ASTROPHYSICAL DYNAMICS:
FROM STARS TO GALAXIES

IAU SYMPOSIUM No. 271

T0336026

COVER ILLUSTRATION:

The cover illustration of this book is a collage of four panels intending to illustrate the fundamental reasons for the existence of the conference. The symposium spanned a wide range of topics in astrophysical dynamics, from stars to galaxies, highlighting the common processes and common scientific approaches to understanding these phenomena. Hence both a star and a galaxy are pictured, both from observations and from simulations. The two simulations chosen are both from work that Prof. Juri Toomre has been involved in, since the symposium was dedicated to honouring his long, illustrious and varied career near the occasion of his 70th birthday.

Upper left panel Numerical simulations of dynamo action realized in a convective spherical envelope by the Toomre group using the ASH code (see e.g. Brun, Miesch & Toomre, 2004, Astrophys. Jou., 614, 1073). Shown is the longitudinal component of the magnetic field with blue/red indicating negative/positive polarity.

Upper right panel Solar event on August 1, 2010 as viewed by the Solar Dynamics Observatory. This multi-wavelength (211, 193 & 171 Angstrom) extreme ultraviolet snapshot from the Solar Dynamics Observatory (SDO) shows the sun's northern hemisphere in mid flare-eruption. Different colors in the image represent different gas temperatures ranging from 1 to 2 million degrees K.
Credit: NASA/SDO/AIA

Lower left panel Simulation of galaxy mergers by Toomre & Toomre 1972 (Astrophys. Jou.,178, 623)

Lower right panel Mice galaxies. NGC 4676, or the Mice Galaxies, are two spiral galaxies in the constellation Coma Berenices, that are presently in the process of colliding and merging. Photographed in 2002 by the Hubble Space Telescope.
Credit: `http://hubblesite.org/newscenter/archive/releases/2002/11/image/d`.

INTERNATIONAL ASTRONOMICAL UNION

UNION ASTRONOMIQUE INTERNATIONALE

ASTROPHYSICAL DYNAMICS: FROM STARS TO GALAXIES

PROCEEDINGS OF THE 271st SYMPOSIUM OF THE INTERNATIONAL ASTRONOMICAL UNION HELD IN NICE, FRANCE JUNE 21–25, 2010

Edited by

NICHOLAS H. BRUMMELL
University of California, Santa Cruz, CA, USA

A. SACHA BRUN
IRFU/SAp, CEA, Saclay, France

MARK S. MIESCH
HAO, NCAR, Boulder, CO, USA

and

YANNICK PONTY
Observatoire de la Côte d'Azur, Nice, France

CAMBRIDGE
UNIVERSITY PRESS

CAMBRIDGE UNIVERSITY PRESS
Cambridge, New York, Melbourne, Madrid, Cape Town,
Singapore, São Paulo, Delhi, Mexico City

Cambridge University Press
The Edinburgh Building, Cambridge CB2 8RU, UK

Published in the United States of America by Cambridge University Press, New York

www.cambridge.org
Information on this title: www.cambridge.org/9780521197397

First published 2011

A catalogue record for this publication is available from the British Library

ISBN 978-0-521-19739-7 Hardback

Table of Contents

DAY 1. THE SUN AND STARS: OBSERVATIONAL CONSTRAINTS, THEORIES AND MODELS

Morning session. Chair: N. Brummell

Afternoon session. Chair: D. Gough

DAY 2. GALAXIES: OBSERVATIONAL CONSTRAINTS, THEORIES AND MODELS

Morning session. Chair: E. Zweibel

Afternoon session. Chair: A.Toomre

DAY 3. NONLINEAR ASTROPHYSICS

Morning session. Chair: S. Brun

DAY 4. COSMIC MAGNETISM

Morning session. Chair: N. Weiss

Afternoon session. Chair: K. Moffatt

DAY 5. ASTROPHYSICAL TURBULENCE

Morning session. Chair: M. Proctor

Preface

The purpose of IAU symposium 271 was to enable interaction, discussion, study, and thereby to enhance our understanding, of some of the important nonlinear dynamical processes present in the Universe. The intention was to pay special attention to those that are ubiquitously present in a great variety of astronomical objects, from stars like our Sun to galaxies.

A wide range of temporal and spatial scales are present in many of the essential dynamical phenomena operating in most celestial bodies, and thus fluid and magnetic instabilities, highly nonlinear states and turbulence play a central role in these systems. Today, understanding the behaviour and evolution of such systems requires high-accuracy, multi-scale astronomical observations and thoughtful analysis of the data garned, coupled with detailed theoretical study via models. Relatively recently, high-performance numerical simulations have become an essential and revealing tool for assessing the often subtle and surprising highly nonlinear regime of such models.

Symposium 271 offered a unique opportunity for world experts with widely varying perspectives to share their knowledge and opinions on the latest advances in the study of the common underlying processes from the field of nonlinear astrophysical dynamics. In the end, more than 110 scientists attended from 35 countries ranging as far afield as China, India, Russia, Egypt, Japan, Australia, South Africa, USA, Columbia, etc.

This conference was particularly special in that it was a fitting occasion to celebrate the long and illustrious career of Professor Juri Toomre close to the date of his 70th birthday. The Symposium theme aptly befits Juri's widespread achievements in many realms of astrophysical dynamics. It was decided to have this meeting in Nice because Juri used to enjoy his visits to the Observatory here to visit Jean-Paul Zahn, who was Director there for some time. Juri's long career and many successes were remembered during the conference and were celebrated with a banquet in his honour.

Nic Brummell, Sacha Brun, Mark Miesch and Yannick Ponty, Organisers
December 2010

THE ORGANIZING COMMITTEE

Scientific

Nicholas Brummell (co-chair, USA)
Sacha Brun (co-chair, France)
Kwing Chang (China)
Paul Charbonneau (Canada)
Joergen Christensen-Daalsgaard (Denmark)
David Galloway (Australia)
Douglas Gough (UK)
Siraj Hasan (India)

Mark Miesch (USA)
Keith Moffatt (UK)
Yannick Ponty (France)
Annick Pouquet (USA)
Steve Tobias (UK)
Nigel Weiss (UK)
Ellen Zweibel (USA)
Jean-Paul Zahn (France)

Local

S. Brun (co-chair, Saclay)
N. Bessolaz
S. Berard
P. Chavegrand
L. Jouve
S. Mathis

Y. Ponty (co-chair, Nice)
R. Pinto
A. Strugarek
S. Szeles
B. Thooris

Acknowledgements

The symposium is sponsored and supported by the IAU Divisions IV (Stars) and by the IAU Commissions No. 10 (Solar Activity), No. 27 (Variable Stars), No. 28 (Galaxies) and No. 35 (Stellar Constitution).

The organisers gratefully acknowledge support from the International Astronomical Union, and further funding from:

Astrosim

CEA France
European Science Foundation
Observatoire de la Côte d'Azur, France

European Network for Computational Astrophysics
CNRS France
European Research Council

CONFERENCE PHOTOGRAPH

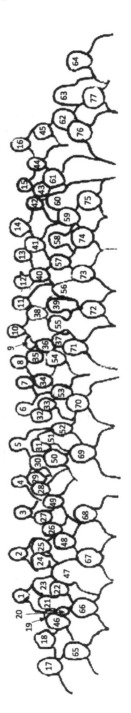

CONFERENCE PHOTOGRAPH INDEX

Participants

Rainer **Arlt**, Astrophysikalisches Institut Potsdam, Germany rarlt@aip.de
David **Arnett**, Steward Observatory, USA wdarnett@gmail.com
Wayne **Arter**, CCFE, UK wayne.arter@ccfe.ac.uk
Kyle **Augustson**, JILA University of Colorado Boulder, USA kyle.augustson@colorado.edu
Nicolas **Bessolaz**, CEA/IRFU/Service d'Astrophysique, Saclay, France nicolas.bessolaz@cea.fr
Lionel **Bigot**, Observatoire de la Côte d'Azur, France lbigot@oca.eu
Adrian **Barker**, DAMTP, University of Cambridge, UK ajb268@cam.ac.uk
Frederic **Bournaud**, CEA/IRFU/Service d'Astrophysique, Saclay, France frederic.bournaud@cea.fr
Axel **Brandenburg**, NORDITA, Stockholm, Sweden brandenb@nordita.org
Benjamin **Brown**, University of Wisconsin Madison, USA bpbrown@astro.wisc.edu
Matthew **Browning**, Canadian Institute for Theoretical Astrophysics, Canada browning@cita.utoronto.ca
Nicholas **Brummell**, University of California Santa Cruz, USA brummell@soe.ucsc.edu
Allan Sacha **Brun**, CEA/IRFU/Service d'Astrophysique, Saclay, France sacha.brun@cea.fr
Paul **Bushby**, Newcastle University, UK paul.bushby@ncl.ac.uk
Benjamin **Byington**, University of California Santa Cruz, USA bbyingto@soe.ucsc.edu
Tao **Cai**, Hong Kong University of Science and Technology, Hong Kong ctust@ust.hk
Francoise **Combes**, Observatoire de Paris, France francoise.combes@obspm.fr
Simon **Candelaresi**, Nordita, Stockholm, Sweden iomsn@physto.se
Massimo **Capaccioli**, University of Naples, Federico II, Italy capaccioli@na.astro.it
Miljenko **Cemeljic**, Academia Sinica Institute of Astronomy and Astrophysics , Taiwan miki@tiara.sinica.edu.tw
Kwing **Chan**, Hong Kong University of Science et Technology, Hong Kong maklchan@ust.hk
Piyali **Chatterjee**, NORDITA, Stockholm, Sweden piyalic@nordita.org
Joergen **Christensen-Dalsgaard**, Univ. of Aarhus, Denmark jcd@phys.au.dk
Remo **Collet**, Max Planck Institute for Astrophysics, Germany remo@mpa-garching.mpg.de
Anna **Curir** , INAF- Astronomical Observatory of Torino, Italy curir@oato.inaf.it
Olivier **Do Cao**, CEA/IRFU/Service d'Astrophysique, Saclay, France olivier.do-cao@cea.fr
Marc **DeRosa**, Lockheed Martin Solar and Astrophysics Laboratory derosa@lmsal.com
Fabio **Del Sordo**, NORDITA, Stockholm, Sweden fadiesis@gmail.com
Jean-François **Donati**, IRAP, Observatoire Midi-Pyrénées, France donati@ast.obs-mip.fr
Bérengère **Dubrulle**, CEA/IRAMIS Saclay, France berengere.dubrulle@cea.fr
Chris **Engelbrecht**, Dept. of Physics, University of Johannesburg, South Africa chrise@uj.ac.za
Adnan **Erkurt**, Istanbul University, Dept of Astronomy & Space Sciences, Turkey adnan.erkurt@ogr.iu.edu.tr
Katia **Ferrière**, IRAP, Observatoire Midi-Pyrénées, France ferriere@ast.obs-mip.fr
Cary **Forest**, University of Wisconsin, Madison, USA cbforest@wisc.edu
Fabio **Frescura**, University Witwatersrand, South Africa fabio.frescura@wits.ac.za
Uriel **Frisch**, Observatoire Nice- Cote d'Azur, France uriel@obs-nice.fr
David **Galloway**, School of Mathematics and Statistics, University of Sydney, Australia dave@maths.usyd.edu.au
Adriana **Gazol**, CRyA, UNAM, Mexico a.gazol@crya.unam.mx
Douglas **Gough**, DAMTP, University of Cambridge, UK douglas@ast.cam.ac.uk
Vitaly **Groppen**, North-Caucasian Institute of Mining and Metallurgy, Russia groppen@mail.ru
Rebecca **Grouchy**, Observatoire de Paris, France rebecca.grouchy@obspm.fr
Gustavo **Guerrero**, NORDITA, Stockholm, Sweden guerrero@nordita.org
Céline **Guervilly**, LGIT Grenoble, France celine.guervilly@obs.ujf-grenoble.fr
Deborah **Haber**, JILA, University of Colorado, USA dhaber@solarz.colorado.edu
Shravan **Hanasoge**, Max-Planck Institute for Solar System Research, Germany hanasoge@princeton.edu
Rajib **Hazarika**, Diphu Govt. College,Assam,India drabrh_dgc5163@rediffmail.com
Frank **Hill**, National Solar Observatory, USA fhill@noao.edu
Bradley **Hindman**, JILA / University of Colorado, USA hindman@solarz.colorado.edu
Alexander **Hubbard**, NORDITA, Stockholm, Sweden hubbard@pas.rochester.edu
David **Hughes**, University of Leeds, UK d.w.hughes@leeds.ac.uk
Neal **Hurlburt**, Lockheed Martin Advanced Technology Center hurlburt@lmsal.com
Nariman **Ismailov** Baku State University, Azerbaijian box1955n@yahoo.com
Laurène **Jouve**, DAMTP Cambridge, UK lj272@cam.ac.uk
Nagendra **Kumar**, M.M.H. College Ghaziabad, (U.P), India nagendrakgk@rediffmail.com
Gareth **Kennedy**, Institut de Cincies del Cosmos, University of Barcelona, Spain gareth.f.kennedy@gmail.com
Volodymyr **Kryvdyk**, Taras Shevchenko National University of Kyiv, Ukrenia kryvdyk@univ.kiev.ua
Katarzyna **Kulpa-Dybel**, Astronomical Observatory of the Jagiellonian University, Poland kulpa@oa.uj.edu.pl
Maurice **Laloum** , CNRS/IN2P3/LPNHE (retired) maurice.laloum@orange.fr
Marian **Lazar**, Plasma Research Dept., Rurh University, Germany mlazar@tp4.ruhr-uni-bochum.de
John **Leibacher**, National Solar Observatory, Tucson, USA jleibacher@nso.edu
Robyn **Levine**, Canadian Institute for Theoretical Astrophysics, Canada levine@cita.utoronto.ca
Lien-Hsuan **Lin**, Institute of Astronomy and Astrophysics, Academia Sinica, Taiwan lhlin@asiaa.sinica.edu.tw
Marjaana **Lindborg**, University of Helsinki, Finland marjaana.lindborg@helsinki.fi
Olga **Lobanova**, Russia legenda223@rambler.ru
Stéphane **Mathis**, CEA/DSM/IRFU/SAp/LDEE AIM Paris-Saclay, France stephane.mathis@cea.fr
Romain **Meyrand**, IAS, France romain.meyrand@ias.fr
Mark **Miesch**, High Altitude Observatory, NCAR, USA miesch@ucar.edu
H. Keith **Moffatt**, DAMTP University of Cambridge, UK h.k.moffatt@damtp.cam.ac.uk
Motahareh **Mohammadpour**, Mazandaran University, Iran mohammadpour@umz.ac.ir
Herbert J. **Muthsam**, Faculty of Mathematics, University of Vienna, Austria herbert.muthsam@univie.ac.at
Evangelia **Ntormousi**, University Observatory Munich, Germany eva@usm.uni-muenchen.de
Christoph **Olczak**, Max Planck Institute for Astronomy, Germany olczak@mpia.de
Ana **Palacios**, Graal, University of Montpellier, France ana.palacios@univ-montp2.fr
Paniveni **Paniveni**, NIEIT, MYSORE, India paniveniudaya@yahoo.co.in
Paolo **Padoan**, ICREA / ICC - University of Barcelona, Spain ppadoan@ucsd.edu
Clare **Parnell**, University of St Andrews, UK clare@mcs.st-and.ac.uk
Rui **Pinto**, CEA/IRFU/Service d'Astrophysique, Saclay, France rui.pinto@cea.fr
Ruth **Peterson**, Astronomy & UCO/Lick, UC Santa Cruz, CA, USA peterson@ucolick.org
Hélène **Politano**, Observatoire de la Cote d'Azur, France politano@oca.eu
Yannick **Ponty**, Observatoire de la Cote d'Azur, France yannick.ponty@oca.eu
Annick **Pouquet**, NCAR, Boulder, USA pouquet@ucar.edu
Michael **Proctor**, DAMTP, University of Cambridge, UK mrep@cam.ac.uk
Walid A. **Rahoma**, Astronomy Dept., Faculty of Science, Cairo University, Egypt walid.rahoma@gmail.com
Erico **Rempel**, Institute of Aeronautical Technology (ITA) rempel@ita.br
Paolo **Repetto**, Instituto de Astronomia UNAM, Mexico prepetto@astroscu.unam.mx
Tamara **Rogers**, University of Arizona Tucson, USA tamirogers@mac.com
Fouad **Sahraoui**, Lab. de Physique des Plasmas, CNRS-Polytechnique, France fouad.sahraoui@lpp.polytechnique.fr
Jesper **Schou**, Solar group, Stanford University, USA schou@sun.stanford.edu

Lara **Silvers**, City University London, UK   lara.silvers.1@city.ac.uk
Antoine **Strugarek**, CEA/IRFU/Service d'Astrophysique, Saclay, France   antoine.strugarek@cea.fr
Joel **Tanner**, Yale University, USA   joel.tanner@yale.edu
Michael **Thompson**, High Altitude Observatory, NCAR, USA   mjt@ucar.edu
Alan **Title**, Lockheed Martin Advanced Technology Center, USA   title@lmsal.com
Steve **Tobias**, University of Leeds, UK   smt@maths.leeds.ac.uk
Juri **Toomre**, JILA & Dept of Astrophysical and Planetary Sciences, USA   jtoomre@solarz.colorado.edu
Alar **Toomre**, MIT, USA   toomre@math.mit.edu
Regner **Trampedach**, JILA, University of Colorado, USA   trampeda@lcd.colorado.edu
Saku **Tsuneta**, National Astronomical Observatory of Japan, Japan   saku.tsuneta@nao.ac.jp
Allard Jan **Van Marle**, K.U Leuven, Netherland   AllardJan.vanMarle@wis.kuleuven.be
Geoffrey **Vasil**, Canadian Institute for Theoretical Astrophysics, Canada   vasil@cita.utoronto.ca
Yu-Ting **Wu**, Department of Physics, National Tsing-Hua University, Taiwan   d9622814@oz.nthu.edu.tw
Ling Jun **Wang**, University of Tennessee at Chattanooga, USA   lingjun-wang@utc.edu
Jörn **Warnecke**, Nordita, Stockholm, Sweden   joern@nordita.org
Nigel **Weiss**, DAMTP, University of Cambridge, UK   now@damtp.cam.ac.uk
Joe **Wolf**, University of California, Irvine, USA   wolfj@uci.edu
Toby **Wood**, University of California Santa Cruz, USA   tsw25@soe.ucsc.edu
Jean-Paul **Zahn**, LUTH, Observatoire de Paris, France   jean-paul.zahn@obspm.fr
Olexandra **Zhukova**, Observatory of Kiev, Ukrenia   a-zhukova@ukr.net
Nataliya **Zubreva**, Ukraine Observatory, Ukrenia   natasha.zubreva@gmail.com
Ellen **Zweibel**, University of Wisconsin-Madison, USA   zweibel@astro.wisc.edu

Local organising committee: events

The local organising committee organised a very nice boat trip to Villefranche-sur-Mer and a wonderful dinner to honour Prof. Juri Toomre's 70th birthday in the middle of the conference.

Address by the Organisers

Firstly, we would like to thank everyone for taking the time and effort to travel from far and wide to IAU Symposium 271. We had participants from all over the world, including most of Europe, the USA, Taiwan, Russia, South Africa, Australia, Egypt, Mexico, Japan, to name but a few of the 35 countries represented.

The aim of the conference was to encourage and enable scientific discussion on a broad range of topics, related to objects as varied as our Sun and other stars to galaxies. The emphasis was on the underlying processes that united these various objects, such as hydrodynamic and hydromagnetic turbulence, and complex nonlinear dynamics in general. A mixture of techniques including observations, basic theory and computation were discussed.

One of the reasons for the breadth of the topics discussed in the Symposium was that a large part of the motivation for the conference stemmed from a desire to recognise and honour the long and illustrious career of Professor Juri Toomre in the year of his 70th birthday.

Juri was born in Estonia in 1940 but immigrated to the USA with his family in 1949. He received both a Bachelors and a Masters degree from Massachusetts Institute of Technology (MIT) by 1963 and then went to Trinity College, Cambridge, England as a Marshall scholar to work with Prof. H. Keith Moffatt in the Department of Applied Mathematics and Theoretical Physics. Juri obtained a Ph.D. in 1967 with a thesis on "Hydromagnetic Jets".

Juri then returned to the USA to work as a postdoctoral scholar at the Department of Mathematics, New York university and the Goddard Institute for Space Studies, New York. During this time, Juri worked with his brother, Alar, a professor of Mathematics at MIT, on models of close encounters of galaxies, and the "tails" and "bridges" that can be formed in their tidal interaction, resulting in a famous paper, Toomre & Toomre, 1972 (Astrophys. Jou., 178, 623). Juri also cemented a strong relationship with a group of peers at this time, which resulted in long-standing collaborative projects that bore fruition in a long series of papers with various combinations of the group members Douglas Gough, Jean Latour, Ed Spiegel, Juri and Jean-Paul Zahn. This group become affectionately known as "the Convective Collective" since the body of work addressed stellar convection theory (see page 339).

In 1975, Juri became a Professor in the (now) Astrophysics and Planetary Science Department and a Fellow of the (then) Joint Institute for Laboratory Astrophysics (now JILA) at the University of Colorado, Boulder, a position that he retains to this today.

Juri's work with the Convective Collective continued into the early '80s, and has expanded along these lines towards more complex anelastic and compressible systems with magnetic fields and in various geometries, with a series of students and postdoctoral scholars ever since. Indeed, Juri became a veritable clearing house for any up and coming person with an interest in stellar fluid dynamics, either spawning students (e.g. David Hathaway, Neal Hurlburt, Anil Deane, Phil Jones, Xin Xie, Mark Rast, Mark Miesch, Matt Browning, Ben Brown) or grooming postdocs (e.g. David Hughes, Fausto Cattaneo, Nic Brummell, Keith Julien, Tom Clune, Sacha Brun), lists that read like a "Who's Who" of mathematical stellar fluid dynamics today. One of Juri's distinct successes was to recognise the potential of high performance computing, and to champion its use in the study of the highly nonlinear systems that are relevant to stellar situations. Juri's stellar convection and MHD group is a major user of the nations supercomputing facilities, and, indeed, has evolved into a new version of the Convective Collective known (also affectionately) as "the ASH Mob" (after the main Anelastic Spherical Harmonic computational code that they use as their main tool).

On a parallel work strand, in the mid-1980s, through his close collaboration with Douglas Gough (Cambridge, England), Juri also became an early pioneer in the subject of helioseismology, the inversion of sound data to infer information about the interior of the Sun. Juri again became the benevolent father and incubator for a series of students and postdocs associated with this breakthrough line of work (e.g. Deborah Haber, Frank Hill, Brad Hindman).

Along the way, Juri has been a major contributor to both service in the academic field and teaching in the University. Juri has been vice-chair of the Solar Observatories Council of Association of Universities for Research in Astronomy (AURA) with oversight for the National Solar Observatory, served several terms on the Observatories Council dealing with National Optical Astronomy Observatory (NOAO), and has been member and chairman of the Space Telescope Institute Council (STIC) which has oversight for the Space Telescope Science Institute (STScI). Juri is currently chair of the scientific advisory committee to the Global Oscillations Network Group (GONG, the major ground-based project in helioseismology), was a Co-I on the Solar Oscillations Investigation (SOI) Michelson Doppler Interferometer (MDI) helioseismology experiment on the Solar and Heliospheric Observatory (SOHO), and is now Co-I on the Helioseismic Magnetic Imager (HMI) experiment on the newly launched Solar Dynamics Observatory (SDO). Juri also recently served on the Astro2010 Astronomy & Astrophysics Decadal Survey central committee. Juri also received the 2010 University of Colorado Hazel Barnes prize, the highest accolade for research and teaching at the university.

Overall, Juri's scientific nose for a good problem, acute awareness of the cutting edge, skill with a turn of phrase, keen eye for a strong selling point, and nuturing nature have made for a deservedly long and extremely successful career. He has established himself as a long term server of the scientific goal and a leader and mentor to others with the same ideals. We therefore dedicated this meeting to honouring Juri's personal achievements and his devotion to these goals.

Nic Brummell, Sacha Brun, Mark Miesch and Yannick Ponty

PAPERS

Astrophysical Dynamics: From Stars to Galaxies
Proceedings IAU Symposium No. 271, 2010
N.H. Brummell, A.S. Brun, M. S. Miesch, & Y. Ponty, eds.

© International Astronomical Union 2011
doi:10.1017/S1743921311017418

Some recent and future helioseismological inferences concerning the solar convection zone

Douglas Gough

Institute of Astronomy & Department of Applied Mathematics and Theoretical Physics,
University of Cambridge; Physics Department, Stanford University
email: douglas@ast.cam.ac.uk

Abstract. Several uncertain helioseismic findings of potential interest to Jüri about the solar convection zone are briefly discussed, along with some personal optimistic hopes for the future.

Keywords. Helioseismology, solar abundances, sun's age, solar magnetic field, meridional flow

---

1. Some Background

Helioseismology can be summoned to yield the seismic structure of the sun: the variation through the sun of those quantities that control the propagation of (the essentially adiabatic) seismic waves, principally density and pressure (which are related by hydrostatics) and the relation between them under adiabatic change, which is characterized by the adiabatic exponent $\gamma_1 = (\partial \ln p / \partial \ln \rho)_{\text{ad}}$; I add to that list macroscopic motion and magnetic field, about which some information can be obtained, although they cannot be determined completely by seismology, even in principle. Any property of the sun that cannot be expressed solely in terms of those quantities is not a purely seismic variable, and cannot be determined by seismology alone. I should add that I shall be assuming in my discussion a knowledge of the mass M (at least GM) and the radius R (which I refrain from even attempting to define here), which are obtained by non-seismic means. I am being explicit about this point to draw attention to the fact that isospectral stellar structures exist with different M and R, a property (amongst others) which renders asteroseismology without M or R less informative than helioseismology.

It is common to represent the (almost spherically symmetrical) hydrostatic structure of the sun by $c^2(\mathbf{r})$ and $\rho(\mathbf{r})$, where c and ρ are sound speed and density; c^2 is preferred to c because for a perfect gas $c^2 = \gamma_1 p / \rho$ which is approximately proportional to $\gamma_1 T / \mu$, where p is pressure and μ is the mean molecular mass (to a first approximation the perfect-gas law is an adequate guide); c^2 therefore resembles temperature T. Because the structures of modern models are quite close to that of the sun, it is expedient to consider the small relative deviations $\delta \ln c^2$ and $\delta \ln \rho$ of the sun from a reference theoretical model, such as Model S of Christensen-Dalsgaard *et al.* (1996), rather than the bare values of c^2 and ρ. Those deviations are typically no greater than 0.2% and 1.5% respectively.

Solar models are typically produced by evolution from the zero-age main sequence, adjusting the initial helium abundance Y_0 for a given initial heavy-element abundance Z_0 (or, equivalently, any combination of initial abundances Y_0 and Z_0), and a scaling factor in an algorithm to model convection – typically a mixing-length parameter – to reproduce the observed luminosity L and radius R. That establishes a relation between Y_0 and Z_0: models with lower Z_0 have lower Y_0. It is worth noting that in the models (*i*) temperature T decreases at fixed radius r as Y_0 decreases, partly because μ is lower

3

globally and partly because there is a greater abundance of hydrogen in the core to fuel the nuclear reactions which therefore provide the observed luminosity L at lower T (and ρ), and (ii) as the star ages, a slight depression in c^2 is produced near $r = 0$ because in the central core nuclear reactions have converted hydrogen into helium and thereby increased μ locally. In the early days of helioseismology the former property played an important role in calibrating Y_0, to a value that was then perceived to be incompatible with neutrino observations (those were the days before neutrino transitions had been detected); nowadays the latter property is used for calibrating the age of solar models.

The most prominent property of $\delta \ln c^2$ is a narrow hump near $x = r/R = 0.65$; it is associated with the tachocline, about which I shall say a few words later. There is also a discrepency in the convection zone, probably due largely to an error in the value adopted (implicitly) for R (which I continue to refrain from discussing – but see Takata & Gough, 2003). Finally, there is a large-scale discrepancy in the radiative interior, which is unexplained. I emphasize that c^2 in some models deviates from that in the sun by only a few tenths per cent (errors in the determination of the solar c^2 are even smaller).

2. On the age of the sun

Guenter Houdek and I (e.g. 2007, 2009) have recently been seismically calibrating solar models to characterize the central dip in c^2, and hence to estimate the main-sequence age $t_\odot$ of the sun. Modes of the lowest degree l must be used, for it is they that penetrate the most deeply into the core. We have confined ourselves to only such modes (having $l \leqslant 3$), with the intention of using our procedure for stars other than the sun.

Were the structure of the sun to be smooth, the cyclic frequencies $\nu_{n,l}$ of the low-degree modes would be given (asymptotically), in terms of $x_{n,l} := \nu_{n,l}/\nu_0$, by

$$x_{n,l} \sim n + \frac{1}{2}l + \varepsilon + \sum_i \sum_{j=0}^i A_{ij} L^{2j} x_{n,l}^{1-2i} \tag{2.1}$$

for large order n, where $L^2 = l(l+1)$ and ε, ν_0 and A_{ij} are functionals of the solar structure, independent of l and n. The most deeply probing terms are the most strongly L-dependent, having coefficients $A_{ii} \simeq \int f_{ii} \mathrm{d}r$ for each i, in which, except where $R - r \ll R$,

$$f_{ii}(r) \propto \left(\frac{1}{r} \frac{\mathrm{d}}{\mathrm{d}r} \right)^i c^{2i-1}. \tag{2.2}$$

The L-independent terms, namely ε and $A_{i0} x_{n,l}^{1-2i}$, depend principally on the surface layers, whose influence on the oscillation frequencies is rendered uncertain by the inadequacy of our understanding of convection. The fiducial frequency ν_0 is the inverse of twice the acoustic radius of the star, and is therefore a global indicator.

A first attempt to calibrate $t_\odot$ was made by fitting formula (2.1) to raw observed frequencies. That suffered from 'contamination' by oscillatory (with respect to n) deviations from the smooth formula (Gough, 2001). There was also the problem of having to calibrate the models with respect to two parameters, namely $t_\odot$ and Y_0, whose influences on the unknown coefficients in the formula were not easily separated. (I must acknowledge that at about the same time Bonanno, Schlattl and Paternò (2002) attempted a similar calibration to determine $t_\odot$; however, they assumed a value for Z_0, and did not allow for the chemical composition to vary, so their one-parameter fit was more straightforward, although, of course, less reliable.) The oscillatory deviations are produced by acoustic glitches, caused partly by the (near) discontinuity in $\mathrm{d}^2 c^2/\mathrm{d}r^2$ at the base of the convection zone and by depressions in γ_1 caused by the ionization of helium (and hydrogen) (Gough, 1990). They have 'frequencies' roughly twice the acoustic depths τ_{g} of the glitches. The amplitude of the γ_1-induced oscillation depends on Y, providing a separate datum for calibrating $t_\odot$ and Y_0, and thereby rendering the calibration more stable. The outcome currently is

$$t_\odot = 4.60 \pm 0.02 \ \mathrm{Ga},$$

with $Y_0 = 0.253 \pm 0.004$ and $Z_0 = 0.016 \pm 0.001$. The present photospheric abundances of the calibrated model are

$$Y_{\mathrm{s}} = 0.227, \quad Z_{\mathrm{s}} = 0.0146.$$

I must emphasize here that this seismically calibrated model, in common with others of its genre (e.g. Turck-Chièze *et al.*, 2001), is not a seismic model, for its structure agrees with only a few limited aspects of the sun, and is not seismically acceptable throughout.

An interesting question that arises naturally from this exercise is how the age of the sun (measured, as we do, from what one can define as the instant of the zero-age main sequence – I shall discuss that instant in an instant) compares with the ages of the oldest meteorites. That can have implications regarding the formation of the solar system, which, it is generally believed, took place over a timespan of some 10^7 years or less (there are some who would say 'more'). Evidently, even if we disregard the systematic modelling errors, we have not yet achieved adequate precision. But the goal is almost in sight. The value quoted above is (perhaps conveniently) not significantly greater than modern determinations of the ages of the oldest meteorites (e.g. von Hippel, Simpson & Manset, 2001).

Finally, a word about an origin of solar time. On the main sequence the characteristic evolution timescale of the sun exceeds the thermal diffusion time by a factor 300 or so. Therefore, once established, the sun is (probably) in thermal balance to quite a good approximation, the rate of generation of nuclear energy in the core equalling the radiant luminosity a the surface. The sun arrived on the main sequence by gravitational (Kelvin-Helmholtz) contraction moderated by thermal diffusion, on a timescale $t_{\mathrm{KH}} \simeq 10^7 \mathrm{a}$. This is comparable, not entirely fortuitously, with the formation timescale of the planetary system. The nuclear generation of heat halted the contraction, but only gradually. So one might wonder whether it is even meaningful to define a precise instant from which to measure main-sequence age. In fact it is, because it turns out, again not entirely fortuitously, that on the main sequence the relative abundance X_{c} of hydrogen ($X = 1 - Y - Z$) at the centre of the sun is very nearly a linearly decreasing function of time (Gough, 1995). Therefore one can extrapolate $X_{\mathrm{c}}(t)$ backwards quite reliably to its initial value X_0 (gravitational settling in the core over a time t_{KH} is tiny) to define an origin.

3. On the heavy-element abundance of the sun

The value $Z_{\mathrm{s}}/X_{\mathrm{s}} = 0.0193$ obtained by the model-fitting procedure described in the previous section is substantially lower than values that were fashionable a while ago – e.g. $Z_{\mathrm{s}}/X_{\mathrm{s}} = 0.0274$ (Anders and Grevesse, 1989); Model S has an abundance ratio $Z_{\mathrm{s}}/X_{\mathrm{s}} = 0.0245$ – although it is somewhat greater than the value promulgated recently by Asplund *et al.* (e.g. 2005), who carried out a new spectroscopic analysis of the (near photospheric) solar atmosphere taking the turbulence produced by convection into account, and recommended $Z_s/X_s = 0.0165$. A brief account of the issue raised by the new analysis is not inappropriate here.

It is normally presumed that the photosphere, which is well mixed in by convection, represents the composition of the radiative envelope beneath the convection zone, aside from a small modification from gravitational settling. This exposes a discrepancy with solar modelling that has exercised many minds in recent years. The adjustment of Z_0 suggested by the new spectroscopic analyses is large: more than 30%. But the effect on the equation of state is small, since heavy elements constitute less than 2% of the solar material. Therefore the effect of lowering Z_0 in a solar model is essentially just to reduce the opacity by a similar amount. That reacts on the temperature gradient required to maintain the heat flux, producing a change in T and a comparable change in c^2, throwing the model out of agreement with the sun. Any model calculated using this low value of Z_0, with $t_\odot$ roughly 4.6 Ga, using generally accepted microscopic physics (equation of state, opacity, nuclear reaction rates) and adopting the usual so-called standard tenets of stellar-structure theory, must necessarily be ruled out by seismic observation. Although that may seem obvious, there has been a spate of publications labouring the point, and presenting numerical examples of the seismic disagreement with low-Z models often without edifying comment; they have been catalogued recently by Basu and Antia (2008). I hope it is hardly necessary for me to point out that the seismological analysis is based on extremely simple and well understood physics (at the level required for the present discussion), and therefore is not open to serious doubt.

The disagreement can be presented in a variety of ways. Perhaps the most acceptable amongst discussions adopting the tenets of standard stellar-structure theory is to retain the equation of state (it surely cannot be wrong by as much as 30%) and, I recommend, the nuclear reaction rates, for they have been studied extensively in the last decades in connexion with the neutrino problem (which is now, at least in its basic form, resolved; I should acknowledge, however, that, as has been pointed out by several critics, the p-p cross-section determining the slowest, controlling, reaction of the chain has been determined only theoretically). Assuming no mixing in the deep interior, one can easily scale Y_0 from a model to estimate $X(r)$ in the sun today with adequate precision for the purpose in hand, hence obtain $T(r)$ from $c^2(r)$, and thereby compute the opacity required for transporting the required amount of heat by radiative transfer. Not surprisingly, the outcome (e.g. Gough, 2004) is close to the value in reference model S of Christensen-Dalsgaard et al. (1996). The problem posed by Asplund and his colleagues is therefore to reconcile that value with the photospheric chemical composition.

Several possibilities for resolving the disagreement come immediately to mind. Perhaps the sun condensed gravitationally in its primordial interstellar gas cloud around a seed giant-planet-like condensation which had already shed some of its hydrogen and helium. Then the sun's radiative interior could have been rich in Z, and the difference, being stable to double-diffusive convection, could have survived to the present day. This is a mechanistic justification for having an appropriate compositional variation in the outer reaches of the sun's envelope, an hypothesis suggested originally by Guzik, Watson and Cox (2005). The density variation caused by the composition variation consequent on this hypothesis, even were it confined to a thin interface, would contribute no more than about 30% to the amplitude of the associated oscillatory signature (see §2) in the eigenfrequencies. This is probably too small to be detected unambiguously with currently available data. But maybe in the future, with a sufficiently sophisticated analysis, it could be disentangled from other aspects of the stratification near the base of the convection zone.

It has been suggested that alternatively there could be mechanisms other than radiative transfer, such as gravity waves, to transport the heat; but wouldn't it be incredible for such processes to mimic the functional form produced by the physics of the radiatively induced atomic transitions that determine opacity? That objection can be levelled against most other suggestions too. So one is, perhaps reluctantly, led to wonder whether the abundance determinations by Asplund and his colleagues are correct. It is interesting to note that recently their value has been revised upwards a little – $Z_s/X_s = 0.0181$ (Asplund *et al.* 2009) – and that Caffau *et al.* (2010) have carried out a parallel spectroscopic analysis, obtaining a somewhat higher value still, namely 0.0211. These two values bracket that obtained by the model calibration reported in §2. However, I reiterate that the calibrated model is not a seismic model.

It behoves us to seek some independent way of determining Z. One might attempt that seismologically, by measuring $W := (r^2/Gm)\mathrm{d}c^2/\mathrm{d}r$, where $m = 4\pi \int \rho r^2 \mathrm{d}r$. In the adiabatically stratified regions of the convection zone, $W \simeq \Theta := 1 - (\partial\ln\gamma_1/\partial\ln\rho)_p - \gamma_1[1 + (\partial\ln\gamma_1/\partial\ln p)_\rho]$ (Gough, 1984), which has humps where γ_1 is lowered by ionization. We first determined Y by that method. I recall announcing the intention to carry out the determination at a meeting in Cambridge in 1985. Donald Lynden-Bell said he thought it was impossible, and wagered that in any case it would not be accomplished within 10 years. He was right. But for the wrong reason. He thought that we would be unable to measure the helium hump in W with adequate precision. But actually we measured it so precisely as to show that it was incompatible with equations of state of the time (Kosovichev *et al.*, 1992), implying that those equations could not be trusted to convert W into a reliable value of Y. However, we were able to refine previous estimates using currently available equations of state, and found Y to be lower, by several per cent, than the value of Y_0 used in typical theoretical models – a finding subsequently corroborated by others (e.g. Serenelli and Basu, 2010) – thereby emphasizing the need to consider the influence of gravitational settling. Christensen-Dalsgaard, Proffitt and Thompson (1993) demonstrated that models incorporating gravitational settling can be enormously closer in structure to the sun than those that do not. Gravitational settling was therefore included in Christensen-Dalsgaard's model S.

So now I suggest that history be repeated. My colleague Katie Mussack and I will try to measure the minute humps in Θ produced by the ionization of principally C, N and O beneath the region of appreciable HeII ionization (Mussack & Gough, 2009). Of course we cannot expect a precision comparable with what can be achieved for helium, but a robust, albeit roughly determined, amplitude of the ionization-induced variation in $W(r)$ about a background – whose value is rendered uncertain by our inadequate understanding of the van der Waals effects from bound species of hydrogen and helium (Baturin *et al.* 2000) – should be achievable; the current debate might then, at least partially, be settled.

4. Adiabatic stratification of the deep convection zone

The first numerical simulations of solar convection, predecessors of calculations for which Jüri is now famous, did not provide a reliable indication of the stratification deep in the convection zone. Yet we all know from laboratory experiments with convection that the lapse rate approaches its neutral value at large Rayleigh number. One expects $\Delta := \gamma_1^{-1} - \Gamma_1^{-1} \simeq \nabla - \nabla_{\mathrm{ad}}$, where $\Gamma_1 = \mathrm{d}\ln p/\mathrm{d}\ln\rho$, to be extremely small, and indeed mixing-length theory predicts values of order 10^{-6} deep in the solar convection zone. However, it is certainly of interest to seek independent, seismological, evidence for the smallness of Δ. That is a difficult, because one cannot measure Δ directly: one must be content with what is essentially a measurement of Γ_1^{-1} subtracted from the corresponding adiabatic value. An upper bound is therefore the best one can expect.

The only attempt of which I am aware was carried out in the early days of helioseismology when the data were much less precise than they are today. The result was

$$\Delta < 0.03$$

(Gough, 1984). It would be interesting to see by how much this bound can be tightened with more modern helioseismic data. What we tend to do today is to assume that Δ is utterly negligible, and use that result to infer the thermodynamic quantity $\Theta \simeq W$. It was principally by measuring the hump in W in the second ionization zone of helium that the helium abundance has been measured (to within the undetermined errors in the equation of state). The adiabatic constraint reduces the function space in which one isolates the thermodynamic diagnostic, thereby eliminating some of the extraneous contaminating properties of the stratification.

5. Stratification of the tachocline

As is well known, the convection zone rotates differentially. Described in very broad terms, the latitudinal variation of the angular velocity Ω observed at the surface persists throughout the convection zone, and is separated from a uniformly rotating radiative interior by a thin shear layer called the tachocline. Spiegel and Zahn (1992) demonstrated that had the convection zone abutted directly onto the radiative interior (presumed to be nonmagnetic) the differential rotation would have burrowed into the interior within the sun's lifetime. They concluded that some mechanism in the tachocline must isolate the interior from the shear. They suggested the presence of a thin layer of horizontally isotropic essentially two-dimensional turbulence, of sufficient vigour to overcome the shear. McIntyre and I (1998) argued subsequently that two-dimensional turbulence in a rotating flow does not behave in that manner, as I have believed for a long time (e.g. Gough and Lynden-Bell, 1968), and we cited some more recent evidence in support. We argued that the only conceivable way that the interior could rotate uniformly is for it to be rigid, held by a large-scale (primordial) magnetic field. I still hold that view, although I hasten to add that it is far from being generally accepted (e.g. Brun and Zahn, 2006), although the conclusion that some agent rigidifies the interior is coming to look more and more likely.

Whatever causes the rigidity of the interior, it is inevitable that gyroscopic pumping in the convection zone must produce a proclivity for a meridional circulation connecting the convection zone with the tachocline, in at least all but the lowest latitudes. Indeed, Spiegel and Zahn (1992) analysed such a flow in their two-dimensionally turbulent tachocline. That flow transports to the convection zone helium that had settled under gravity, homogenizing the tachocline with the convection zone above. The outcome is to reduce the mean molecular mass in the tachocline, and thereby raise the sound-speed. That process is no doubt the cause of the sound-speed anomaly beneath the convection zone which I mentioned at the end of my introductory background discussion. Julian Elliott and I (1999) attempted to calibrate the thickness of the tachocline by fitting the anomaly to a solar model with an artificially mixed layer, obtaining a value $0.02R_\odot$.

The reason I say we attempted (rather than succeeded in) performing the calibration is that although the final model that we obtained deviated from model S with an anomaly essentially identical to that observed, it was not quite in the right place. What we failed to point out is that simply moving it to the right place by adjusting the depth of the convection zone would have produced a large-scale deviation in sound speed throughout the radiative interior. This is a phenomenon that had been known for a long time (e.g. Christensen-Dalsgaard *et al.*, 1985) and is no doubt why Brun, Turck-Chièze and Zahn (1999) had had trouble fitting their evolved solar models to the seismic inferences.

An investigation by Takata and Shibahashi (2003) and a more recent unpublished investigation by Jørgen Christensen-Dalsgaard and myself have failed to produce a seismically acceptable spherically symmetrical model with a partially mixed layer that resides completely beneath the convection zone. The implications are unclear at present, although the result may be evidence for tachocline asphericity (although I hasten to add that the essentially hydrostatic balance of forces implies that at least the base of the tachocline, as denoted by the molecular-mass gradient, must be very nearly spherical).

6. Solar-cycle variation of the stratification of the convection zone

There is much discussion at this conference on the dynamics of the solar cycle. Are there seismological consequences that could be used to test the theories? Libbrecht and Woodard (1990) have presented seismic frequency changes of low- and intermediate-degree modes during the rising phase of cycle 22. They found that the changes were approximately inversely proportional to the inertiae of the modes, indicating that the predominant structural variations are confined to the near-surface layers of the sun. But there might be another component of the variation, an oscillatory component barely discernible to the eye, which could be indicative of a localized temporally varying acoustic glitch. However, it may not be real, and indeed Antia and Basu have declared it to be insignificant (e.g. Gough 2002). Goldreich *et al.* (1991) suggested that it might be due to a thin sheet of horizontal magnetic field buried somewhere in the sun, as had been discussed by Gough and Thompson (1988, 1990) and Vorontsov (1988). The oscillatory feature would therefore be expected to be greatest at sunspot maximum. Subsequently I measured the frequency of the oscillation (Gough, 1994), and found it to to be about 700s, corresponding roughly to the depth of the HeII ionization zone and therefore locating the glitch in the convection zone. It seems quite unlikely that the integrity of a magnetic sheet could be maintained against the disruptive influence of the turbulent convection; and indeed numerical simulations by Tobias and his collaborators (2001) have supported that view. Instead, it is more plausible that a magnetic field in the convection zone would be more evenly distributed, on a vertical length scale greater than the helium glitch. Therefore an increase in the intensity of the field, which might be expected at solar maximum, would actually dilute the glitch and thereby reduce the amplitude, Γ, of the oscillatory feature, not augment it. (A tangled field would act similarly.)

Whether the evidence for a variation in Γ is significant or not, an upper bound can be set on its magnitude from Libbrecht and Woodard's observations: $\Delta \ln \Gamma \lesssim 0.025$. That corresponds to a variation in the horizontal magnetic field given by $\sqrt{(\Delta B^2)} \lesssim 2.5$T. Were that bound to be achieved, the associated magnetic energy variation would exceed the local energy density in the convective motion by nearly a factor 10, a result which, as Jüri and his collaborators (e.g. these proceedings) have demonstrated, is not dynamically impossible. The strength of a tangled field would be yet greater.

The variation in cycle 23 was rather different from its predecessor. Basu and Mandel (2004) studied fourth differences (with respect to order n) of seismic frequencies, from which they claimed to have found the first evidence for structural changes with solar activity. In keeping with earlier discussions, they wrote that they believed the changes to be caused by a magnetic field, although they made no attempt to estimate its magnitude. Soon afterwards, Verner, Chaplin and Elsworth (2006) obtained a qualitatively similar result from raw frequencies of low-degree modes. The magnitude of the frequency variations imply $\sqrt{\left(\Delta B^2\right)} \simeq 10\mathrm{T}$, assuming the field to be predominantly horizontal. There has been much said and much written about the anomalies of the last solar cycle – or at least about the long delay between its decline and the onset of the new cycle. Here is another difference, although, because we have no pertinent seismic data prior to cycle 22, we do not know whether it indicates an anomaly in cycle 23 or one in cycle 22.

An important consequence of these investigations is that the quite substantial temporal variations of the oscillatory component of the seismic frequencies must be taken into account when trying to infer helium abundance. That was not done in the model calibrations by Houdek and myself discussed in §2, nor in the original calibrations of the HeII hump (e.g. Däppen *et al.*, 1988). It appears, therefore, that Y has been underestimated (and with it Z), implying that the sun is closer to typical standard models than we have recently surmised.

7. Deep meridionial circulation and magnetic field

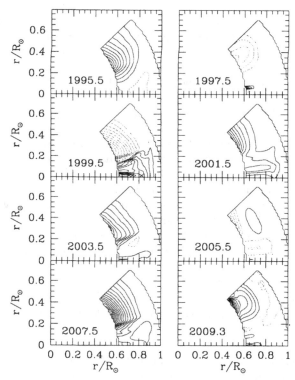

Figure 1. Lines of meridional magnetic field in a quadrant of the sun at different epochs as inferred by Antia, Chitre and Gough (unpublished) from GONG and SOI/MDI helioseismic data ignoring advection by the meridional flow. Continuous curves indicate anticlockwise field loops, dashed curves clockwise.

Direct helioseismological measurement of deep meridional flow is difficult because the effects on global seismic modes of north-south advection – or indeed the effects of any radial flow or zonal flow with zero longitudinal average – do not perturb the frequencies to leading order in the local Doppler shift. Flow in the outermost layers of the sun can be detected by localized Doppler measurements of high-degree waves that are damped within a circumundulation time and do not cohere to form standing waves; but deeply penetrating modes live longer, and cannot be analysed in that way. One must adopt a procedure for measuring the distortion of the eigenfunctions, either from leakage under projection onto putative undistorted wave forms (Schou *et al.*, 2009), variations in temporal phase (Gough and Hindman, 2010), or directly by some technique such as telechronoseismology (e.g. Duvall and Hanasoge, 2009); no valid seismic detection has yet been reported.

The interior magnetic field is also difficult to measure with confidence, because corresponding to any (at least axisymmetric) magnetic configuration is an isospectral density and sound-speed configuration (Zweibel and Gough, 1995). It is necessary to augment the seismic data with nonseismic information, or assumption, to draw any inference. For example, it is hardly possible for the solar-cycle HeII-glitch variation discussed in the previous section to have been produced by a thermal anomaly, and certainly not by a change in the chemical composition or the equation of state.

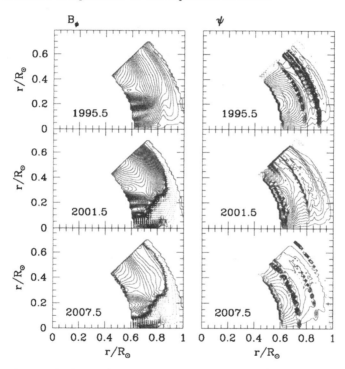

Figure 2. The right panels depict first iterates to determine the streamlines of meridional flow associated with a selection of the magnetic-field configurations illustrated in Fig. 1; continuous curves indicate anticlockwise flow, dashed curves clockwise. Panels on the left depict contours of constant B_ϕ, separated by 0.02T, such that continuous curves circle the axis of rotation positively, dashed curves negatively.

I conclude my discussion by reporting briefly on an indirect procedure currently being carried out by Antia, Chitre and myself to use seismically determined angular-velocity variations to infer the axisymmetric component of a putative magnetic field and associated meridional flow. By necessity the procedure is fraught with assumption, but we believe that it points to a way forward. The idea is to find that magnetic field and meridional flow that are consistent with magnetic induction, assuming, for the moment, that the angular-velocity variations are produced entirely by Maxwell stresses. Thus the meridional components of the momentum equation and the thermal energy equation –

the difficult equations that are at the centre of convection and dynamo theory – are not addressed, but are instead replaced by seismic observations of the angular velocity $\Omega(r, \theta, t)$, and, to provide boundary conditions for the analysis, direct measurements of the flow and the line-of-sight magnetic field in the photosphere. I refrain from burdening you with the details of how we are doing it, except to say that we are progressing slowly from a procedure with many assumptions which we wish to shed one by one, or, failing that, whose influence on the robustness of the results we shall try to ascertain. To date we have ignored microscopic and turbulent diffusion of magnetic field and momentum.

Our first experiment was to ignore meridionial flow entirely, and determine merely the field that produces a Maxwell stress compatible with the angular-velocity variation. Results are illustrated in Fig. 1. They extend from the declining phase of cycle 22 essentially to the present. We have not yet succeeded in obtaining the field close to the poles.

In Fig. 2 we present a few examples of the meridional flow associated with that field, which we estimated as a linear perturbation, ignoring its distorting effect on the field. One might note with cautious interest that at some latitudes the flow reverses beneath the depth at which direct seismic probing has yet been possible. But we warn that even the solution of this simple idealization is not yet complete. One might note that the azimuthal component B_ϕ of the magnetic field reaches values of a few tenths Tessla, which is much lower than the values inferred from the seismic frequency variations reported by Basu and Mandel, and Verner, Chaplin and Elsworth. We are now trying to learn how to iterate to a fully consistent solution in which the field is properly advected by the flow. Maybe, if we ever get close enought to reality, we'll be able to address some aspects of the superb simulations that Jüri and his colleagues have produced.

I am grateful to the Leverhulme Foundation for an Emeritus Fellowship, and to P. Younger for typing the manuscript.

References

Anders, E. & Grevesse, N., 1989, *Geochim. Cosmochim. Acta*, **53**, 197

Asplund, M., Grevesse, N. & Sauval, A. J., 2005, *Cosmic Abundances as Records of Stellar Evolution and Nucleosynthesis in honor of David L. Lambert*, ASP Conf. Ser., **336**, 25

Asplund, M., Grevesse, N., Sauval, A. J., & Scott, P., 2009, *Ann. Rev. Astron. Astrophys.*, **47**, 481

Basu, S. & Antia, H. M., 2008, *Phys. Rep.*, **457**, 217

Basu, S. & Mandel, A., 2004, *Astrophys. J.*, **617**, L155

Baturin, V. A., Däppen, W., Gough, D. O., & Vorontsov, S. V., 2000, *Mon. Not. R. Astron. Soc.*, **316**, 71

Bonanno, A., Schlattl, H., & Paternò, L., 2002, *Astron. Astrophys.*, **390**, 1115

Brun, A. S. & Zahn, J.-P., 2006, *Astron. Astrophys.*, **457**, 665

Brun, A. S., Turck-Chièze, S., & Zahn, J.-P., 1999, *Astrophys. J.*, **525**, 1032

Caffau, E. *et al.*, 2010, *Astron. Astrophys.*, **514**, 92

Christensen-Dalsgaard, J., Duvall, T. L., Jr, Gough, D. O., Harvey, J. W., & Rhodes, E. J., Jr, 1985, *Nature*, **315**, 378

Christensen-Dalsgaard, J., Proffitt, C. R., & Thompson, M. J., 1993, *Astrophys. J.*, **403**, L75

Christensen-Dalsgaard, J. *et al.*, 1996 *Science*, **272**, 1286

Däppen, W., Gough, D. O., & Thompson, M. J., 1988, *Seismology of the Sun and Sun-like stars* (ed. E.J. Rolfe, ESA SP-286 Noordwijk), 505

Duvall, T. L., Jr & Hanasoge, S. M., 2009, *GONG 2008/SOHO 21 ASP Conf. Ser.* **416**, 103

Elliott, J. R. & Gough, D. O., 1999, *Astrophys. J.*, **516**, 475

Goldreich, P., Murray, N., Willette, G., & Kumar, P., 1991, *Astrophys. J.*, **370**, 752

Gough, D. O., 1984, *Mem. Soc. Astron. Italiana*, **55**, 13

Gough, D. O., 1990, *Progress of seismology of the Sun and stars*, (ed. Y. Osaki & H. Shibahashi, Springer, Heidelberg), *Lecture Notes in Physics*, **267**, 283

Gough, D. O., 1994, *The Sun as a variable star*, (ed. J.M. Pap, C. Fröhlich, H.S. Hudson & S.K. Solanki, Cambridge Univ. Press, Cambridge), *Proc. IAU Colloq.* **143**, 252

Gough, D. O., 2001, *Astrophysical Ages and Time Scales*, (ed. T. von Hippel, C. Simpson & N. Manset), *Astron. Soc. Pacific Conf. Ser.*, **245**, 31

Gough, D. O., 2002, *Proc. SOHO 11 Symposium*, (ed. A. Wilson, ESA SP-508, Noordwijk), 577

Gough, D. O., 2004, *Proc. SOHO 17: 10 years of SOHO and beyond*, (ed. H. Lacoste & L. Ouwehand, ESA SP-617, Noordwijk), 1

Gough, D. O. & Hindman, B., 2010, *Astrophys. J.*, **714**, 960

Gough, D. O. & Lynden-Bell, D., 1968, *J. Fluid Mech.* **32**, 437

Gough, D. O. & McIntyre, M. E., 1998, *Nature*, **394**, 755

Gough, D. O. & Thompson, M. J., 1988, *Advances in helio- and asteroseismology* (ed. J. Christensen-Dalsgaard & S. Frandsen, *Proc. IAU Symp.* **123**, Reidel, Dordrecht), 175

Gough, D. O. & Thompson, M. J., 1990, *Mon. Not. R. Astron. Soc.*, **242**, 25

Guzik, J. A., Watson, L. S., & Cox, A. N., 2005, *Astrophys. J.*, **627**, 1049

von Hippel, T., Simpson, C., & Manset, N. (ed.), 2001, *Astrophysical Ages and Time Scales*, *Astron. Soc. Pacific Conf. Ser.*, **245**

Houdek, G. & Gough, D. O., 2007, *Mon. Not. R. Astron. Soc.*, **375**, 861

Houdek, G. & Gough, D. O., 2009, *Comm. Asteroseismology*, **159**, 27

Kosovichev, A. G. *et al.*, 1992, *Mon. Not. R. Astron. Soc.*, **259**, 356

Libbrecht, K. G. & Woodard, M. F., 1990, *Lecture Notes in Phys*, **367**, 145

Mussack, K. & Gough, D. O. 2009, *Proc. GONG 2008 / SOHO XXI, ASP Conf. Ser.*, **416**, 203

Schou, J., Woodard, M. F., & Birch, A. C., 2009, *Bull. Am. Astron. Soc.*, **41**, 813

Serenelli, A. M. & Basu, S., 2010, *Astrophys. J.*, **719**, 865

Spiegel, E. A. & Zahn, J.-P., 1992, *Astron. Asstrophys.*, **265**, 106

Takata, M. & Gough, D. O., 2003, *Local and global helioseismology: the present and future, Proc. SOHO12/GONG+2002*, (ed. A. Wilson, ESA SP-517, Noordwijk), 397

Takata, M. & Shibahashi, H., 2003, *Publ. Astron. Soc. Japan*, **55**, 1015

Tobias, S. M., Brummell, N. H., Clune, T., & Toomre, J., 2001, *Astrophys. J.*, **549**, 1183

Turck-Chièze, S. *et al.*, 2001, *Astrophys. J.*, **555**, L69

Verner, G. A., Chaplin, W. J., & Elsworth, Y., 2006, *Astrophys. J.*, **640**, L95

Vorontsov, S. V., 1988, *Advances in helio- and asteroseismology* (ed. J. Christensen-Dalsgaard & S. Frandsen, *Proc. IAU Symp.* **123**, Reidel, Dordrecht), 151

Zweibel, E. G. & Gough, D. O., 1995, *Proc. Fourth SOHO Workshop: Helioseismology*, (ed. J.T. Hoeksema, V. Domingo, B. Fleck & B. Battrick, European Space Agency SP-376, Noordwijk), vol 2, p.73

Discussion

THOMPSON: You talked about the 700s oscillatory signal in the $\nu_{max} - \nu_{min}$ solar-cycle variations, and said that a diffuse magnetic field in the HeII ionization zone would dilute the signal of the HeII glitch. So what is the phase of the oscillation?

GOUGH: Because magnetic field dilutes the acoustic glitch, the phase of the oscillatory variation from sunspot minimum to sunspot maximum deviates from the phase of the mean signal, as depicted by Verner, Chaplin and Elsworth, by π. It would deviate likewise from the fourth differences plotted by Basu and Mandel, were it not for the frequency variation of the amplitude of the signal which produces an additional small deviation. This phase is consistent with that of the variations reported by Basu and Mandel, and Verner, Chaplin and Elsworth.

HILL: Have you estimated the magnitude of the meridional flow as a function of r?

GOUGH: Not really. It would be dangerous to make inferences from an unconverged iteration (although it does appear that the velocity increases with depth immediately beneath the photosphere). I showed the picture of that iteration merely to whet my (and, I hope, others') appetites.

ZWEIBEL: Would the convection-zone magnetic field affect the stratification and thereby the baroclinic terms in the equation for Ω, producing evolution of Ω in addition to that produced by Maxwell stresses?

GOUGH: The magnetic field may have a significant influence on baroclinicity in the upper layers of the convection zone – perhaps the outer 15% by radius – where the density is relatively low, thereby adding to the complexity of the dynamics of the meridional flow which advects Ω. One should not forget that anisotropic Reynolds stresses are no doubt also important. As you know, our investigation is in a very early stage, and so far we have ignored those processes; but they are on our list of matters that we intend to investigate. Indeed, the purpose of our exercise is not merely to produce field and flow configurations that might plausibly reflect those in the sun, but primarily to understand the mechanisms that generate them.

TOOMRE: You side-stepped the heavy-element issue by saying that the low Z values deduced from surface observations appear to gradually increase with recent reanalyses. So possibly there is no real problem with the theoretical structure models.

GOUGH: At the moment it is difficult, even for the executors, to judge the accuracy of abundance 'determinations'. Yes, I did point out that reported values of Z have tended to increase with time since Martin Asplund's first announcement, but that was due mainly, although not entirely, to new independent investigators entering the fray. On the whole we believe that the precision of scientific measurements increases with the passing time, and we hope that the accuracy does too. However, the latter is not always the case; indeed, Martin Asplund's original work in this area exemplified that. I suspect that the value of the photospheric Z will settle down soon, and that it will end up being lower than the value of Z that seismic models elaborated with the tenets of standard stellar-evolution theory require of the radiative interior.

Astrophysical Dynamics: From Stars to Galaxies
Proceedings IAU Symposium No. 271, 2010
N. Brummell, A.S. Brun, M. S. Miesch & Y. Ponty, eds.
© International Astronomical Union 2011
doi:10.1017/S174392131101742X

Helioseismic Observations of Solar Convection Zone Dynamics

Frank Hill, Rachel Howe, Rudi Komm, Irene González Hernández, Shukur Kholikov, and John Leibacher

National Solar Observatory, Tucson, Arizona, USA 85726

Abstract. The large-scale dynamics of the solar convection zone have been inferred using both global and local helioseismology applied to data from the Global Oscillation Network Group (GONG) and the Michelson Doppler Imager (MDI) on board SOHO. The global analysis has revealed temporal variations of the "torsional oscillation" zonal flow as a function of depth, which may be related to the properties of the solar cycle. The horizontal flow field as a function of heliographic position and depth can be derived from ring diagrams, and shows near-surface meridional flows that change over the activity cycle. Time-distance techniques can be used to infer the deep meridional flow, which is important for flux-transport dynamo models. Temporal variations of the vorticity can be used to investigate the production of flare activity. This paper summarizes the state of our knowledge in these areas.

Keywords. Sun: helioseismology, Sun: interior, Sun: activity

---

1. Introduction

Since the advent of helioseismology in 1961, substantial advances have been made in our understanding of the large-scale dynamics in the solar convection zone. The modern era of helioseismology started in 1995 with the deployment of the Global Oscillation Network Group (GONG) set of ground-based instruments and the launch of the Solar and Heliospheric Observatory (SOHO) carrying three helioseismology instruments (MDI, GOLF and VIRGO). Now, with 15 years of data we can observe a number of features of the flows in the solar convection zone. Here we will briefly describe some of the latest observations pertaining to the torsional oscillation and solar cycle timing. We will also discuss deep and shallow meridional flows and divergence observations. Finally the relationship between vorticity, active regions and flares will be briefly reviewed.

2. Torsional Oscillation and Solar Cycle Timing

Howard & LaBonte (1980) discovered a plasma stream on the solar surface that was slightly faster than the average surface rotation rate and that migrated from the solar poles to the equator over the course of the solar cycle. They named this east-west zonal flow the "torsional oscillation" since it had some of the characteristics of the process postulated by Walén (1944) as the basis of the solar cycle. Solar activity forms at the high-latitude boundary of the equatorward torsional oscillation stream, suggesting a close link between the flow and the activity. Long-term observations from Mt. Wilson show asymmetric patterns of the flow in the north and south solar hemispheres (Ulrich & Boyden 2005). A similar pattern is observed in coronal emissions (Altrock *et al.* 2008).

Helioseismology has revealed that the torsional oscillation is not confined to the solar surface, but extends downward to depths of at least 60 Mm, half way through the solar convection zone (Howe *et al.* 2000, Vorontsov *et al.* 2002). It also shows that there is

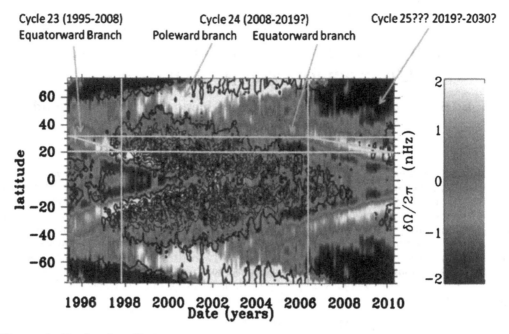

Figure 1. Torsional oscillation at a depth of 7 Mm as a function of latitude and date. The magnetic field is overlaid as contours of 5 G. The labels at the top indicate the flows associated with cycles 23 and 24, and where the first sign of cycle 25, which is apparently late, should appear.

a variation in the solar rotation rate near the poles that can be interpreted as either a poleward branch of the torsional oscillation or as an alternating polar acceleration and deceleration. Fig. 1 shows the pattern of the torsional oscillation at a depth of 7 Mm as a function of latitude and date determined from global inversion of GONG and SOHO/MDI data, with a contour overlay of the surface magnetic field.

It now appears that the timing of the solar cycle is well-correlated with the temporal evolution of the torsional oscillation (Howe *et al.* 2009). For example, the rate of migration of the equatorward branch is apparently related to the length of the minimum. Fig. 1 demonstrates that, for cycle 24, the equatorward migration took 1.5 years longer to reach a latitude of $20 - 25°$, the point at which the activity ramped up in cycle 23. The recent minimum was also about 1.5 years longer. Fig. 1 also shows that, when the equatorward branch reached the $20 - 25°$ latitude range for cycle 23, the surface activity increased rapidly. The same behavior is now being seen for cycle 24. Howe *et al.* (2009) pointed out that the position of the flow belts by mid-2009 corresponded to that at the epoch in the previous cycle when solar activity was just about to start rising – and indeed, by the end of 2009 the new cycle was finally underway. The equatorward branch appears several years prior to the surface magnetic field. For cycle 24, the equatorward branch appeared at the end of 2002, some 7 years prior to the activity. It is thus an early indication of the cycle timing. Finally, the poleward branch appears even earlier, about when the surface activity for the prior cycle increases rapidly. This was approximately 12 years in the case of cycle 24. Note that the cycle 25 poleward branch has not yet appeared. Does this mean that cycle 25 will be very weak or even non-existent? Or is it simply delayed?

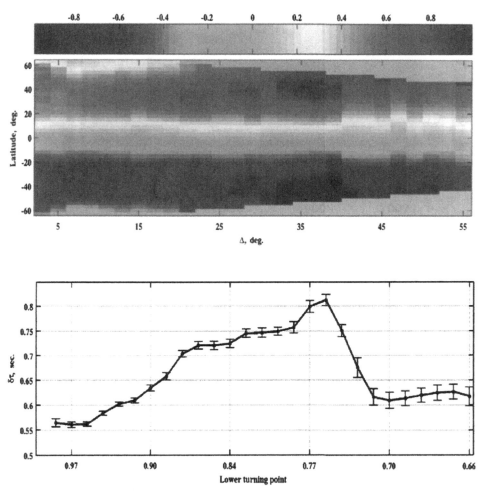

Figure 2. Deep meridional flow travel-time differences. Top: Average of 15 years of GONG meridional flow measurements as a function of the separation between the two latitudes that are correlated. The lower turning point for waves with travel distance of about 45° reaches the tachocline region. Bottom: mean time difference averaged over both hemispheres for the latitude range 20° − 40° as a function of propagation depth. The time difference increases until just before the bottom of the convective zone, then it sharply drops.

3. Meridional Flows and Divergence

According to flux-transport dynamo models, the north-south meridional flow plays an important role in determining the characteristics of the solar cycle (Dikpati & Gilman 2009). In particular, the magnitude of the deep return flow (thought to be located at the base of the convection zone) from the pole to the equator sets the strength and timing of the activity. From estimates of the signal-to-noise ratio for measurements focused near the base of the convection zone, Braun & Birch (2008) concluded that helioseismic measurements needed to span a solar cycle in order to detect the 1-2 m/s return flow required by mass conservation. Fig. 2 shows the travel-time differences for 15 years of GONG data derived from near-sectoral modes that are primarily sensitive to the meridional flow. There is a distinct and significant change in the nature of the differences at

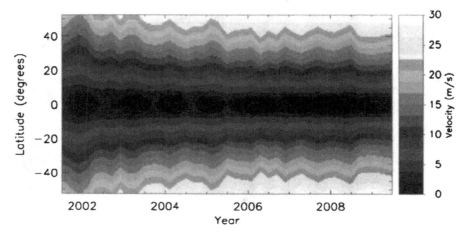

Figure 3. Temporal variation of a fitted polynomial to meridional circulation observations inferred at a depth of 5.8 Mm. Positive velocities are taken towards each respective pole.

the bottom of the convection zone, corresponding to a jump of about 2 m/s with errors of about 0.2 m/s (Leibacher, Kholikov & Hill 2010). This indicates that these measurements have adequate signal-to-noise to resolve the predicted return flow amplitude. Determining the actual flow requires an inversion to cancel out near-surface effects. Preliminary inversions surprisingly indicate that there are multiple meridional cells in depth, and that the return flow may actually lie in the radiative envelope below the base of the convection zone. More work must be done to verify these results.

The meridional flow immediately below the surface contributes to the evolution of activity by transporting the surface magnetic field to the poles, canceling and reversing the polarity of the polar field. Since the strength of the polar field is correlated with the maximum level of activity in the subsequent cycle, variations in the near-surface meridional flow also play a role in setting the characteristics of the cycle. Fig. 3 shows that the near-surface meridional flow has been increasing in magnitude over the declining phase of cycle 23 (González Hernández *et al.* 2010). Hathaway & Rightmire (2010) postulate that this increase of speed is one of the factors contributing to the peculiar minimum between cycles 23 and 24, and that cycle 24 will turn out to be a weak one.

The ring-diagram analysis of local helioseismology provides the horizontal flow field as a function of depth and heliographic position. Using the continuity equation and assuming incompressibility on large spatial scales allows us to derive the vertical velocity field, the divergence, and the vorticity (Komm 2007). Fig. 4 shows the divergence near the surface as a function of depth and latitude during maximum and minimum activity. The pattern of divergence indicates the presence of multiple cells in the north-south meridional direction, with a convergence at the active latitudes. The pattern, though substantially weaker, is also present at solar minimum.

It thus seems that there is considerable structure in the solar meridional flow in the form of multiple cells in both latitude and depth, and it is likely that the characteristics of these structures will vary in time. These aspects of the solar dynamics must be incorporated into dynamo models if the models are to explain the solar cycle. Indeed, recent simulations (e.g. Muñoz-Jaramillo, Nandy, & Martens 2009, Dikpati *et al.* 2010) are beginning to include these new results into the calculations.

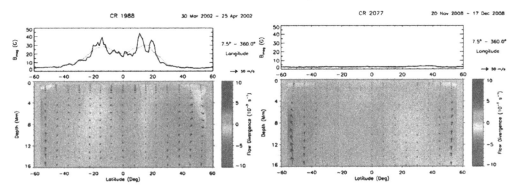

Figure 4. The divergence of flows as a function of depth and latitude for maximum (left) and minimum (right) activity levels. A multi-cellular pattern in latitude is visible that is more prominent during maximum.

4. Subsurface Vorticity, Active Regions and Flares

As stated, the ring diagram method can provide measurements of the vorticity below the photosphere. Mason *et al.* (2006) found that there is a specific pattern of vorticity below virtually every active region in the declining phase of cycle 23. Fig. 5 shows the pattern, which appears as four regions of oppositely directed horizontal vorticity. These can be interpreted as cross-sections of two vortex rings stacked on top of each other, with oppositely directed circulation in each ring. This is consistent with a downdraft from the solar surface caused by the cooling effect of the magnetic field meeting an upflow at a depth of about 10 Mm.

Further work found a correlation between the flaring activity in an active region, the strength of the subsurface vorticity, and the magnitude of the surface magnetic field (Komm & Hill 2009). Fig. 6 shows that active regions with both strong vorticity and strong surface field are very likely to produce many strong flares. This suggests that the routine measurement of vorticity below active regions would provide a useful space weather forecast tool. In addition, the temporal variation of the subsurface vorticity has been demonstrated to be able to predict the occurrence of flares as much as three days in advance (Reinard *et al.* 2010). The physical mechanism is that the kinetic energy contained in the vortex rings is transformed into a twisting of the magnetic field that eventually overcomes the repulsion of like polarity and explosively reconnects.

5. Conclusion

Helioseismology has now matured to the point that it is revealing a number of new aspects of the relationship between internal solar dynamics and surface activity. The characteristics of the torsional oscillation, the meridional flow, and the vorticity below active regions seem to play major roles in determining the solar cycle and the evolution of the surface magnetic field on both long and short time scales. It thus seems that models of the surface activity must incorporate the sub-surface flow fields in order to accurately capture the evolution of the magnetic field in the photosphere and the corona.

6. Acknowledgments

This work was partially supported by NASA GIP Grant #NNG08EI54I and NASA NNH07AG21I. This work utilizes data obtained by the Global Oscillation Network Group

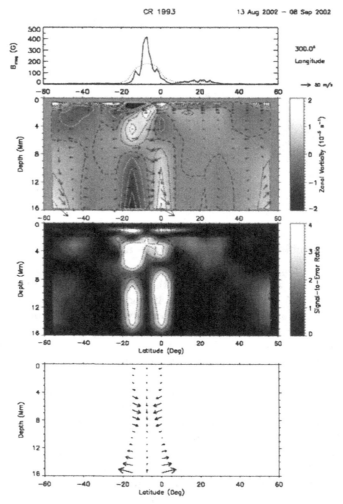

Figure 5. Zonal vorticity at 300° longitude in CR 1993 as a function of latitude and depth. Top: Gross magnetic flux (solid line) and binned over 15° (dotted line). Second panel: the x-component of vorticity. The arrows represent the meridional and vertical velocity components with the vertical magnitude increased by a factor of ten for visibility. Third panel: The signal-to-error ratio. Bottom panel: Idealized schematic of flows below a strong active region (with arbitrary amplitudes). From Mason *et al.* (2006), reproduced by permission of the AAS.)

(GONG) project, managed by the National Solar Observatory, a division of the National Optical Astronomy Observatories, which are operated by AURA, Inc., under a cooperative agreement with the National Science Foundation. The data were acquired by instruments operated by the Big Bear Solar Observatory, High Altitude Observatory, Learmonth Solar Observatory, Udaipur Solar Observatory, Instituto de Astrofísica de Canarias, and Cerro Tololo Inter-American Observatory.

On a personal note, since this Symposium was in honor of Juri Toomre, FH would like to thank Juri for all of the support and guidance he has provided to himself and to the GONG project over the years.

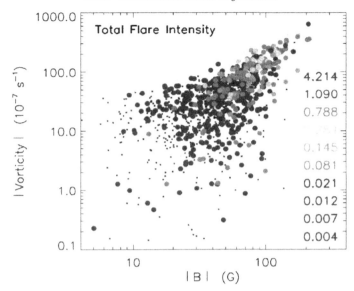

Figure 6. The location of approximately 1000 active regions from 2001 to 2006 in a (vorticity, surface magnetic field) plane. The size and gray scale of the points indicate the total flare productivity in units. The most flare-productive active regions have the highest values of both vorticity and surface magnetic field strength. From Komm & Hill (2009), reproduced by permission of the AGU.

References

Altrock, R., Howe, R., & Ulrich, R. K. 2008, in: R. Howe, R. W. Komm, K. S. Balasubramaniam & G. J. D. Petrie (eds.), *Subsurface and Atmospheric Influences on Solar Activity* (San Francisco: Astronomical Society of the Pacific), p. 335

Braun, D. C. & Birch, A. C. 2008, *ApJ* (Letters), 689, L161

Dikpati, M. & Gilman, P. A. 2009, *Space Sci. Revs*, 144, 67

Dikpati, M., Gilman, P. A., de Toma, G., & Ulrich, R. K. 2010, *Geophys. Res. Lett* 37, L14107

González Hernández, I., Howe, R., Komm, R. W., & Hill, F. 2010, *ApJ* (Letters), 713, L16

Hathaway, D. H. & Rightmire, L. 2010, *Science*, 327, 1350

Howard, R. & LaBonte, B. J. 1980, *ApJ* (Letters), 239, L33

Howe, R., Christensen-Dalsgaard, J., Hill, F., Komm, R. W., Larsen, R. M., Schou, J., Thompson, M. J., & Toomre, J. 2000, *ApJ* (Letters), 533, L163

Howe, R., Christensen-Dalsgaard, J., Hill, F., Komm, R. W., Schou, J., & Thompson, M. J. 2009, *ApJ* (Letters), 701, L87

Komm, R. W. 2007, *AN*, 328, 269

Komm, R. & Hill, F. 2009, *JGR*, 114, A06105

Leibacher, J. W., Kholikov, S., & Hill, F. 2010, *BAAS*, 42, 823

Mason, D. Komm, R. Hill, F. Howe, R. Haber, D. A., & Hindman, B. W. 2006, *ApJ*, 645, 1543

Muñoz-Jaramillo, A., Nandy, D., & Martens, P. C. H.. 2009, *ApJ*, 698, 461

Reinard, A. A., Henthorn, J. Komm, R., & Hill, F. 2010, *ApJ* (Letters), 710, L121

Ulrich, R. K. & Boyden, J. E. 2005, *ApJ* (Letters), 620, L123

Vorontsov, S. V., Christensen-Dalsgaard, J., Schou, J., Strakhov, V. N., & Thompson, M. J. 2002, *Science*, 296, 101

Walén, C. 1944, *Arkiv f. Mat., Astron. o. Fys.*, 30A (15), 1

Discussion

BRANDENBURG: The torsional oscillation pattern from helioseismology assumes north-south symmetry. How does this compare with the observed torsional oscillation at the surface?

HILL: The surface measurements show a difference between the zonal flow in the north and south hemispheres. The measurements from global helioseismology are necessarily symmetric across the equator due to the spherical harmonic decomposition of the data. However, we have also determined the torsional oscillation pattern from the ring diagram method of local helioseismology and these results show a north-south asymmetric pattern.

TOOMRE: What are the areas of emphasis for GONG science, now in its mature phase, for the coming decade?

HILL: The helioseismic section of GONG will focus on observations of the sub-surface flows and their relationship to the solar cycle and the surface activity. Since the solar cycle occurs on decadal time scales, we need continual, consistent, and long-term measurements of the interior dynamics to understand it. GONG will also develop space weather forecast tools based on helioseismology, investigate the relationship of the flows and the direction of the magnetic field using vector magnetograms, and seek to detect sub-surface magnetic fields.

THOMPSON: You showed the positive correlation between inferred subsurface vorticity and surface magnetic field. I still have a nagging concern that these measurements are made in regions of strong magnetic field where we don't fully understand the forward problem of wave propagation. Do you agree that there may be systematic errors and the vorticity could be an artifact?

HILL: It seems to me that the ring-diagram analysis is less susceptible to magnetic contamination in comparison with the time-distance approach. The major concern for the rings is the suppression of the amplitude of the oscillations in active regions. We can test this by artificially suppressing the p-mode amplitudes in quiet-sun control areas, and reanalyzing the data to see if the vorticity pattern appears.

Astrophysical dynamics: from stars to galaxies
Proceedings IAU Symposium No. 271, 2010
N. Brummell, A.S. Brun, M.S. Miesch & Y. Ponty, eds.

Large-scale magnetic fields of low-mass dwarfs: the many faces of dynamo

J.-F. Donati

CNRS/Université de Toulouse, Observatoire Midi-Pyrérées, Toulouse, France
email: donati@ast.obs-mip.fr

Abstract. Magnetic field are ubiquitous to low-mass stars and can potentially impact their evolution and their internal structure; yet the physical processes (called dynamo) that succeed at generating them in the stellar convective zones of cool dwarfs are still enigmatic. Although theoretical modelling and numerical simulations of stellar dynamo action showed breathtaking progress in the last decade, they are not yet in the state of accurately predicting the various magnetic topologies that different low-mass stars can generate.

Thanks to the advent of new-generation instruments, spectropolarimetric observations can now reveal the large-scale magnetic topologies of cool dwarfs, from the brown dwarf threshold (spectral type M8) up to the limit beyond which outer convective zones get vanishingly small (spectral type F5). In particular, they can reconstruct through tomographic methods the poloidal and toroidal components of the large-scale field, hence offering a fresh option for guiding dynamo theories to a more mature state.

We review here the latest observational advances, showing in particular that magnetic topologies of low-mass dwarfs can drastically vary with mass and rotation rate, and discuss their implications for our understanding of dynamo processes.

Keywords. stars: magnetic fields, stars: imaging, stars: rotation, stars: late-type, techniques: polarimetry

1. Introduction

Most cool stars exhibit a large number of solar-like activity phenomena; dark spots are present at the their surfaces (e.g., Berdyugina 2005), where they come and go on timescales ranging from days (as they are carried in and out of the observer's view by the star's rotation) to months (as they appear and disappear over a typical spot lifetime) and years or decades (with spots fluctuating in number and location throughout activity cycles). Prominences are also detected in cool stars, both as absorption and emission transients (e.g., in Balmer lines) tracing magnetically confined clouds (e.g., Cameron & Robinson 1989) either transiting the stellar disc (and scattering photons away from the observer, as for dark filaments on the Sun) or seen off-limb (and scattering photons towards the observer, as for bright prominences on the Sun). Cool stars are also surrounded by low-density coronal plasma at MK temperatures showing up at various wavelengths in the spectrum (e.g., radio, X-ray and optical line emission) and associated with frequent flaring, recurrent coronal mass ejections, and winds. Activity phenomena in cool stars scale up with faster rotation and later spectral types (e.g., Hall 2008).

The current understanding is that activity phenomena are a by-product of the magnetic fields that cool stars generate within their convective envelopes through dynamo processes, involving cyclonic turbulence and rotational shearing. In the particular case of the Sun, dynamo processes are presumably concentrating in a thin interface layer (the so-called tachocline) confined at the base of the convective zone (CZ) and where rotation gradients are supposedly largest. The spectacular images of the Sun collected with

TRACE and (more recently) HINODE demonstrate that the activity of the Sun very tightly correlates with the presence of magnetic field emerging from the surface, either in the form of large closed loops (mostly at medium latitudes) or open field lines (mostly at high latitudes); the exact process through which magnetic fields succeed at heating the tenuous outer atmosphere to MK temperatures is however still unclear.

Cool stars are assumed to behave similarly. This is supported by observations showing that activity scales up with rotation rate (at any given spectral type), as suggested by dynamo theories. One of the key parameter for measuring the efficiency of magnetic field generation is the Rossby number Ro, i.e., the ratio of the rotation period of the star to the convective turnover time. It describes how strongly the Coriolis force is capable of affecting the convective eddies, with small Ro values indicating very active stars rotating fast enough to ensure that the Coriolis force strongly impacts convection. The observation that activity correlates better with Ro than with rotation (e.g., Noyes et al. 1984, Mangeney & Praderie 1984) or equivalently, that cooler stars are relatively more active at a given rotation rate, agrees well with the theoretical expectation that convective turnover times increase with decreasing stellar luminosities.

Magnetic fields are also responsible for slowing down cool stars through the braking torque of winds magnetically coupled to the stellar surface (Schatzman 1962; Mestel 1999). This is qualitatively compatible with the fact that most cool stars rotate slowly (like the Sun itself), with the exception of close binaries (whose spin angular momentum is constantly refueled from the orbital reservoir through tidal coupling) and young stars (which have not had time yet to dissipate their initial load of angular momentum). Magnetised wind models yield a good match to the observed distribution of rotation periods in young open clusters of ages ranging from several tens to several hundreds of Myr, further confirming that magnetic fields are likely what triggers the spinning down of cool stars as they arrive on the main sequence.

The main lesson from the solar paradigm is thus that dynamo processes are essentially ubiquitous in all cool stars with outer convective layers and generate magnetic fields with a high degree of temporal variability at all timescales. Extrapolating the solar analogy much further is potentially hazardous; in particular, assuming that conventional dynamo models (entirely tailored to match observations of the Sun) also apply to cool stars with very different convective depths and rotation rates is subject to caution. In very active stars rotating 100 times faster than the Sun for instance, the magnetic feedback onto the convection pattern may be strong enough to distort theoretical dynamo patterns; similarly, very-low-mass fully-convective stars obviously lack the interface layer where conventional dynamo processes are expected to concentrate, but are nevertheless strongly active. Magnetic studies of low-mass stars are therefore our best chance for exploring the various faces of dynamo processes over a large range of masses and rotation rates.

2. Magnetic properties of cool stars

The very first estimates of magnetic fields in cool stars other than the Sun were obtained by measuring the differential broadening of spectral lines as a function of their magnetic sensitivities (Robinson et al. 1980), making it possible to derive the first trends on the magnetic properties of low-mass stars (e.g., Saar 2001). These studies find that the average surface magnetic strength is, in most cases, roughly equal to the equipartition field, i.e., the field whose magnetic pressure balances the thermal pressure of the surrounding gas; only very active stars with rotation periods lower than about 5 d (among which fully-convective M dwarfs and young low-mass protostars) strongly deviate from this relation (e.g., Johns-Krull & Valenti 1996) A similar behaviour is observed in the

Sun, where fields of moderately active plages are close to equipartition while those of active sunspots are stronger by a factor of 2 or more. This suggests that magnetic regions at stellar surfaces progressively evolve from a plage-like to a spot-like structure, with flux tubes having increasingly larger sizes or being more tightly packed, as activity increases.

These studies also find that the average magnetic flux at the surfaces of cool stars increases more or less linearly with $1/Ro$ until it saturates at $Ro \simeq 0.1$ (corresponding to a rotation period of about 2 d for a Sun-like star), with most of the increase being attributable to the fractional area covered with fields (at least in moderately active stars). The detection of a saturation regime, confirmed with new magnetic flux measurements from molecular lines in M dwarfs (Reiners & Basri 2008), supports the idea that magnetic fields are eventually capable of modifying, if not controlling, the convective motions through some feedback mechanism; this may potentially explain in particular why magnetic regions at low and high activity levels are morphologically different.

The first detections of Zeeman polarisation signatures from solar-type stars (Donati *et al.* 1997) and their tomographic modelling with stellar surface imaging tools such as Zeeman-Doppler Imaging (or ZDI, e.g., Donati *et al.* 1992) opened up an alternative option for studying dynamo processes. In particular, the medium- and large-scale magnetic fields accessible through ZDI, though energetically less important than magnetic fluxes derived from Zeeman broadening, are nevertheless optimally suited for checking topological predictions of dynamo models on global fields and their potentially cyclic variations, to which other methods are insensitive.

Initial studies, concentrating on a few very active rapidly rotating stars in the saturated-dynamo regime brought surprising results. In particular, they demonstrated that strong toroidal fields can show up directly at the stellar surface, in the form of monopolar regions of dominantly azimuthal fields or even complete rings encircling the star at various latitudes (Donati *et al.* 1992, Donati & Cameron 1997, Donati 1999, Donati *et al.* 2003a); while tori of strong azimuthal fields are likely present in the Sun at the base of the CZ (e.g., to account for the non-stochastic arrangement of surface sunspots, known as Hale's polarity law), they do not build up at the surface of the Sun - hence the surprise. The poloidal components detected on the active stars observed in these exploratory studies consist mainly of a significant non-axisymmetric term with alternating patterns of opposite radial field polarities. Other studies confirmed and amplified these initial results, reporting the presence of strong and often dominant toroidal fields at photospheric level (Dunstone *et al.* 2008), even in less active stars with longer rotation periods (Petit *et al.* 2005) or earlier spectral types (Marsden *et al.* 2006). A recent study focussing on main-sequence Sun-like stars with different rotation periods suggests that significant surface toroidal fields are detected whenever the rotation period is lower than $\simeq$20 d (Petit *et al.* 2008), i.e., $\simeq$25% shorter than the rotation period of the Sun.

ZDI observations also demonstrated that large-scale magnetic topologies of active stars are latitudinally sheared by surface differential rotation at a level comparable to that of the Sun (Donati & Cameron 1997), with the equator lapping the pole by one complete rotation cycle about every 100 d. This conclusion agrees with previous results derived from indirect tracers of differential rotation (photometric monitoring, e.g., Hall 1991). Differential rotation displays a steep increase with earlier spectral types, reaching values of 10 times the solar shear or more in late F stars (e.g., Marsden *et al.* 2006, Donati *et al.* 2008a). This trend is independently confirmed from observations of spectral line shapes (Reiners 2006) and suggests that F stars with shallow convective zones (CZs) are departing very strongly from solid-body rotation.

The major improvement in instrumental sensitivity brought by the twin new-generation spectropolarimeters ESPaDOnS@CFHT and NARVAL@TBL made it possible to start surveying the magnetic topologies of cool stars, from mid F to late M stars. It allowed in particular the large-scale field properties of M dwarfs to be investigated for the first time on both sides of the full convection threshold (presumably occurring at spectral type M4, i.e., at a mass of 0.35 $M_\odot$, Baraffe *et al.* 1998). Spectropolarimetric monitoring of the rapidly rotating M4 dwarf V374 Peg revealed that the star hosts a strong large-scale mostly-poloidal, mainly axisymmetric field despite its very short period (0.44 d), high activity level and low Ro (Donati *et al.* 2006, Morin *et al.* 2008a); additional observations of active mid-M dwarfs further confirmed that dynamo processes in fully-convective stars with masses of about 0.3 $M_\odot$ are apparently very successful at generating strong poloidal fields with simple axisymmetric configurations (Morin *et al.* 2008b).

Comparing to partly-convective early M dwarfs reveals that the transition in the large-scale field properties is fairly sharp and located at a mass of about 0.4 to 0.5 $M_\odot$ (Donati *et al.* 2008b), i.e., slightly above the 0.35 $M_\odot$ theoretical full-convection threshold. This sharp transition also coincides with a strong decrease in surface differential rotation (with photospheric shears smaller by a factor of 10 or more than that of the Sun) and with a strong increase in the lifetime of large-scale fields. Preliminary results on very-low mass stars (< 0.2 $M_\odot$) suggest that the situation is even more complex, with some stars hosting very strong and simple large-scale fields (like those of mid-M dwarfs) and some others with much weaker and complex magnetic topologies (resembling those of early-M dwarfs). Observations of a larger sample are needed to clarify the situation but the preliminary results already demonstrate that at least some very-low-mass stars are capable of producing a strong large-scale axisymmetric poloidal field. This conclusion is independently confirmed by the detection of highly-polarised rotationally-modulated radio emission from late M and early L dwarfs attributable to intense large-scale magnetic fields (e.g., Berger *et al.* 2005) through electron cyclotron maser instability (Hallinan 2008).

Figure 1 presents graphically the main results obtained so far in the framework of the ongoing survey effort, aimed at identifying which stellar parameters mostly control the field topology. To make it more synthetic, the plot focusses only on a few basic properties of the reconstructed magnetic topologies, namely the reconstructed magnetic energy density e (actually the integral of B^2 over the stellar surface), the fractional energy density p in the poloidal field component, and the fractional energy density a in mostly axisymmetric modes (i.e., with $m < \ell/2$, m and ℓ being the order and degree of the spherical harmonic modes describing the reconstructed field). Each selected star is shown in the plot at a position corresponding to its mass and rotation period, with a symbol depicting these three characteristics of the recovered large-scale fields, i.e., e (symbol size), p (symbol colour) and a (symbol shape). This plot clearly illustrates the two main transitions mentioned above:

• below $Ro \simeq 1$, stars more massive than 0.5 $M_\odot$ succeed at producing a substantial (and sometimes even dominant) toroidal component with a mostly non-axisymmetric poloidal component;

• below 0.5 $M_\odot$, stars (at least very active ones) apparently manage to trigger strong large-scale fields that are mostly poloidal and axisymmetric.

At very low-masses and high rotation rates, dwarfs seem capable of generating either strong aligned dipoles or very weak non-axisymmetric large-scale fields (Morin et al. 2010).

Long-term monitoring of large-scale magnetic topologies can potentially reveal whether the underlying dynamo processes are cyclic like in the Sun (with the field switching its overall polarity every 11 yr), constant or chaotic. Initial studies carried over a decade demonstrated indeed that both the field topologies and the differential rotation patterns

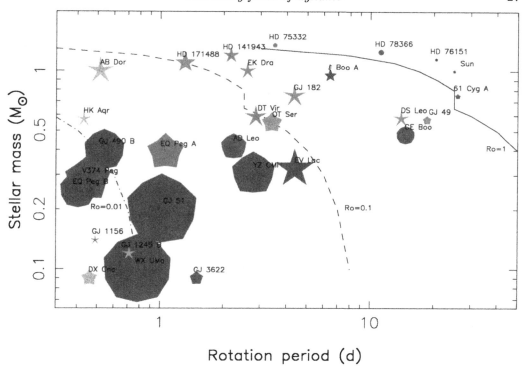

Figure 1. Basic properties of the large-scale magnetic topologies of cool stars, as a function of stellar mass and rotation rate. Symbol size indicates relative magnetic energy densities e, symbol colour illustrates field configurations (blue and red for purely toroidal and purely poloidal fields respectively) while symbol shape depicts the degree of axisymmetry of the poloidal field component (decagon and stars for purely axisymmetric and purely non-axisymmetric poloidal fields respectively). The full, dashed and dash-dot lines respectively trace where the Rossby number Ro equals 1, 0.1 and 0.01. The smallest and largest symbols correspond to mean large-scale field strengths of 3 G and 1.5 kG respectively (updated from Donati & Landstreet 2009).

are variable on long-timescales (e.g., Donati *et al.* 2003b) but have failed to catch stars in the process of switching their global magnetic polarities, suggesting that their dynamos (if cyclic) do not reverse much more often than that of the Sun; similar conclusions are obtained from long-term monitoring of solar-type stars using indirect proxies like overall brightness or chromospheric emission (Hall 2008). Very recently, first evidence for global polarity switches was reported in a star other than the Sun, namely the Jupiter-hosting F8 star τ Boo (Donati 2008a). During repeated spectropolarimetric monitoring (see Fig. 2), two successive polarity switches of τ Boo were recorded within about 2 yr, suggesting an activity cycle about 10 times faster than that of the Sun (Fares *et al.* 2009); although still fragmentary, observations already show that the poloidal and toroidal field components do not vary in phase across the cycle period.

3. Testing dynamo models with observations

Observational evidence that magnetic fields of cool stars are generated through dynamo processes is very strong. As recalled above, magnetic fields are ubiquitous to all stars with significant outer convection (i.e., spectral type later than mid F), and direct spectroscopic estimates demonstrate that magnetic fluxes scale up with rotation rate (and more tightly with $1/Ro$) until they saturate - in agreement with what conventional dynamo theories

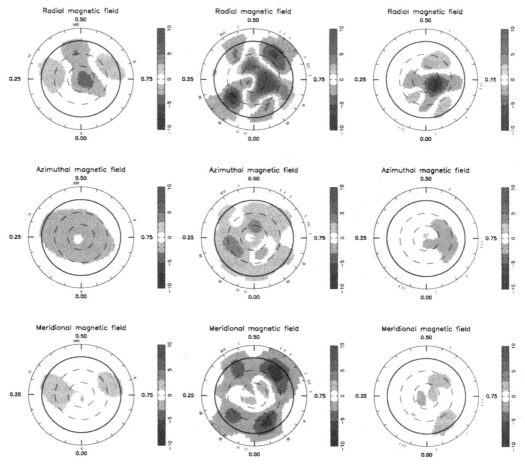

Figure 2. Large-scale magnetic topology of the F7 planet-hosting star τ Boo derived with Zeeman-Doppler imaging in 2006 June (left panel), 2007 June (middle panel) and 2008 June (right panel). In each panel, the 3 components of the field in spherical coordinates are displayed (from top to bottom) with magnetic fluxes labelled in G, the star being shown in flattened polar projection down to a latitude of $-30°$. Radial ticks around each plot indicate phases of observations. Both poloidal and toroidal fields globally switched polarities between successive epochs (from Fares *et al.* 2009).

predict. The other option, i.e., that these fields would be fossil remnants from a prior evolutionary stage, finds little support either from observations or theory; while models predict fossil fields to be dissipated by convection (as a result of the very high turbulent magnetic diffusivity) in as little as 1000 yr, observations indicate that fields are very often highly variable (both locally and globally) on timescales of months to decades and thus cannot reasonably result from an evolutionary process that ended at least tens of Myr before. Magnetic field measurements on cool stars thus bring, at first order, strong and independent support to generic dynamo models.

Dynamo models have undergone considerable progress in recent years; mean-field models are now implementing more physics (e.g., the presence of an interface layer or the effect of meridional circulation) while direct numerical simulations are now able to reach strongly turbulent regimes capable of producing intense mean magnetic fields (Brun *et al.* 2004, Browning 2008). However, despite such progress, there is still a large number of open questions, some of them concerning the very basic physics of dynamo processes, e.g., the identification of the primary mechanism through which the large-scale poloidal component is regenerated (Charbonneau *et al.* 2005). Above all, dynamo models are almost completely tailored for the Sun, with all model parameters finely tuned to reproduce solar observations as well as possible; checking them against observations of other stars with different masses and rotation rates in particular is a mandatory validation test that they yet have to undergo. The growing body of published results on large-scale magnetic topologies of cool stars should provide the opportunity for doing this in the near future. Meanwhile, we will summarize here the main topics on which the recent results provide new insight into dynamo processes.

The presence of toroidal fields at the surface of partly-convective stars with $Ro < 1$ is undoubtedly a surprising discovery, leading some to conclude that dynamo processes in very active stars must be operating either throughout the whole CZ or at the very least within a subphotospheric layer (e.g., Donati & Cameron 2007, Donati 1999, Donati *et al.* 2003a) rather than just at the base of the CZ (as usually assumed in conventional dynamo theories). Interestingly, a similar idea - a distributed dynamo shaped by near-surface shear - was recently invoked and investigated theoretically in the particular case of the Sun as an alternative to conventional interface dynamo models (Brandenburg 2005); in particular, this new model can potentially solve a number of long-standing issues (e.g., the large number of toroidal flux belts produced by interface dynamos) if further validated by new simulations. Very recent spectropolarimetric observations of the Sun with the HINODE spacecraft revealed that the quiet inter-network regions (i.e., the inner regions of supergranular cells of the quiet Sun) are pervaded by transient mainly-horizontal magnetic flux (Lites et al. 2008), giving still further support for a non-conventional distributed and/or near-surface dynamo in the Sun. Admittedly, surface toroidal fields detected in cool active stars are more stable than those seen on the Sun and participate to the large-scale field; they could however share a similar origin and scale up in strength and size with $1/Ro$, being only visible at low ZDI-like spatial resolutions for stars with $Ro < 1$. In turn, this may suggest that the newly-discovered horizontal fields of the Sun also participate to the activity cycle.

Despite their high level of activity, fully-convective stars obviously lack the interface layer where dynamo processes presumably operate; understanding their magnetism thus represents a major challenge for theoreticians. A wide range of predictions has been made about the kind of fields that such dynamos can produce; while early studies speculate that they generate small-scale fields only (e.g., Durney *et al.* 1993), newer models find that they can potentially trigger purely non-axisymmetric large-scale fields (e.g., Chabrier & Küker 2006) with CZs rotating as solid bodies (Küker & Rüdiger 1997). The latest simulations show that axisymmetric poloidal fields can also be produced with significant differential rotation (Dobler *et al.* 2006), but that toroidal fields are usually dominant and differential rotation rather weak whenever Ro is low enough (Browning 2008). In this context, the recent discovery (from both spectropolarimetry and radio observations) that fully-convective stars are able to generate strong and simple large-scale mostly-axisymmetric poloidal fields while rotating almost as solid bodies (Donati *et al.* 2006, Morin *et al.* 2008a, 2008b) is very unexpected and hard to reconcile with any of the existing models.

The first results of the spectropolarimetric survey indicate that the sharp transition in the large-scale magnetic topologies and surface differential rotation of M dwarfs more or less occurs where the internal structure of the star drastically changes with mass (the inner radiative zone shrinking in radius from $0.5~R_\star$ to virtually nothing when $M_\star$ decreases from $0.5~M_\odot$ to $0.4~M_\odot$, e.g., Baraffe *et al.* 1998). It is also worthwhile noting that X-ray luminosities of M dwarfs (relative to their bolometric luminosities) are roughly equal (at similar Ro) on both sides of the full-convection threshold, while the strengths of their large-scale fields feature a clear discontinuity (at a mass of about $0.4\,M_\odot$, Donati et al. 2008b, Morin et al. 2008b). All this suggests that dynamo processes become much more efficient at producing large-scale mainly-axisymmetric poloidal fields essentially as a response to the rapid growth in convective depths with decreasing stellar masses; this is qualitatively compatible with the idea that the geometry (and in particular the aspect ratio) of the CZ may control the kind of dynamo wave that a cosmic body can excite (Goudard & Dormy 2008). The latest finding that dwarfs stars seem capable, at very low-masses and high rotation rates, of generating either strong aligned dipoles or very weak non-axisymmetric large-scale fields (Morin et al. 2010) is equally intriguing, and may suggest a bistable behaviour of dynamo processes in a restricted range of stellar parameters.

The first detection of global magnetic polarity switches in a star other than the Sun is a major first step towards a better understanding of activity cycles of low-mass stars. Looking at, e.g., how cycle periods vary with stellar mass and rotation rate, or how poloidal and toroidal fields fluctuate with time across the cycle period, should ultimately reveal what physical processes mostly control the cycle. Results using a Babcock-Leighton flux transport dynamo model on the Sun suggest that meridional circulation is a crucial parameter (e.g., Dikpati & Charbonneau 1999); while meridional circulation is difficult to estimate directly in stars other than the Sun, its relation to rotation and differential rotation can potentially be tracked back from how cycle periods vary with stellar parameters. The geometry of the CZ is potentially also important (Goudard & Dormy 2008).

4. Conclusion

Spectropolarimetric observations have the potential to reveal the drastically different magnetic topologies that dynamos of low-mass stars can generate for various stellar masses and rotation rates. They demonstrated in particular that large-scale fields of cool stars change abruptly close to the full-convection threshold, with dwarfs on the low-mass side being much more successful at producing strong, mostly axisymmetric, dipole-like magnetic configurations; partly convective stars also undergo a noticeable change around the $Ro = 1$ limit. Provided generous allocation of telescope time, spectropolarimetric techniques can even unveil properties of stellar magnetic cycles (and the details of the underlying polarity switches of both poloidal and toroidal field components) as well as how they depend on mass and rotation rates.

Large-scale coordinated observing campaigns organised within the framework of the MagIcS international program (aimed at investigating stellar magnetic fields throughout the HR diagram) should continuously provide new advances along these lines in the future. Extending observations to near-infrared (nIR) wavelengths is of obvious interest, both for reaching very-low-mass stars and brown dwarfs (too faint to be observed in the visible domain) and for improving the overall sensitivity to magnetic fields (thanks to the increased Zeeman splitting); we thus propose to build a new nIR cryogenic spectropolarimeter (called SPIRou) operating from $1 - 2.5\ \mu m$ as a next-generation instrument for CFHT in 2015.

Time is also ripe for a thorough statistical investigation of theoretical and numerical dynamos; working out in particular whether existing models and codes are able to reproduce convincingly the abrupt changes in the large-scale field properties that observations reveal sounds like an obvious goal for the near future, both potentially fruitful and technically feasible.

References

Baraffe I., Chabrier G., Allard F., & Hauschildt P. H., 1998, *A&A* 337, 403

Berdyugina S. V., 2005, *Living Reviews in Solar Physics* 2, 8

Berger E., Rutledge R. E., Reid I. N., et al., 2005, *ApJ* 627, 960

Brandenburg A., 2005, *ApJ* 625, 539

Browning M. K., 2008, *ApJ* 676, 1262

Brun A. S., Browning M. K., & Toomre J., 2005, *ApJ* 629, 461

Cameron A. C. & Robinson R. D., 1989, *MNRAS* 236, 57

Chabrier G. & Küker M., 2006, *A&A* 446, 1027

Charbonneau P., 2005, *Living Reviews in Solar Physics* 2, 2

Dikpati M. & Charbonneau P., 1999, *ApJ* 518, 508

Dobler W., Stix M., & Brandenburg A., 2006, *ApJ* 638, 336

Donati J.-F., 1999, *MNRAS* 302, 457

Donati J.-F., Brown S. F., Semel M., et al., 1992, *A&A* 265, 682

Donati J.-F. & Cameron A. C., 1997, *MNRAS* 291, 1

Donati J.-F., Cameron A. C., Semel M., et al., 2003a, *MNRAS* 345, 1145

Donati J.-F., Cameron A. C., & Petit P., 2003b, *MNRAS* 345, 1187

Donati J.-F., Forveille T., Cameron A. C., et al., 2006, *Science* 311, 633

Donati J.-F. & Landstreet J. D., 2009, *ARA&A* 47, 333

Donati J.-F., Morin J., Petit P., et al., 2008b, *MNRAS* 390, 545

Donati J.-F., Moutou C., Fares R., et al., 2008a, *MNRAS* 385, 1179

Donati J.-F., Semel M., Carter B. D., Rees D. E., & Collier Cameron, A. 1997, *MNRAS*, 291, 658

Dunstone N. J., Hussain G. A. J., Cameron A. C., et al., 2008, *MNRAS* 387, 1525

Durney B. R., De Young D. S., & Roxburgh I. W., 1993, *Sol. Phys.* 145, 207

Fares R., Donati J.-F., Moutou C., et al., 2009, *MNRAS* 398, 1383

Goudard L. & Dormy E., 2008, *Europhysics Letters* 83, 59001

Hallinan G., Antonova A., Doyle J. G., et al., 2008, *ApJ* 684, 644

Hall J. C., 2008, *Living Reviews in Solar Physics* 5, 2

Hall D. S., 1991, in LNP 380, 353 (Berlin Springer Verlag)

Johns-Krull C. M. & Valenti J. A., 1996, *ApJ* 459, L95

Küker M. & Rüdiger G., 1997, *A&A* 328, 253

Lites B. W., Kubo M., Socas-Navarro H., et al., 2008, *ApJ* 672, 1237

Mangeney A. & Praderie F., 1984, *A&A* 130, 143

Marsden S. C., Donati J.-F., Semel M., Petit P. & Carter, B. D., 2006, *MNRAS* 370, 468

Mestel L., 1999, Stellar magnetism (Oxford : Clarendon)

Morin J., Donati J.-F., Forveille T., et al., 2008a, *MNRAS* 384, 77

Morin J., Donati J.-F., Petit P., et al., 2008b, *MNRAS* 390, 567

Morin J., Donati J.-F., Petit P., et al., 2010, *MNRAS* (in press)

Noyes R. W., Hartmann L. W., Baliunas S. L., Duncan D. K., & Vaughan A. H., 1984, *ApJ* 279, 763

Petit P., Donati J.-F., Aurière M., et al., 2005, *MNRAS* 361, 837

Petit P., Dintrans B., Solanki S. K., et al., 2008, *MNRAS* 388, 80

Reiners A., 2006, *A&A* 446, 267

Reiners A. & Basri G., 2008, *ApJ* 684, 1390

Robinson R. D., Worden S. P., & Harvey J. W., 1980, *ApJ* 236, L155

Saar S. H., 2001, in: ASP Conf Series 223, 292

Schatzman E., 1962, *AnAp* 25, 18

Astrophysical Dynamics: from Stars to Galaxies
Proceedings IAU Symposium No. 271, 2010
N. Brummell, A.S. Brun, M.S. Miesch & Y. Ponty, eds.

© International Astronomical Union 2011
doi:10.1017/S1743921311017443

Stellar hydrodynamics caught in the act: Asteroseismology with CoRoT and Kepler

Jørgen Christensen-Dalsgaard[1] and Michael J. Thompson[2,3]

[1] Department of Physics and Astronomy, Aarhus University, 8000 Aarhus C, Denmark
email: `jcd@phys.au.dk`

[2] School of Mathematics & Statistics, University of Sheffield, Sheffield, S3 7RH, UK

[3] High Altitude Observatory, National Center for Atmospheric Research, P.O. Box 3000, Boulder, CO 80307-3000, USA
email: `mjt@ucar.edu`

Abstract. Asteroseismic investigations, particularly based on data on stellar oscillations from the CoRoT and *Kepler* space missions, are providing unique possibilities for investigating the properties of stellar interiors. This constitutes entirely new ways to study the effects of dynamic phenomena on stellar structure and evolution. Important examples are the extent of convection zones and the associated mixing and the direct and indirect effects of stellar rotation. In addition, the stellar oscillations themselves show very interesting dynamic behaviour. Here we discuss examples of the results obtained from such investigations, across the Hertzsprung-Russell diagram.

Keywords. hydrodynamics, asteroseismology, space vehicles, stars: interiors, stars: evolution, stars: oscillations, planetary systems

1. Introduction

Stellar interiors are obvious sites for interesting dynamical phenomena, with strong potential effects of convection and other forms of instabilities, rotation and its evolution, as well as magnetic fields, forming a fascinating playing field for theoretical investigations. However, 'classical' observations of stellar surfaces provide only limited observational tests of the resulting models. Observations of stellar oscillations, on the other hand, are sensitive to many aspects of the structure and dynamics of stellar interiors. In addition, the oscillations are themselves interesting dynamical phenomena. The diagnostic potential of oscillations has been evident for more than a decade in the solar case, where helioseismic investigations have yielded very detailed information about the structure and rotation of the solar interior (e.g., Gough & Toomre 1991; Christensen-Dalsgaard 2002) and the detailed properties of the solar near-surface region through local helioseismology (for a review, see Gizon *et al.* 2010). For a comprehensive review of global helio- and asteroseismology, see also Aerts *et al.* (2010).

The extension of such detailed investigations to other stars has been eagerly sought since the beginning of helioseismology, given the expected general presence of the oscillations in all cool stars. There is strong evidence that the solar modes are excited stochastically by the near-surface convection, and hence similar modes are expected in all stars with such vigorous convective motions (Christensen-Dalsgaard & Frandsen 1983). However, the predicted amplitudes of a few parts per million in relative intensity or at most tens of centimeters per second in velocity have made their detection extremely challenging. Ground-based observations have had some success for a limited number of stars (for early examples, see Kjeldsen *et al.* 1995; Bouchy & Carrier 2001, 2002; Frandsen *et al.*

2002), but the required efforts in terms of manpower and valuable telescope time have been very considerable. However, in the last few years observations of stellar oscillations, and other types of variability, have made tremendous progress, largely as a result of the CoRoT and *Kepler* space missions, with significant contributions also from the Canadian MOST satellite (Walker *et al.* 2003; Matthews 2007). Here we provide a brief overview of some of the results from these missions which promise to revolutionize the study of stellar internal structure and dynamics.

2. CoRoT and Kepler

Both CoRoT and *Kepler* carry out photometric observations, with the dual purpose of detecting extra-solar planets (exoplanets) using the transit technique and making asteroseismic investigations. In fact, the observational requirements for these two types of investigation are very similar. A central goal of the exoplanet studies is to characterize the population of Earth-like planets in orbit around Sun-like stars. The corresponding transit corresponds to a reduction in stellar intensity by around 10^{-4}. This level of sensitivity allows the detection of solar-like oscillations in sufficiently extended observations. Also, both types of investigations require very long and continuous observations. Finally, asteroseismology has the potential to characterize the properties of the central stars in planetary systems detected by the transit technique. This has proven to be very useful in a few cases for the *Kepler* mission and is central to the PLATO mission proposed to ESA.

The CoRoT satellite (Baglin *et al.* 2006, 2009; Michel *et al.* 2008a) was launched on 27 December 2006 into an orbit around the Earth. The satellite has an off-axis telescope with a diameter of 28 cm, and a focal plane with 4 CCD detectors, two of which (defining the exoplanet field) are optimized for studying planet transits and the other two, with slightly defocussed images, optimized for asteroseismology (the asteroseismic field). The field of view of each CCD is $1.3° \times 1.3°$. Given the low-Earth orbit, great care was taken to minimize effects of scattered light. Further details on the design and operations of the mission were provided by Auvergne *et al.* (2009).

CoRoT's orbit allows extended observations, up to 5 months, in two regions with a diameter of around 20°, near the Galactic plane and the direction of the Galactic centre and anticentre, respectively; these are known as the 'CoRoT eyes'. The observed fields are selected within these regions, such that the asteroseismic detectors contain a suitable sample of relatively bright asteroseismic targets, up to a total of 10 targets, with an adequate density of stars in the corresponding exo field. It should be noted that while the observations in the latter are probably not sufficiently sensitive to study solar-like oscillations in main-sequence stars, they do provide very extensive data on other types of stellar variability, including solar-like oscillations in red giants (see Section 8). The photometric analysis is carried out on the satellite, with only the photometric signal being transmitted to the ground; a typical cadence for the astero field is 32 s.

The mission suffered a loss of two of the four CCD detectors (one each in the exo and astero fields) on 8 March 2009† but is otherwise performing flawlessly. Early results on solar-like oscillations from CoRoT data were presented by Michel *et al.* (2008b). The mission has now been extended until March 2013.

The *Kepler* mission (Borucki *et al.* 2009; Koch *et al.* 2010) was launched on 7 March 2009 into an Earth-trailing heliocentric orbit, with a period of around 53 weeks. This provides a very stable environment for the observations, in terms of stray light and

---

† Coincidentally a day after the launch of the *Kepler* mission!

other disturbances, although with the drawback of a gradually decreasing rate of data transmission with the increasing distance from Earth. Even so, it is hoped to keep the mission operating well beyond the current nominal lifetime of $3\frac{1}{2}$ years. The *Kepler* photometer consists of a Schmidt telescope with a corrector diameter of 95 cm and a 16° diameter field of view. The detector at the curved focal plane has 42 CCDs with a total of 95 megapixels. This provides a field of around 105 square degrees. The data are downlinked in the form of small images around each of the target stars, with the detailed photometry being carried out on the ground. This allows up to 170,000 targets to be observed at a 30-minute cadence, while up to 512 stars can be observed at a 1-minute cadence; the latter are the prime targets for asteroseismology, although for more slowly varying stars the long-cadence data are also extremely valuable. The spacecraft is rotated by 90° four times per orbit to keep the solar arrays pointed towards the Sun. Thus the observations are naturally divided into quarters. New target lists are uploaded for each quarter, although targets can be observed for as little as one month each; typically, most targets are in fact observed for several quarters, in many cases throughout the mission. For further details on the spacecraft and the operations, see Koch *et al.* (2010).

A detector module, corresponding to two of the 42 CCD detectors, failed in January 2010. Otherwise, the mission has been operating successfully, reaching very close to the expected performance. Borucki *et al.* (2010) provided an early overview of *Kepler* results on exoplanets.

Kepler observes a fixed field in the region of the constellations of Cygnus and Lyra, centred 13.5° above the Galactic plane and chosen to secure a sufficient number of targets of the right type while avoiding excessive problems with background confusion. A very detailed characterization of the field was carried out before launch, resulting in the Kepler Input Catalog (KIC; Brown, T. M., *et al.* 2011). To avoid problems with highly saturated trailing images, the field is located such that the brightest stars are placed in gaps between the CCDs. In addition, the CCDs are positioned such that a star is located at approximately the same point on a CCD following each quarterly rotation.

Kepler asteroseismology (e.g., Christensen-Dalsgaard *et al.* 2008) is carried out in the Kepler Asteroseismic Science Consortium (KASC), which at the time of writing has around 450 members. The members are organized into 13 working groups, generally dealing with different classes of pulsating stars. The KASC is responsible for proposing targets, and for the analysis and publication of the results. Data for the KASC are made available through the Kepler Asteroseismic Science Operations Centre (KASOC) in Aarhus, Denmark, which also organizes the sharing and discussion of manuscripts before publication. The structure of the Kepler Asteroseismic Investigation (KAI) was presented by Kjeldsen *et al.* (2010).

In the early phases of the KAI a survey was made of a very large number of stars, to characterize their oscillation properties and provide the basis for selecting targets for more extended observations. Initial results of this survey phase were discussed by Gilliland *et al.* (2010).

3. Pulsating stars

As a background for the discussion below of specific types of pulsating stars we provide a brief overview of the properties of stellar pulsations. For more detail, see Aerts *et al.* (2010). We restrict the discussion to slowly rotating stars and oscillations of modest amplitude. In this case the oscillations can, to leading order, be characterized as small perturbations around a spherically symmetric equilibrium structure; individual modes depend on co-latitude θ and longitude ϕ as spherical harmonics $Y_l^m(\theta, \phi)$, where l

measures the total number of nodal lines on the stellar surface and m the number of nodal lines crossing the equator, with $|m| \leqslant l$. Modes with $l = 0$ are typically described as *radial* oscillations. For each l, m the star has a set of modes distinguished by the radial order n.

From a dynamic point of view there are two basic types of stellar modes†: acoustic modes (or p modes) where the restoring force is pressure, and internal gravity waves (or g modes) where the restoring force is buoyancy variations across spherical surfaces. Thus g modes are only found for $l > 0$. Being acoustic the properties of the p modes predominantly depend on the sound-speed variation within the star. In many cases the result is that the frequencies ν approximately scale with the inverse dynamical time scale, or

$$\nu \propto \left(\frac{GM}{R^3}\right)^{1/2} \propto \langle\rho\rangle^{1/2} , \tag{3.1}$$

where M and R are the mass and radius of the star, G is the gravitational constant and $\langle\rho\rangle$ is the mean density. The g-mode frequencies depend on the variation of the gravitational acceleration and density gradient throughout the star, the latter in turn being very sensitive to the variation of composition.

In unevolved stars typical g-mode frequencies are lower than the frequencies of p modes. However, as the star evolves, the core contracts, leading to a very high local gravitational acceleration; in addition, strong composition gradients are built up by the nuclear burning and possibly convective mixing. In this case the local g-mode frequency may become large, and as a result the modes can take on a mixed character, with p-mode behaviour in the outer parts of the star and g-mode character in the deep interior. We discuss examples of this in Sections 7 and 8.

An important measure of the properties of a mode of oscillation is its normalized mode inertia

$$E = \frac{\int_V \rho|\boldsymbol{\delta r}|^2 \mathrm{d}V}{M|\boldsymbol{\delta r}_\mathrm{s}|^2} , \tag{3.2}$$

where $\boldsymbol{\delta r}$ is the displacement vector, $\boldsymbol{\delta r}_\mathrm{s}$ the surface displacement, and the integral is over the volume V of the star. Acoustic modes have their largest amplitude in the outer layers of the star and hence typically have relatively low inertias, whereas g modes are generally confined in the deep interiors of stars, with correspondingly high inertia.

Energetically, two fundamentally different mechanisms may excite the oscillations. Some stars function as a heat engine where the relative phase of compression and heating in a critical layer of the star is such that thermal energy is converted into mechanical energy, contributing to driving of the oscillations, and dominating over the other parts of the star which have the opposite effect. This is typically associated with opacity variations of specific elements; an important example is the effect of the second ionization of helium (Cox & Whitney 1958), which causes instability in stars where the corresponding region in the star is located at the appropriate depth beneath the stellar surface. The driving leads to initially exponential growth of the oscillations, with so far poorly understood mechanisms setting in to control the limiting amplitudes of the modes.

In stars where the oscillations are not self-excited in this manner the modes may be excited stochastically, through driving from other dynamical phenomena in the star. This is likely the case in the Sun where near-surface convection at nearly sonic speed is a strong source of acoustic noise which excites the resonant modes of the star (e.g., Goldreich &

---

† In addition, modes corresponding to *surface gravity waves* can be distinguished, but at degrees so far only relevant for spatially resolved observations of the Sun.

Keeley 1977; Houdek *et al.* 1999). In this case the oscillation amplitudes result from a balance between the stochastic energy input and the damping of the modes. Such excitation is expected in all stars with a significant outer convection zone, i.e., stars with effective temperature below around 7000 K. In principle, it excites all modes in a star; however, typically only modes with low inertia (cf. Eq. 3.2), i.e., acoustic modes of rather high radial order, are excited to sufficiently high amplitudes to be readily observable.

The stochastic excitation of acoustic oscillations leads to a characteristic bell-shaped distribution of mode amplitudes (see Fig. 9 below). It has been found (Brown *et al.* 1991; Brown & Gilliland 1994; Kjeldsen & Bedding 1995; Bedding & Kjeldsen 2003; Stello

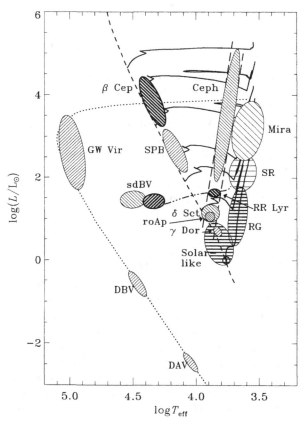

Figure 1. Schematic location of classes of pulsating stars in the Hertzsprung-Russell diagram. The diagonal dashed line marks the main sequence, where stars undergo central hydrogen burning. Evolution tracks following the main sequences for a few stellar masses are shown by solid lines, the triple-dot-dashed line indicates the location of horizontal-branch stars with central helium burning, and the dotted line sketches the white dwarf cooling track, after stars have stopped nuclear burning. The hatching indicates the excitation mechanism: slanted lines from lower right to upper left for heat-engine excitation of p modes, slanted lines from lower left to upper right for heat-engine excitation of g modes, and horizontal lines for stochastic excitation. The two nearly vertical dashed lines mark the limits of the Cepheid instability strip, where stars are generally excited by an opacity-driven heat engine operating in the second helium ionization zone. Stars discussed in the present paper are shown with bolder lines: RR Lyrae stars (RR Lyr; Section 4), massive main-sequence stars (β Ceph; Section 5), long-period subdwarf B variables (sdBV; Section 6), solar-like pulsators (Section 7) and red giants (RG; Section 8).

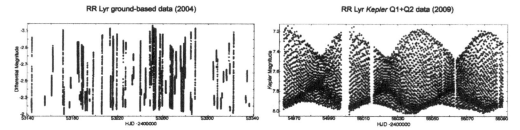

Figure 2. Lightcurves for RR Lyrae lightcurves. The left panel shows the combined results of observations with 6 ground-based telescopes (Kolenberg *et al.* 2006), compared with the first two quarters of *Kepler* data (Kolenberg *et al.* 2011).

et al. 2008) that the frequency ν_{max} at maximum power scales as the acoustic cut-off frequency (Lamb 1909), leading to

$$\nu_{\mathrm{max}} \propto M R^{-2} T_{\mathrm{eff}}^{-1/2} , \qquad (3.3)$$

where T_{eff} is the effective temperature. This relation so far lacks a solid theoretical underpinning (see, however Belkacem *et al.* 2011), but it has proven to be very useful in characterizing stars observed to have solar-like oscillations (see Section 8).

As illustrated in Fig. 1 pulsating stars are found throughout the Hertzsprung-Russell diagram, in all phases of stellar evolution. Thus there are excellent possibilities for learning about a broad range of stars. Most of these classes have been observed by CoRoT and *Kepler*. In the following we discuss a few important examples.

4. RR Lyrae stars

The RR Lyrae stars are amongst the 'classical' pulsating stars that have been studied extensively from the ground, since the discovery in 1900 of the variability of RR Lyr, the prototype of the class which happens to be in the *Kepler* field (see Smith 1995, for more information on these stars). They are low-mass stars with a low content of heavy elements, in the core helium-burning phase of evolution. As they have relatively well-defined luminosities they serve a very useful purpose as distance indicators to nearby galaxies, such as the Magellanic Clouds. Their pulsations are excited by the heat-engine mechanism, operating as a result of opacity variations in the second helium ionization zone. They oscillate predominantly in one or two low-order radial modes, with amplitudes large enough to allow observations with even very modest equipment.

What makes these stars interesting in the context of space asteroseismology are the very interesting, and poorly understood, dynamical properties of the oscillations in a substantial fraction of the class. Blažko (1907)† discovered in one member of the class that the maximum amplitude varied cyclically with a period of 40.8 d. This phenomenon has since been found in a number of RR Lyrae stars, including RR Lyr itself (Kolenberg *et al.* 2006). A centennial review, including a discussion of the so far questionable attempts at explaining the effect, was provided by Kolenberg (2008).

The continuity, duration and precision of space-based observations offer obvious advantages in investigating such long-term phenomena. This is illustrated in Fig. 2 which

---

† In the discovery paper the original Cyrillic name was written as 'Blažko'. However, traditionally the name of the effect is now written as 'the Blazhko effect' with a slightly different transliteration; we follow that tradition here.

compares results of an extensive ground-based campaign on RR Lyr with *Kepler* observations. The latter obviously provide a far better phase coverage throughout the Blazhko
cycle. Phase plots at maximum and minimum amplitude in the cycle are illustrated in
Fig. 3. Similar results have been obtained by CoRoT (Poretti *et al.* 2010). An early survey of 28 RR Lyrae stars by *Kepler* (Kolenberg *et al.* 2010) found the Blazhko effect in
40 % of the stars, a rather higher fraction than previously suspected.

One may hope that these vastly improved data will bring us closer to an understanding
of this enigmatic phenomenon. As an interesting piece of evidence Szabó *et al.* (2010),
using *Kepler* data, found that three Blazhko RR Lyrae stars, including RR Lyr itself,
showed period doubling in certain phases of the Blazhko cycle, with slight variations in
the maximum amplitude between alternating pulsation cycles. Also, from CoRoT observations Chadid *et al.* (2011) investigated cycle-to-cycle changes in the Blazhko modulation. Such results evidently provide constraints on, and inspiration for, the attempts at
theoretical modelling of these stars.

5. Massive main-sequence stars

The pulsations in hot main-sequence stars, the so-called β Cephei stars, have been
known for a century, but the cause of the pulsations was only definitely identified around
1990, when substantial revisions in opacity calculations produced opacities which allowed
excitation of the observed modes through the heat-engine mechanism operating through
the opacity from iron-group elements (Moskalik & Dziembowski 1992). This causes

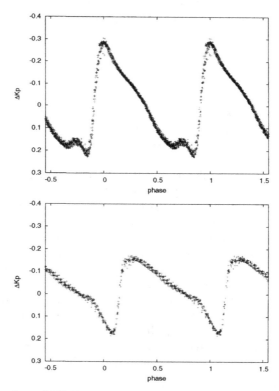

Figure 3. Phase plots of RR Lyr at maximum and minimum amplitude, from *Kepler*
observations. Figure courtesy of R. Szabó.

excitation of both p and g modes, with a tendency towards g modes at lower effective temperature and hence mass, and a transition to the slowly pulsating B stars (SPB stars) dominated by high-order g modes with very long periods. An excellent overview of the excitation of oscillations in the B stars was provided by Pamyatnykh (1999).

These massive stars are of very considerable astrophysical interest as precursors for the core-collapse supernovae. Their structure and evolution depend strongly on the properties of the convective core, including additional mixing outside the convectively unstable region caused by overshoot or 'semiconvection', as well as other dynamical phenomena associated, for example, with rotation. Thus the potential for asteroseismic investigation is very valuable, provided that adequate data can be obtained. Particularly useful diagnostics can be obtained from observations of g modes. High-order g modes are approximately uniformly spaced in period, with a period spacing $\Delta\Pi$ given, to leading order, by

$$\Delta\Pi = \frac{2\pi^2}{\sqrt{l(l+1)}} \left(\int_{r_1}^{r_2} N \frac{\mathrm{d}r}{r} \right)^{-1} \tag{5.1}$$

(Tassoul 1980); here N is the buoyancy frequency and $[r_1, r_2]$ is the interval where the modes are trapped, with $N^2 > 0$. Assuming an ideal gas the buoyancy frequency is determined by

$$N^2 \simeq \frac{g^2\rho}{p}(\nabla_{\mathrm{ad}} - \nabla + \nabla_\mu), \tag{5.2}$$

where g is the local gravitational acceleration, ρ is density and p is pressure; also, following the usual convention,

$$\nabla = \frac{\mathrm{d}\ln T}{\mathrm{d}\ln p}, \qquad \nabla_{\mathrm{ad}} = \left(\frac{\partial \ln T}{\partial \ln p} \right)_{\mathrm{ad}}, \qquad \nabla_\mu = \frac{\mathrm{d}\ln \mu}{\mathrm{d}\ln p}, \tag{5.3}$$

where T is temperature, μ is the mean molecular weight and the derivative in ∇_{ad} is taken corresponding to an adiabatic change. In a detailed analysis Miglio *et al.* (2008) pointed out the diagnostic potential of *departures* from the uniform period spacing. Owing to the presence of the term in ∇_μ in Eq. (5.2) the buoyancy frequency is very sensitive to the detailed composition profile, such as may result outside a convective core. The resulting sharp features in the buoyancy frequency introduce perturbations to $\Delta\Pi$ with a characteristic oscillatory behaviour, the details of which depend strongly on conditions at the edge of the core.

As for the RR Lyrae stars (Section 4) the oscillations can readily be detected in ground-based observations. The difficulty is to obtain adequate frequency resolution and precision, given the long periods and generally fairly dense spectra. Substantial successes have been achieved with coordinated multi-site observations over several months (Handler *et al.* 2004), leading to interesting information about convective core overshoot (Ausseloos *et al.* 2004; Pamyatnykh *et al.* 2004) and internal rotation (Aerts *et al.* 2003; Dziembowski & Pamyatnykh 2008). However, it is evident that such massive campaigns can only be carried out in very special cases, and even then they do not provide the full desired data continuity or sensitivity to low-amplitude modes.

Observations from CoRoT and *Kepler* have the potential to secure very long continuous observations of these stars (Degroote *et al.* 2010a; Balona *et al.* 2011). A very interesting case was discussed by Degroote *et al.* (2010b), for the star HD 50230. This is a massive main-sequence star, of spectral type B3V, which was observed by CoRoT for 137 days. The resulting power spectrum showed a large number of g-mode frequencies, with periods up to a few days, in addition to several high-frequency p modes. In the long-period part of the spectrum the authors were able to identify a group of eight modes with almost constant period spacing, arguing that such a sequence is very unlikely to be found by chance. As illustrated in Fig. 4 these period spacings showed a highly regular variation with period, of precisely the form predicted by Miglio *et al.* (2008) to result from a sharp feature in the buoyancy frequency. As pointed out by Miglio *et al.* the decrease in the amplitude with increasing period is a sensitive diagnostic of the properties of the feature. A more detailed interpretation of the results will require more stringent characterization of other properties of the star, through 'classical' observations as well as more extensive asteroseismic analyses of the rich spectrum. However, the result clearly demonstrates the potential of such observations for characterizing the properties of convective cores in massive main-sequence stars.

6. Subdwarf B stars

The subdwarf B stars (sdB stars) are very hot core helium burning stars, at the blue end of the horizontal branch (for a review, see Heber 2009). The high effective temperature is the result of the stars having lost most of their hydrogen envelope, through processes that are so far not fully understood. Pulsations were first observed in such stars by Kilkenny *et al.* (1997), with very high frequency. That the stars might be unstable to acoustic modes was found in parallel, and independently, by Charpinet *et al.* (1996). The driving arises from the heat-engine mechanism operating on opacity from the iron-group elements. Subsequently Green *et al.* (2003) also observed long-period oscillations in somewhat cooler sdB stars, corresponding to g modes of high order. A detailed analysis of the excitation was carried out by Fontaine *et al.* (2003). To be effective, the iron-group opacity must be enhanced through enrichment of the elements in the critical region through radiative levitation (see also Fontaine *et al.* 2006); owing to the high gravitational acceleration such processes of settling and levitation are quite efficient in these stars.

As in the previous cases discussed, major ground-based efforts have been made to study these pulsations, involving coordinated observations between several observatories over extended periods (e.g., Randall *et al.* 2006a,b; Baran *et al.* 2009). The difficulties of such observations are particularly severe for the g-mode pulsators, with periods of order hours and amplitudes of only a few parts per thousand. Yet these modes are particularly

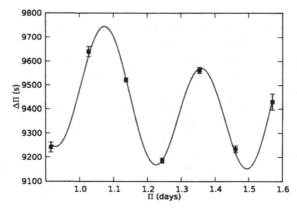

Figure 4. Period spacings in the B3V star HD 50230, from CoRoT observations. The variations in $\Delta\Pi$, here fitted with a decaying sinusoid, reflect the properties of the buoyancy frequency just outside the convective core in the star. Adapted from Degroote *et al.* (2010b).

interesting as diagnostics of the deep interiors of the stars. Thus the sdB variables have been prime targets for asteroseismology with CoRoT and *Kepler*. The *Kepler* initial survey has included a substantial number of sdB stars and so far led to the discovery of 14 pulsating stars, all except one of which are long-period pulsators, with substantial numbers of g modes of intermediate and high order (Østensen *et al.* 2010, 2011; Baran *et al.* 2011). Thus the periods may be expected to satisfy the asymptotic relation (5.1). Reed *et al.* (2011) showed that this is indeed the case and noted the importance for mode identification.

As a specific example of the power of these data we consider a detailed analysis of the star KPD 1943+4058 based on the initial *Kepler* data. Here Reed *et al.* (2010) detected 21 modes, with periods between 0.7 and 2.5 h, and in addition three modes in the p-mode region. Similar observational results were obtained by Van Grootel *et al.* (2010a), who in addition carried out a fit of the g-mode periods to models of the star. The models were described by their total mass, the mass in the thin envelope and an assumed fully mixed core, and the composition of the core, characterized by the combined abundance of carbon and oxygen. In addition, the models were constrained to be consistent with the spectroscopically inferred effective temperature and surface gravity. The fit to the observations measured by a merit function S defined by

$$S^2 = \sum_i (\Pi_i^{(\mathrm{obs})} - \Pi_i^{(\mathrm{mod})})^2 \,, \qquad (6.1)$$

where $\Pi_i^{(\mathrm{obs})}$ and $\Pi_i^{(\mathrm{mod})}$ are the observed and model periods, respectively, and the identification of the computed modes with the observed modes is part of the fitting procedure. Some results are illustrated in Fig. 5. The analysis provided very precise determinations of the properties of the star, including strong evidence for a mixed core region significantly larger than the expected convectively unstable region, indicating substantial core overshoot.

It should be noted that the periods of the best-fitting model did not agree with the observed periods to within their observational uncertainty. This evidently shows that further improvements of the modelling, beyond the parameters included in the fit, are needed. It seems likely that the number of observed periods is sufficient that a formal

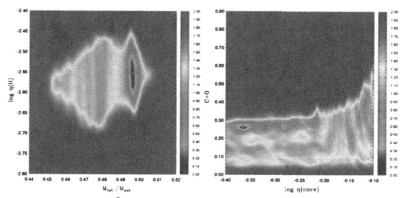

Figure 5. The merit function S^2 (cf. Eq. 6.1) in a fit of the observed *Kepler* periods for the star KPD 1943+4058 to a set of models; the colour scale for $\log S^2$ is indicated to the right. The left panel shows the fit in terms of the total mass of the star, in solar units, and the logarithm of the fraction of the mass in the hydrogen-rich envelope. The right panel plots the merit as a function of the logarithm of the mass fraction outside the mixed core and the combined abundance, by mass, of carbon and oxygen in the core. From Van Grootel *et al.* (2010a).

inversion of the differences can be attempted, as has been applied with great success in the solar case (e.g., Gough *et al.* 1996). The results may provide a direct indication of the aspects of the models where improvements are needed.

Similar analyses will be possible for the remaining stars for which extensive g-mode data have been obtained with *Kepler*, and they will evidently improve as the stars continue to be observed. Also, an sdB star showing extensive g-mode pulsations has been observed by the CoRoT mission (Charpinet *et al.* 2010). Model fitting to the resulting periods by Van Grootel *et al.* (2010b) yielded results rather similar to those discussed above.

7. Solar-like oscillations in main-sequence stars

Solar-like oscillations are predominantly acoustic in nature and excited by turbulent convection in the star's outer convective envelope. As already noted, although this broadband excitation mechanism excites all modes in principle, because of their low mode inertias it tends to be the high-order p modes that are excited to observable amplitude. The first star in which such oscillations were detected was of course the Sun, and the study of the Sun's oscillations has led to the rich field of helioseismology, in which Juri Toomre has played a leading role (see, *e.g.*, Gough & Toomre 1991; Christensen-Dalsgaard 2002).

The asteroseismic investigation of solar-type stars has taken a major step forward thanks to the *Kepler* mission (Chaplin *et al.* 2010). To date, *Kepler* has yielded clear detections of oscillations in 500 solar-type stars (Chaplin *et al.* 2011). A plot of the distribution in the HR diagram of *Kepler* stars with detected solar-like oscillations is shown in Fig. 6. This represents an increase by more than a factor of 20 of the number of known stars on and near the main sequence which show solar-like oscillations.

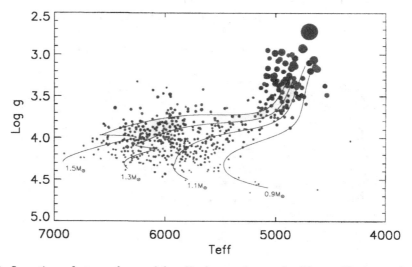

Figure 6. Location of stars observed by *Kepler* to show solar-like oscillations, plotted as a function of effective temperature and the logarithm of the surface gravity. The size of the symbols measures the oscillation amplitude, while the colour (in the electronic version) indicates the apparent brightness of the stars, with red for the brightest stars. For comparison, evolution tracks at the indicated masses are also shown. Adapted from Chaplin *et al.* (2011) and Verner *et al.* (2011). Figure courtesy of C. Karoff.

The high-order, low-degree p modes in solar-type stars occupy resonant acoustic cavities that extend from the surface to a region close to the stellar core. Their cyclic frequencies ν_{nl} satisfy a simple expression:

$$\nu_{nl} \simeq \Delta\nu \left(n + \frac{l}{2} + \epsilon\right) - d_{nl} . \tag{7.1}$$

Here

$$\Delta\nu = \left(2 \int_0^R \frac{dr}{c}\right)^{-1} , \tag{7.2}$$

where c is the adiabatic sound speed and the integral is over the distance r to the centre of the star; also

$$d_{nl} = \frac{l(l+1)\Delta\nu}{4\pi^2\nu_{nl}} \left[\frac{c(R)}{R} - \int_0^R \frac{dc}{dr}\frac{dr}{r}\right] \tag{7.3}$$

(Tassoul 1980; Gough 1986). Accordingly, the frequencies of such modes of the same degree are separated by *large separations*

$$\Delta\nu_{nl} = \nu_{nl} - \nu_{n\,l-1} \approx \Delta\nu , \tag{7.4}$$

while the small correction d_{nl} gives rise to *small separations*

$$\delta\nu_{nl} = \nu_{nl} - \nu_{n-1\,l+2} \approx (4l+6)\frac{\Delta\nu}{4\pi^2\nu_{nl}} \left[\frac{c(R)}{R} - \int_0^R \frac{dc}{dr}\frac{dr}{r}\right] \tag{7.5}$$

between the frequencies of modes that differ in degree by 2 and in order by 1. Finally ϵ is a slowly varying function of frequency which is predominantly determined by the properties of the near-surface region.

The quantity $\Delta\nu$ is a measure of the acoustic radius of the star. It shares the scaling (Eq. 3.1) of the frequencies with the mean density, and hence so too do the large separations. For main-sequence stars d_{nl}, and thus the small separations, are mainly determined by the central regions of the star, being sensitive in particular to the sound-speed gradient in the core, and hence they provide a measure of the star's evolutionary state. Thus measuring the large and small frequency separations gives a measure of the mean density and evolutionary state of the star. A useful seismic diagnostic is the asteroseismic HR diagram, in which the star's average large separation is plotted against its average small separation: this is illustrated in Fig. 7. For main-sequence stars, the asteroseismic HR diagram allows the mass and age of the star to be estimated, assuming that other physical inputs (such as initial chemical composition and the convective mixing-length parameter) are known.

The existence of the large separation also motivates another diagnostic, which is to plot the frequencies of the star in a so-called *échelle diagram*. Here the frequencies ν_{nl} are reduced modulo $\Delta\nu$ and $\bar\nu_{nl} \equiv \nu_{nl} \bmod \Delta\nu$ is plotted on the x-axis while $\nu_{nl} - \bar\nu_{nl}$ is plotted on the y-axis. If the spacing of frequencies according to asymptotic expression (7.1) were exact, the modes of like degree would be aligned as nearly vertical lines in the échelle diagram, with the lines corresponding to $l = 0, 2$ being separated from one another by the small separation and the line corresponding to $l = 1$ being offset from those by an amount corresponding to half the large separation. Deviations from such a simple picture reveal deviations from the simple asymptotic relation and contain physically interesting information about the star. An example of an échelle diagram for star KIC 11026764, is shown in Fig. 8. The ridges corresponding to $l = 0, 2$ are evident. The ridge corresponding

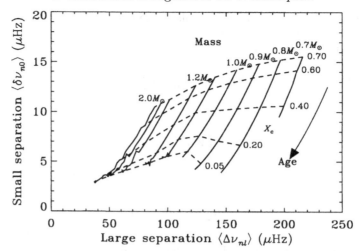

Figure 7. Asteroseismic HR diagram for a homogeneous set of stellar models of different masses and ages. The solid lines show how a star of given mass evolves in terms of its large and small frequency separations. The dashed lines connect stars which have aged by the same fraction of their total main-sequence lifetime.

to $l = 1$ is more irregular: this is due to avoided crossings in this relatively evolved star, an issue which is discussed further below.

Even without a measurement of the small separations, it is possible to make useful seismic estimates of stellar masses and radii using observational measures of $\Delta\nu$ and of the frequency ν_{max} of maximum mode amplitude. From Eq. (7.2) the former scales as M/R^3, whereas by Eq. (3.3) the latter scales as M/R^2: hence a measurement of these two yields an estimate of both M and R. This has been applied to an ensemble of 500 stars in the *Kepler* field by Chaplin *et al.* (2011). This paper by Working Group 1 (WG1) of the KASC concludes that while the estimated radii of the 500 stars are similar to those expected from stellar population models, the observed distribution of masses is wider at its peak than the model distribution, and is offset towards slightly lower masses.

Chaplin *et al.* (2010) published observations of three bright solar-type stars, which were monitored during the first month of *Kepler* science operations. This paper was the first to establish the asteroseismic potential of *Kepler* observations of solar-type stars: about 20 modes were distinguished in each star, and the frequencies and frequency separations allowed radii, masses and ages of the stars to be estimated. The three stars that were the objects of the study, KIC 6603624, KIC 3656476 and KIC 11026764, were given the working names Saxo, Java and Gemma† by the WG1 members. One of these stars, Gemma, was revealed to have evolved off the main sequence and proved more challenging to model and constrain asteroseismologically: this interesting case is discussed further now.

Gemma is one of the best-studied solar-like stars to be investigated thus far with *Kepler* data. The observed power spectrum, as obtained by Chaplin *et al.* (2010), is shown in Fig. 9. The analysis of the frequencies of the star was the subject of the paper by Metcalfe *et al.* (2010). Gemma is 10-20 per cent more massive than the Sun and also somewhat older. The core of Gemma is therefore more chemically evolved than is the core of the

† The working group allocated the names of pet cats to the stars that have been the early objects of study: this ideosyncracy is due to the WG1 lead, Bill Chaplin.

Sun; the models indicate that the star has a small compact helium core, surrounded by a hydrogen-burning shell. This leads to interesting behaviour of the star's frequencies which provides a powerful diagnostic of the star's evolutionary state. As a solar-like star evolves at the end of its main-sequence life, it continues to grow in radius while forming a strong gradient in mean-molecular weight at the edge of its core. Also, the core contracts, increasing the central condensation and the gravitational acceleration in the deep interior of the star. These effects in turn cause a strong peak to form in the buoyancy frequency (cf. Eq. 5.2), which supports g modes at frequencies which are representative for the stochastically excited solar-like oscillations. Thus at such frequencies the star has two resonant cavities supporting nonradial modes: one in the deep envelope of the star where the modes behave like p modes, increasingly confined to the outer regions of the star with increasing degree, and one in the core where the modes behave like g modes. These two regions are separated by an intermediate region where the modes are evanescent.

With increasing age the star undergoes an overall expansion which causes the frequencies of the p modes to decrease, while the increase in the central condensation and hence the buoyancy frequency causes the frequencies of g modes to increase. Although the g modes are not in general themselves observable directly, at times in the star's evolution the frequencies of a g mode and a p mode get sufficiently close for a strong coupling between the two modes to be possible, giving rise to a so-called mixed mode. This evolution of the frequency spectrum is illustrated in Fig. 10 for a representative stellar model of Gemma. This shows how the radial ($l = 0$) and $l = 1$ p-mode frequencies change as a function of age of the star. By their nature, g modes are non-radial; and hence the radial modes cannot couple to them: thus the overall expansion of the star simply causes the $l = 0$ frequencies to decrease monotonically with increasing age. The $l = 1$ frequencies also tend to decrease with age; but occasionally a given $l = 1$ mode approaches the frequency of an $l = 1$ g mode: at that point, a strong coupling between the two modes occurs if the evanescent region between their two resonant cavities is not too large. The frequencies of the two modes never actually cross: instead, the modes undergo an avoided crossing, which results in the observable mode increasing in frequency as the star evolves,

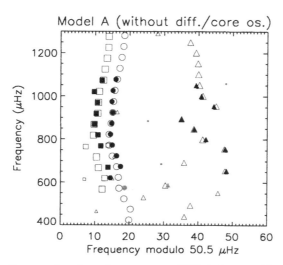

Figure 8. An échelle diagram for the solar-type star KIC 11026764 (Gemma) indicating both the observed frequencies (filled symbols) and those of a stellar model (open symbols). Modes of degree $l = 0, 1, 2$ are denoted respectively by circles, triangles and squares. From Metcalfe *et al.* (2010).

for the duration of the strong mode coupling. The evolving frequencies of the physical
g modes can be discerned in Fig. 10 as the loci of the avoided crossings that take place
for the $l = 1$ p modes.

The frequency spectrum of $l = 0$ and $l = 1$ modes at any particular stellar age can
be read off Fig. 10 by taking a vertical cut through the ridges: an example at an age of
about 5.98 Gyr is indicated. It is evident that the $l = 0$ modes will be essentially evenly
spaced in frequency, consistent with the asymptotic expression (Eq. 7.1), whereas the
series of avoided crossings will cause the $l = 1$ modes to be nonuniform. Moreover, the
location in frequency space where avoided crossings are "caught in the act" is strongly
dependent on the age of the star and so can enable the age of the star to be determined
rather precisely.

This behaviour is consistent with measured frequencies of Gemma, illustrated in the
échelle diagram in Fig. 8. The $l = 0$ and $l = 2$ frequencies are approximately uniformly
spaced, whereas the $l = 1$ frequencies are more irregularly spaced, consistent with avoided
crossings. (The $l = 2$ and higher-degree p modes are affected much less by coupling to
the g modes than are the $l = 1$ modes, because their lower turning points are further
from the core and hence the evanescent region between the resonant cavities of the p and
g modes is wider.)

Metcalfe *et al.* (2010) modelled Gemma, fitting to the individual measured frequencies
and hence exploiting in particular the non-uniform distribution of the $l = 1$ modes.
The frequencies of one of their resulting models are shown in the échelle diagram. They
found that two families of solutions, one with stellar masses around $1.1\,M_\odot$ and the
other with stellar masses around $1.2\,M_\odot$, that fitted the observed frequencies equally
well. Notwithstanding this 10% ambiguity in mass, the radius and age of the star were
determined with a precision of about 1%, and an estimated accuracy of about 2% for the
radius and about 15% for the age. The mass ambiguity would be resolved if the range of

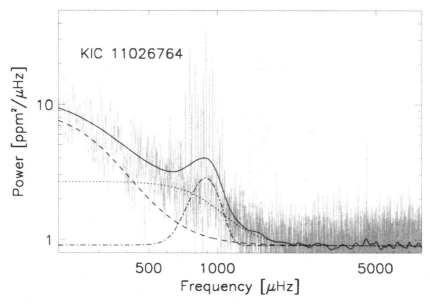

Figure 9. Observed power spectrum of the star KIC 11026764, known as Gemma, based on early
Kepler observations. The solid line shows the smoothed spectrum, separated into the oscillation
signal (dash-dotted line) and background components from granulation and faculae (dashed and
dotted lines, respectively). Figure, adapted from Chaplin *et al.* (2010), courtesy of C. Karoff.

measured frequencies could be extended to higher frequencies, at which point the model frequencies not only of the $l = 1$ modes but also of the $l = 0, 2$ modes diverge between the two families of models.

The properties of the modes in the vicinity of the avoided crossings is illustrated in more detail in terms of the mode inertia (cf. Eq. 3.2) in Fig. 11. When the dipolar modes behave predominantly as acoustic modes, with frequencies decreasing with increasing age, their inertia is close to that of the neighbouring radial mode. It increases when a mode behaves predominantly as a g mode; at the point of closest approach in an avoided crossing the two modes have the same inertia, intermediate between the g-mode and the p-mode behaviour. As discussed below for red giants, the inertia has an important influence on the mode visibility. For the dipolar modes in the Gemma models the contrast between the p- and g-mode behaviour is modest and the modes are readily observable, even when they are most g-mode-like. On the other hand, for modes of degree $l = 2$ and higher the intermediate evanescent region is substantially broader and the distinction between the p- and g-mode behaviour correspondingly stronger. Thus it is less likely to observed mixed modes at these higher degrees, as confirmed by the interpretation of the observed spectrum.

Apart from the intrinsic interest in the asteroseismic studies of solar-like stars, these studies provide an important possibility for characterizing stars that host extra-solar planetary systems. When a planet is detected using the transit technique the variation in the detected stellar brightness depends on the ratio between the diameters of the planet and the central star. A reliable determination of the planet radius, of great importance to the characterization of the nature of the planet, therefore depends on determining the stellar radius. As discussed above this can be provided by the asteroseismic analysis of the oscillations of the star; indeed, the target stars for the *Kepler* search for extra-solar planets are generally in a range of parameters where solar-like oscillations are expected, although most stars are too faint for reliable observations to be possible. However, the potential of the technique was demonstrated by Christensen-Dalsgaard *et al.* (2010) who analysed *Kepler* asteroseismic data for a known exoplanet host. More recently, astero-

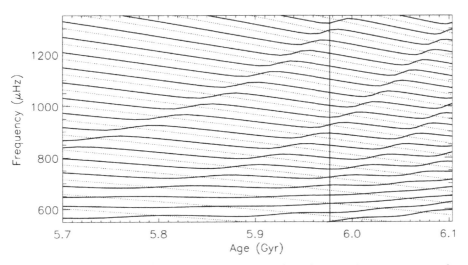

Figure 10. Evolution of the $l = 0$ (dotted) and $l = 1$ (solid) mode frequencies as a function of age for a representative stellar model of KIC 11026764 (Gemma). The vertical line indicates the age of 5.77 Gyr of one good-fitting model to Gemma's observed frequencies. From Metcalfe *et al.* (2010).

seismic analysis was used in the characterization of the first rocky planet detected by *Kepler* (Batalha *et al.* 2011).

We finally note that the frequencies of solar-like oscillation are sensitive to stellar magnetic activity. In the solar case this has been studied extensively (e.g., Woodard & Noyes 1985; Libbrecht & Woodard 1990). Some evidence for such variation was found by Metcalfe *et al.* (2007) for the star β Hyi. As discussed by Karoff *et al.* (2009) the long observing sequences possible with CoRoT and in particular *Kepler* provide a rich possibility for detecting similar effects in other stars. In fact, in a solar-like star observed by CoRoT García *et al.* (2010) detected variations in oscillation frequencies and amplitudes which appeared to be the result of a rather short stellar activity cycle.

8. Red giants

Assuming that the solar oscillations are excited stochastically by convection (Goldreich & Keeley 1977) one would expect that all stars with vigorous outer convection zones exhibit such oscillations. A rough estimate of the amplitudes (Christensen-Dalsgaard & Frandsen 1983) suggested that the amplitude increases with increasing luminosity. Thus red giants are obvious targets for the search for, and investigation of, solar-like oscillations.

The first definite detection of individual modes in a red giant was obtained by Frandsen *et al.* (2002) in the star ξ Hya. The frequency spectrum showed very clearly a series of

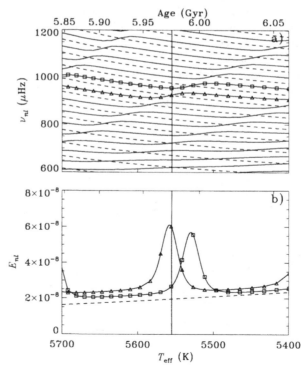

Figure 11. Panel a) shows a blow-up of Fig. 10 around the model illustrated in Fig. 8. Panel b) shows mode inertias for two of the $l = 1$ modes (solid lines, identified in both panels by triangles and squares, respectively), and the intervening radial mode (dashed line). The increase in the inertia, relative to the radial mode, is an indication of a predominant g-mode character of the modes.

uniformly spaced peaks with a separation $\Delta\nu \simeq 7\,\mu$Hz. Simple modelling, given also the location in the HR diagram, strongly suggested that only radial modes had been observed; the alternative identification, of alternating $l = 0$ and $l = 1$ modes, would correspond to a true large separation twice as big and hence a radius smaller by roughly a factor 1.3, entirely inconsistent with the observed luminosity and effective temperature.

Evidence for solar-like oscillations in giants had also been obtained from more statistical analyses. Christensen-Dalsgaard *et al.* (2001) noted that the relation between the standard deviation and mean of the amplitude variations in the so-called semi-regular variables, based on visual observations carried out by the American Association of Variable Star Observers (AAVSO), was consistent with the expectations for stochastically excited oscillations. The solar-like nature of the oscillations of selected semi-regular variables was confirmed by Bedding (2003) through analysis of their power spectra. Also, Kiss & Bedding (2003, 2004) analysed large sets of OGLE† observations of red giants, obtaining clear indication of several modes of oscillation which may likely be identified as solar-like. Detailed analyses of the OGLE data were carried out by Soszyński *et al.* (2007) and Dziembowski & Soszyński (2010), confirming the solar-like nature of the observed oscillations. These investigations extend the realm of solar-like oscillations to stars with a luminosity of up to $10\,000\,L_\odot$ and periods of several months.

The red-giant phase (see Salaris *et al.* 2002, for a review) follows after the phase exemplified by Gemma, discussed in Section 7 above. The stars ascend the Hayashi region at almost constant effective temperature and strongly increasing radius and luminosity, with a very compact helium core and an extended, mostly convective, envelope. The energy production takes place through hydrogen fusion in a thin shell around the helium core. The tip of the red-giant branch is defined by the ignition of helium near the centre. Stars with central helium fusion are located in the so-called 'red clump' in the HR diagram (see Fig. 15); even for these, however, most of the energy is produced by the hydrogen shell. The strongly centralized helium fusion gives rise to a small convective core, although the bulk of the helium core remains radiative. In both the ascending red giant and the clump phase the small extent of the core gives rise to a very high gravitational acceleration and hence buoyance frequency, further amplified by the presence of strong composition gradients (cf. Eq. 5.2). Thus all nonradial modes have the character of high-order g modes in the core. The resulting mixed nature of the modes, and the high density of modes of predominantly g-mode character, are illustrated in Fig. 12 which shows the mode inertia E (cf. Eq. 3.2) for a typical model on the ascending red-giant branch. Most of the modes with $l = 1$ and 2 clearly have much higher inertias than the radial modes, and hence are predominantly g modes. However, there are resonances where the modes are largely trapped in the outer acoustic cavity. These p-dominated modes have inertias close to the inertia of the radial modes, reflecting their small amplitudes in the core.

Dziembowski *et al.* (2001) considered the excitation and damping of modes in red giants. They found that the very high radial order of the nonradial modes in the core led to substantial radiative damping, even for modes that were predominantly acoustic. On this basis Christensen-Dalsgaard (2004) concluded that nonradial oscillations were unlikely to be observed in red giants. (This would be consistent with the results of Frandsen *et al.* (2002) on ξ Hya where apparently only radial modes were found.) Fortunately this conclusion was wrong: nonradial modes are indeed observed in red giants and provide fascinating diagnostics of their interiors.

A first indication of the presence of nonradial oscillations in red giants came from line-profile observations by Hekker *et al.* (2006). However, the major breakthrough came

† Optical Gravitational Lensing Experiment

with the observation of a substantial number of red giants by CoRoT, as presented by De Ridder *et al.* (2009). A selection of the resulting power spectra are shown in Fig. 13. The presence of solar-like oscillations is obvious, the peaks shifting to lower frequency with increasing radius (cf. Eq. 3.3). Also, De Ridder *et al.* (2009) showed an example of an échelle diagram which beyond any doubt identified modes of degree 0, 1 and 2.

The potential visibility of nonradial modes in red giants was made very much clearer by Dupret *et al.* (2009), following an analysis by Chaplin *et al.* (2005). The observational visibility of a mode is determined by the peak height H in the power spectrum. This is related to the total observed power P of the mode by $P \propto H\Delta$, where Δ is the width of the peak. If the mode is observed for much longer than the natural damping time, the width is given by the damping rate, i.e., the imaginary part $|\omega_i|$ of the frequency. If the damping is dominated by the near-surface layers, as is often the case, at a given frequency ω_i is related to the mode inertia E by

$$\omega_i \propto E^{-1} \, . \tag{8.1}$$

Thus those modes that are predominantly g modes, with high inertia (cf. Fig. 12) have much smaller widths than the p-dominated modes. The power in the mode is determined by a balance between the energy input from the stochastic noise near the stellar surface and the damping. Assuming again Eq. (8.1) the outcome is that $P \propto E^{-1}$ at fixed frequency. It follows that the peak height H is independent of E at a given frequency and hence that the g-dominated modes should be observed at the same height as the p-dominated modes.

This, however, assumes that the duration $\mathcal{T}$ of the observation is much longer than the lifetime $|\omega_i|^{-1}$ of the mode. If this is not the case, the peaks are broader and the height consequently smaller. As an approximate scaling of this dependence Fletcher *et al.* (2006) proposed

$$H \propto \frac{P}{|\omega_i| + 2/\mathcal{T}} \, ; \tag{8.2}$$

for $\mathcal{T} \ll |\omega_i|^{-1}$, in particular, $H \propto P \propto E^{-1}$ and the g-dominated modes are essentially

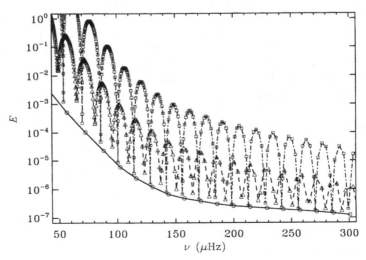

Figure 12. Mode inertias (cf. Eq. 3.2) against cyclic frequency in a red-giant model of mass $1.4\,M_\odot$ and radius $5R_\odot$. Modes of degree $l = 0$ (circles connected by a solid line), $l = 1$ (triangles connected by a dashed line) and $l = 2$ (squares connected by a dot-dashed line) are illustrated.

invisible. As is clear from Fig. 12 the g-dominated modes in practice often have inertias much higher than the p-dominated modes and hence correspondingly longer lifetimes; thus they may be expected have small observed peak heights, unless observations of very long duration are analysed. However, modes of mixed character, particularly those with $l = 1$, may have damping times comparable with or shorter than the very long observations made available by CoRoT and *Kepler* and hence may be visible. Concerning ξ Hya, the apparent absence of nonradial modes in the observations was probably caused by the relatively short observing run of around one month, compared with the five-month observations by De Ridder *et al.* (2009). Even the most p-mode-like dipolar modes have somewhat higher mode inertia and hence lifetimes than the radial modes; thus the peak height of these modes was likely suppressed in the observations by Frandsen *et al.* (2002).

Very extensive results on red-giant oscillations have been obtained by CoRoT and *Kepler* (e.g., Hekker *et al.* 2009; Bedding *et al.* 2010; Mosser *et al.* 2010; Kallinger *et al.* 2010a,b; Stello *et al.* 2010; Hekker *et al.* 2011). These confirm the acoustic nature of the observed spectra, with a clear detection of modes of degree 0, 1 and 2. This is illustrated in Fig. 14 (see also Gilliland *et al.* 2010), for a large sample of stars observed with *Kepler* during the first 16 months of the mission; the observations are shown as stacked spectra ordered according to decreasing large frequency separation $\Delta\nu$ and hence increasing radius and luminosity. This is characterized at the right-hand edge of the figure by the ratio between luminosity and mass, estimated from the oscillation parameters (see below). Stellar 'noise' from granulation is evident in the low-frequency region. The frequencies

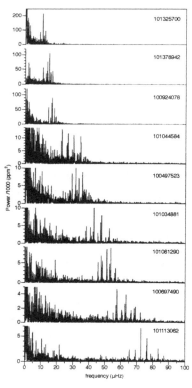

Figure 13. Power spectra of solar-like oscillations in red giants, from five months of observations with CoRoT. The stars are identified by their CoRoT identification number, with radius increasing towards the top. From De Ridder *et al.* (2009).

approximately satisfy an asymptotic relation similar to Eq. (7.1), with a closely spaced pair of bands of $l = 0, 2$ modes and an intermediate band of $l = 1$ modes. However, since the acoustic propagation region is generally confined to the convection zone or the region just beneath it the small separation between $l = 0$ and 2 is not directly related to the properties of the stellar core, let alone the age of the star, unlike the situation on the main sequence. Montalbán et al. (2010a) carried out an extensive analysis of the overall properties of the oscillation frequencies for a large sample of red-giant models. They noted that the outer convective zone in red-clump phase is not quite as deep as for the stars ascending the red-giant branch; the extent of the acoustic propagation region beyond the convective envelope was found to have a potentially measurable effect on the small frequency separations.

The mean large frequency separation $\Delta\nu$ and the frequency $\nu_{\max}$ at maximum power satisfy the scaling relations (3.1) and (3.3). Thus the stellar properties can be characterized by these quantities. This is used in Fig. 15 to plot a 'Hertzsprung-Russell' diagram of red giants observed with Kepler, replacing the luminosity by $\nu_{\max}$ as a measure of radius and hence luminosity. The observations are compared with evolution tracks for a range of masses, using scaling from the solar value of $\nu_{\max}$. The distribution of stars clearly shows the higher density in the region of the helium-burning red clump. The scaling relations can also be used to determine the stellar parameters from the observed $\Delta\nu$, $\nu_{\max}$ and T_{eff} (e.g. Kallinger et al. 2010a). This provides a unique possibility for population studies of red giants (e.g., Miglio et al. 2009; Kallinger et al. 2010b; Mosser et al. 2010; Hekker et al. 2011). The CoRoT results are particularly interesting in this regard, given that they allow a comparison of the populations in the centre and anti-centre directions of the Galaxy (Miglio et al., in preparation) and hence provide information about the evolution and dynamics of the Galaxy. A more precise determination of the stellar parameters can be obtained with the so-called grid-based methods, where stellar modelling is used to relate the effective temperature, mass and radius (e.g., Gai et al. 2011). This was used by Basu et al. (2011) to investigate the properties of two open clusters in the Kepler field.

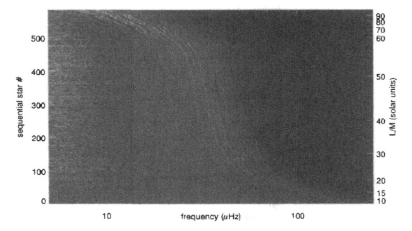

Figure 14. Stacked power spectra of red giants observed with Kepler, with the frequency of maximum power decreasing towards the top. The numbers at the right edge provide an estimate of the corresponding ratio between luminosity and mass, in solar units. Figure courtesy of T. Kallinger.

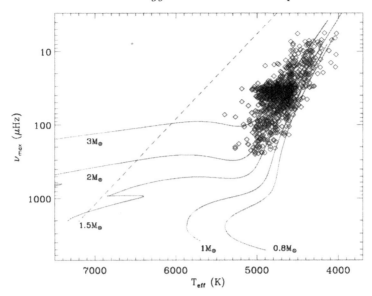

Figure 15. 'Hertzsprung-Russell diagram' of red giants observed with *Kepler*, using the frequency $\nu_{\max}$ at maximum power as a proxy for luminosity. The curves show evolution tracks for models at the indicated ages. Figure courtesy of D. Huber (see Huber *et al.* 2010).

To the extent that the modes are trapped in the convective envelope, whose structure is very similar amongst different stars apart from a scaling depending on the stellar mass and radius, one would expect a corresponding similarity between the oscillation frequencies. This is confirmed by the so-called 'universal red-giant oscillation pattern', a term introduced by Mosser *et al.*, of the oscillations (Huber *et al.* 2010; Mosser *et al.* 2011). To illustrate this, Fig. 16 shows a normalized and stacked échelle diagram. Here collapsed échelle diagrams for the individual stars, normalized by the large separation, have been stacked after taking out the variation with stellar parameters of the average ϵ (cf. Eq. 7.1). Clearly there is very little variation with $\nu_{\max}$ and hence stellar properties in the location of the ridges and hence the scaled small separations. This is emphasized by the collapsed version of the diagram in the lower panel; it should be noticed that this also, as indicated, provides a weak indication of modes of degree $l = 3$. A lower limit to the width of the ridges is provided by the natural width of the peaks, corresponding to the lifetime of the modes. For $l = 0$ and 2 Huber *et al.* (2010) found a width of around $0.2\,\mu$Hz, essentially independent of the stellar parameters and corresponding to a mode lifetime† of around 18 d. Similar results were obtained by Baudin *et al.* (2011) based on CoRoT observations. Interestingly, this value of the lifetime is similar to the estimate obtained by Houdek & Gough (2002) from modelling of ξ Hya.

In Fig. 16 it is evident that the ridge for $l = 1$ appears substantially wider than for $l = 0$ and 2. This can be understood from the analysis of Dupret *et al.* (2009), discussed above, which showed that several dipolar modes may reach observable amplitudes in the vicinity of an acoustic resonance (see also Montalbán *et al.* 2010b). In a major breakthrough in red-giant asteroseismology Beck *et al.* (2011) and Bedding *et al.* (2011) demonstrated that the frequencies in such groups of peaks showed clear g-mode-like behaviour; this allowed a determination of the uniform g-mode period spacing (cf. Eq. 5.1).

† defined as the e-folding time of the displacement

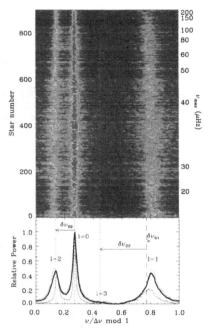

Figure 16. The upper panel shows stacked collapsed, rescaled and shifted échelle diagrams for red giants observed with *Kepler*. These have been collapsed in the lower panels, where the peaks corresponding to $l = 0 - 3$ are indicated. The thick lines correspond to the full set, and the thin blue and red lines (in the electronic version) correspond to stars with $\nu_{\max} > 100\,\mu$Hz and $\nu_{\max} < 50\,\mu$Hz, respectively. Figure courtesy of D. Huber (see Huber *et al.* 2010).

It was further demonstrated by Bedding *et al.* (2011) that the inferred value of $\Delta\Pi$ allowed to distinguish between stars on the ascending red-giant branch and stars in the helium-burning clump phase. With further analyses these observations will undoubtedly provide very valuable diagnostics of the central properties of red giants.

The frequencies of purely acoustic modes also contain information beyond the basic parameters $\Delta\nu$ and $\nu_{\max}$. Sharp features in the sound speed lead to systematic departures from the simple asymptotic behaviour in Eq. (7.1), with characteristic properties which provide information about the location and strength of the feature (e.g., Gough 1990). An important example is the effect on the sound speed of the localized reduction in the adiabatic exponent caused by the second ionization of helium, which provides information about the helium abundance (e.g., Vorontsov *et al.* 1991; Monteiro & Thompson 2005; Houdek & Gough 2007). Carrier *et al.* (2010) and Miglio *et al.* (2010) found the signature of this effect in the red giant HR 7349 observed by CoRoT. Miglio *et al.* noted that the inferred location of the second helium ionization zone provided a strong constraint on the properties of the star. Such analyses will undoubtedly be possible for a large number of red giants observed by CoRoT and *Kepler*.

9. Concluding remarks

The last few years have seen amazing progress in the availability of data for asteroseismic investigations of stellar interiors. The full scientific exploitation of these data is just starting. The community is realizing the challenges provided by actual observations, compared with the simulations that preceded the missions, and we are hard at work at optimizing the techniques for the data analysis and interpretation, with the goal of addressing specific aspects of stellar interiors. The ability to study solar-like oscillations in hundreds of stars with *Kepler* has been a positive surprise, allowing comparative studies in *ensemble asteroseismology*† (Chaplin *et al.* 2011); however, we still need to extend the investigations to unevolved stars of lower masses, as will surely be possible as ever longer timeseries for individual stars are accumulated. The present authors, at least, had not foreseen the possibility of detailed analyses of the mixed modes in subgiants, providing archaeological information about the properties of the mixed cores during the main-sequence phase (Deheuvels & Michel 2010). The very recent detection and analysis of features in the pulsations directly related to the g modes in the cores of red giants (Beck *et al.* 2011; Bedding *et al.* 2011) have also been a major surprise, with huge potentials both for characterizing the evolutionary state of the stars and for investigating the properties of the deep interiors of these stars. The detection in the *Kepler* field of a dozen subdwarf B stars showing long-period oscillations, with additional cases being found by CoRoT, is providing tight constraints on the overall properties of stars in this very late phase of evolution and there is certainly a potential for much more detailed investigations. And the list goes on.

We are fortunate currently to have access to three largely complementary space missions with asteroseismic potentials. The MOST mission is perhaps somewhat overshadowed by CoRoT and *Kepler*, but it continues to provide excellent data on a variety of bright stars with the possibility of selecting targets over a large part of the sky; the recent combined analysis of MOST data and ground-based radial-velocity data for Procyon (Huber *et al.* 2011) demonstrates the potential of MOST even in the area of solar-like pulsations. CoRoT has fully demonstrated the required sensitivity to study solar-like oscillations in main-sequence stars. The mission has the very substantial advantage of being able to observe both in the direction towards and away from the Galactic centre which allows comparative studies of stellar populations. Also, the stars observed in the asteroseismology field are relatively bright, facilitating the ground-based support observations of these targets, and the CoRoT 'eyes' contain a very broad sample of interesting targets of most types. Finally, *Kepler* can observe stars for the duration of the mission, optimizing the precision and sensitivity of the observations and allowing the uninterrupted study of potential changes in stellar properties. The heliocentric orbit provides a more stable and quiet environment than the low-earth orbit of CoRoT, in terms of scattered light and magnetospheric disturbances. Also, the *Kepler* field has been found to be extremely rich in a variety of interesting stars, now to a large extent characterized through the KAI survey phase.

Even so, there is a continued need to improve the observational situation and strong prospects that this will happen. The BRITE Constellation mission, under development by Austria, Canada and Poland (Kuschnig *et al.* 2009) will fill a very important niche by carrying out photometric observations of the brightest stars across the sky in two colours. On a longer timescale the ESA PLATO mission, if selected, will greatly extend the *Kepler* results (Catala *et al.* 2011). As *Kepler*, PLATO has the dual purpose of exoplanet investigations and asteroseismology. However, PLATO will look at substantially brighter stars in much larger fields. This is important for the crucial ground-based follow-up observations to confirm the detection of a planet in an apparent transit, particularly for earth-size planets which will be a key goal for PLATO as it is for *Kepler*. Also, as a result PLATO will allow asteroseismic characterization of a substantial fraction of the

† or, according to D. O. Gough, perhaps better *synasteroseismology*

stars around which potential planets are found, unlike *Kepler* where this is an exception. PLATO will be placed in an orbit around the L_2 point which shares the advantage, in terms of stability, of *Kepler's* heliocentric orbit. The planned observing schedule consists of continuous observations of two fields for two or three years each, followed by a 'stop-and-stare' phase where each field is observed for a few months. The latter part of the mission will allow investigation of a substantial fraction of the sky, providing a survey of a far more wide-ranging and varied set of stellar pulsations and other types of stellar variability than has been possible even with *Kepler*.

The great advances provided by the space asteroseismic observations should not blind us to the continued usefulness of ground-based observations. In photometry that allows study of rare objects that may not be available to the space observatories. Also, it provides greater flexibility, e.g., in carrying out observations in several colours, of importance to mode identification. Even more important is the use of ground-based radial-velocity observations, particularly for solar-like oscillations. Observations of solar oscillations from the SOHO mission have clearly demonstrated that the solar background, from granulation and activity, is substantially higher relative to the oscillations for photometric observations than for radial-velocity observations, as already noted by Harvey (1988) (see also Grundahl *et al.* 2007). This background is evident in the *Kepler* observations of Gemma illustrated in Fig. 9. This puts a natural limit to the precision and mode selection possible in photometric observations. Thus radial-velocity observations of carefully selected stars are still required to reach the ultimate precision and level of detail in asteroseismic investigations. Such observations can been carried out from the ground, as has been done successfully for a small sample of stars (e.g., Bouchy & Carrier 2002; Bedding *et al.* 2004; Bazot *et al.* 2005; Kjeldsen *et al.* 2005; Bedding *et al.* 2007; Arentoft *et al.* 2008); to reduce the gaps in the data they have been carried out in a coordinated fashion, involving two or more observatories. However, the required state-of-the-art instrumentation is only available for limited periods and certainly not for the month-long observations from several observatories that are needed to secure adequate frequency resolution. This is the motivation for the development of the Stellar Observations Network Group (SONG) network (Grundahl *et al.* 2009, 2011). SONG is planned to consist of 7 – 8 nodes with a suitable geographical distribution in the northern and southern hemisphere. Each node will consist of a 1 m telescope, equipped with a high-resolution spectrograph for Doppler-velocity observations and a so-called lucky-imaging camera for photometry in crowded fields. With the use of an iodine cell as reference, and with careful optimization of the optics, it is estimated that SONG will be able to study velocity oscillations in a star as the Sun at magnitude 6. The lucky-imaging camera is designed for characterization of exoplanet systems through gravitational micro-lensing (e.g., Dominik *et al.* 2010). At present a prototype SONG node is under construction with Danish funding, to be placed at the Izaña Observatory on Tenerife, with expected deployment and start of operations in 2011. A Chinese node is being designed and is expected to be operational in 2013, and funding and support for further nodes will be sought through a network of international collaborations.

The data from these projects will provide excellent possibilities for testing present stellar evolution calculations and indicating where improvements should be made. Such improvements are certainly required, particularly when it comes to the treatment of stellar internal dynamics. Impressive progress is being made in the application of complex, yet unavoidably highly simplified, treatments of the interplay between rotation, circulation, and magnetic fields, including also the evolution of stellar internal angular velocity (e.g., Palacios *et al.* 2003, 2006; Mathis & Zahn 2004, 2005) (see also Maeder 2009). Indeed, full stellar evolution calculations will undoubtedly require such simplifications in the foreseeable future. However, it is equally important that these simplified treatments be tested by detailed simulations of the hydrodynamical phenomena, albeit for conditions that often do not fully reflect the stellar internal conditions. An important example is the simulation of near-surface convection, where computations under reasonably realistic conditions are in fact possible (e.g., Nordlund *et al.* 2009; Trampedach 2010). Simulations of the deeper parts of the convective envelope and the region below (see Miesch 2005; Miesch & Toomre 2009, for reviews) unavoidably require simplification, but are providing deep insights into the interaction between convection and rotation, and the generation of magnetic fields (Brun *et al.* 2004; Browning *et al.* 2006; Miesch *et al.* 2008). Such simulations are now being extended to stars other than the Sun (Brun & Palacios 2009; Brown, B. P., *et al.* 2011), including the dynamics of stellar convective cores (Brun *et al.* 2005; Featherstone *et al.* 2009). The observations by CoRoT, *Kepler* and projects to follow provide excellent prospects for testing the results of such modelling and hence improve our understanding of stellar internal dynamics.

Acknowledgement: We wish to take this occasion to thank Juri for many years of enjoyable collaboration, as well as for his inspiration and constant friendship. We are very grateful to P. Degroote, J. De Ridder, G. Doğan, D. Huber, T. Kallinger, C. Karoff, K. Kolenberg, R. Szabó and V. Van Grootel for the provision of, or help with, the figures, and to Travis Metcalfe for comments that helped improve the paper. We thank Sacha Brun and Nick Brummell for the excellent organization of an exciting conference, and for their patience with the present authors. The National Center for Atmospheric Research is sponsored by the National Science Foundation (NSF).

References

Aerts, C., Thoul, A., Daszyńska, J., Scuflaire, R., Waelkens, C., Dupret, M. A., Niemczura, E., & Noels, A. 2003, *Science*, 300, 1926

Aerts, C., Christensen-Dalsgaard, J., & Kurtz, D. W. 2010, *Asteroseismology*, Springer, Heidelberg

Arentoft, T., Kjeldsen, H., Bedding, T. R., Bazot, M., Christensen-Dalsgaard, J., et al. 2008, *ApJ*, 687, 1180

Ausseloos, M., Scuflaire, R., Thoul, A., & Aerts, C. 2004, *MNRAS*, 355, 352

Auvergne, M., Bodin, P., Boisnard, L., Buey, J.-T., Chaintreuil, S., et al. 2009, *A&A*, 506, 411

Baglin, A., Michel, E., & Auvergne, M. and the CoRoT team 2006, in *Proc. SOHO 18 / GONG 2006 / HELAS I Conf. Beyond the spherical Sun*, ed. K. Fletcher, ESA SP-624, ESA Publications Division, Noordwijk, The Netherlands

Baglin, A., Auvergne, M., Barge, P., Deleuil, M., & Michel, E. and the CoRoT Exoplanet Science Team 2009, in *Proc. IAU Symp. 253, Transiting Planets*, eds F. Pont, D. Sasselov & M. Holman, IAU and Cambridge University Press, p. 71

Balona, L. A., Pigulski, A., De Cat, P., Handler, G., Gutiérrez-Soto, J., et al. 2011, *MNRAS*, in the press

Baran, A., Oreiro, R., Pigulski, A., Pérez Hernández, F., Ulla, A., et al. 2009, *MNRAS*, 392, 1092

Baran, A. S., Kawaler, S. D., Reed, M. D., Quint, A. C., O'Toole, S. J., et al. 2011, *MNRAS*, in the press. [arXiv:1103.1666v1]

Basu, S., Grundahl, F., Stello, D., Kallinger, T., Hekker, S., et al. 2011, *ApJ*, 729, L10

Batalha, N. M., Borucki, W. J., Bryson, S. T., Buchhave, L. A., Caldwell, D. A., et al. 2011, *ApJ*, 729, 27

Baudin, F., Barban, C., Belkacem, K., Hekker, S., Morel, T., et al. 2011, *A&A*, 529, A84

Bazot, M., Vauclair, S., Bouchy, F., & Santos, N. C. 2005, *A&A*, 440, 615

Beck, P. G., Bedding, T. R., Mosser, B., Stello, D., Garcia, R. A., et al. 2011, *Science*, 332, 205.

Bedding, T. R. 2003, *Ap&SS*, 284, 61

Bedding, T. R. & Kjeldsen, H. 2003, *PASA*, 20, 203

Bedding, T. R., Kjeldsen, H., Butler, R. P., McCarthy, C., Marcy, G. W., O'Toole, S. J., Tinney, C. G., & Wright, J. T. 2004, *ApJ*, 614, 380

Bedding, T. R., Kjeldsen, H., Arentoft, T., Bouchy, F., Brandbyge, J., *et al.* 2007, *ApJ*, 663, 1315

Bedding, T. R., Huber, D., Stello, D., Elsworth, Y. P., Hekker, S., *et al.*, 2010, *ApJ*, 713, L176

Bedding, T. R., Mosser, B., Huber, D., Montalbán, J., Beck, P., *et al.* 2011, *Nature*, 471, 608

Belkacem, K., Goupil, M. J., Dupret, M. A., Samadi, R., Baudin, F., Noels, A., & Mosser, B. 2011, *A&A*, in the press. [arXiv:1104.0630v2]

Blažko, S. 1907, *AN*, 175, 325

Borucki, W., Koch, D., Bathalha, N., Caldwell, D., Christensen-Dalsgaard, J., *et al.* 2009, in *Proc. IAU Symp. 253, Transiting Planets*, eds F. Pont, D. Sasselov & M. Holman, IAU and Cambridge University Press, p. 289

Borucki, W. J., Koch, D., Basri, G., Batalha, N., Brown, T., *et al.* 2010, *Science*, 327, 977

Bouchy, F. & Carrier, F. 2001, *A&A*, 374, L5

Bouchy, F. & Carrier, F. 2002, *A&A*, 390, 205

Brown, B. P., Miesch, M. S., Browning, M. K., Brun, A. S., & Toomre, J. 2011, *ApJ*, 731, 69

Brown, T. M. & Gilliland, R. L. 1994, *ARAA*, 32, 37

Brown, T. M., Gilliland, R. L., Noyes, R. W., & Ramsey, L. W. 1991, *ApJ*, 368, 599

Brown, T. M., Latham, D. W., Everett, M. E., & Esquerdo, G. A. 2011, submitted to *ApJ*. [arXiv:1102.0342]

Browning, M. K., Miesch, M. S., Brun, A. S., & Toomre, J. 2006, *ApJ*, 648, L157

Brun, A. S. & Palacios, A. 2009, *ApJ*, 702, 1078

Brun, A. S., Miesch, M. S., & Toomre, J. 2004, *ApJ*, 614, 1073

Brun, A. S., Browning, M. K., & Toomre, J. 2005, *ApJ*, 629, 461

Carrier, F., De Ridder, J., Baudin, F., Barban, C., Hatzes, A. P., *et al.* 2010, *A&A*, 509, A73

Catala, C. & Appourchaux, T. and the PLATO Mission Consortium 2011, in *Proc. GONG-SoHO 24: A new era of seismology of the sun and solar-like stars*, ed. T. Appourchaux, *J. Phys.: Conf. Ser.*, 271, 012084

Chadid, M., Perini, C., Bono, G., Auvergne, M., Baglin, A., Weiss, W. W., & Deboscher, J. 2011, *A&A*, 527, A146

Chaplin, W. J., Houdek, G., Elsworth, Y., Gough, D. O., Isaak, G. R., & New, R. 2005, *MNRAS*, 360, 859

Chaplin, W. J., Appourchaux, T., Elsworth, Y., García, R. A., Houdek, G., *et al.*, 2010, *ApJ*, 713, L169

Chaplin, W. J., Kjeldsen, H., Christensen-Dalsgaard, J., Basu, S., Miglio, A., *et al.* 2011, *Science*, 332, 213

Charpinet, S., Fontaine, G., Brassard, P., & Dorman, B. 1996, *ApJ*, 471, L103

Charpinet, S., Green, E. M., Baglin, A., van Grootel, V., & Fontaine, G., *et al.* 2010, *A&A*, 516, L6

Christensen-Dalsgaard, J. 2002, *Rev. Mod. Phys.*, 74, 1073

Christensen-Dalsgaard, J. 2004, *Solar Phys.*, 220, 137

Christensen-Dalsgaard, J. & Frandsen, S. 1983, *Solar Phys.*, 82, 469

Christensen-Dalsgaard, J., Kjeldsen, H., & Mattei, J. A. 2001, *ApJ*, 562, L141

Christensen-Dalsgaard, J., Arentoft, T., Brown, T. M., Gilliland, R. L., Kjeldsen, H., Borucki, W. J., & Koch, D. 2008, in *Proc. HELAS II International Conference: Helioseismology, Asteroseismology and the MHD Connections*, eds L. Gizon & M. Roth, *J. Phys.: Conf. Ser.*, 118, 012039

Christensen-Dalsgaard, J., Kjeldsen, H., Brown, T. M., Gilliland, R. L., Arentoft, T., Frandsen, S., Quirion, P.-O., Borucki, W. J., Koch, D., & Jenkins, J. M. 2010, *ApJ*, 713, L164

Cox, J. P. & Whitney, C. 1958, *ApJ*, 127, 561

Degroote, P., Aerts, C., Samadi, R., Miglio, A., Briquet, M., Auvergne, M., Baglin, A., Baudin, F., Catala, C., & Michel, E. 2010a, *AN*, 331, 1065

Degroote, P., Aerts, C., Baglin, A., Miglio, A., Briquet, M., *et al.* 2010b, *Nature*, 464, 259

Deheuvels, S. & Michel, E. 2010, *AN*, 331, 929

De Ridder, J., Barban, C., Baudin, F., Carrier, F., Hatzes, A. P., *et al.* 2009, *Nature*, 459, 398

Dominik, M., Jørgensen, U. G., Rattenbury, N. J., Mathiasen, M., Hinse, T. C., *et al.* 2010, *AN*, 331, 671

Dupret, M.-A., Belkacem, K., Samadi, R., Montalban, J., Moreira, O., *et al.* 2009, *A&A*, 506, 57

Dziembowski, W. A. & Pamyatnykh, A. A. 2008, *MNRAS*, 385, 2061

Dziembowski, W. A. & Soszyński, I. 2010, *A&A*, 524, A88

Dziembowski, W. A., Gough, D. O., Houdek, G., & Sienkiewicz, R. 2001, *MNRAS*, 328, 601

Featherstone, N. A., Browning, M. K., Brun, A. S., & Toomre, J. 2009, *ApJ*, 705, 1000

Fletcher, S. T., Chaplin, W. J., Elsworth, Y., Schou, J., & Buzasi, D. 2006, *MNRAS*, 371, 935

Fontaine, G., Brassard, P., Charpinet, S., Green, E. M., Chayer, P., Billères, M., & Randall, S. K. 2003, *ApJ*, 597, 518

Fontaine, G., Brassard, P., Charpinet, S., & Chayer, P. 2006, *Mem. Soc. Astron. Ital.*, 77, 49

Frandsen, S., Carrier, F., Aerts, C., Stello, D., Maas, T., *et al.* 2002, *A&A*, 394, L5

Gai, N., Basu, S., Chaplin, W. J., & Elsworth, Y. 2011, *ApJ*, 730, 63

García, R. A., Mathur, S., Salabert, D., Ballot, J., Régulo, C., Metcalfe, T. S., & Baglin, A. 2010, *Science*, 329 1032,

Gilliland, R. L., Brown, T. M., Christensen-Dalsgaard, J., Kjeldsen, H., Aerts, C., *et al.*, 2010, *PASP*, 122, 131

Gizon, L., Birch, A. C., & Spruit, H. C. 2010, *ARAA*, 48, 289

Goldreich, P. & Keeley, D. A. 1977, *ApJ*, 212, 243

Gough, D. O. 1986, in *Hydrodynamic and magnetohydrodynamic problems in the Sun and stars*, ed. Y. Osaki, Department of Astronomy, University of Tokyo, p. 117

Gough, D. O. 1990, in *Progress of seismology of the sun and stars, Lecture Notes in Physics*, vol. 367, eds Y. Osaki & H. Shibahashi, Springer, Berlin, p. 283

Gough, D. O. & Toomre, J. 1991, *ARAA*, 29, 627

Gough, D. O., Kosovichev, A. G., Toomre, J., Anderson, E. R., Antia, H. M., *et al.* 1996, *Science*, 272, 1296

Green, E. M., Fontaine, G., Reed, M. D., Callerame, K., Seitenzahl, I. R., *et al.* 2003, *ApJ*, 583, L31

Grundahl, F., Kjeldsen, H., Christensen-Dalsgaard, J., Arentoft, T., & Frandsen, S. 2007, *Comm. in Asteroseismology*, 150, 300

Grundahl, F., Christensen-Dalsgaard, J., Kjeldsen, H., Jørgensen, U. G., Arentoft, T., Frandsen, S., & Kjærgaard, P. 2009, in *Proc. GONG2008/SOHO21 meeting: Solar-stellar Dynamos as revealed by Helio- and Asteroseismology*, eds M. Dikpati, T. Arentoft, I. González Hernández, C. Lindsey & F. Hill, *ASP Conf. Ser.*, 416, San Francisco, p. 579

Grundahl, F., Christensen-Dalsgaard, J., Jørgensen, U. G., Kjeldsen, H., Frandsen, S., & Kjærgaard Rasmussen, P. 2011, in *Proc. GONG-SoHO 24: A new era of seismology of the sun and solar-like stars*, ed. T. Appourchaux, *J. Phys.: Conf. Ser.*, 271, 012083

Handler, G., Aerts, C. (and an international team of 50 astronomers) 2004, in *Proc. IAU Colloq. No 193: Variable Stars in the Local Group*, eds D. W. Kurtz & K. Pollard, ASP Conf. Ser., 310, San Francisco, p. 221

Harvey, J. W. 1988, in *Proc. IAU Symposium No 123, Advances in helio- and asteroseismology*, eds J. Christensen-Dalsgaard & S. Frandsen, Reidel, Dordrecht, p. 497

Heber, U. 2009, *ARAA*, 47, 211

Hekker, S., Caerts, C., De Ridder, J., & Carrier, F. 2006, *A&A*, 458, 931

Hekker, S., Kallinger, T., Baudin, F., De Ridder, J., Barban, C., Carrier, F., Hatzes, A. P., Weiss, W. W., & Baglin, A. 2009, *A&A*, 506, 465

Hekker, S., Gilliland, R. L., Elsworth, Y., Chaplin, W. J., De Ridder, J., Stello, D., Kallinger, T., Ibrahim, K. A., Klaus, T. C., & Li, J. 2011, *MNRAS*, in the press. [arXiv:1103.0141]

Houdek, G., Balmforth, N. J., Christensen-Dalsgaard, J., & Gough, D. O. 1999, *A&A*, 351, 582

Houdek, G. & Gough, D. O. 2002, *MNRAS*, 336, L65

Houdek, G. & Gough, D. O. 2007, *MNRAS*, 375, 861

Huber, D., Bedding, T. R., Arentoft, T., Gruberbauer, M., Guenther, D. B., Houdek, G., Kallinger, T., Kjeldsen, H., Matthews, J. M., Stello, D., & Weiss, W. W. 2011, *ApJ*, 94

Huber, D., Bedding, T. R., Stello, D., Mosser, B., Mathur, S., *et al.* 2010, *ApJ*, 723, 1607

Kallinger, T., Weiss, W. W., Barban, C., Baudin, F., Cameron, C., Carrier, F., De Ridder, J., Goupil, M.-J., Gruberbauer, M., Hatzes, A., Hekker, S., Samadi, R., & Deleuil, M. 2010a, *A&A*, 509, A77

Kallinger, T., Mosser, B., Hekker, S., Huber, D., Stello, D., *et al.* 2010b, *A&A*, 522, A1

Karoff, C., Metcalfe, T. S., Chaplin, W. J., Elsworth, Y., Kjeldsen, H., Arentoft, T., & Buzasi, D. 2009, *MNRAS*, 399, 914

Kilkenny, D., Koen, C., O'Donoghue, D., & Stobie, R. S. 1997, *MNRAS*, 285, 640

Kiss, L. L. & Bedding, T. R. 2003, *MNRAS*, 343, L79

Kiss, L. L. & Bedding, T. R. 2004, *MNRAS*, 347, L83

Kjeldsen, H. & Bedding, T. R. 1995, *A&A*, 293, 87

Kjeldsen, H., Bedding, T. R., Viskum, M., & Frandsen, S. 1995, *AJ*, 109, 1313

Kjeldsen, H., Bedding, T. R., Butler, R. P., Christensen-Dalsgaard, J., Kiss, L. L., McCarthy, C., Marcy, G. W., Tinney, C. G., & Wright, J. T. 2005, *ApJ*, 635, 1281

Kjeldsen, H., Christensen-Dalsgaard, J., Handberg, R., Brown, T. M., Gilliland, R. L., Borucki, W. J., & Koch, D. 2010, *AN*, 331, 966

Koch, D. G., Borucki, W. J., Basri, G., Batalha, N. M., Brown, T. M., *et al.* 2010, *ApJ*, 713, L79

Kolenberg, K., Smith, H. A., Gazeas, K. D., Elmaslı, A., Breger, M., *et al.* 2006, *A&A*, 459, 577

Kolenberg, K. 2008, in *Proc. HELAS II International Conference: Helioseismology, Asteroseismology and the MHD Connections*, eds L. Gizon & M. Roth, *J. Phys.: Conf. Ser.*, 118, 012060

Kolenberg, K., Szabó, R., Kurtz, D. W., Gilliland, R. L., Christensen-Dalsgaard, J., *et al.* 2010, *ApJ*, 713, L198

Kolenberg, K., Bryson, S., Szabó, R., Kurtz, D. W., Smolec, R., *et al.* 2011, *MNRAS*, 411, 878

Kuschnig, R., Weiss, W. W., Moffat, A., & Kudelka, O. 2009, in *Proc. GONG2008/SOHO21 meeting: Solar-stellar Dynamos as revealed by Helio- and Asteroseismology*, eds M. Dikpati, T. Arentoft, I. González Hernández, C. Lindsey & F. Hill, *ASP Conf. Ser.*, 416, San Francisco, p. 587

Lamb, H. 1909, *Proc. London Math. Soc.*, 7, 122

Libbrecht, K. G. & Woodard, M. F. 1990, *Nature*, 345, 779

Maeder, A. 2009, *Physics, formation and evolution of rotating stars*, Springer, Berlin

Mathis, S. & Zahn, J.-P. 2004, *A&A*, 425, 229 (Erratum: *A&A*, 462, 1063; 2007)

Mathis, S. & Zahn, J.-P. 2005, *A&A*, 440, 653

Matthews, J. M. 2007, *Comm. in Asteroseismology*, 150, 333

Metcalfe, T. S., Dziembowski, W. A., Judge, P. G., & Snow, M. 2007, *MNRAS*, 379, L16

Metcalfe, T. S., Monteiro, M. J. P. F. G., Thompson, M. J., Molenda-Żakowicz, J., Appourchaux, T., *et al.* 2010, *ApJ*, 723, 1583

Michel, E., Baglin, A., Weiss, W. W., Auvergne, M., Catala, C., *et al.* 2008a, *Comm. in Asteroseismology*, 156, 73

Michel, E., Baglin, A., Auvergne, M., Catala, C., Samadi, R., *et al.* 2008b, *Science*, 322, 558

Miesch, M. S. 2005, *Living Rev. Solar Phys.*, 2, 1. URL (cited on 29/3/11): http://www.livingreviews.org/lrsp-2005-1

Miesch, M. S. & Toomre, J. 2009, *ARFM*, 41, 317

Miesch, M. S., Brun, A. S., DeRosa, M. L., & Toomre, J. 2008, *ApJ*, 673, 557

Miglio, A., Montalbàn, J., Noels, A., & Eggenberger, P. 2008, *MNRAS*, 386, 1487

Miglio, A., Montalbán, J., Baudin, F., Eggenberger, P., Noels, A., Hekker, S., De Ridder, J., Weiss, W., & Baglin, A. 2009, *A&A*, 503, L21

Miglio, A., Montalbán, J., Carrier, F., De Ridder, J., Mosser, B., Eggenberger, P., Scuflaire, R., Ventura, P., D'Antona, F., Noels, A., & Baglin, A. 2010, *A&A*, 520, L6

Montalbán, J., Miglio, A., Noels, A., Scuflaire, R., & Ventura, P. 2010a, *ApJ*, 721, L182

Montalbán, J., Miglio, A., Noels, A., Scuflaire, R., & Ventura, P. 2010b, *AN*, 331, 1010

Monteiro, M. J. P. F. G. & Thompson, M. J. 2005, *MNRAS*, 361, 1187

Moskalik, P. & Dziembowski, W. A. 1992, *A&A*, 256, L5

Mosser, B., Belkacem, K., Goupil, M.-J., Miglio, A., Morel, T., *et al.* 2010, *A&A*, 517, A22

Mosser, B., Belkacem, K., Goupil, M. J., Michel, E., & Elsworth, Y., *et al.* 2011, *A&A*, 525, L9

Nordlund, Å., Stein, R. F., & Asplund, M. 2009, *Living Rev. Solar Phys.*, 6, 2. URL (cited on 29/3/11): http://www.livingreviews.org/lrsp-2009-2

Østensen, R. H., Silvotti, R., Charpinet, S., Oreiro, R., & Handler, G., *et al.* 2010, *MNRAS*, 409, 1470

Østensen, R. H., Silvotti, R., Charpinet, S., Oreiro, R., & Bloemen, S., *et al.* 2011, *MNRAS*, in the press. [arXiv:1101.4150]

Palacios, A., Talon, S., Charbonnel, C., & Forestini, M. 2003, *A&A*, 399, 603

Palacios, A., Charbonnel, C., Talon, S., & Siess, L. 2006, *A&A*, 453, 261

Pamyatnykh, A. A. 1999, *AcA*, 49, 119

Pamyatnykh, A. A., Handler, G., & Dziembowski, W. A. 2004, *MNRAS*, 350, 1022

Poretti, E., Paparó, M., Deleuil, M., Chadid, M., & Kolenberg, K., *et al.* 2010, *A&A*, 520, A108

Randall, S. K., Green, E. M., Fontaine, G., Brassard, P., & Kilkenny, D., *et al.* 2006a, *ApJ*, 643, 1198

Randall, S. K., Green, E. M., Fontaine, G., Brassard, P., Terndrup, D. M., Brown, N., Fontaine, M., Zacharias, P., & Chayer, P. 2006b, *ApJ*, 645, 1464

Reed, M. D., Kawaler, S. D., Østensen, R. H., Bloemen, S., Baran, A., *et al.* 2010, *MNRAS*, 409, 1496

Reed, M. D., Baran, A., Quint, A. C., Kawaler, S. D., O'Toole, S. J., *et al.* 2011, *MNRAS*, in the press. [arXiv:1102.4286 [astro-ph.SR]]

Salaris, M., Cassisi, S., & Weiss, A. 2002, *PASP*, 114, 375

Smith, H. A. 1995, *RR Lyrae stars*, Cambridge Astrophysics Series, Cambridge University Press

Soszyński, I., Dziembowski, W. A., Udalski, A., Kubiak, M., Szymański, M. K., Pietrzyński, G., Wyrzykowski, L., Szewczyk, O., & Ulaczyk, K. 2007, *AcA*, 57, 201

Stello, D., Bruntt, H., Preston, H., & Buzasi, D. 2008, *ApJ*, 674, L53

Stello, D., Basu, S., Bruntt, H., Mosser, B., Stevens, I. R., *et al.* 2010, *ApJ*, 713, L182

Szabó, R., Kolláth, Z., Molnár, L., Kolenberg, K., Kurtz, D. W., *et al.* 2010, *MNRAS*, 409, 1244

Tassoul, M. 1980, *ApJ Suppl.*, 43, 469

Trampedach, R. 2010, *Ap&SS*, 328, 213

Van Grootel, V., Charpinet, S., Fontaine, G., Brassard, P., Green, E. M., *et al.* 2010a, *ApJ*, 718, L97

Van Grootel, V., Charpinet, S., Fontaine, G., Green, E. M., & Brassard, P. 2010b, *A&A*, 524, A63

Verner, G. A., Elsworth, Y., Chaplin, W. J., Campante, T. L., Corsaro, E., *et al.* 2011, submitted to *MNRAS*

Vorontsov, S. V., Baturin, V. A., & Pamyatnykh, A. A. 1991, *Nature*, 349, 49

Walker, G., Matthews, J., Kuschnig, R., Johnson, R., Rucinski, S., *et al.* 2003, *PASP*, 115, 1023

Woodard, M. F. & Noyes, R. W. 1985, *Nature*, 318, 449

Astrophyiscal Dynamics: From Galaxies to Stars
Proceedings IAU Symposium No. 271, 2010
N. Brummell, A.S. Brun, M. S. Miesch & Y. Ponty, eds.

Toroidal Field Reversals

T. M. Rogers[1]

[1]Department of Planetary Sciences, The University of Arizona
1629 E. University Blvd. Tucson AZ 85721
email: tami@lpl.arizona.edu

Abstract. I present axisymmetric numerical simulations of the solar interior, with differential rotation imposed in the convection zone and tachocline and a dipolar poloidal field confined to the radiative interior. In these simulations toroidal field reversals which are equator-ward propagating are driven in the absence of a dynamo. These reversals are driven in the tachocline and are seen at the top of the convection zone. While not solar-like in many ways, these reversals do show some solar-like properties not previously seen in full MHD simulations.

Keywords. Sun: interior, magnetic fields

1. Introduction

One of the main constraints on models and simulations of the solar dynamo process is the sunspot cycle, marked by equator-ward propagation of sunspot pairs. The basic picture of the sunspot cycle is that magnetic field is pumped into the tachocline by overshooting convection, where it is stretched and amplified by the strong radial and latitudinal differential rotation there. The toroidal field, thus amplified and strong enough to traverse the convection zone without being destroyed, becomes magnetically buoyant and rises through the convection zone, manifesting itself at the surface of the Sun as sunspot pairs. In flux-transport dynamo models the toroidal field is advected equator-ward by the meridional circulation at the base of the convection zone. Of course, there are many uncertainties in this picture of the solar dynamo, including how well the field is pumped, organized, amplified in the tachocline, as well as how it becomes buoyant, remains in tact in its traverse through the convection zone, etc. Dynamo models usually focus on addressing the gross features of the sunspot cycle, such as the equator-ward propagation of toroidal field which reverses in time. Such solutions are easy to achieve in mean-field models, but have proven more difficult in more physical, less-tunable, magneto-hydrodynamic (MHD) simulations of the solar convection zone. Simulations of dynamos in the solar convection zone from two decades ago (Glatzmaier (1985)) and more recent simulations (Brown (2010), Ghizaru et. al. (2010)) have produced reversing toroidal field but with generally pole-ward or no latitudinal propagation.

It was speculated early on (Glatzmaier (1985)) that equator-ward propagation could be achieved in the adjacent stable region, if the reversals and propagation were due to a dynamo wave (Parker (1955)) and obeyed the Parker-Yoshimura sign rule. Unfortunately, three dimensional calculations including an adjacent stable region are not abundant. The most recent of such simulations (Browning et. al. (2006)) showed that magnetic field was easily stored and amplified in an imposed tachocline, but saw no reversals. Here I show here that toroidal field reversals can be instigated by the Tayler instability and that equator-ward propagation occurs by advection of the toroidal field by meridional circulation, similar to flux-transport models.

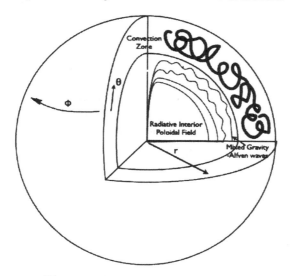

Figure 1. Schematic of model setup.

2. Model Setup

I emphasize that the simulations presented here are axisymmetric and therefore, not dynamo simulations. The original purpose of the simulations was to study the uniform rotation of the solar radiative interior†. For that reason the models simulate a large fraction of the solar radius extending from 0.10 $R_\odot$ to 0.90 $R_\odot$ and accurately represents the stable stratification of the radiation zone (in order to better represent internal gravity waves). Because the model is axisymmetric I am not able to accurately simulate the differential rotation of the convection zone. Therefore, I artificially impose the observed differential rotation profile of the convection zone. After running a purely hydrodynamic model for some time I add a dipolar poloidal field to the radiative interior. The initial field configuration is such that the field lines close within the radiative interior but overlap the tachocline. A schematic of the model setup is shown in Figure 1.‡

I solve the full MHD equations in the anelastic approximation:

$$\nabla \cdot \mathbf{B} = 0 \tag{2.1}$$

$$\nabla \cdot (\bar{\rho}\mathbf{u}) = 0 \tag{2.2}$$

$$\frac{\partial \mathbf{u}}{\partial t} + (\mathbf{u} \cdot \nabla)\mathbf{u} = -\nabla P - C\bar{g}\hat{r} + 2(\mathbf{u} \times \Omega) + \frac{1}{\rho}(\mathbf{J} \times \mathbf{B})$$

$$+ \bar{\nu}(\nabla^2 \mathbf{u} + \frac{1}{3}\nabla(\nabla \cdot \mathbf{u})) \tag{2.3}$$

$$\frac{\partial T}{\partial t} + (\mathbf{v} \cdot \nabla)T = -v_r\left(\frac{d\bar{T}}{dr} - (\gamma - 1)\bar{T}h_\rho\right)$$

$$+ (\gamma - 1)Th_\rho v_r + \gamma\bar{\kappa}[\nabla^2 T + (h_\rho + h_\kappa)\frac{\partial T}{\partial r}] \tag{2.4}$$

† Which I did discuss at the meeting but is not included in this proceeding
‡ Since the time of the meeting this work has been submitted to The Astrophysical Journal, figures in this conference proceeding are the same as some of those depicted in that paper.

$$\frac{\partial \mathbf{B}}{\partial t} = \nabla \times (\mathbf{u} \times \mathbf{B}) + \eta \nabla^2 \mathbf{B} \tag{2.5}$$

where equations 1 and 2 ensure magnetic flux and mass conservation. Equation 3 is the momentum equation, 4 is the energy equation and 5 is the magnetic induction equation. The numerical method is similar to that in Glatzmaier (1984), except here we use a finite difference scheme in the radial direction as opposed a Chebyshev expansion. This allows us more flexibility in allocation of radial resolution. The resolution of the model presented is 1500 radial zones by 512 latitudinal zones. In radius, 500 zones are devoted to the radiative interior and 1000 are devoted to the tachocline and convection zone.

3. Results

The model is initiated with a purely poloidal magnetic field. Toroidal field is quickly generated by stretching of that field by the differential rotation in the tachocline. The induction of toroidal field is described by the toroidal component of the induction equation, given by:

$$\frac{\partial B_\phi}{\partial t} = r B_r \frac{\partial}{\partial r}\left(\frac{u_\phi}{r}\right) - r u_r \frac{\partial}{\partial r}\left(\frac{B_\phi}{r}\right) + \frac{\sin\theta B_\theta}{r}\frac{\partial}{\partial\theta}\left(\frac{u_\phi}{\sin\theta}\right) - \frac{\sin\theta u_\theta}{r}\frac{\partial}{\partial\theta}\left(\frac{B_\phi}{\sin\theta}\right) + \eta\nabla^2 B_\phi \tag{3.1}$$

Initially, toroidal field induction is dominated by the first term on the right hand side of equation 3.1 and as such the toroidal field is oppositely signed at high and low latitudes,

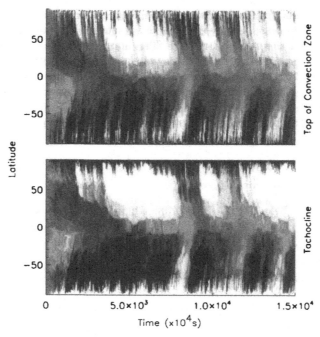

Figure 2. Toroidal field as a function of time and latitude. Blue (black) represents negative toroidal field, while white represents positive. The top panel shows the field at the top of the convection zone, while the bottom panel shows the field in the tachocline.

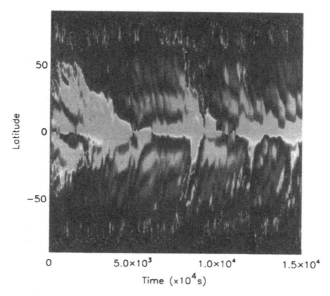

Figure 3. Tayler instability criterion, as described by equation 3.2, in the tachocline, as a function of time and latitude. Blue represents times and regions where the model is stable, Red represents times and positions of instability. Instability clearly corresponds to times of "reversals" as seen in figure 2.

anti-symmetric about the equator. This configuration quickly becomes unstable, leaving one dominant sign in the northern hemisphere and the opposite sign in the southern hemisphere (the sign dictated by the third term on the rhs of 3.1 and the initial poloidal field configuration). Throughout most of the simulation we find that the dominant induction term is the advection of toroidal field by latitudinal velocity (the fourth term on the rhs of 3.1). The dominant sign in each hemisphere remains fixed in time in the radiative interior. However, the sign of the toroidal field in the tachocline and convection zone changes in time, as can be seen in Figure 2. Clearly, there is a dominant toroidal field sign in each hemisphere (which I will refer to as the dominant sign) and a weaker field sign in each hemisphere (which I refer to as the minority sign). In the northern hemisphere, positive toroidal field is the dominant sign; in the southern hemisphere negative toroidal field is the dominant sign. Occasionally, the dominant sign toroidal field decays rapidly and the minority sign becomes dominant, leading to the appearance of a reversal. The cause of rapid toroidal field decay can be traced to the axisymmetric Tayler instability (Tayler (1975), Spruit (1999)). The basic criterion for the this instability (Spruit (1999)), is given by:

$$\cos\theta \frac{\partial}{\partial\theta}\left(\ln\left(\frac{B^2}{\sin^2\theta}\right)\right) > 0 \tag{3.2}$$

In Figure 3 we show the lhs of equation 3.2 as a function of time and latitude at a height within the tachocline, with red (white) representing values larger than zero and blue (black) representing values less than zero. One can clearly see that reversals seen in Figure 2 are associated with times when the axisymmetric Tayler instability criterion is satisfied. In Figure 3 one can see that the criterion is satisfied much of the time at high latitudes, but reversals coincide with those times when the criterion is satisfied at low latitudes as well.

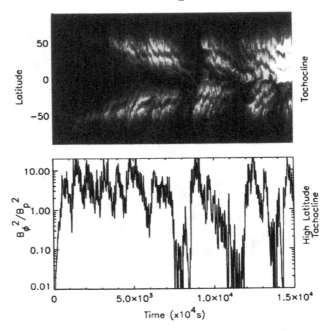

Figure 4. Ratio of toroidal to poloidal field strength, in the tachocline, as a function if time and latitude (top) and at high latitudes (bottom). Note that this ratio is larger at low latitudes due to the fact that the poloidal field is much larger at high latitudes than at low latitudes.

The condition in 3.2 is for a purely toroidal field and as shown by Wright (1973) and Braithwaite & Spruit (2004), the addition of a poloidal field can have a stabilizing effect. Both showed that with mixed field components the instability criterion then also depends on the relative strengths and geometries of the toroidal and poloidal field components. I similarly find that the toroidal field must satisfy the gradient condition expressed in 3.2, but in addition, its strength must also overcome the stabilizing effect of the poloidal field, which occurs when the local ratio B_ϕ^2/B_p^2 is $\approx$10-20. Once these criterion are satisfied the toroidal field rapidly decays, as seen in Figure 4. In Figure 4 I show the ratio of the toroidal field energy to the poloidal field energy as a function of time and latitude (top panel), while the bottom panel just shows this ratio at a chosen (high) latitude. Two things are immediately obvious in this figure. First, it is clear in the top panel of figure 4 that the ratio of field energy, B_ϕ^2/B_p^2, is significantly higher at low latitudes than at high latitudes, which is predominantly due to the poloidal field in the tachocline being substantially larger at high latitudes than at low latitudes.[†] Second, the ratio of toroidal to poloidal magnetic energy is rather oscillatory at high latitudes, and occasionally after reaching a strong peak, there is rapid destruction of toroidal field energy. During these times, the Tayler instability criterion as described in 3.2, is satisfied and the minority signed toroidal field is advected equator-ward by meridional circulation. Looking at Figure 2 one can estimate the timescale for the minority field to be swept to the equator. Comparing this to the advection time calculated using the mean latitudinal velocity at the base of the convection zone ($\approx 3 \times 10^3$ cm/s) one finds that the advection time is similar, confirming the dominance of the advection of toroidal field by latitudinal flow. Initially

† This implies that the field is not as well confined at high latitudes than at low latitudes, which has some implications for the ability of field to enforce uniform rotation in the solar interior. This matter will be discussed in a forthcoming paper.

4. Implications

These simulations are clearly not good solar analogues. The short comings of this model are many: axisymmetry (lack of dynamo), artificially imposed differential rotation and large diffusion coefficients are the first to come to mind (there are others). Nevertheless, there are some interesting features. First, this model is the first MHD simulation which gets equator-ward propagating and reversing toroidal field, and does so without a dynamo. Second, this model demonstrates that the equator-ward propagation can easily be obtained by meridional circulation which is driven by the differential rotation in the solar convection zone, a key component of flux-transport mean field models. Finally, these simulations give another measure of stability for mixed field configurations.

References

Braithwaite, J. & Spruit, H. C. 2007, *ApJ*, 469, 275
Brown, B. P., Browning, M. K., Brun, A. S., Miesch, M., & Toomre, J. 2010, *ApJ*, 711, 424
Browning, M. K., Miesch, M., Brun, A. S., & Toomre, J. 2006, *ApJ*, 648, L157
Ghizaru, M., Charbonneau, P., & Smolarkiewicz, P. K. 2010, *ApJL*, 715, L133
Glatzmaier, G. 1985, *GAFD*, 30, 490
Glatzmaier, G. 1985, *Journal of Computational Physics*, 55, 461
Parker, E. N. 1955, *ApJ*, 122, 293
Spruit, H. C. 1999, *A&A*, 349,189
Tayler, R. J. 1975, *Bulletin of the American Astronomical Society*, 7,252
Wright, G. A. E. 1973, *MNRAS*, 162,339

Discussion

BRANDENBERG: How do you model the radiation conductivity? Did you use a rescaled Kramers opacity?

ROGERS: The thermal diffusivity is modeled as the solar radiative conductivity, divided by ρc_p, multiplied by some factor for numerical stability.

$$\kappa(r) = f \frac{16\sigma T^3}{3\rho^2 \chi c_p} \tag{4.1}$$

where σ is the Stefan-Boltzmann constant, and χ is the opacity.

BRANDENBERG: How realistic is the overshoot you get?

ROGERS: I don't know if we know what "realistic" overshoot is, given we don't have a direct measurement. I think these are some of the best simulations of overshoot in the Sun since they have the stable stratification right and the gradient of the thermal diffusivity is right. The overshoot I get is very small, maybe 10% of a pressure scale height, depending on how it is defined.

BRANDENBERG: Do the g-modes restrict the timestep, i.e. how useful is it to use the anelastic approximation?

ROGERS: Yes, the gravity waves restrict the timestep, or probably more accurately the timestep can be limited by the angular velocity generated by the dissipation of gravity waves. This is particularly true at the center of the simulations where a small amount of angular momentum deposition could lead to somewhat large angular velocities compounded by the fact that the horizontal grid spacing there is very fine.

BRUN: What is the main angular momentum transport process at the base of the convection zone in your simulation?

ROGERS: Note to reader: this question refers to the part of the talk where I discussed angular momentum transport in the radiative interior. This part of the talk is not included in this conference proceeding. I don't really know the answer to this question. Just below the convection zone the angular momentum transport is dominated by overshoot and waves, although referring to them as waves in this region is probably not accurate. In the models I have run the magnetic field has little influence in that region

BRUN: How does the presence of magnetic field change the balance of angular momentum and the properties of the internal waves (propagation, group velocity...)?

ROGERS: In the axisymmetric models presented here the answer is I don't know. However, in models and simulations of IGW interacting with a *purely* toroidal field the answer is that magnetic fields can reflect the waves creating a duct and enhanced angular momentum deposition in the region between the field and the base of the convection zone. The magnetic field then filters some waves, preventing them from reaching the deep interior, but also acts as an amplification mechanism on other waves. The net is that the field causes the wave transport in the deep radiative interior to be highly dependent in time and space, because the reflection/amplification depends on the local field strength (which varies in time).

Astrophysical Dynamics from Stars to Galaxies
Proceedings IAU Symposium No. 271, 2010
N. Brummell, A.S. Brun, M. S. Miesch & Y. Ponty, eds.

© International Astronomical Union 2011
doi:10.1017/S1743921311017467

Differential Rotation and Magnetism in Simulations of Fully Convective Stars

Matthew K. Browning[1]

[1]Canadian Institute for Theoretical Astrophysics,
University of Toronto, 60 St. George St, Toronto, Canada
email: `browning@cita.utoronto.ca`

Abstract. Stars of sufficiently low mass are convective throughout their interiors, and so do not possess an internal boundary layer akin to the solar tachocline. Because that interface figures so prominently in many theories of the solar magnetic dynamo, a widespread expectation had been that fully convective stars would exhibit surface magnetic behavior very different from that realized in more massive stars. Here I describe how recent observations and theoretical models of dynamo action in low-mass stars are partly confirming, and partly confounding, this basic expectation. In particular, I present the results of 3–D MHD simulations of dynamo action by convection in rotating spherical shells that approximate the interiors of 0.3 solar-mass stars at a range of rotation rates. The simulated stars can establish latitudinal differential rotation at their surfaces which is solar-like at "rapid" rotation rates (defined within) and anti-solar at slower rotation rates; the differential rotation is greatly reduced by feedback from strong dynamo-generated magnetic fields in some parameter regimes. I argue that this "flip" in the sense of differential rotation may be observable in the near future. I also briefly describe how the strength and morphology of the magnetic fields varies with the rotation rate of the simulated star, and show that the maximum magnetic energies attained are compatible with simple scaling arguments.

Keywords. convection, MHD, stars: low-mass, stars: magnetic fields, stars: rotation, turbulence

---

1. Introduction: Puzzles of Low-Mass Stellar Magnetism

Magnetic fields are ubiquitous in low-mass stars, and in at least some cases those magnetic fields exhibit a remarkable amount of spatial and temporal organization. The most famous example is the Sun's cyclical magnetism: sunspots appear on the solar disk first at mid-latitudes, then progressively nearer the equator over the course of a roughly 11-year cycle; the number and polarity of the spots varies in the same way (see., e.g., Ossendrijver 2003). These organized magnetic fields are widely (though not universally) thought to be generated partly in the tachocline of shear at the base of the solar convective envelope – in part because it is a site of strong differential rotation, but also because the stable stratification below the convection zone might allow magnetic fields to be amplified enormously before becoming susceptible to magnetic buoyancy instabilities (see, e.g., Parker 1993; see Spruit 2010 for a different view).

But not all stars have a tachocline. Moving down the main sequence to lower masses and cooler temperatures, the convective envelope deepens and the radiative core shrinks. Stars of less than about a third a solar mass (corresponding to spectral types of roughly M3.5 or later) are thought to be convective throughout their interiors, and so do not possess an internal boundary layer akin to the solar tachocline. Although these low-mass stars might in principle still possess differential rotation – like the latitudinal shear present throughout the solar convection zone – a widespread theoretical expectation had been that they would exhibit magnetic fields that differed appreciably from those realized in more massive solar-like stars (e.g., Durney *et al.* 1993).

Observations of stellar magnetism paint a somewhat murkier picture. On the one hand, there have been recent suggestions (particularly from spectropolarimetric observations) that the surface magnetic topologies of stars with a small radiative core *do* differ appreciably from those of stars that are fully convective (e.g., Morin *et al.* 2010; Reiners & Basri 2009). Fully convective stars (with late-M spectral types) also appear to spin down (through angular momentum loss via a magnetized stellar wind) much more slowly than early-M dwarfs (e.g., Browning *et al.* 2010; Reiners *et al.* 2009), though it is unclear whether this is due to changes in the strength or morphology of surface magnetic fields or to changes in the mass loss rate. On the other hand, it also seems clear that some fully convective stars can build magnetic fields with remarkably strong large-scale components (Donati *et al.* 2006), with the strength of those fields sensitive to rotation at some level (e.g., Mohanty & Basri 2003). Indeed, some recent comparisons between line-of-sight and unsigned field measurements (extracted from Stokes V and Stokes I observations) suggest that some fully convective stars actually harbor *more* organized fields than slightly more massive stars with a small radiative core (Reiners & Basri 2009). Puzzlingly, some of these stars appear to show large-scale magnetic field organization but no evident surface differential rotation (e.g., Donati *et al.* 2006).

The most central question raised by these observations is, simply, how are spatially organized fields realized in fully convective stars? But this basic issue is tightly linked to a whole set of other questions: what is the nature of the differential rotation in these stars, and what role does it play in the dynamo process? Can these stars ever drive solar-like differential rotation at observable levels? How do the strength and morphology of the dynamo-generated fields vary with rotation rate? How does the magnetism modify the transport of energy and angular momentum throught the stellar interior? These questions motivate the work described in this paper.

Here, I describe the results of a series of 3-D simulations of convection and magnetism in rotating spherical domains that are intended to represent fully convective stars of 0.3 solar masses at various rotation rates. In §2 I describe the basic setup of these simulations and the numerical methods employed. The convective flow patterns and some aspects of energy transport are detailed in §3. I describe the differential rotation that arises in these models in §4. In §5 I describe the strength and morphology of the magnetic fields realized at various rotation rates.

2. Model Formulation and Numerical Methods

The simulations described here are highly idealized representations of 0.3 solar-mass stars rotating between one-tenth and ten times as rapidly as the Sun (0.1–$10 \times \Omega_0 = 2.6 \times 10^{-6}$ s^{-1}). They were all carried out using the Anelastic Spherical Harmonic (ASH) code, which solves the 3–D Navier-Stokes equations with magnetism in the anelastic approximation (Clune *et al.* 1999; Miesch *et al.* 2000; Brun, Miesch & Toomre 2004). The setup at each rotation rate is essentially identical to that described in Browning (2008); here I summarize only the most important details. Before delving into them, note that the ASH code is well-suited to this problem because it correctly captures the global spherical geometry of the star, thereby allowing the study of intrinsically large-scale processes like differential rotation, meridional circulation, and global-scale dynamo action. The tradeoff, of course, is that because we can only resolve a finite range of spatial and temporal scales in the simulation, including the largest possible length scales (the radius of the star) in our modeling implies that the smallest lengths resolved are still quite large – typically about 1 Mm. Thus I focus here on the dynamics of the largest scales, while recognizing that at some level of detail these must be influenced by smaller-scale dynamics that these simulations cannot resolve.

The spherical computational domain typically extends from 0.06-0.96R, where R is the overall stellar radius of 2.07×10^{10} cm, thus excluding both the surface boundary layer and the innermost portions of the star. I exclude the inner few percent of the star from these calculations both because the coordinate systems employed in ASH are singular there, and because the small numerical mesh sizes at the center of the star would require impractically small timesteps. The initial stratifications of the mean density, energy generation rate, gravity, radiative diffusivity, and entropy gradient dS/dr are adopted from a 1-D stellar model (I. Baraffe, private communication, after Baraffe & Chabrier 2003). These thermodynamic quantities are updated throughout the course of the simulation as the evolving convection modifies the spherically symmetric mean state. Variables are expanded in terms of spherical harmonic basis functions $Y_l^m(\theta, \phi)$ in the horizontal directions and Chebyshev polynomials $T_n(r)$ in the radial. As with all numerical simulations, the eddy viscosities and diffusivities employed are vastly greater than their counterparts in actual stars; here I have taken these to be constant in radius, and adopted a Prandtl number $\nu/\kappa = 0.25$ and a magnetic Prandtl number $Pm = \nu/\eta$ that varies between 0.25 and 8 depending on the simulation. At a detailed level, the flows and magnetic fields attained in the simulations are sensitive to the values of these non-dimensional numbers. Because of this, I focus here on the broad trends these simulations exhibit: the types of flows and magnetic fields they can achieve as the basic parameters of the problem are varied, rather than the precise values of magnetic energy, zonal wind velocity, etc, that are attained. I particularly concentrate here on the role that changes in the overall stellar rotation rate play, since this turns out to have an outsized influence on the flows and magnetic fields that are achieved.

3. Convective Flows and Energy Transport

The convective flows in these simulations possess structure on many scales. Although many small-scale features continually emerge, propagate, and survive only for a short while, there are also large-scale organized motions that can persist for extended intervals. A sampling of this behavior is provided by Figure 1, which shows volume renderings of the radial velocity in two simulations, one rotating at the solar angular velocity and the other ten times slower. (The opacity mapping used in that figure is such that only motions near

Figure 1. Radial velocity in simulations rotating at the solar angular velocity (*b, c*) and tenfold slower (*a*). Upflows are reddish (light), downflows are blueish (dark). Panel *c* shows a cutaway of one hemisphere.

the outer boundary of the simulation are visible in Fig. 1a, 1b; Figure 1c shows a cutaway of one hemisphere in the more rapidly rotating simulation, to highlight the radial extent of the motions.) Rotation has a strong effect on the convective patterns that are realized: when rotation is very slow, the convection is more or less isotropic on each spherical surface, with a network of upflows and downflows that meanders in orientation over the sphere. Increasing the rotation rate breaks this symmetry and imposes a preferred direction, leading (in some cases) to convective rolls like those shown in Figure 1b, 1c. (If the eddy viscosity in the simulations is decreased, these rolls become less pronounced and can break up, but there is still a tendency towards alignment along the rotation axis.) These roll-like structures are a well-known feature of convection in rotating spherical shells (see, e.g., Busse 2002; Gilman 1977).

One other feature worth highlighting is the strong asymmetry between upflows and downflows realized near the top of the simulated stars: the downflows are stronger and narrower than the former, mainly because of the strong density stratification. (Downflows are cool and contract; upflows are hot and expand.) Deeper in the interior, the flows are weaker and of somewhat larger physical scale: motions can span large fractions of a hemisphere and extend radially for some distance. Flow amplitudes also vary appreciably with depth, with typical rms velocities declining by a factor of about ten in going from the surface to the center.

The variation in flow amplitude with radius is linked to both the density stratification and to radial variations in the amount of energy that must be transported by convection. Convection ultimately arises because of a need to transport heat outwards: if more energy has to be carried by convection, the convective velocity will generally be higher. Although the interior is unstably stratified everywhere, radiation still carries some of the energy at small radii. This is because the end state of efficient convection is an interior stratification that is approximately isentropic, *not* isothermal as in unstratified convection: thus there is still a non-zero radiative flux. Together with the overall increase of the total luminosity with radius (out to the point where nuclear energy generation stops), this implies that the total luminosity carried by convection peaks at large radii (around $r = 0.80R$). Thus the convective velocity is appreciably greater near the surface than at depth. Another important effect arises because of the asymmetry between upflows and downflows: this implies a negative (inward) kinetic energy flux, which in a steady state must be compensated for by an increased outward enthalpy (convective) flux. This effect, too, depends on depth (since the star is more strongly stratified near the surface), again leading to more vigorous convection near the top of the simulation domain.

4. Differential Rotation: Solar or Anti-Solar

In addition to transporting heat, the convective flows also transport angular momentum. One might naively expect that parcels of fluid would *individually* conserve angular momentum – so that (for instance) a fluid element moving outward would tend to slow down (move retrograde relative to the frame), and a fluid element moving latitudinally from equator to pole would tend to spin up. This would imply anti-solar differential rotation at the surface, with a slow equator and fast poles, and longitudinal velocities that decrease with distance from the rotation axis. The fact that the Sun in fact has a fast equator and slow pole is enough to suggest that there is more to the story than this: that convection, acting in concert or conflict with meridional circulations and magnetic fields, can redistribute angular momentum in surprising ways. Here I describe briefly the types of differential rotation that are achieved in these simulations at varying rotation rates, while deferring an in-depth analysis of the angular momentum transport to forthcoming work (Browning & Miesch 2010).

All of the MHD calculations described here had hydrodynamical progenitor simulations. These all began in a state of uniform rotation, but convection quickly established interior rotation profiles that varied with radius and latitude. The resulting differential rotation, displayed in Figure 2 for two *hydrodynamic* cases, depends on the overall stellar rotation rate: in Figure 2*a*, from a simulation rotating at one-tenth the solar angular velocity, the rotation profile is "anti-solar" at the surface, with a slow equator and fast poles; in Figure 2*b*, from a simulation rotating at the solar angular velocity, the rotation profile is solar-like at the surface. Both cases also show angular velocity contrasts in radius, with angular velocity decreasing with depth in the more rapidly rotating case and decreasing with depth in the slower rotator. In the simulation rotating at the solar rate, the interior angular velocity profile is largely constant on cylinders, reflecting the strong Proudman-Taylor constraint; in the more slowly rotating case this cylindrical alignment is not evident.

Similar transitions in the nature of angular momentum transport and differential rotation have been noted by several previous authors in other contexts. In particular, Gilman (1977) noted that his simulations of a solar-like convection zone exhibited solar-like equatorial acceleration in some regimes and anti-solar rotation profiles in others; he identified the transition between these two regimes with a transition from a convective Rossby number (essentially the ratio of buoyancy driving to Coriolis forces) greater than unity to less than unity. When the rotational influence on the convection was strong relative to buoyancy driving, he attained equatorial acceleration; when it was weak, the equator rotated more slowly than the poles. Similar results were obtained in simulations of core convection in A-type stars (Browning *et al.* 2004). The simulations here exhibit qualitatively similar behavior. Convection influenced by rotation (that is, with a Rossby number roughly less than unity) tends to drive solar-like differential rotation; when rotation is slower, the differential rotation profile is anti-solar.

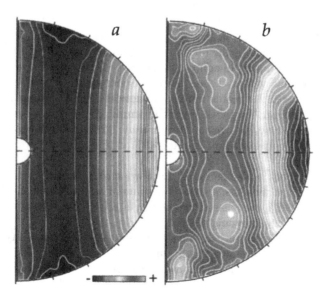

Figure 2. Differential rotation achieved in sample hydrodynamic simulations rotating at the solar angular velocity (*a*) and ten times slower (*b*). Light tones are prograde, dark ones retrograde.

The interior rotation profiles can be quite different in the presence of strong dynamo-generated magnetic fields. In MHD simulations, the magnetic fields react back strongly upon the flows, acting to quench and in some instances essentially eliminate the differential rotation. An example of this for a simulation rotating at the solar rate was explored in Browning (2008). The extent to which the differential rotation is reduced depends on the overall magnetic field strength, which in turn (as discussed below) depends somewhat on the overall stellar rotation rate. Thus, there is a "sweet spot" for differential rotation: when the simulations rotate very slowly, they drive weak anti-solar differential rotation; when the rotation rate is somewhat larger they drive strong solar-like differential rotation; and when rotation is more rapid still they build magnetic fields so strong that the differential rotation is partly quenched. These transitions occur at fairly slow overall rotation rates: because the luminosities of M-dwarfs are so low, their convective velocities are also low, and so they are strongly influenced by rotation even at solar-like angular velocities. Only the very slowest rotators (those with rotation periods much longer than the solar value) drive anti-solar differential rotation.

It seems plausible that these basic predictions could be testable in the near future by photometric monitoring with, e.g., Kepler. The exact rotation periods at which stars might transition from anti-solar to solar-like differential rotation, and then again to no evident differential rotation (because it has been wiped out by the dynamo-generated Maxwell stresses), are probably not reliably predictable by these calculations. They depend on the values of viscosity, magnetic diffusivity, etc, adopted in the modeling. But the basic existence of these different regimes, and their rough dependence on rotation rate, might well be robust. The main challenge observationally will probably be to measure differential rotation reliably in the slowest rotators, which are presumably less active and so less spotted.

5. Dynamo Action Achieved

The flows in each simulation act as a magnetic dynamo, amplifying a tiny seed field by many orders of magnitude and sustaining it against Ohmic decay. The magnetic energy (ME) grows exponentially until it reaches a steady state. In the most rapid rotators (which, again, are not really rotating all that rapidly in absolute terms), the final magnetic energy density is approximately in equipartition with the flows; in slower rotators it tends to be smaller. The exact values of the magnetic energy are sensitive at some level to the values of the magnetic Reynolds number, magnetic Prandtl number, etc, that I have adopted in each calculation. To give a specific example: in an evolved calculation rotating at the solar angular velocity, ME was approximately 120% of the total kinetic energy KE (relative to the rotating frame) and about 140% of the convective (non-axisymmetric) kinetic energy (CKE).

As the fields grow, they react back on the flows through the Lorentz force. Thus KE begins to decline once ME reaches a threshold value of about 5% of KE; here this decline is associated mainly with a decrease in the energy of differential rotation DRKE, whereas CKE is less affected by the growing fields.

Like the flows that build them, the magnetic fields possess both intricate small-scale structure and substantial large-scale components. The typical length scale of the field increases with depth, partly tracking the radial variation in the size of typical convective flows. The smallest field structures are typically on finer scales than the smallest flow fields, partly because I have adopted a magnetic Prandtl number Pm greater than unity. By decomposing the magnetism into its azimuthal mean (TME), and fluctuations around that mean (FME), we can gain a coarse estimate of the typical size of field structures: if the field is predominantly on small scales, only a small signal will survive this azimuthal averaging. In these simulations, TME accounts for at most about 20% of the total magnetic energy in the bulk of the interior; it is smallest near the surface (where TME is typically less than 5 % ME), and largest (as a fraction of ME) at depth.

The fraction of energy in the mean (axisymmetric) components increases with increasing rotation rate. The ratio of the toroidal mean energy to the poloidal mean also changes: in the very slowest rotators the two components are comparable, while in more rapidly rotating cases with strong differential rotation, TME exceeds the poloidal mean energy (PME) by a factor of a few.

The mean (axisymmetric) fields realized in some of the simulations are remarkably strong and long-lived. Mean toroidal field strengths can exceed 10 kG in some locations; some prominent field structures persist for thousands of days. The overall field polarity is stable over long intervals (decades), in sharp contrast to some simulations of solar convection without a tachocline (Brun *et al.* 2004), in which the field polarity flipped at irregular intervals of less than 600 days.

6. Closing Thoughts

The overall picture that emerges from these simulations is that fully convective stars can act effectively as magnetic dynamos, building fields that have structure on both large and small spatial scales. The large-scale fields can be remarkably strong and long-lived. The convection drives differential rotation, in a manner that depends on the overall rotation rate: the very slowest rotators would appear anti-solar at their surface, while more rapid rotators establish a fast equator and slow pole. That differential rotation is, however, largely quenched by Maxwell stresses in cases that build strong magnetic fields.

The maximum magnetic energy densities attained in these calculations are of order equipartition with the kinetic energy density relative to the rotating frame. Although stronger magnetic fields might be possible in some instances (see, e.g., Featherstone *et al.* 2009), it is worth noting that the assumption of equipartition yields magnetic field estimates broadly in line with those recently predicted on somewhat different grounds by Christensen and collaborators (see, e.g., Christensen *et al.* 2009; Reiners *et al.* 2009). Specifically, Reiners et al. (2009) argue, based on an energy flux scaling derived empirically from a series of planetary dynamo calculations, that the magnetic flux in stars should scale approximately as

$$Bf \propto M^{1/6} L^{1/3} R^{-7/6} \qquad (6.1)$$

with M, L, and R the stellar mass, luminosity and radius, and where Bf is the surface magnetic flux. This turns out to be the same scaling one derives by assuming that 1. magnetic fields are in approximate equipartition with the convective kinetic energy and 2. that convective energy is given roughly by mixing-length arguments, so that it is proportional to the heat flux to the one-third power. To order of magnitude, the convective luminosity is given by

$$L \sim \frac{\text{convective energy}}{\text{convective overturning time}} \sim \frac{\rho v^2 (\frac{4}{3}\pi d^3)}{l_c/v} \qquad (6.2)$$

with d a lengthscale characterizing the depth of the convection zone and l_c the length characterizing convective eddies. This in turn implies that $v \sim (\frac{3L}{4\pi\rho}l_c)^{1/3}\frac{1}{d}$. Meanwhile, equipartition of kinetic and magnetic energy densities implies that $B \sim \rho^{1/2}v$. If we then take $\rho \sim M/R^3$, and assume that to order of magnitude all lengthscales are comparable

$(l_c \sim d \sim R)$, we obtain that $v \sim L^{1/3} R^{1/3} M^{-1/3}$, so that finally $B \sim L^{1/3} M^{1/6} R^{-7/6}$, as in the Reiners *et al.* (2009) scaling. This is not to suggest that greater or lesser field energies are not possible, but does indicate that objects deviating greatly from the Reiners *et al.* (2009) scaling are breaking one of the assumptions I made above. Either equipartition does not hold, or the relevant flow velocity is not related to the background heat flux in the way assumed here.

It is a pleasure to acknowledge many helpful conversations about this and related puzzles in stellar rotation and magnetism with Juri and the other members of the "ASH mob."

References

Browning, M. K., 2008, *ApJ*, 676, 1262

Browning, M. K., Basri, G., Marcy, G. W., West, A. A., & Zhang, J., 2010, *AJ*, 139, 504

Brun, A. S., Miesch, M. S., & Toomre, J. 2004, *ApJ*, 614, 1073

Christensen, U. R., Holzwarth, V., & Reiners, A., 2009, *Nature*, 457, 167

Donati, J. F., Forveille, T., Cameron, A. C., Barnes, J. R., Delfosse, X., Jardine, M. M., & Valenti, J. A., 2006, *Science*, 311, 633

Durney, B. R., De Young, B. S., & Roxburgh, I. W., 1993, *Sol. Phys.*, 145, 207

Featherstone, N. A., Browning, M. K., Brun, A. S., & Toomre, J., 2009, *ApJ*, 706, 1000

Gilman, P. A., 1977, *GAPFD*, 8, 93

Miesch, M. S., Elliott, J. R., Toomre, J., Clune, T. L., Gl atzmaier, G. A., & Gilman, P. A., 2000, *ApJ*, 532, 593

Mohanty, S. & Basri, G., 2003, *ApJ*, 583, 451

Morin, J., et al., 2010, *MNRAS* 1077

Ossendrijver, M. 2003, *Astron. Astrophys. Rev.*, 11, 287

Parker, E. N. 1993, *ApJ*, 408, 707

Reiners, A. & Basri, G., 2009, *A&A*, 496, 787

Reiners, A., Basri, G., & Christensen, U. R., 2009, *ApJ*, 697, 373

Spruit, H. C. 2010, arXiv:1004.4545

Discussion

A. BRANDENBURG: Is the dependence of rotation rate on spectral type real and how does this depend on age?

M. BROWNING: The conclusion that very low-mass stars take longer to spin down than somewhat higher-mass ones appears to me to be quite robust. But there is certainly an age effect: if you look at young enough clusters, many stars of all spectral types are still rotating quite rapidly. In the field, though – i.e., looking at much older stars – you tend to find that very few early-M stars are detectably rotating (see, e.g., Browning *et al.* (2010), where we used Keck HIRES spectra to look for signs of rotation in dozens of field M-dwarfs). A significantly larger fraction of late-M stars (or L-dwarfs) are detectably rotating.

K. MOFFAT: Your conclusion that rotation plus convection are conducive to the growth of large-scale fields is hardly new! These ingredients imply large-scale helicity and so a corresponding α-effect. Have you been able to interpret your simulations from this "mean-field" point of view?

M. BROWNING: The short answer is no, these simulations do not seem to be particularly well-described by mean-field theory. There is certainly large-scale helicity of the predicted sense, but it doesn't seem to be especially well-linked to the growth of the large-scale field. The role of rotation in these simulations seems to be partly about increasing the correlation length of the convection to something on the order of the domain size, rather than just imparting a preferred sign to the helicity in each hemisphere. That said, we certainly have more work to do in trying to make connections between these results and mean-field theory.

K. FERRIERE: Why in the Sun does the equator rotate faster than the poles?

M. BROWNING: I would say that, honestly, we still have a "description" of how this happens rather than a first-principles theory. (See Miesch *et al.*, these proceedings, for some of that description.) Ultimately there are Reynolds stresses arising from correlations in the fluctuating velocity components, and the sense of those stresses is such that you break the tendency for individual parcels of fluid to just conserve angular momentum as they move outward or inward. But baroclinic effects, meridional circulations, etc, all appear to play roles as well.

Astrophysical Dynamics: From Stars to Galaxies
Proceedings IAU Symposium No. 271, 2010
N.H.Brummell, A.S. Brun, M.S. Miesch & Y. Ponty

© International Astronomical Union 2011
doi:10.1017/S1743921311017479

Global-scale wreath-building dynamos in stellar convection zones

Benjamin P. Brown[1,2], Matthew K. Browning[3], Allan Sacha Brun[4], Mark S. Miesch[5] and Juri Toomre[6]

[1] Dept. Astronomy, University of Wisconsin, Madison, WI 53706-1582
email: bpbrown@astro.wisc.edu

[2] Center for Magnetic Self Organization in Laboratory and Astrophysical Plasmas, University of Wisconsin, Madison, WI 537066-1582

[3] Canadian Institute for Theoretical Astrophysics, University of Toronto, Toronto, ON M5S3H8 Canada

[4] DSM/IRFU/SAp, CEA-Saclay and UMR AIM, CEA-CNRS-Université Paris 7, 91191 Gif-sur-Yvette, France

[5] High Altitude Observatory, NCAR, Boulder, CO 80307-3000

[6] JILA and Dept. Astrophysical & Planetary Sciences, University of Colorado, Boulder, CO 80309-0440

Abstract. When stars like our Sun are young they rotate rapidly and are very magnetically active. We explore dynamo action in rapidly rotating suns with the 3-D MHD anelastic spherical harmonic (ASH) code. The magnetic fields built in these dynamos are organized on global-scales into wreath-like structures that span the convection zone. Wreath-building dynamos can undergo quasi-cyclic reversals of polarity and such behavior is common in the parameter space we have been able to explore. These dynamos do not appear to require tachoclines to achieve their spatial or temporal organization. Wreath-building dynamos are present to some degree at all rotation rates, but are most evident in the more rapidly rotating simulations.

Keywords. convection, MHD, stars: interiors, stars: magnetic fields, stars: rotation

1. Introduction

When stars like the Sun are young, they rotate quite rapidly. Observations of these young Suns indicate that they have strong surface magnetic activity and can undergo global-scale polarity reversals similar to the 22-year solar cycle. The magnetic fields observed at the surface of these stars are thought to originate in stellar dynamos driven in their convective envelopes. There, plasma motions couple with rotation to generate global-scale magnetic fields. Though correlations between the rotation rate of stars and their magnetic activity are observed (e.g., Pizzolato *et al.* 2003) it is at present unclear how the stellar dynamo process depends in detail on rotation.

Motivated by this rich observational landscape, we have explored the effects of more rapid rotation on 3-D convection and dynamo action in simulations of stellar convection zones. These simulations have been conducted using the anelastic spherical harmonic (ASH) code, a tool developed by a team of postdocs and graduate students working with Juri Toomre to study global-scale magnetohydrodynamic convection and dynamo action in stellar convection zones (e.g., Clune *et al.* 1999; Miesch *et al.* 2000; Brun *et al.* 2004, and contribution by Miesch in these proceedings).

We began our explorations of convection in rapidly rotating suns by exploring hydrodynamic simulations at a variety of rotation rates (Brown *et al.* 2008). These simulations capture the convection zone only, spanning from $0.72 R_\odot$ to $0.97 R_\odot$, and take solar values for luminosity and stratification but the rotation rate is more rapid. The total density contrast across such shells is about 25. In those simulations we found that the differential rotation generally becomes stronger as the rotation rate increases, while the meridional circulations appear to become weaker and multi-celled in both radius and latitude.

In this paper we review the dynamos we have found in our simulations of more rapidly rotating solar-type stars. These wreath-building dynamos form surprisingly organized structures in their convection zones (§2) and some even undergo quasi-cyclic magnetic reversals (§3). Wreath-building dynamos appear throughout the parameter space we have surveyed to date (§4). We close by reflecting on the challenges that lie ahead (§5).

2. A Dynamo with Magnetic Wreaths

Our first simulation discussed here, case D3, is of a star rotating three times faster than the Sun (Brown *et al.* 2010a). Vigorous convection in this simulation drives a strong differential rotation and achieves sustained dynamo action at relatively low magnetic Prandtl number; here $Pm = \nu/\eta$ is 0.5, where ν is the viscosity and η is the magnetic diffusivity.

The magnetic fields created in this dynamo are organized on global-scales into banded wreath-like structures. These are shown for case D3 in Figure 1a. Two such wreaths are visible in the equatorial region, spanning the depth of the convection zone and latitudes from roughly $\pm 30°$. The dominant component of the magnetism is the longitudinal field B_ϕ, and the two wreaths have opposite polarities. Here the wreath in the northern hemisphere has negative polarity B_ϕ while the wreath in the southern hemisphere is positive in sense. The wreaths are not isolated flux structures; instead, magnetic fields meander in and out of each wreath, connecting them across the equator and to higher latitudes (Fig. 1b). The lack of visible magnetism in the polar regions reflects the relatively low magnetic Reynolds number associated with the convection (average fluctuating Rm' is roughly 50 at mid-convection zone).

It has been a great surprise that such structures can exist in the convection zone of this simulation. Generally, it has been expected that convection should shred such structures or pump them downwards into a stable tachocline at the base of the convection zone. Here the entire domain is convectively unstable, and no such tachocline is present. The wreaths persist for long intervals in time, with the mean (longitudinally averaged) magnetic fields showing relatively little variation in time. The convection leaves its imprint on the wreaths, with the strongest downflows dragging the field towards the bottom of the convection zone. This is visible in the distinct waviness apparent in Figure 1. On the poleward edges the wreaths are wound up into the vortical convection there, and this appears to play an important role in regenerating the poloidal field.

The magnetic wreaths are built by both the global-scale differential rotation and by the emf arising from correlations in the turbulent convection. Generally, the mean longitudinal magnetic field $\langle B_\phi \rangle$ in the wreaths is generated by the Ω-effect: the stretching of mean poloidal field by the shear of differential rotation into mean toroidal field. Production of $\langle B_\phi \rangle$ by the differential rotation is balanced by turbulent shear and advection, and by ohmic diffusion on the largest scales.

The mean poloidal field is generated by the turbulent emf $\mathcal{E}_{FI} = \langle \boldsymbol{u'} \times \boldsymbol{B'} \rangle$, where the fluctuating velocity is $\boldsymbol{u'} = \boldsymbol{u} - \langle \boldsymbol{u} \rangle$ and the fluctuating magnetic fields are $\boldsymbol{B'} = \boldsymbol{B} - \langle \boldsymbol{B} \rangle$. In these simulations, $\mathcal{E}_{FI}$ is generally strongest at the poleward edge of the wreaths, centered at approximately $\pm 20°$ latitude, whereas the Ω-effect and $\langle B_\phi \rangle$ peak at roughly $\pm 15°$ latitude. This spatial offset between $\mathcal{E}_{FI}$ and $\langle B_\phi \rangle$ means that the turbulent emf is

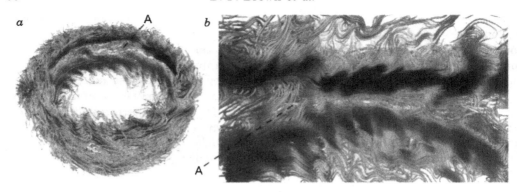

Figure 1. Magnetic wreaths in case D3. (*a*) Full volume rendering of magnetic wreaths, show-ing entire simulation. Lines trace the vector magnetic field with color denoting amplitude and polarity of the longitudinal magnetic field B_ϕ (red or light tones, positive; blue or darker tones, negative). Rather than being simple flux surfaces, magnetic fields thread in and out of each wreath, connecting the wreaths across the equator and linking them to the polar regions. (*b*) Zoom in view of region A showing cross-equatorial connectivity.

not generally well represented by a simple α-effect description, e.g.,

$$E_{\mathrm{FI}} = \langle \boldsymbol{u'} \times \boldsymbol{B'} \rangle|_\phi \neq \alpha \langle B_\phi \rangle \tag{2.1}$$

when α is a scalar quantity. This is true even when α is estimated from the kinetic and magnetic helicities present in the simulation. More sophisticated mean-field models may do much better at matching the observed emf E_{FI}, and other terms in the mean-field expansion may play a significant role; in particular, the gradient of $\langle B_\phi \rangle$ is large on the poleward edges of the wreaths where E_{FI} is significant.

3. A Cyclic Dynamo in a Stellar Convection Zone

We turn now to a more rapidly rotating dynamo simulation, case D5, rotating five times faster than the Sun (Brown *et al.* 2010b). As in case D3, strong global-scale mag-netic wreaths are built in the convection zone. Now however, the wreaths begin to show significant time-variation and undergo quasi-regular polarity reversals.

One such reversal is illustrated in Figure 2. Before the reversal (Fig. 2*a*), the wreaths look similar to those found in case D3, though here magnetism permeates the entire convection zone, including the polar regions where relic wreaths from the previous re-versal are visible. The equatorial region shows significantly more connectivity and large fluctuations of B_ϕ, with small knots of alternating polarity visible throughout. This cross-equatorial connectivity appears to play an important role in the reversal process. The magnetic fields built in case D5 attain somewhat larger amplitudes than those realized in case D3: here at mid-convection zone B_ϕ can reach $\pm 40\,\mathrm{kG}$, while in case D3 the peak amplitudes were closer to $\pm 26\,\mathrm{kG}$.

During the reversal (Fig. 2*b*), new wreaths of opposite polarity form near the equator while the old wreaths propagate towards the poles. This poleward propagation appears to be a combination of a nonlinear dynamo wave, arising from systematic spatial offsets between the generation terms for mean poloidal and toroidal magnetic field, and possi-bly a poleward-slip instability arising from magnetic stresses within the wreaths. After a reversal (Fig. 2*c*) the new magnetic wreaths grow in strength and dominate the equato-rial region. In the polar regions the wreaths from the previous cycle begin to dissipate,

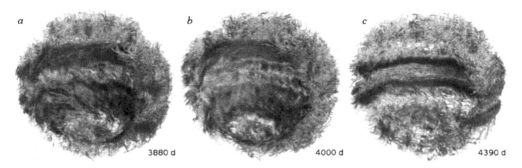

a 3880 d *b* 4000 d *c* 4390 d

Figure 2. Global-scale magnetic reversal in case D5. (*a*) Magnetic wreaths shown in field line tracing shortly before a polarity reversal, with positive polarity wreath above equator and negative polarity below. Volume shown spans slightly more than a full hemisphere with both polar caps visible where relic wreaths from the previous cycle remain visible. (*b*) During a reversal, new wreaths with opposite polarity form at the equator while old wreaths migrate toward the poles. (*c*) When the reversal completes, the polarity of the wreaths have flipped, with negative polarity wreath above the equator and positive below. Cancellation occurs in the polar regions and the old wreaths slowly dissipate. Times of snapshots are labeled.

reconnecting with the pre-existing flux there and being shredded by the turbulent convection there.

Generally, the dynamo processes identified in case D3 serve to build the wreaths of case D5. As in case D3, before a reversal the turbulent emf E_{FI} that contributes to the mean poloidal field is generally largest on the poleward edge of the wreaths at latitudes above $\pm 20°$, while the Ω-effect and $\langle B_\phi \rangle$ peak at roughly latitude $\pm 15°$. During a reversal both E_{FI} and the production of $\langle B_\phi \rangle$ associated with the Ω-effect surf on the poleward edge of the wreaths as those structures move poleward. This systematic phase shift appears to contribute to that propagation.

4. Wreath-building Dynamos

The two simulations we have explored here, cases D3 and D5, are part of a much larger family of simulations that we have conducted exploring convection and dynamo action in younger suns. The properties of this broad family are summarized in Figure 3*a*. Indicated here are 26 simulations at rotation rates ranging from $0.5\,\Omega_\odot$ to $15\,\Omega_\odot$. At individual rotation rates (e.g., $3\Omega_\odot$), further simulations explore the effects of lower magnetic diffusivity η and hence higher magnetic Reynolds numbers. Some of these follow a path where the magnetic Prandtl number Pm is fixed at 0.5 (triangles) while others sample up to Pm=4 (diamonds). The most turbulent simulations have fluctuating magnetic Reynolds numbers of about 500 at mid-convection zone. Wreath-building dynamos are achieved in most simulations (17), though a smaller number do not successfully regenerate their mean poloidal fields (9, indicated with crosses). Very approximate regimes of dynamo behavior are indicated, based on the time variations shown by the different classes of dynamos.

Near the onset of wreath-building dynamo action we generally find little time variation in the axisymmetric magnetic fields associated with the wreaths. This is illustrated for case D3 in Figure 3*b*, showing the mean longitudinal $\langle B_\phi \rangle$ at mid-convection zone over an interval of about 4000 days. Though small variations are visible on a roughly 500 day timescale, the two wreaths retain their polarities for the entire time simulated (more than 20,000 days), which is significantly longer than the convective overturn time

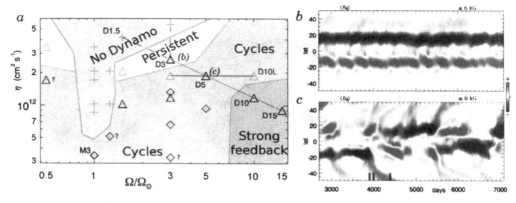

Figure 3. Wreath-building dynamos. (a) Parameter space showing variety of wreath-building dynamos currently explored. Magnetic diffusivity η and rotation rate Ω are shown for dynamo simulations at rotation rates ranging sampling 0.5–15$\Omega_\odot$, with very approximate dynamo regimes shown. In some regions, magnetic Reynolds numbers are too low to sustain dynamo action, while in other regions persistent magnetic wreaths form which do not show evidence for cycles. At higher magnetic Reynolds numbers (occurring here at low η or high Ω), wreaths typically undergo quasi-cyclic reversals. At the highest rotation rates the Lorentz force can substantially modify the differential rotation, but dynamo action is still achieved. Cases with question marks show significant time-variation but have not been computed for long enough to definitively establish cyclic behavior. (b) Time-latitude plot of mean (axisymmetric) B_ϕ at mid-convection zone in persistent case D3 (Brown *et al.* 2010a). (c) Cyclic case D5 shown for same span of time (Brown *et al.* 2010b). Three reversals are visible here, occuring on roughly 1500 day periods. Times of snapshots shown in Figure 2 are indicated.

(roughly 10–30 days), the rotation period (9.3 days), or the ohmic diffusion time (about 1300 days at mid-convection zone). We refer to the dynamos in this regime as persistent wreath-builders.

Generally, we find that wreath-building dynamos begin to show large time dependence as the magnetic diffusivity η decreases and as the rotation rate Ω increases. In many cases this leads to quasi-regular global-scale reversals of magnetic polarity, as discussed for case D5 in §3. We illustrate three of these cycles occuring in case D5 in Figure 3b. In this simulation, reversals occur with a roughly 1500 day timescale, though during some intervals the dynamo can fall into other states. For reference, the ohmic diffusion time in this simulation is about 1800 days, while the rotation period is 5.6 days. Similar cycles occur in other simulations, but the period of reversals varies with both η and Ω. Cycles appear to become shorter as η decreases, opposite to what might be expected if the ohmic time determined the cycle period. The dependence of cycle period on Ω is less certain. At present, determining why cycles are realized in many of these dynamos remains difficult. The phenomena appears to be at least partially linked to the magnetic Reynolds number of the differential rotation and possibly to that of the fluctuating convection.

In these wreath-building dynamos the major reservoir of kinetic energy that feeds the generation of magnetism is the axisymmetric differential rotation, and this global-scale shear is strongly reduced in the dynamo simulations. Individual convective structures are largely unaffected by the magnetic wreaths except when the fields reach very large amplitudes; in case D5 this occurs when B_ϕ exceeds values of roughly 35 kG at mid-convection zone. At the highest rotation rates the Lorentz force of the axisymmetric magnetic fields becomes strong enough to substantially modify the differential rotation, largely wiping out the latitudinal and radial shear (e.g., cases D10 and D15 in Fig. 3a).

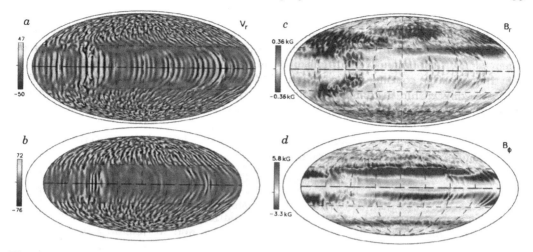

Figure 4. Nests of convection in dynamo case D10L. Convective patterns are shown by snapshots in Mollweide projection of the radial velocities v_r (a) near the surface (0.95 $R_\odot$) and (b) at mid-convection zone (0.85 $R_\odot$). Two active nests of convection are clearly visible. (c) Near the surface, the radial magnetic field B_r is concentrated in the stronger nest. This patch of radial field propagates with the nest. (d) At mid-convection zone the active nests leave their imprint on the magnetic wreaths, shown here by longitudinal field B_ϕ.

In these cases, wreath-like structures can still form though they typically have more complex structure and are less axisymmetric.

Some of these rapidly rotating dynamos have held further surprises. One such simulation, case D10L rotating ten times faster than the Sun, is shown in Figure 4. Magnetic wreaths form in this dynamo, but so do strong localized nests of convection. Two such nests are visible near the equatorial region in the convective patterns of radial velocity (Fig. 4a, b). At certain longitudes the convection is significantly stronger, and these nests span the convection zone. The nests leave their imprint on the magnetism, concentrating the radial magnetic field B_r into patches near the surface (Fig. 4c) and interacting strongly with the toroidal field B_ϕ at depth (Fig. 4d). These active nests appear to be distinct dynamical structures, and they propagate at a different rate than either individual convective cells or the local differential rotation. If such nests persist in stellar convection zones, they are likely to strongly affect inferences of the global-scale magnetic field. We have found active nests in rapidly rotating hydrodynamic simulations (Brown *et al.* 2008), but to our knowledge this is the first time such structures have been found in a fully saturated dynamo state.

We have found wreath-like structures in simulations rotating at the solar rotation rate as well (one example appears in the contribution by Miesch, these proceedings). Achieving such structures in solar simulations is challenging, as the angular velocity contrast in the Sun is smaller than that realized in the rapidly rotating dynamos. Generally the solar simulations require low values of η to build wreaths, which in turn calls for high resolutions and that exacts a large computational cost.

The magnetic boundary conditions adopted near the base of the convection zone also play an important role: here we have explored simulations with perfectly conducting bottom boundaries. Previous simulations of the solar dynamo (case M3, Brun *et al.* 2004) used a potential field bottom boundary and did not find magnetic wreaths like those shown here. When the boundary conditions are changed in case M3 to the perfect conducting boundaries used here, the axisymmetric magnetic fields grow in strength, become more organized and form wreaths near the base of the convection zone. A variant on that solar dynamo simulation is shown in the contribution by Miesch in these proceedings. Conversely, magnetic wreaths are difficult to achieve in the rapidly rotating dynamos when we use the boundary conditions of Brun *et al.* (2004). We take some comfort in the fact that wreaths continue to form in simulations which include a portion of the stable radiative zone within the domain. In the simulations with model tachoclines, the wreaths of magnetism are essentially unmodified from those found in the simulations which only capture the convection zone. This suggests that a perfectly conducting bottom boundary may mimic the presence of a highly-conductive radiative zone below the convection zone better than a potential field extrapolation does.

5. Overview

Stellar convection spans a vast range of spatial and temporal scales which remain well beyond the grasp of direct numerical simulation, and we remain humbled by the complexities posed by highly turbulent convection on global-scales in rotating, stratified, and magnetized plasmas. Stellar dynamo studies must drastically simplify the physics of the stellar interior: as an example, molecular values of η in the solar convection zone range from roughly 10^2–10^5 cm^2/s as one moves from the tachocline to the near surface regions, while the molecular viscosity ν is of order 10 cm^2/s there. In contrast, our simulations employ values of η and ν that are of order 10^{12} cm^2/s; this large value is more similar to simple estimates of turbulent diffusion associated with granulation at the surface where $\nu_t \sim V_t L_t \sim 10^{11}$ cm^2/s given $V_t \sim 1$ km/s and $L_t \sim 1$ Mm. Despite this daunting separation in parameter space, it is striking that coherent magnetic structures can arise at all in the midst of turbulent convection. We find the combination of global-scale spatial organization and cyclic behavior fascinating, as these appear to be the first self-consistent 3D convective stellar dynamos to achieve such behavior in the bulk of the convection zone.

A variety of wreath-building dynamos have been found in these simulations of rapidly rotating suns: some build persistent wreaths, while others undergo significant time variations including quasi-cyclic reversals of global-scale magnetic polarity. The parameter space is complex, and some simulations show cycles in one hemisphere but not the other. Cyclic cases like case D5 can even wander into and then back out of non-cycling states. The role of tachoclines in stellar dynamos remains a matter of great debate. These simulations suggest that, at least in rapidly rotating stars, tachoclines may not play as crucial a role in the organization and storage of the global-scale magnetic field as in the solar dynamo. A major step forward will be to explore simulations that couple wreath-building dynamos in the convection zone, through a tachocline of shear, to the stable radiative interior below; those simulations are ongoing now. We are also exploring how convection and dynamo action may be different in other solar-type stars, including K-type dwarfs and the fully convective M-type stars, which present their own mysteries (see Browning 2008, and contribution by Browning, these proceedings). The future of 3D dynamo simulations is very bright, and cyclic solutions are beginning to appear in a variety of situations (e.g., Ghizaru *et al.* 2010; Käpylä *et al.* 2010; Mitra *et al.* 2010). These are exciting times indeed for stellar dynamo theorists!

This research is supported by NASA through Heliophysics Theory Program grants NNG05G124G and NNX08AI57G, with additional support for Brown through the NASA GSRP program by award number NNG05GN08H and NSF Astronomy and Astrophysics postdoctoral fellowship AST 09-02004. CMSO is supported by NSF grant PHY 08-21899. Miesch was supported by NASA SR&T grant NNH09AK14I. NCAR is sponsored by the National Science Foundation. Browning is supported by research support at CITA. Brun was partly supported by the Programme National Soleil-Terre of CNRS/INSU (France), and by the STARS2 grant from the European Research Council. The simulations were carried out with NSF PACI support of PSC, SDSC, TACC and NCSA.

References

Brown, B. P., Browning, M. K., Brun, A. S., Miesch, M. S., & Toomre, J. 2008, *ApJ*, 689, 1354
—. 2010a, *ApJ*, 711, 424
Brown, B. P., Miesch, M. S., Browning, M. K., Brun, A. S., & Toomre, J. 2010b, *ApJ*, submitted
Browning, M. K. 2008, *ApJ*, 676, 1262
Brun, A. S., Miesch, M. S., & Toomre, J. 2004, *ApJ*, 614, 1073
Clune, T. L., Elliott, J. R., Glatzmaier, G. A., Miesch, M. S., & Toomre, J. 1999, Parallel Computing, 25, 361
Ghizaru, M., Charbonneau, P., & Smolarkiewicz, P. K. 2010, *ApJ*, 715, L133
Käpylä, P. J., Korpi, M. J., Brandenburg, A., Mitra, D., & Tavakol, R. 2010, Astronomische Nachrichten, 331, 73
Miesch, M. S., Elliott, J. R., Toomre, J., Clune, T. L., Glatzmaier, G. A., & Gilman, P. A. 2000, *ApJ*, 532, 593
Mitra, D., Tavakol, R., Käpylä, P. J., & Brandenburg, A. 2010, *ApJ*, 719, L1
Pizzolato, N., Maggio, A., Micela, G., Sciortino, S., & Ventura, P. 2003, *A&A*, 397, 147

Discussion

T. ROGERS: Could the poleward propagation of field be likened to the dynamo wave, in which the sign of propagation depended on helicity associated with convection versus radiation zones?

B. BROWN: Maybe. Here we clearly see signs that the poleward propagation is partly due to Maxwell stresses and a resulting poleward slip, and this phenomena occurs essentially unmodified in simulations that include a tachocline and convective penetration into a stable radiative zone. But there may be dynamo wave aspects to the reversals as well. [Note: subsequent work indicates that a non-linear dynamo wave does play an important role in the reversal process; see Brown *et al.* (2010b).]

E. ZWEIBEL: Why is the "failed dynamo" region on the η–Ω plane localized near the solar rotation rate $(\Omega = \Omega_\odot)$? (see Figure 3*a*)

B. BROWN: On either side of the solar rotation rate, the latitudinal shear is strong. Simulations spinning slower than the Sun have strong anti-solar differential rotation (retrograde equators, prograde poles) while the more rapidly rotating simulations have stronger solar-like differential rotation (prograde equators, retrograde poles). Both cases lead to higher magnetic Reynolds numbers associated with the latitudinal shear. Simulations at the solar rotation rate have weaker differential rotation, and need lower values of η to achieve dynamo action in these wreath-building dynamos.

Astrophysical Dynamics: From Stars to Galaxies
Proceedings IAU Symposium No. 271, 2010
N.H.Brummell, A.S.Brun, M.S. Miesch & Y.Ponty, eds.

© International Astronomical Union 2011
doi:10.1017/S1743921311017480

Characterizing the Quiet Sun Scale Magnetic Field

Alan Title[1]

[1]Lockheed Martin Advanced Technology Center, 3251 Hanover Street, Palo Alto, 94304, USA

Abstract. Observations with the Solar Optical Telescope on Hinode indicate that the Quiet Sun magnetic field occurs on every scale of convection including granulation. Data reported here show that, regardless of the position on the disk, the polarity in the inner network regions are balanced to 1 part in 72. This is consistent with both local dynamo processes or the creation of surface features by the granulation downflows.

Keywords. Sun: magnetic fields

1. Introduction

For a number of years there has been evidence of the presence of magnetic fields virtually everywhere on the surface of the Quiet Sun (QS). This has suggested that flux was appearing at the surface on the scale of granulation in a manner similar to the now well established "magnetic carpet" which maintains the kilogauss field elements observed in the supergranule boundaries. MDI data had shown that the carpet is sustained by continuous emergence of "ephemeral regions", magnetic structures with an average flux a few times 1018 Mx. High sensitivity magnetograph and spectro-polarimeter (SP) measurements have long shown the existence of "internet-work fields" structures with fluxes as low as 1015 Mx. While there is general agreement that the network and ephemeral fields have strengths on the order of a kilogauss, there has remained some controversy about the strength of the internet-work fields. Their estimated strengths have ranged between a few hundred gauss and a kilogauss.

Careful measurements of MDI time sequences have revealed the presence and rapid evolution of "magnetic fragments". Fragments are different from ephemeral regions in that they are not easily associated with an opposite polarity mate or mates. The number of fragments increases with increased spatial resolution of magnetic maps. This has led to the statement that more flux seems to appear on the Sun in "unassociated" fragments than recognizable bipoles. The abundance of fragments can not indicate that there are magnetic monopoles, but rather that the emergence of flux on the surface can take complex and not easily recognized forms. Numerical simulations of flux emergence and local dynamo action exhibit patterns of appearance in which bipole emergence would be impossible to recognize at the resolution of current telescopes.

Many new insights into the characteristics of the QS magnetic field have arisen from the observations made with the Solar Optical Telescope (SOT) on the Hinode. These insights have their origin in the uniform image quality allowed from the seeing-free space platform; the lack of the need for extensive image processing techniques to increase resolution that tend to diminish or eliminate low contrast transient features; the combination of a very sensitive SP and a filter magnetograph; and a set of powerful and efficient software tools for construction and analysis of movies and SP maps.

The first observations made with both the birefringent filter magnetograph and SP revealed magnetic field virtually everywhere in the QS. High signal-to-noise spectra showed that granule boundaries were nearly everywhere filled with mixed polarity near vertical magnetic fields, while horizontal fields spread over much of the entire surface. The presence of a pervasive sea of flux on the granular scale has immediate implications because the granular pattern has a lifetime on the order of ten minutes. A consequence of the granular flows is that flux is continuously swept into the granule boundaries on the granule turn over time. In order to always observe flux on the spatial scale of granulation, flux must be continuously emerging on the granule temporal scale. Figure 1 shows a continuum image and a map of the vertical field in a region of very QS. Magnetogram contours overlaid on the continuum image illustrate that the majority of the magnetic flux is in the granule boundaries. Nevertheless, there are also clear examples of magnetic signal in granules. This should be expected both in the case that all flux does not emerge in boundaries or when during the initial stages of emergence there is a significant vertical component over the center of granules.

 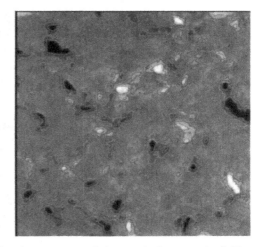

Figure 1. Continuum image (left) overlaid by the contours of the vertical magnetic field (right). The image and the magnetic map were constructed from a SP scan.

Hinode filter magnetograph movies show that active regions emerge in a very complex manner. Between the flux concentrations that eventually contract into the leading and trailing parts of the active region there emerges a sea of smaller bipoles roughly aligned with the two growing regions of opposite polarity (see figure 2). This has been seen previously and even semi-empirical models had suggested that a phenomena, O-loops, could explain the observation that sunspots seem to grow nearly in place. What the new Hinode observations have shown is that many of the same features seen in emergence of active regions are repeated in emergence events that are much smaller (see figure 3). The Hinode data suggests that much of the flux that appears on the solar surface does not initially emerge as a single bipole, but as a sea of small flux structures each of which is oriented in roughly the same direction. The smaller the total flux in an emergence event the more random the orientation of the components and the shorter their lifetime on the solar surface. Because of the complexity of the emergence patterns it is difficult to impossible to associate a particular flux element with the opposite polarity feature or features with which it emerged. In part this is caused by merging that has occurred before the flux elements have reached the surface. Given the Hinode observations, it is not surprising that more fragments have been observed than bipole pairs in earlier data.

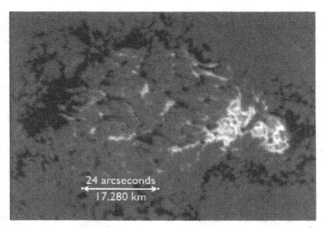

Figure 2. An evolving active region on 11 December 2007 showing the emergence of intermediate bipoles between the leading and following flux concentrations.

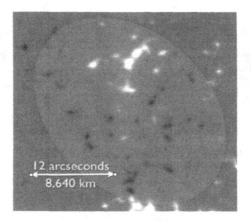

Figure 3. An example of a complex emergence of an array of small magnetic flux elements. The magnetogram is from a frame of a movie of QS on 5 November 2007. It is difficult to identify bipoles in this complex pattern either in this single frame or in the movie.

Numerical models of how a flux tube started 5000 km below the visible surface emerges exhibit features that have a remarkable similarity to the Hinode observations described above. The rapid decrease in the atmospheric pressure near the surface requires that the magnetic pressure in the tube decreases in order to maintain hydrostatic equilibrium. As a result a uniform circular flux tube that started 5000 km below the surface expands perpendicular to its axis as it rises forming a relatively thin flat oval of horizontal flux just below the surface. Examples of the start and the evolution of such a numerical simulation is shown in figure 4. As the "pond" of flux forms near the surface, strong granular downflows drag the magnetic field lines downward creating local bipoles. Because the pond can have dimensions larger than the spacing of the convective downflows, the total (unsigned) flux that emerges on the surface can be greater than the total flux in a cross section of the original emerging flux tube. In figure 5 a pair of 3D slices show the flux pattern at 5000 km and just below below the surface. A few field lines are shown that illustrate the formation of bipoles that are created by the convective downflows.

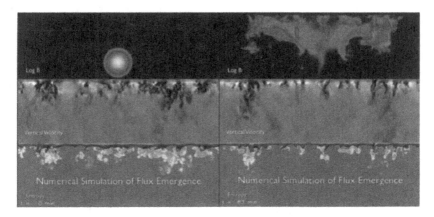

Figure 4. Cross section of a rising originally circular flux tube together with cross sections of the vertical velocity and entropy in the computational volume. The frame on the left is the initial condition while the frame on the right sampled the simulation when 10% of the flux in the original tube has reached the surface.

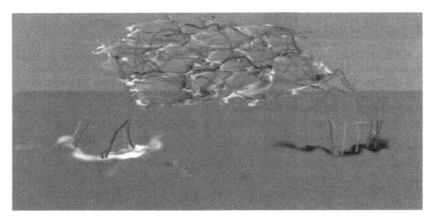

Figure 5. 3D slices of the magnetic flux at 5000 km and just below the surface together with a few example field lines.

The pond emergence phenomena coupled with the realization that an appreciable fraction of the magnetic field has to be appearing and disappearing on the time scale of tens of minutes seems to argue against feature recognition and tracking schemes as a good methodologies for characterizing the majority of the ubiquitous QS fields that can now be observed at the sub-arc second level. Below some distinguishing characteristics of QS fields are described that are based on characteristics of the flux distributions determined pixel by pixel.

2. Data Analysis

As a first step in trying to develop a diagnostic to characterize the magnetic fields in different solar regions, histograms of the integrated V (IV) signal were created. These histograms are good approximations of the distribution of LOS flux values in the sampled regions. Because of the wide range in the number of points with a given signal, distribution functions were the main focus of the initial analysis. Figure 6 shows an example magnetic map and the histogram of pixel IV signal values. The core of the histogram, the values near zero signal, are well fitted by Gaussians in the example shown as well as all the other maps measured. A lower limit to the amount of area covered by detectable field in the maps can be estimated by subtracting the number of pixels in the Gaussian fit to the core from the total number of pixels in the entire map. In the case illustrated, 24% of the area of the map have signals outside of the Gaussian core and 71% of the total unsigned signal results from pixels outside of the Gaussian core.

 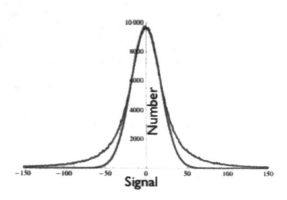

Figure 6. SOT SP (left) on 3/05/2009 centered at (4.1,8.1) arc seconds with respect to disk center. The map was 768×768 with 0.149 arc seconds pixels. The right panel shows in blue the histogram of the magnetic signals in the map. Shown in red is the best fit Gaussian to the core of the histogram. 24% of the points in the map are outside the Gaussian core and they represent 71% of the total unsigned signal in the map.

The map shown in figure 6 is roughly 115×115 arc seconds so should contain several supergranulation cells. It is possible to identify the cell boundaries with a mask that isolates the higher signal regions. By dilating the high flux map and inverting it a map is created that isolates the cell interiors (see figure 7). This map shows an essentially uniform mixture of polarities. On a quick glance it is even difficult to note the blank areas caused by the removal of the network flux.

With the separation of the high and lower values points distribution functions can be created from the entire map as well as the network and cell interiors separately. Figure 8 shows distribution functions for the entire map and those of the high value positive points. The power law indices of the distribution functions are −2.29, −2.34, and −2.34 for the negative, positive, and positive high value points, respectively.

Figures 9 through 11 summarize the major results of the set of measurements on magnetic maps in QS over a range of radial positions. Figure 9 shows the variation of histogram slope with radial position. In this figure the magnitude of the indices are shown (all measured indices are negative). A quadratic increase in the slope toward both the polar and equatorial limbs is observed. Figure 10 shows the total (unsigned) and mean

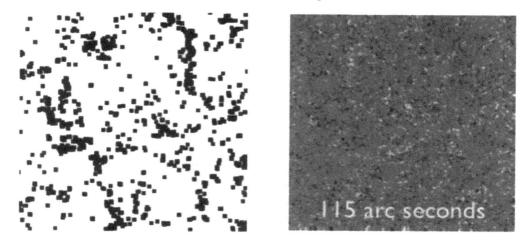

Figure 7. The left hand panel is a mask obtained by dilating and then inverting the high signal map, while the right hand panel is the result of applying the mask to the magnetogram.

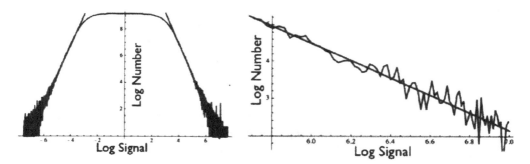

Figure 8. In the left hand panel is the distribution function of the full magnetogram overlaid with the best fit lines to the indices of the positive and negative map points, which are −2.29 and −2.34 for the negative and positive points, respectively. The right hand panel has an expanded scale distribution function of the high value positive points isolated by the high signal mask. Here the index is also −2.34.

flux as a function of radial position for ONLY those points in the inner network. This figure demonstrates that the average signal in the inner network is near zero for all radial positions. Figure 11 is a similar figure for the value of mean and total flux for high flux points – the network points. Here there is significantly more variation in the value of the mean flux, while the total flux decreases quadratically but only weakly toward the limbs.

3. Summary of QS Observations

A notable feature of the distribution functions is that the power law index decreases from −2.3 at disk center to −4 at the limb. If the distribution of the amplitude of individual features was Gaussian such large negative indices would not occur. In order to create the observed power law for an ensemble of magnetic features requires that the peak amplitude of the features has a power law distribution of −0.5 ± 0.2. The drop in

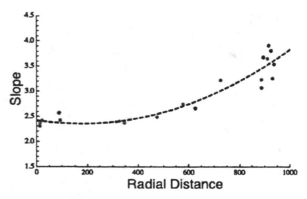

Figure 9. Plot of average slope of distribution functions versus radial position on the disk. The red dots are from equatorial locations. The dashed curve is the best fit to the data to all points. The best fit is of the form $a + cx + br^2$, where a,b,c= $(2.4, -7.4 \times 10^{-5}, 2.2 \times 10^{-6})$.

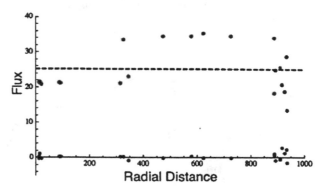

Figure 10. Plots of the unsigned flux in the regions of the quiet sun well outside of the network (green) versus radial position. In red are points near the equator. Shown in purple are the values of the average signal. In all regions the average signal is near zero which shows that in the inner network regions the polarity is well balanced. The ratio average to the flux unsigned flux is 72.6. The dashed line is the best fit to the unsigned flux data it shows essentially no change in the unsigned flux from disk center to the limbs.

the indices toward the limb indicates that the high amplitude features are more vertical that those of lower amplitude.

More interesting is the observation that the internetwork fields are very well balanced on the scale of the sample boxes that are 115 arc seconds square, or that contain a few supergranules, independent of position on the solar disk. This is true even if the supergranular boundaries that they are internal to are not nearly as well balanced. This property could be caused by local dynamo action or equally likely just a consequence of the stitching cause by the granulation downflows into a rising pond of horizontal flux.

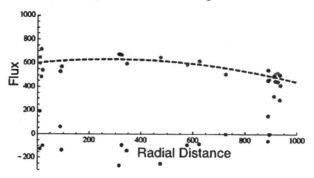

Figure 11. Plots of unsigned flux (green) and of the average flux (purple) versus radial position for network regions. Shown in large black points are total fluxes for the equatorial regions, The dashed curve is a quadratic fit to the unsigned flux data.

Astrophysical Dynamics: From Stars to Galaxies
Proceedings IAU Symposium No. 271, 2010
N. H. Brummell, A. S. Brun, M. S. Miesch & Y. Ponty, eds.

© International Astronomical Union 2011
doi:10.1017/S1743921311017492

Dipolar and Quadrupolar Magnetic Field Evolution over Solar Cycles 21, 22, and 23

M. L. DeRosa[1], A. S. Brun[2] and J. T. Hoeksema[3]

[1]Lockheed Martin Solar and Astrophysics Laboratory, 3251 Hanover St. B/252, Palo Alto, CA 94304 USA
email: derosa@lmsal.com

[2]Laboratoire AIM Paris-Saclay, CEA/Irfu Université Paris-Diderot CNRS/INSU, F-91191 Gif-sur-Yvette, France

[3]Hansen Experimental Physics Laboratory, Stanford University, 466 Via Ortega, Stanford, CA 94305 USA

Abstract. Time series of photospheric magnetic field maps from two observatories, along with data from an evolving surface-flux transport model, are decomposed into their constituent spherical harmonic modes. The evolution of these spherical harmonic spectra reflect the modulation of bipole emergence rates through the solar activity cycle, and the subsequent dispersal, shear, and advection of magnetic flux patterns across the solar photosphere. In this article, we discuss the evolution of the dipolar and quadrupolar modes throughout the past three solar cycles (Cycles 21–23), as well as their relation to the reversal of the polar dipole during each solar maximum, and by extension to aspects of the operation of the global solar dynamo.

Keywords. Sun: magnetic fields

1. Introduction

Understanding the solar dynamo is one of the longstanding problems in the field of solar physics. Much of what we know about the solar dynamo arises from observations of the patterns of magnetic flux that emerge onto the photosphere and their subsequent evolution (Hathaway 2010), combined with numerical modeling efforts that focus on gaining a theoretical perspective on the interplay between the large-scale flows, the tachocline and upper shearing layers, convection, and magnetic fields within the solar interior (Charbonneau 2005; Browning *et al.* 2006). The most useful numerical models capture many of the observed behaviors of magnetic flux, such as (for example) the spectrum of features visible on the photosphere, or the timing and phases of dipole reversals.

As a result, long-term observations of photospheric magnetic fields over multiple solar activity cycles are extremely useful as boundary conditions for modeling, and/or as independent validation of models, or for simply gaining intuition. In this article, we take a spectral approach and analyze the solar photospheric magnetic field from the perspective of its spherical harmonic decomposition, determining the harmonic coefficients from three different time series of the photospheric magnetic field, with a special emphasis on dipolar and quadrupolar fields and their phase relationship over the past three solar cycles. The three datasets are outlined in §2, followed by a discussion of the dipolar and quardupolar modes in §3 and §4, respectively. The energy spectra are presented in §5, followed by a brief discussion and concluding remarks in §6.

2. Spherical Harmonic Decomposition of the Data

Time series of maps of the radial magnetic field covering the entire solar photosphere (i.e., the full 360° of longitude and 180° of latitude) were decomposed into their constituent spherical harmonic modes. This process involves finding the coefficients $B_{\ell,m}(t)$ for a time series of photospheric radial magnetic field maps $B_r(\theta, \phi, t)$ such that

$$B_r(\theta, \phi, t) = \sum_{\ell, m} B_{\ell,m}(t)\, Y_{\ell,m}(\theta, \phi),$$

where the spherical harmonic functions $Y_{\ell,m}(\theta, \phi)$ are characterized by the quantum numbers degree ℓ and order m. The magnetic data originate from three different sources, as follows:

(*a*) Diachronic maps, sampled once per Carrington rotation (CR), were assembled from full-disk magnetograms observed by the ground-based Wilcox Solar Observatory (WSO) in Stanford, CA. The WSO maps span CR 1642–2089, containing data from 1976-May-27 through 2009-Nov-9, and are available at `http://wso.stanford.edu/synopticl.html`.

(*b*) Diachronic maps, sampled once per Carrington rotation, were assembled from full-disk magnetograms observed by the Michelson Doppler Imager (MDI; Scherrer *et al.* 1995) on board the Solar and Heliospheric Observatory (SOHO) spacecraft. The MDI maps span CR 1910–2088, containing data from 1996-Jul-1 through 2009-Oct-13, and are available at `http://soi.stanford.edu/magnetic/index6.html`.

(*c*) Instantaneous synoptic maps from an evolving-flux model of the solar photosphere (Schrijver 2001) were sampled once every two weeks. The model is constructed by inserting flux from the 96-minute time series of full-disk MDI magnetograms into the model (Schrijver & De Rosa 2003). During each 6h time step, the population of flux concentrations is advected horizontally according to empirically-based prescriptions for differential rotation, poleward meridional flows, and dispersal due to supergranulation. Neighboring flux concentrations are allowed to coalesce or partially cancel (depending on their polarities) if they become separated by a distance less than 4200 km. The data used in this study span much of the SOHO era, ranging from 1996-Jul-1 through 2009-Dec-31, and can be downloaded using the `pfss` package available through SolarSoft.

3. Evolution of Solar Dipole

The solar dipolar magnetic field can be analyzed in terms of its polar and equatorial harmonic components. The polar component, corresponding to the $\ell=1$, $m=0$ spherical harmonic function and which has a mode amplitude of $B_{1,0}$, is observed to be strongest during solar minimum when there is a significant amount of magnetic flux located near the polar regions of the sun. These polar caps are the result of small amounts of trailing-polarity flux "escaping" from each active region throughout the course of a sunspot cycle, followed by their advection poleward by the meridional flow pattern. On average, flux from trailing polarities is more likely to reach the poles, from which it follows that the same amount of leading-polarity flux must cross the equator and cancel with leading-polarity flux from the opposite hemisphere. In contrast, the equatorial dipole components, having mode amplitudes $B_{1,\pm1}$, are strongest during solar maximum and result from the presence of active regions during the sunspot activity cycle.

Because the WSO data span multiple sunspot activity cycles, a historical perspective on the evolution of the dipole can be gained, as shown in Figures 1 and 2. For example, the upper panel of Figure 1 illustrates that the (unsigned) magnitude of the solar-minimum polar dipole is much lower at the present time, following the most recent sunspot cycle (Cycle 23), than following either of the two preceding sunspot cycles (Cycles 21 and 22).

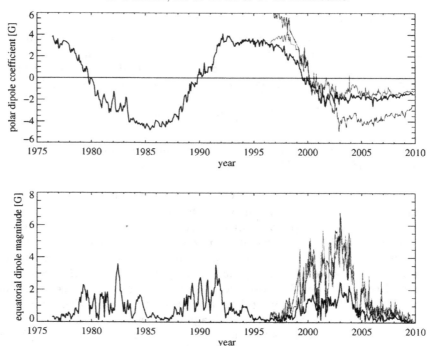

Figure 1. The temporal evolution of the (signed) polar dipole coefficient $B_{1,0}$ (upper panel), and the (unsigned) magnitude of the equatorial dipole coefficient $(|B_{1,-1}|^2 + |B_{1,1}|^2)^{1/2}$ (lower panel) for each of the three datasets. The horizontal axis encompasses the past three sunspot cycles. In each plot, the thick solid line corresponds to the WSO maps, the dashed line corresponds to the MDI maps, and the dotted line corresponds to the evolving-flux model. [In the color version of this article (available online), the thick solid line is black, the dotted line is green, and the dashed line is magenta.]

Similarly, the equatorial dipole strength was lower during the Cycle 23 maximum than during the maxima of either Cycle 21 or 22, as shown in the lower panel of Figure 1. We infer that such cycle-to-cycle trends are likely not unusual, especially if one were to consider the variation in the sunspot index (a broad-brush measure of magnetic activity) over time as determined, for example, by the Royal Observatory of Belgium (available at http://www.sidc.be/sunspot-index-graphics/sidc_graphics.php). Interestingly, the range over which the ratio of the energies contained in the equatorial versus the polar dipole components has remained about the same over this time, as shown in the lower panel of Figure 2.

During the course of a sunspot cycle, the polar dipole reverses sign as the polar caps established during the previous cycle are eroded away and are reformed with additional opposite-polarity flux that is being advected poleward. Visualizations of the dipole axis computed from data from Cycle 23 show that this reversal process took several years to complete. During this time, and especially when $B_{1,0}$ was near zero, the reduced energy in the polar dipole was partially offset by an increase in energy in the equatorial dipole, resulting in a reduction of the total energy in all dipolar modes only by about an order of magnitude above its solar-minimum value, as shown in the upper panel of Figure 2. When the polar dipolar component is weak, the axis of the equatorial dipolar component is observed to precess in longitude. These dynamics occur because the longitude of the dipole axis is set by the strongest active regions on the photosphere at any given time.

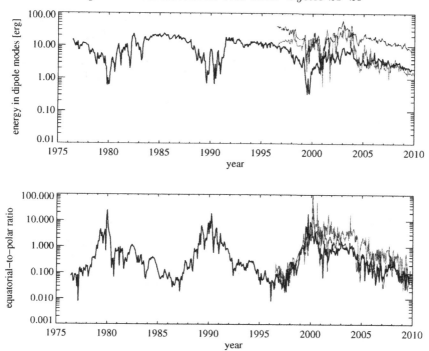

Figure 2. The temporal evolution of the total energy in the dipolar modes $\sum_m |B_{1,m}|^2$ (upper panel), and the ratio between the energy in the equatorial and polar dipole modes $|B_{1,0}|^2/(|B_{1,-1}|^2 + |B_{1,1}|^2)$ (lower panel). The solid, dashed, and dotted lines are as in Fig. 1.

As older active regions decay and newer sources of flux appear on the photosphere, this longitude can (and does!) change in a continuous fashion in response to the ever-evolving configuration of active-region flux on the solar surface.

4. Evolution of Solar Quadrupole

The evolution of the energy contained in the quadrupolar ($\ell=2$) modes exhibit much more variation than the dipolar energy, as can be seen by comparing the upper panel of Figure 2 with the upper panel of Figure 3, in which are illustrated the energy in the dipolar and quadrupolar modes, respectively. As with the equatorial dipole components, the quadrupolar modes have more power in them during maximal activity levels than during solar minima. During such times when there is a large amount of activity, it is possible for the energy in the quadrupolar modes to be greater than the energy in the dipolar modes. The ratio between these two groups of modes is shown in the lower panel of Figure 3, from which it is evident that during each of the past three sunspot cycles there have been periods of time when the quadrupolar energy has exceeded the dipolar energy by as much as a factor of 10. An example of one such period is illustrated in the coronal field model of Figure 4, where the solar magnetic field appears to be dominated by a quadrupolar mode having an axis of symmetry in the equatorial plane oriented perpendicular to the line of sight.

5. Energy Spectra

One property of the spherical harmonic functions $Y_{\ell,m}(\theta, \phi)$ is that the degree ℓ is equal to the number of node lines (i.e., lines where $Y_{\ell,m}=0$) on the $r=R_\odot$ surface. In other words, the spatial scale represented by any harmonic mode (i.e., the distance between

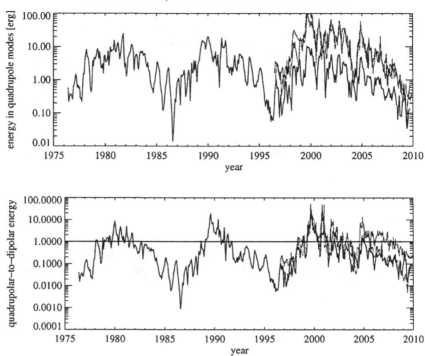

Figure 3. The temporal evolution of the total energy in the quadrupolar modes $\sum_m |B_{2,m}|^2$ (upper panel), and the ratio between the energy in the equatorial and polar dipole modes $\sum_m |B_{1,m}|^2 / \sum_m |B_{2,m}|^2$ (lower panel). The solid, dashed, and dotted lines are as in Fig. 1.

neighboring node lines) is determined by its spherical harmonic degree ℓ. As a result, the range of ℓ values containing the greatest amount of energy indicates of the dominant spatial scales of the magnetic field. To this end, we have averaged the power spectra from each of the three datasets both over time and over m, and have displayed the result in Figure 5. In the figure, it can be seen that the magnetic power spectra form a broad peak with a maximum degree occurring at about $\ell=25$. This size scale corresponds to about the same size as a typical active region, indicating that much of the magnetic energy can be found (perhaps not surprisingly) on the spatial scale of active regions.

The upper panel of Figure 5 illustrates the dependence of the energy spectra on the spatial resolution of the magnetograph. The WSO curve (the solid black line in the figure) does not show the same broad peak at $\ell=25$ as the curves from the other datasets (both of which utilize MDI data). This is because the significantly lower spatial resolution of the WSO magnetograph (which has $180''$ pixels, and is stepped by $90''$ in the east-west direction and $180''$ in the north-south direction) versus MDI (which has a plate scale of $2''$ in full-disk mode) does not allow modes even as high as $\ell=25$ to be adequately resolved. It is likely that as longer time series of data from newer, higher-resolution magnetograph instrumentation are assembled, energy spectra (such as those shown in Fig. 5) may change, especially at the higher end of the ℓ spectrum due to the better observations of finer scales of magnetic field.

Of interest too is the exponent of the power-law dropoff of these average spectra for ℓ values greater than the value of ℓ for which the energy peaks. We can see from the lower panel of Figure 5 that the energy spectra from the MDI diachronic map dataset and the evolving-flux synoptic-map dataset have different slopes (i.e., they have different

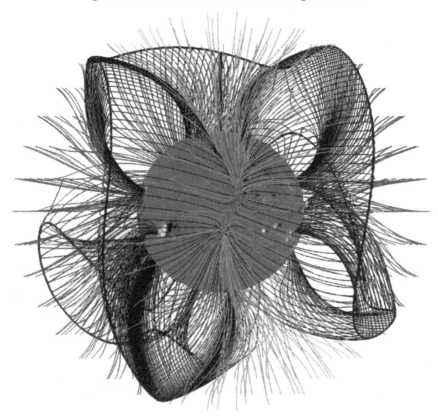

Figure 4. A potential-field source-surface (PFSS) model of the coronal magnetic field for 2000-Oct-11, as an example of a time when the photospheric (and thus coronal) magnetic field has a quadrupolar component that dominates the dipolar component. The thin black lines correspond to high-arching closed field lines that abut the source-surface neutral line (thick black line). Some collections of such high-arching fieldlines, when viewed from a certain perspective, appear as "tunnels" that generally correspond to the location of helmet streamers seen in coronagraph imagery (e.g., Wang *et al.* 2007). Fieldlines intersecting the upper boundary of the model (which are assumed to open into the heliosphere) are colored either light or dark gray, depending on polarity.

power-law exponents), which must be an indication of the differences in the dynamics by which energy gets transferred from larger spatial scales to ephemeral-region scales in each of the two cases. Interestingly, the evolving-flux model was shown to have approximately the same flux distribution as a MDI full-disk magnetogram both at both low and high activity levels (see Fig. 2 of Schrijver 2001), making the differences between the model and the sun less obvious.

6. Discussion and Concluding Remarks

We have performed a spherical harmonic expansion on time series of photospheric magnetic field data assembled from multiple datasets, and illustrated several interesting features of the large-scale photospheric magnetic field. We find that both the polar and equatorial dipole components were much reduced during Cycle 23 than during the two preceding sunspot activity cycles. In contrast, the ratio of energies in the dipolar versus

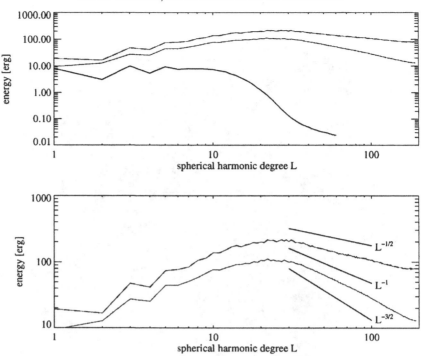

Figure 5. The energy spectra (upper panel) as a function of spherical harmonic degree ℓ as averaged over the full temporal extent of each dataset. The lower panel shows the same data for a portion of the vertical axis used in the upper panel. The solid, dashed, and dotted lines are as in Fig. 1.

the quadrupolar modes is similar, varying with about a factor of 10 between maximal and minimal activity levels during all three cycles.

During the past three sunspot activity cycles when the polar dipole is reversing sign, the quadrupolar modes were observed to predominate over the (both equatorial and polar) dipolar modes. Whether this feature plays a dynamical role in the reversal of the dipole, or whether it is simply a consequence of the polar dipole component being small (allowing higher-degree modes to dominate), remains at issue. We note in passing that similar dominance by the quadrupolar over the dipolar modes is observed during reversals of the geomagnetic field, and indeed interplay between these low-degree modes have been hypothesized to trigger polarity reversals for the earth's magnetic field (McFadden *et al.* 1991; Leonhardt & Fabian 2007).

We also must keep in mind that the most detailed observations of photospheric magnetic fields occurred during the most recent sunspot cycle (Cycle 23), and that this particular sunspot activity cycle appears somewhat unusual (though not necessarily unprecedented) in many respects (Hoeksema 2010). For instance, the asymmetric nature of the distribution of active regions during this cycle resulted in one of the polar caps forming about one year prior to the other, and such dynamics may affect the type of large-scale diagnostics discussed in this article. Additionally, the upcoming Cycle 24 has also been slow to start up, resulting in an extended decline phase of Cycle 23, a fact that has confounded efforts to predict the length and strength of Cycle 24 (see, for example, the range of predictions compiled by Pesnell 2008). It is clear that improved observations over the course of upcoming cycles will help to explain some of the unsolved mysteries presented here.

Acknowledgement

A.S.B. acknowledges funding by the ERC through grant STARS2 #207430.

References

Browning, M. K., Miesch, M. S., Brun, A. S., & Toomre, J. 2006, *ApJL*, 648, L157
Charbonneau, P. 2005, *Liv. Rev. Solar Phys.*, 2, 2
Hathaway, D. H. 2010, *Liv. Rev. Solar Phys.*, 7, 1
Hoeksema, J. T. 2010, in A. G. Kosovichev, A. H. Andrei, & J.-P. Roelot (eds.), *Solar and Stellar Variability: Impact on Earth and Planets*, Proc. IAU Symposium 264 (Cambridge: Cambridge Univ. Press), p. 222
Leonhardt, R. & Fabian, K. 2007, *Earth Planet. Sci. Lett.*, 253, 172
McFadden, P. L., Merrill, R. T., McElhinny, M. W., & Lee, S. 1991, *JGR*, 96, 3923
Pesnell, W. D. 2008, *Solar Phys.*, 252, 209
Scherrer, P. H. *et al.* 1995, *Solar Phys.*, 162, 129
Schrijver, C. J. 2001, *ApJ*, 547, 475
Schrijver, C. J. & De Rosa, M. L. 2003, *Solar Phys.*, 212, 165
Wang, Y., Biersteker, J. B., Sheeley, Jr., N. R., Koutchmy, S., Mouette, J., & Druckmüller, M. 2007, *ApJ*, 660, 882

Discussion

BRANDENBURG: If the quadrupole becomes dominant over the dipole, this should mean that Hale's polarity law is completely violated.

DEROSA: Axel, I believe you are referring to the fact that the $Y_{2,0}$ mode, when aligned with the axis of rotation, suggests that both poles should have flux possessing the same polarity, which would seem to violate the aspect of Hale's polarity law that indicates that the northern and southern hemispheres should be antisymmetric. However, I note that the quadrupole does not necessarily have to be aligned with the axis of rotation, and can in fact be tilted 90° so as to be oriented in the equatorial plane. This is exactly what is occurring at the time there is more energy in the quadrupolar modes than in the dipolar modes [c.f., Fig. 4 in this article].

J. TOOMRE: What may be the relation between active longitudes and the predominance of quadrupole vs dipole modes?

DEROSA: The active longitudes, depending where they are, play a prominent role in determining which modes predominate, especially if they are regularly spaced. If they are separated by, say, 180° and there is activity in the northern hemisphere at one of the active longitudes, and in the southern hemisphere at the other active longitude, then this puts power in the dipolar mode, for example. If there are more than two active longitudes, I would expect there to be persistent power in the higher-order spherical harmonics as a result.

Astrophysical Dynamics: From Stars to Galaxies
Proceedings IAU Symposium No. 271, 2010
N.H. Brummell, A.S. Brun, M. S. Miesch & Y. Ponty, eds.
© International Astronomical Union 2011
doi:10.1017/S1743921311017509

On the formation of ring galaxies

Yu-Ting Wu[1] and Ing-Guey Jiang[2]

[1]Department of Physics, National Tsing-Hua University,
Hsinchu, Taiwan 30013, R. O. C.
email: d9622814@oz.nthu.edu.tw

[2]Department of Physics and Institute of Astronomy, National Tsing-Hua University,
Hsinchu, Taiwan 30013, R. O. C.
email: jiang@phys.nthu.edu.tw

Abstract. The formation scenario of ring galaxies is addressed in this paper. We focus on the P-type ring galaxies presented in Madore, Nelson & Petrillo (2009), particularly on the axis-symmetric ones. Our simulations show that a ring can form through the collision of disc and dwarf galaxies, and the locations, widths, and density contrasts of the ring are well determined. We find that a ring galaxy such as AM 2302-322 can be produced by this collision scenario.

Keywords. galaxies: formation, galaxies: interactions, galaxies: kinematics and dynamics

1. Introduction

Ring galaxies, a peculiar class of galaxies, contain various ring-like structures with or without clumps, nuclei, companions or spokes. For example, the famous Cartwheel galaxy was first discovered by Zwicky (1941) and shows an outer ring, an inner ring, a nucleus and spokes (Theys & Spiegel 1976; Fosbury & Hawarden 1977; Higdon 1995). According to the morphology of ring galaxies, Few & Madore (1986) studied 69 ring galaxies in the southern hemisphere and developed a classification scheme. The ring galaxies are classified into two main classes, the O-type and P-type galaxies. The O-type galaxies have a central nucleus and smooth regular ring, while P-type systems often contain an offset nucleus and a knotty ring.

As more and more observational data on ring galaxies have been obtained (Arp & Madore 1987; Bushouse & Standford 1992; Marston & Appleton 1995; Elmegreen & Elmegreen 2006; Madore et. al. 2009), three dominant theories have been proposed to explain the formation and evolution of ring galaxies: the collision scenario, the resonance scenario and the accretion scenario. In the collision scenario suggested by Lynds & Toomre (1976), ring galaxies are formed after a head-on collision between an intruder galaxy and a disk galaxy. The formation of the Cartwheel galaxy is thought to be a prototype of this scenario. Furthermore, from the observational statistics by Few & Madore (1986), P-type galaxies have an excess of companions and can be considered to be formed by the collision scenario. In the resonance scenario, ring-like patterns are formed by gas accumulation at Lindblad resonances which respond to external perturbations, such as a bar or an oval. O-type galaxies with central nucleus and no obvious companions are thought to experience the resonance formation process, for instance, IC 4214 (Buta *et al.* 1999). For the third scenario, accretion scenario, it was proposed to explain the origin of the polar ring galaxies. The polar ring galaxies contain the host galaxies and the outer rings with gas and stars which orbit nearly perpendicular to the plane of the host galaxies. This type of galaxy is believed to be formed when the material from another galaxy or intergalactic medium is accreted onto the host galaxy. A possible prototype of this scenario is the formation of NGC 4650A (Bournaud & Combes 2003).

Recently, a catalog and imaging atlas of 104 P-type galaxies are presented in Madore, Nelson & Petrillo (2009). Our goal is to examine the formation of P-type ring galaxies shown in that catalog. We will determine the locations, widths, and density contrasts of rings and study the evolution of these quantities. In this project, we focus on axis-symmetric rings as in AM 2302-322.

2. The Model

In this study, the target and the intruder are assumed to be a disc galaxy and a dwarf galaxy respectively. We investigate the response of collisions between the target disc galaxy, consisting of the stellar disc and the dark matter halo, and the less massive dwarf galaxy, containing the dark matter halo and the stellar component. The stellar disc and the dark matter halo of the target galaxy have the same density profiles as in Hernquist (1993). The intruder dwarf galaxy comprises the dark matter halo and stellar part with Plummer spheres (Binney & Tremaine 1987; Read *et al.* 2006). The simulation are carried out with the parallel tree-code GADGET (Springel, Yoshida & White 2001) and the softening length are set to be 0.15 kpc and 0.09 kpc for the target disc galaxy and the intruder dwarf galaxy, respectively.

The initial positions of particles can be easily determined according to the above given density profiles. However, assigning initial velocities to each particle will need more complicated techniques. For the spherically symmetric systems, such as the dark matter halo of the disc galaxy and both components of the dwarf galaxy, the initial velocities are calculated from the phase-space distribution functions which can be obtained from Eddington's formula (Binney & Tremaine 1987). For non-spherically symmetric systems, such as the stellar disc, the velocities are determined from the moments of the collisionless Boltzmann equation as in Hernquist (1993) because the analytical phase-space distribution function is not available.

Throughout this paper, the following system of units is used: the gravitational constant G is 43,007.1, the unit of length is 3 kpc, the unit of mass is $10^{10} M_\odot$ and the unit of time is 5.09×10^9 yr. Using the above units, for the disc galaxy, the disc mass, M_d, is 5.6, the disc radial scale length, h, is 3.5, the disc vertical scale length, z_0, is 0.7, the halo mass, M_h, is 32.48, the halo core radius, r_c, is 3.5 and the halo tidal radius, r_t, is 35.0. The disc galaxy has a total of 340,000 particles, i.e. 290,000 dark matter particles and 50,000 stellar particles in the disc. The dynamical time, T_{dyn}, can be defined by the velocity, $v_{1/2}$, of a test particle at disc's half-mass radius, $R_{1/2} = 5.95$, that is $T_{dyn} \equiv 2\pi R_{1/2}/v_{1/2} = 0.174$, where $v_{1/2} = 214.77$.

We combine the dark matter halo and the stellar disc to set up the disc galaxy after constructing these two components independently. These two components of the disc galaxy influence each other and then approach to a new equilibrium. According to the virial theorem, when the disc galaxy is in equilibrium, the value of $2K/|U|$ should be around one, where K and U are total kinetic energy and total potential energy, respectively. Hereafter, $2K/|U|$ is called the *virial ratio*. The disc galaxy approaches a new equilibrium at $t = 15T_{dyn}$ and the energy conservation is fulfilled because the total energy variation is 0.082 per cent. Hence, the disc galaxy at $t = 15T_{dyn}$ will be used to represent the target disc galaxy at the beginning of the collision simulation.

For the dwarf galaxy, the total mass is 9.52, which is a quarter of the mass of the disc galaxy, and the mass to light ratio is 5. The scale lengths of the dark matter halo and the stellar component are 3.0 and 1.5 respectively. The total number of particles is 85,000, containing 68,000 dark matter particles and 17,000 stellar particles. The mass of each particle in our simulation is the same. Because the dark matter halo and the stellar component of the dwarf galaxy are both spherically symmetric and can be set up together, the whole dwarf galaxy with two components is in equilibrium initially. This dwarf galaxy will be the intruder galaxy in the collision simulation.

3. The Evolution

The target disc galaxy and the dwarf galaxy are set up as mentioned previously. The initial separation of these two galaxies is 200, which is far enough to make sure two galaxies are well separated, and the initial relative velocity $v_i = 286.2$. To fix the center of mass of the collision system at the origin, the disc galaxy is located at $(x, y, z) = (0, 0, -40)$ with $(v_x, v_y, v_z) = (0, 0, 57.24)$ and the dwarf galaxy is located at $(x, y, z) = (0, 0, 160)$ with $(v_x, v_y, v_z) = (0, 0, -228.96)$.

To have more detail about the evolution of the galaxies in the simulation, the time interval between each snapshot is set to be $T_s = T_{dyn}/2$. The virial ratio, $2K/|U|$, of the collision system as a function of time during $t = 0$ - $80T_s$ is shown in Fig. 1 (a). Owing to the strong interaction between the disc galaxy and the dwarf galaxy during $t = 6$ - $22T_s$, the virial ratio goes away from one. The first peak and the second peak of $2K/|U|$ are around $t = 7T_s$ and $t = 21T_s$, respectively, while the minimum of $2K/|U|$ is around $t = 15T_s$.

In addition, Fig. 1 (b) shows the centers of mass of both galaxies during the collision. The disc and dwarf galaxies have a close encounter at $t = 7T_s$ and then separate at $t = 15T_s$. After $6T_s$, i.e. at $t = 21T_s$, two galaxies approach one another again because of the gravitational force. In a comparison of Fig. 1 (a) and (b), it is clear that two peaks of $2K/|U|$ at $t = 7T_s$ and $t = 21T_s$ in Fig. 1 (a) imply two encounters between the disc galaxy and the dwarf galaxy at these times. Moreover, the minimum of $2K/|U|$ at $t = 15T_s$ between two peaks of $2K/|U|$ infers that there is a separation between two encounters.

Fig. 2 shows the time evolution of the stellar disc and the stellar component of the dwarf galaxy in the edge-on view. Initially, two galaxies are separated by 200. During the early phases of the encounter, from $t = 0$ to $t = 6T_s$, two galaxies approach each other and then encounter at $t = 7T_s$. Later on, the gravitational force from the dwarf galaxy warps the stellar disc upward and downward between $t = 8T_s$ to $t = 21T_s$. Furthermore, since many particles of the dwarf galaxy escape from the center of the dwarf galaxy, the dwarf galaxy starts to expand after the encounter. Consequently, the stellar disc becomes a layered appearance at $t = 24T_s$ and its thickness increases with time, e.g. at $t = 27T_s$. Most particles of the dwarf galaxy are concentrated around the stellar disc, but some of them extend to a distance about 300.

4. The Ring

Fig. 3 (a) shows the surface density of all stellar particles projected along the collision axis (i.e. onto x-y plane) in the collision system and the fitted disc surface density profile at $t = 0$ and after the encounters, at $t = 7T_s$, $t = 8T_s$ and $t = 9T_s$. To obtain an analytic curve to fit the surface density profiles, we first integrate the density profile of the disc to yield the surface density as

$$\Sigma(R) = \frac{M_d}{2\pi h^2} \exp\left(\frac{-R}{h}\right). \tag{4.1}$$

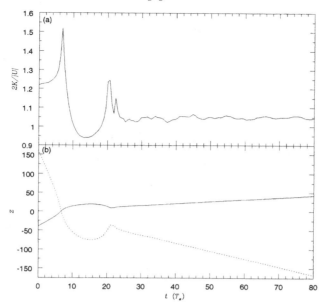

Figure 1. The evolution of the collision simulation. (a) The virial ratio $2K/|U|$ of the collision system as a function of time. (b) The center of mass of each galaxy in the collision simulation during the whole simulation time. The solid line and the dotted line are for the locations of the disc galaxy and the dwarf galaxy, respectively. The unit of t is $T_s = T_{dyn}/2$.

We further modify it to be

$$\Sigma(R) = \alpha \frac{M_d}{2\pi h^2} \exp\left(\frac{-\beta R}{h}\right),\tag{4.2}$$

where two additional fitting parameters, α and β, are introduced. The best-fitting parameter set (α, β) is determined by minimizing χ^2 (Wall & Jenkins 2003) as

$$\chi^2 = \sum_{i=1}^{k} \frac{(Q_i - E_i)^2}{E_i}\tag{4.3}$$

where Q_i is the surface density at a radius R_i and E_i is the $\Sigma(R_i)$ with the fitting parameter set (α, β). In the χ^2 fitting procedure, we neglect the surface density beyond the radius R_{end}, where the surface density Q_{end} is zero.

In Fig. 3 (a), the filled circles, open circles, crosses and open triangles represent the surface density at $t = 7$, $t = 8T_s$, $t = 9T_s$ and $t = 10T_s$, respectively. The solid, dot, short-dashed and long-dashed lines are the best-fitting profiles with the parameter set $(\alpha, \beta) = (3.2, 1.5)$, $(2.1, 1.1)$ and $(1.7, 1.0)$, $(1.4, 0.9)$ for filled circles, open circles, crosses and open triangles, respectively. Fig. 3 (b) shows the same thing on a logarithmic scale. From Fig. 3 (a) and (b), it is clear that after the encounter, a ring-like feature is evident at $t = 7T_s$ and then propagates outward as an expanding ring after $t = 7T_s$. In addition, the rings are also produced at $t = 12T_s$ to $t = 15T_s$ and $t = 21T_s$ to $t = 22T_s$ due to the perturbation after the first encounter at $t = 7T_s$ and the second encounter at $t = 21T_s$. The best-fitting parameter sets, (α, β), at these times are shown in Table 1.

To determine the position of the ring, the density contrast is defined as $\frac{\Delta\Sigma(R)}{\Sigma(R)}$, where $\Delta\Sigma(R)$ is the difference between the surface density and the best-fitting profile at a radius R, i.e. $\Sigma(R)$. The $\Delta\Sigma(R)$ and the density contrast, $\frac{\Delta\Sigma(R)}{\Sigma(R)}$, as a function of radius R at

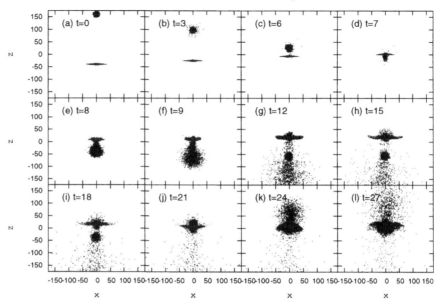

Figure 2. The time evolution of the stellar disc and the stellar component of the dwarf galaxy in the edge-on view. The gray dots and the black dots represent the stellar disc and the stellar part of the dwarf galaxy, respectively. The unit of time is $T_s = T_{dyn}/2$

Table 1. The best-fitting parameter set (α, β) of the surface densities.

	(α, β)	a	b	L	W
$t = 7T_s$	(3.2, 1.5)	3.55	4.35	3.95	0.80
$t = 8T_s$	(2.1, 1.1)	14.05	17.25	15.65	3.20
$t = 9T_s$	(1.7, 1.0)	21.35	27.25	24.30	5.90
$t = 10T_s$	(1.4, 0.9)	27.45	37.25	32.35	9.80
$t = 12T_s$	(1.3, 0.9)	17.25	20.05	18.65	2.80
$t = 13T_s$	(1.3, 0.9)	19.75	23.05	21.40	3.30
$t = 14T_s$	(1.4, 1.0)	21.85	29.45	25.65	7.60
$t = 15T_s$	(1.4, 1.0)	23.15	30.85	27.00	7.70
$t = 21T_s$	(2.1, 1.1)	9.55	12.15	10.85	2.60
$t = 22T_s$	(2.0, 1.0)	19.95	27.15	23.55	7.20

$t = 7T_s$, $t = 8T_s$, $t = 9T_s$ and $t = 10T_s$ are shown in Fig. 3 (c) and (d), respectively. The different symbols (i.e. the filled circles, open circles, crosses and open triangles) represent the $\Delta\Sigma(R)$ and the density contrast, $\frac{\Delta\Sigma(R)}{\Sigma(R)}$, at different times as described in Fig. 3 (a).

From the density contrast as shown in Fig. 3 (d), we can determined the region around a ring-like feature in which all the density contrasts are larger than zero. Then, the average of density contrasts in this region, $(\frac{\Delta\Sigma}{\Sigma})_{av}$, can be calculated and the corresponding radius a and b are determined at which the density contrast reaches $(\frac{\Delta\Sigma}{\Sigma})_{av}$. Therefore, the boundaries of the ring are defined to be at radius a and b. The location (L) of the ring is $L = \frac{a+b}{2}$ and the width (W) of the ring is $W = |b - a|$.

The boundaries, the location, and the width of the ring at different simulation times are shown in Table 1.

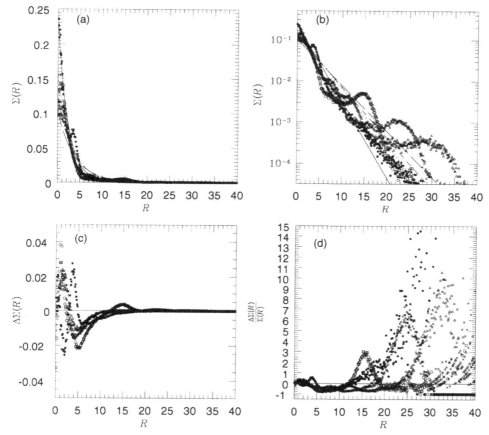

Figure 3. (a) The surface density profiles and the best-fitting profiles at the different times. The filled circles, open circles, crosses and open triangles represent the surface density profiles at $t = 7T_s$, $t = 8T_s$, $t = 9T_s$ and $t = 10T_s$, respectively. The solid, dot, short-dashed and long–dashed lines are the best-fitting profiles for the above four different kinds of points, respectively. (b) The same thing as shown in (a) on a logarithmic scale. (c) $\Delta\Sigma(R)$, which is the difference between the surface density and the best-fitting profile, as a function of radius at $t = 7T_s$ (filled circles), $t = 8T_s$ (open circles), $t = 9T_s$ (crosses) and $t = 10T_s$ (open triangles). (d) The density contrast as a function of radius at $t = 7T_s$ (filled circles), $t = 8T_s$ (open circles), $t = 9T_s$ (crosses) and $t = 10T_s$ (open triangles).

Fig. 4 shows the characteristics of the ring at different simulation times. The average of the density contrast in the ring region as a function of time and the location of the ring as a function of time are plotted in Fig. 4 (a) and (b). Fig. 4 (b) shows three generation rings form at $t = 7T_s$, $t = 12T_s$ and $t = 21T_s$, and then move outward. As the ring keeps going outward, the average of density contrast increases as shown in Fig. 4 (a). In addition, the width of the ring as a function of time is displayed in Fig. 4 (c) and shows that the width of the ring is wider when the density contrast is higher. When the ring is cut with lines along radial directions into 360 equal-size angular bins, the average number of particles in angular bins is shown in Fig. 4 (d), where error bars show the variation

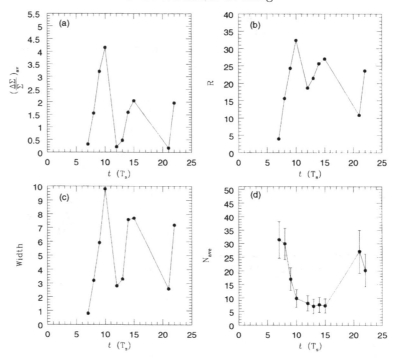

Figure 4. (a) The average of the density contrast in the ring region as a function of time. (b) The location of the ring as a function of time. (c) The width of the ring as a function of time. (d) The average number of particles in angular bins. The error bar shows the variation in different angular bins.

in different angular bins. This panel confirms that the structure of the ring at different times is smooth without a large clump.

5. Concluding Remarks

A catalog and imaging atlas of P-type galaxies are presented by Madore, Nelson & Petrillo (2009) recently. Most P-type ring galaxies have been interpreted as the result of a head-on collision between a disc galaxy and an intruder galaxy, i.e. the collision scenario. We have performed the simulation of the collision between a disc galaxy and an intruder dwarf galaxy to test the collision scenario, and developed a method to determine the location, width and density contrast of a ring structure. Our simulation shows that a ring-like feature could form after the encounters between two galaxies. Because the dwarf galaxy gets destroyed, we produce a ring galaxy without companions. In comparing our simulation to the AM 2302-322 ring galaxy (Madore, Nelson & Petrillo 2009), we find that the density contrast of the ring in our simulation at $t = 7T_s$ is in the same order as the density contrast of AM 2302-322. We conclude that a P-type ring galaxy, such as AM 2302-322, can be formed through the galactic collisions.

References

Arp H. C. & Madore B. F. 1987, *A Catalogue of Southern Peculiar Galaxies and Associations*, Cambridge Univ. Press, Cambridge

Binney J. & Tremaine S. 1987, *Galactic Dynamics (Princeton Univ. Press)*

Bournaud F. & Combes F. 2003, *A&A*, 401, 817

Bushouse H. A. & Stanford S. A. 1992, *ApJS*, 79, 213

Buta R., Purcell G. B., Cobb M. L., Crocker D. A., Rautiainen P., & Salo H. 1999, *AJ*, 117, 778

Elmegreen D. M. & Elmegreen B. G. 2006, *ApJ*, 651, 676

Few J. M. A. & Madore B. F. 1986, *MNRAS*, 222, 673

Fosbury R. A. E. & Hawarden T. G. 1977, *MNRAS*, 178, 473

Hernquist L. 1993, *ApJS*, 86, 389

Higdon J. L. 1995, *ApJ* 455, 524

Lynds R. & Toomre A. 1976, *ApJ*, 209, 382

Madore B. F., Nelson E., & Petrillo K. 2009, *ApJS*, 181, 572

Marston A. P. & Appleton P. N. 1995, *AJ*, 109, 1002

Read, J. I., Wilkinson, M. I., Evans, N. W., Gilmore, G., & Kleyna, J. T. 2006, *MNRAS*, 367, 387

Springel V., Yoshida N., & White S. D. M.. 2001, *NewA*, 6, 79

Theys J. C. & Spiegel E. A. 1976, *ApJ*, 208, 650

Wall J. V. & Jenkins C. R. 2003, *Practical Statistics for Astronomers, Cambridge Univ. Press, Cambridge*

Zwicky F. 1941, *Applied Mechanics (von Kármán volume)*, p.137

Discussion

F. COMBES: In your ring simulations, should you take into account the star formation triggered by the ring wave? the contrast of the ring could be much stronger.

Y.-T. WU: If we consider the star formation in the simulations, it might increase the density contrast of the ring. However, the simulations here can already explain some observational data.

Astrophysical Dynamics: From Stars to Galaxies
Proceedings IAU Symposium No. 271, 2010
N. Brummell, A.S. Brun, M. S. Miesch & Y. Ponty, eds.
© International Astronomical Union 2011
doi:10.1017/S1743921311017510

Modeling mass independent of anisotropy

Joe Wolf

Center for Cosmology, Department of Physics & Astronomy
University of California, Irvine, CA 92697
wolfj@uci.edu

Abstract. By manipulating the spherical Jeans equation, Wolf *et al.* (2010) show that the mass enclosed within the 3D deprojected half-light radius $r_{1/2}$ can be determined with only mild assumptions about the spatial variation of the stellar velocity dispersion anisotropy as long as the projected velocity dispersion profile is fairly flat near the half-light radius, as is typically observed. They find $M_{1/2} = 3\,G^{-1}\,\langle\sigma_{los}^2\rangle\,r_{1/2} \simeq 4\,G^{-1}\,\langle\sigma_{los}^2\rangle\,R_e$, where $\langle\sigma_{los}^2\rangle$ is the luminosity-weighted square of the line-of-sight velocity dispersion and R_e is the 2D projected half-light radius. This finding can be used to show that all of the Milky Way dwarf spheroidal galaxies (MW dSphs) are consistent with having formed within a halo of mass approximately $3\times10^9\,M_\odot$, assuming a ΛCDM cosmology. In addition, the dynamical I-band mass-to-light ratio $\Upsilon_{1/2}^I$ vs. $M_{1/2}$ relation for dispersion-supported galaxies follows a U-shape, with a broad minimum near $\Upsilon_{1/2}^I \simeq 3$ that spans dwarf elliptical galaxies to normal ellipticals, a steep rise to $\Upsilon_{1/2}^I \simeq 3{,}200$ for ultra-faint dSphs, and a more shallow rise to $\Upsilon_{1/2}^I \simeq 800$ for galaxy cluster spheroids.

Keywords. galaxies: formation, kinematics and dynamics, Local Group

1. Introduction

Mass determinations for dispersion-supported galaxies based on only line-of-sight velocity measurements suffer from an uncertainty associated with not knowing the intrinsic 3D velocity dispersion. The difference between tangential and radial velocity dispersions is quantified by the stellar velocity dispersion anisotropy, β. Many questions in galaxy formation are affected by our ignorance of β, including the ability to quantify the dark matter content in the outer parts of elliptical galaxies [Romanowsky *et al.* (2003), Dekel *et al.* (2005)], to measure the mass profile of the Milky Way from stellar halo kinematics [Battaglia *et al.* (2005), Dehnen *et al.* (2006)], and to infer accurate mass distributions in dwarf spheroidal galaxies (dSphs) [Gilmore *et al.* (2007), Strigari *et al.* (2007b)].

Wolf *et al.* (2010) (hereafter W10) used the spherical Jeans equation to show that for each dispersion-supported galaxy, there exists one radius within which the integrated mass as inferred from the line-of-sight velocity dispersion is largely insensitive to β, and that for a wide range of stellar distributions, this radius is approximately the 3D deprojected half-light radius $r_{1/2}$:

$$M_{1/2} \equiv M(r_{1/2}) \simeq 3\,G^{-1}\,\langle\sigma_{los}^2\rangle\,r_{1/2},$$
$$\simeq 4\,G^{-1}\,\langle\sigma_{los}^2\rangle\,R_e,$$
$$\simeq 930\,\left(\frac{\langle\sigma_{los}^2\rangle}{\mathrm{km^2\,s^{-2}}}\right)\,\left(\frac{R_e}{\mathrm{pc}}\right)\,M_\odot. \qquad (1.1)$$

$\langle\sigma_{los}^2\rangle$ is the luminosity-weighted square of the line-of-sight velocity dispersion. In the second line we have used $R_e \simeq (3/4)\,r_{1/2}$ for the 2D projected half-light radius. This approximation is accurate to better than 2% for exponential, Gaussian, King, Plummer, and Sérsic profiles (see Appendix B of W10 for useful fitting formulae).

2. The Spherical Jeans Equation

Given the relative weakness of the scalar virial theorem as a mass estimator (see Section 2 of W10), we turn to the spherical Jeans equation. It relates the tracer velocity dispersion and tracer number density $n_\star(r)$ of a spherically symmetric, dispersion-supported, collisionless stationary system to its total gravitating potential $\Phi(r)$, under the assumption of dynamical equilibrium with no streaming motions:

$$-n_\star \frac{d\Phi}{dr} = \frac{d(n_\star \sigma_r^2)}{dr} + 2\frac{\beta \, n_\star \sigma_r^2}{r}. \tag{2.1}$$

$\sigma_r(r)$ is the radial velocity dispersion of the stars/tracers and $\beta(r) \equiv 1 - \sigma_t^2/\sigma_r^2$ measures the velocity anisotropy, where the tangential velocity dispersion $\sigma_t = \sigma_\theta = \sigma_\phi$. It is informative to rewrite the implied total mass profile as

$$M(r) = \frac{r \, \sigma_r^2}{G} \left(\gamma_\star + \gamma_\sigma - 2\beta \right), \tag{2.2}$$

where $\gamma_\star \equiv -d \ln n_\star / d \ln r$ and $\gamma_\sigma \equiv -d \ln \sigma_r^2 / d \ln r$. Given line-of-sight kinematics, the only term on the right-hand side of Equation 2.2 that can be determined by observations is $\gamma_\star$, which follows from the projected surface brightness profile under an assumption about how it is related to the projected stellar number density $\Sigma_\star(R)$. Via an Abel inversion $n_\star$ is mapped in a one-to-one manner with the spherically deprojected observed surface brightness profile by assuming that $n_\star$ traces the light density.

Line-of-sight kinematic data provides the projected velocity dispersion profile $\sigma_{\rm los}(R)$. As first shown by Binney & Mamon (1982), in order to use the Jeans equation one must relate $\sigma_{\rm los}$ to σ_r:

$$\Sigma_\star \sigma_{\rm los}^2(R) = \int_{R^2}^{\infty} n_\star \sigma_r^2(r) \left[1 - \frac{R^2}{r^2} \beta(r) \right] \frac{dr^2}{\sqrt{r^2 - R^2}}. \tag{2.3}$$

It is clear then that a significant degeneracy associated with using the observed $\Sigma_\star(R)$ and $\sigma_{\rm los}(R)$ profiles exists in trying to determine an underlying mass profile $M(r)$ at any radius, as uncertainties in β will affect both the relationship between $M(r)$ and σ_r in Equation 2.2 and the mapping between σ_r and $\sigma_{\rm los}$ in Equation 2.3.

The technique of W10 for handling the β degeneracy in order to provide a fair representation of the allowed mass profile given a set of observables is to consider general parameterizations for $M(r)$ and $\beta(r)$ and then to undertake a maximum likelihood analysis to constrain all possible parameter combinations. By using such a strategy, W10 derive meaningful mass likelihoods for a number of dispersion-supported galaxies with line-of-sight velocity data sets.

The stellar velocity dispersion anisotropy can be modeled as a three-parameter function

$$\beta(r) = (\beta_\infty - \beta_0) \frac{r^2}{r^2 + r_\beta^2} + \beta_0, \tag{2.4}$$

and the total mass density distribution can be described using the six-parameter function

$$\rho_{\rm tot}(r) = \frac{\rho_s \, e^{-r/r_{\rm cut}}}{(r/r_s)^\gamma [1 + (r/r_s)^\alpha]^{(\delta - \gamma)/\alpha}}. \tag{2.5}$$

For their marginalization, W10 adopt uniform priors over the following ranges: $\log_{10}(r_{1/2}/5) < \log_{10}(r_\beta) < \log_{10}(r_{\rm lim})$; $-10 < \beta_\infty < 0.91$; $-10 < \beta_0 < 0.91$; $\log_{10}(r_{1/2}/5) < \log_{10}(r_s) < \log_{10}(2\,r_{\rm high})$; $0 < \gamma < 2$; $3 < \delta < 5$; and $0.5 < \alpha < 3$, where $r_{\rm lim}$ is the truncation radius for the stellar density. The variable $r_{\rm cut}$ allows the dark matter halo

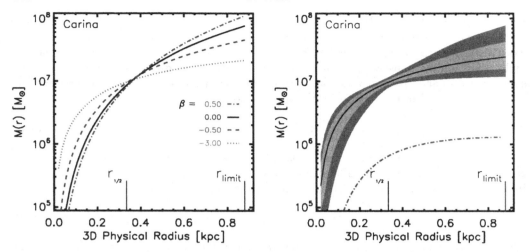

Figure 1. Figure 1 from W10. See the text for details. *Right:* The green dot-dashed line represents the contribution of mass from the stars, assuming a stellar V-band mass-to-light ratio of 3 $M_\odot/L_\odot$.

profile to truncate beyond the stellar extent and the uniform prior $\log_{10}(r_{\mathrm{lim}}) < \log_{10}(r_{\mathrm{cut}}) < \log_{10}(r_{\mathrm{high}})$ is adopted. For distant galaxies W10 use $r_{\mathrm{high}} = 10\, r_{\mathrm{lim}}$ and for satellite galaxies of the Milky Way they set r_{high} equal to the Roche limit for a $10^9\, M_\odot$ point mass. In practice, this allowance for r_{cut} is not important when focusing on integrated masses within the stellar radius.

W10 apply their marginalization procedure to resolved kinematic data for MW dSphs, MW globular clusters, and elliptical galaxies. Since MW dSphs and globular clusters are close enough for individual stars to be resolved, the joint probability of obtaining each observed stellar velocity given its observational error and the predicted line-of-sight velocity dispersion from Equations 2.1 and 2.3 is considered. In modeling the line-of-sight velocity distribution for any system, the observed distribution is a convolution of the intrinsic velocity distribution, arising from the distribution function, and the measurement uncertainty from each individual star. Given that line-of-sight velocity distributions of dispersion-supported systems are often well-described by a Gaussian, the probability of obtaining a set of line-of-sight velocities $\mathcal{V}$ given a set of model parameters $\mathcal{M}$ is described by the likelihood

$$P(\mathcal{V}|\mathcal{M}) = \prod_{i=1}^{N} \frac{1}{\sqrt{2\pi(\sigma_{\mathrm{th.i}}^2 + \epsilon_i^2)}} \exp\left[-\frac{1}{2}\frac{(\mathcal{V}_i - \bar{v})^2}{\sigma_{\mathrm{th.i}}^2 + \epsilon_i^2}\right]. \tag{2.6}$$

The product is over the set of N stars, where $\bar{v}$ is the average velocity of the galaxy. As expected, the total error at a projected position is a sum in quadrature of the theoretical intrinsic dispersion, $\sigma_{\mathrm{th.i}}(\mathcal{M})$, and the measurement error ϵ_i. The posterior probability distribution for the mass at any radius can be generated by multiplying the likelihood by the prior distribution for each of the nine $\rho_{\mathrm{tot}}(r)$ and $\beta(r)$ parameters as well as the observationally derived parameters and associated errors that yield $n_\star(r)$ for each galaxy, including uncertainties in distance. We then integrate over all model parameters, including $\bar{v}$, to derive a likelihood for mass. Following Martinez *et al.* (2009), a Markov Chain Monte Carlo technique is used in order to perform the required ten to twelve dimensional integral.

For elliptical galaxies that are located too far for individual stellar spectra to be obtained, the resolved dispersion profiles are analyzed with the likelihood

$$P(\mathcal{D}|\mathcal{M}) = \prod_{i=1}^{N} \frac{1}{\sqrt{2\pi}\epsilon_i} \exp\left[-\frac{1}{2}\frac{(\mathcal{D}_i - \sigma_{th,i})^2}{\epsilon_i^2}\right], \tag{2.7}$$

where the product is over the set of N dispersion measurements $\mathcal{D}$, and ϵ_i is the reported error of each measurement.

3. Minimizing the Anisotropy Degeneracy

As discussed in W10, the degeneracy between the anisotropy and the integrated mass will be minimized at an intermediate radius within the stellar distribution. Such an expectation follows from considering the relationship between σ_r and σ_{los}.

At the observed center of a spherical, dispersion-supported galaxy ($R = 0$), line-of-sight velocities project onto the radial component with $\sigma_{los} \sim \sigma_r$, while at the edge of the system ($R = r_{lim}$), line-of-sight velocities project onto the tangential component with $\sigma_{los} \sim \sigma_t$. As an example, consider an intrinsically isotropic galaxy ($\beta = 0$). If this system is analyzed using line-of-sight kinematics under the false assumption that $\sigma_r < \sigma_t$ ($\beta < 0$) at all radii, then the total velocity dispersion at $r \simeq 0$ would be overestimated while the total velocity dispersion at $r \simeq r_{lim}$ would be underestimated. Conversely, if one were to analyze the same galaxy under the assumption that $\sigma_r > \sigma_t$ ($\beta > 0$) at all radii, then the total velocity dispersion would be underestimated near the center and overestimated near the galaxy edge. Thus, some intermediate radius should exist where attempting to infer the enclosed mass from only line-of-sight velocities is minimally affected by the unknown value of β.

These qualitative expectations are quantitatively displayed in Figure 1, where inferred mass profiles for the Carina dSph galaxy for several choices of constant β are shown. The right-hand panel shows the same data analyzed using the complete likelihood analysis, where the fairly general $\beta(r)$ profile presented in Equation 2.4 was marginalized over. The dataset is discussed in W10, where the average velocity error is $\sim 3\,\mathrm{km\,s^{-1}}$. Each line in the left panel of Figure 1 shows the median likelihood of the cumulative mass value at each radius for the value of β indicated. The 3D half-light radius and the limiting stellar radius are marked for reference. As expected, forcing $\beta < 0$ produces a systematically higher (lower) mass at a small (large) radius compared to $\beta > 0$. Thus, this requires that every pair of $M(r)$ profiles analyzed with different assumptions about β cross at some intermediate radius. Somewhat remarkable is that there exists a radius where every pair approximately intersects. We see that this radius is very close to the deprojected 3D half-light radius $r_{1/2}$. The right-hand panel in Figure 1 shows the full mass likelihood as a function of radius, with the shaded bands illustrating the 68% and 95% likelihood contours. The likelihood contour also pinches near $r_{1/2}$, as the data preferentially constrain this mass value.

By examining each of the well-sampled dSph kinematic data sets in more detail, W10 finds that the measurement errors, rather than the β uncertainty, always dominate the errors on the mass near $r_{1/2}$, while the mass errors at *both* smaller and larger radii are dominated by the β uncertainty (and thus are less affected by measurement error). To analytically describe this effect, let us briefly examine the Jeans equation in the context of observables.

Consider a dispersion-supported stellar system where $\Sigma_\star(R)$ and $\sigma_{los}(R)$ are determined accurately by observations, such that any viable solution will keep the quantity

$\Sigma_\star(R)\,\sigma^2_{los}(R)$ fixed to within allowable errors. With this in mind, W10 show that Equation 2.3 can be rewritten in a form that is invertible, isolating into a kernel the integral's R-dependence:

$$
\begin{aligned}
\Sigma_\star \sigma^2_{los}(R) &= \int_{R^2}^{\infty} n_\star \sigma^2_r(r)\left[1 - \frac{R^2}{r^2}\beta(r)\right]\frac{dr^2}{\sqrt{r^2 - R^2}} \\
&= \int_{R^2}^{\infty} \frac{n_\star \sigma^2_r}{r^2}\frac{(1-\beta)r^2 + \beta(r^2 - R^2)}{\sqrt{r^2 - R^2}}dr^2 \\
&= \int_{R^2}^{\infty} \frac{n_\star \sigma^2_r(1-\beta)}{\sqrt{r^2 - R^2}}dr^2 - \left(\sqrt{r^2 - R^2}\int_{r^2}^{\infty}\frac{\beta n_\star \sigma^2_r}{\tilde{r}^2}d\tilde{r}^2\right)\Bigg|_{R^2}^{\infty} \\
&\quad + \int_{R^2}^{\infty}\left(\int_{r^2}^{\infty}\frac{\beta n_\star \sigma^2_r}{\tilde{r}^2}d\tilde{r}^2\right)\frac{1}{2}\frac{dr^2}{\sqrt{r^2 - R^2}} \\
&= \int_{R^2}^{\infty}\left[\frac{n_\star \sigma^2_r}{(1-\beta)^{-1}} + \int_{r^2}^{\infty}\frac{\beta n_\star \sigma^2_r}{2\tilde{r}^2}d\tilde{r}^2\right]\frac{dr^2}{\sqrt{r^2 - R^2}},
\end{aligned} \tag{3.1}
$$

where an integration by parts was utilized to achieve the third equality. Note that the second term on the third line must be zero under the physically-motivated assumption that the combination $\beta n_\star \sigma^2_r$ falls faster than r^{-1} at large r. †

Because Equation 3.1 is invertible, the fact that the left-hand side is an observed quantity and independent of β implies that the term in brackets must be well determined regardless of a chosen β. This allows one to equate the isotropic integrand with an arbitrary anisotropic integrand:

$$
n_\star \sigma^2_r\big|_{\beta=0} = n_\star \sigma^2_r[1 - \beta(r)] + \int_r^{\infty}\frac{\beta n_\star \sigma^2_r d\tilde{r}}{\tilde{r}}. \tag{3.2}
$$

After taking a derivative with respect to $\ln r$ and subtracting Equation 2.1 the following result is obtained

$$
M(r;\beta) - M(r;0) = \frac{\beta(r)\,r\,\sigma^2_r(r)}{G}\left(\gamma_\star + \gamma_\sigma + \gamma_\beta - 3\right). \tag{3.3}
$$

We remind the reader that $\gamma_\star \equiv -d\ln n_\star/d\ln r$ and $\gamma_\sigma \equiv -d\ln \sigma^2_r/d\ln r$. Likewise, $\gamma_\beta \equiv -d\ln\beta/d\ln r = -\beta'/\beta$, where $'$ denotes a derivative with respect to $\ln r$. The possibility of a radius r_{eq} is revealed by Equation 3.3, where the term in parentheses goes to zero, such that the enclosed mass $M(r_{eq})$ is minimally affected by not knowing $\beta(r)$:

$$
\gamma_\star(r_{eq}) = 3 - \gamma_\sigma(r_{eq}) - \gamma_\beta(r_{eq}). \tag{3.4}
$$

W10 go through additional detailed arguments to justify that $r_{eq} \simeq r_{1/2}$ for systems with relatively flat observed dispersion profiles. We will utilize this finding (Equation 1.1) to perform tests of galaxy formation.

4. Milky Way dwarf spheroidal galaxies

As an example of the utility of $M_{1/2}$ determinations, Figure 2 presents $M_{1/2}$ vs. $r_{1/2}$ for MW dSph galaxies. Relevant parameters for each of the galaxies are provided in Table 1 of W10. The symbol types labeled on the plot correspond to three luminosity bins that span almost five orders of magnitude in luminosity. It is interesting to compare the data

† Appendix A of W10 shows how an Abel inversion can be used to solve for $\sigma_r(r)$ and $M(r)$ in terms of directly observable quantities.

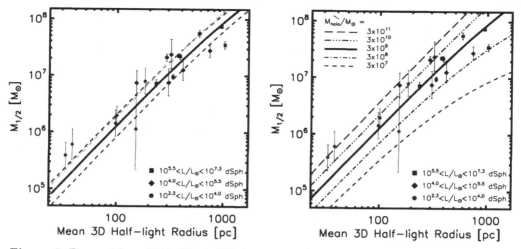

Figure 2. Figure 3 from W10. The half-light masses of the Milky Way dSphs plotted against $r_{1/2}$. See the text for descriptions.

points in Figure 2 to the integrated mass profile $M(r)$ predicted for ΛCDM dark matter field halos of a given M_{halo} mass, which is defined as the halo mass corresponding to an overdensity of 200 compared to the critical density. In the limit that dark matter halo mass profiles $M(r)$ map in a one-to-one way with their M_{halo} mass, then the points on this figure may be used to estimate an associated halo mass for each galaxy.

The solid line in the left panel of Figure 2 shows the mass profile for a NFW dark matter halo [Navarro *et al.* (1997)] at $z = 0$ with a halo mass $M_{halo} = 3 \times 10^9$ $M_\odot$. The median concentration ($c = 11$) predicted by the mass-concentration model [Bullock *et al.* (2001)] updated for WMAP5 ΛCDM parameters [Macciò et al. (2008)] is used. The dashed lines indicate the expected 68% scatter about the median concentration at this mass. The dot-dashed line shows the expected $M(r)$ profile for the same mass halo at $z = 3$ (corresponding to a concentration of $c = 4$), which provides an idea of the spread that would result from the scatter in infall times. An interesting result is that each MW dSph is consistent with inhabiting a dark matter halo of mass $\sim 3 \times 10^9$ $M_\odot$ [Strigari *et al.* (2008)].

The right panel in Figure 2 shows the same data plotted with the median mass profiles for different halo masses. Clearly, the MW dSphs are also consistent with populating dark matter halos of a wide range in M_{halo} above $\sim 3 \times 10^8$ $M_\odot$. This result informs of a very stringent limit on the fraction of the baryons converted to stars in these systems. More importantly, there is no systematic correlation between luminosity and the M_{halo} mass profile that each dSph most closely intersects†. The ultra-faint dSph population (circles) with $L_V < 10^4$ $L_\odot$ is equally likely to be associated with the more massive dark matter halos as are classical dSphs that are more than three orders of magnitude brighter (squares). A simple interpretation of the right-hand panel of Figure 2 shows that the two least luminous satellites (which also have the smallest $M_{1/2}$ and $r_{1/2}$ values) are associated with halos that are *more massive* than any of the classical MW dSphs‡. This behavior is difficult to explain in models constructed to reproduce the Milky Way

† There are hints by Kalirai *et al.* (2010) that this result does not hold for the M31 dSphs.
‡ Martinez *et al.* (2010), Minor *et al.* (2010), and Simon *et al.* (2010) have explored the effects of the inflation of the observed velocity dispersion due to binaries, but find that the faintest system, Segue 1, is still extremely dark-matter dominated.

satellite population, which typically predict a trend between dSph luminosity and halo infall mass. It is possible that a new scale in galaxy formation exists [Strigari *et al.* (2008)] or that there is a systematic bias that makes it difficult to detect low luminosity galaxies which sit within low mass halos [Bullock *et al.* (2010)].

5. Global population of dispersion-supported systems

Figure 3 examines the relationship between the half-light mass $M_{1/2}$ and the half-light luminosity $L_{1/2} = 0.5\,L_I$ for the full range of dispersion-supported stellar systems in the universe: globular clusters, dSphs, dwarf ellipticals, ellipticals, brightest cluster galaxies, and extended cluster spheroids. †

There are several noteworthy aspects to Figure 3, each highlighted in a different way in the three panels. In the middle and right panels the half-light mass-to-light ratios of spheroidal galaxies in the universe demonstrate a minimum at $\Upsilon^I_{1/2} \simeq 2-4$ that spans a broad range of luminosities $L_I \simeq 10^{8.5-10.5}\,L_\odot$ and masses $M_{1/2} \simeq 10^{9-11}\,M_\odot$. It is interesting to note the offset in the average mass-to-light ratios between L_* ellipticals and globular clusters, which may suggest that even within $r_{1/2}$, dark matter may constitute the majority of the mass content of L_* elliptical galaxies. Nevertheless, it seems that dark matter plays a dominant dynamical role ($\Upsilon^I_{1/2} \gtrsim 5$) within $r_{1/2}$ in only the most extreme systems. The dramatic increase in half-light mass-to-light ratios at both larger and smaller luminosity and mass scales is indicative of a decrease in galaxy formation efficiency in the smallest‡ and largest dark matter halos. It is worth mentioning that if ΛCDM is to explain the luminosity function of galaxies a similar trend in the M_{halo} vs. L relationship must exist. While a different mass variable is presented in Figure 3, the resemblance between the two relationships is striking, and generally encouraging for ΛCDM.

One may gain some qualitative insight into the physical processes that drive galaxy formation inefficiency in bright vs. faint systems by considering the $L_{1/2}$ vs. $M_{1/2}$ relation (left panel) in more detail. There exist three distinct power-law regimes $M_{1/2} \propto L^{\aleph}_{1/2}$ with $\aleph > 1$, $\aleph \simeq 1$, and $\aleph < 1$ as mass decreases. Over the broad middle range of galaxy masses, $M_{1/2} \simeq 10^{9-11}\,M_\odot$, mass and light track each other closely with $\aleph \simeq 1$, while faint galaxies obey $\aleph \simeq 1/2$, and bright elliptical galaxies are better described by $\aleph \simeq 4/3$ transitioning to $\aleph \gg 1$ for the most luminous cluster spheroids. One may interpret the transition from $\aleph > 1$ in bright galaxies to $\aleph < 1$ in faint galaxies as a transition between luminosity-suppressed galaxy formation to mass-suppressed galaxy formation. That is, for faint galaxies ($\aleph < 1$), there does not seem to be a low-luminosity threshold in galaxy formation, but rather behavior closer to a threshold (minimum) mass with variable luminosity. For brighter spheroids with $\aleph > 1$, the increased mass-to-light ratios are driven more by increasing the mass at fixed luminosity, suggestive of a maximum luminosity scale (W10).

Going a step further, Figure 3 provides a useful empirical benchmark against which theoretical models can compare. It will be challenging to explain how two of the least luminous MW dSphs have the highest mass-to-light ratios $\Upsilon^I_{1/2} \simeq 3,200$ of any the collapsed structures shown, including intra-cluster light spheroids, which reach values of

---

† These findings were expanded upon by Tollerud *et al.* (2010) to show that all systems which sit embedded within dark matter halos lie along a one dimensional relation within three dimensional space.

‡ McGaugh & Wolf (2010) recently showed that correlations exist when one explores additional observable properties of dSphs.

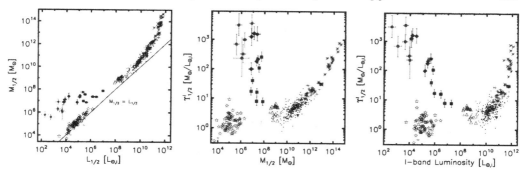

Figure 3. See the text for descriptions. The symbols are linked to galaxy types as follows: Milky Way dSphs (squares, diamonds, circles), galactic globular clusters (stars), dwarf ellipticals (triangles), ellipticals (inverted triangles), brightest cluster galaxies (plus signs), and cluster spheroids (asterisks). See Figure 4 of W10 for references.

$\Upsilon^I_{1/2} \simeq 800$. Not only are the ultra-faint dSphs the most dark matter dominated objects known, given that have even lower baryon-to-dark matter fractions $f_b \sim \Omega_b/\Omega_{dm} \lesssim 10^{-3}$ than galaxy clusters $f_b \simeq 0.1$, W10 points out that ultra-faint dSphs also have higher mass-to-visible light ratios within their stellar extents than even the well-studied galaxy cluster spheroids.

References

Battaglia, G. *et al.* 2005, *MNRAS*, 364, 433

Binney, J. & Mamon, G. A. 1982, *MNRAS*, 200, 361

Bullock, J. S., Kolatt, T. S., Sigad, Y., Somerville, R. S., Kravtsov, A. V., Klypin, A. A., Primack, J. R., & Dekel, A. 2001, *MNRAS*, 321, 559

Bullock, J. S., Stewart, K. R., Kaplinghat, M., Tollerud, E. J., & Wolf, J. 2010, *ApJ*, 717, 1043

Dehnen, W., McLaughlin, D. E., & Sachania, J. 2006, *MNRAS*, 369, 1688

Dekel, A., Stoehr, F., Mamon, G. A., Cox, T. J., Novak, G. S., & Primack, J. R. 2005, *Nature*, 437, 707

Gilmore, G., Wilkinson, M. I., Wyse, R. F. G., Kleyna, J. T., Koch, A., Evans, N. W., & Grebel, E. K. 2007, *ApJ*, 663, 948

Kalirai, J. S. *et al.* 2010, *ApJ*, 711, 671

Łokas, E. L. 2009, *MNRAS*, 394, L102

Macciò, A. V., Dutton, A. A., & van den Bosch, F. C. 2008, *MNRAS*, 391, 1940

Martinez, G. D., Bullock, J. S., Kaplinghat, M., Strigari, L. E., & Trotta, R. 2009, *JCAP*, 06, 014

Martinez, G. D., Minor, Q. E., Bullock, J., Kaplinghat, M., Simon, J. D., & Geha, M. 2010, arXiv1008.4585

McGaugh, S. S. & Wolf, J. 2010, arXiv1003.3448

Minor, Q. E, Martinez, G., Bullock, J., Kaplinghat, M., & Trainor, R. 2010, arXiv1001.1160

Navarro, J. F., Frenk, C. S., & White, S. D. M. 1997, *ApJ*, 490, 493

Romanowsky, A. J., Douglas, N. G., Arnaboldi, M., Kuijken, K., Merrifield, M. R., Napolitano, N. R., Capaccioli, M., & Freeman, K. C. 2003, *Science*, 301, 1696

Simon, J. D. et al. 2010, arXiv1007.4198

Strigari, L. E., Bullock, J. S., Kaplinghat, M., Diemand, J., Kuhlen, M., & Madau, P. 2007b, *ApJ*, 669, 676

Strigari, L. E., Bullock, J. S., Kaplinghat, M., Simon, J. D., Geha, M., Willman, B., & Walker, M. G. 2008, *Nature*, 454, 1096

Tollerud, E. J., Bullock, J. S., Graves, G. J., & Wolf, J. 2010, arXiv1007.5311

Wolf, J., Martinez, G. D., Bullock, J. S., Kaplinghat, M., Geha, M., Muñoz, R. R., Simon, J. D., & Avedo, F. F. 2010, *MNRAS*, 406, 1220

Discussion

E. NTORMOUSI: How would a different dark matter (DM) model (e.g. warm dark matter (WDM)) change the number and size of dwarf galaxies?

J. WOLF: By increasing the temperature of the dark matter particle, one erases substructure on the smallest scales. Thus, there will be less dwarf galaxies, and they would also be less concentrated. If we used a WDM prior instead of a ΛCDM prior to determine field halo masses, then the MW dSphs would have had more massive progenitors.

A. S. BRUN: Is the presence of a DM halo the key in distinguishing between dwarf galaxies and globular clusters (GCs)?

J. WOLF: As shown in Figure 3, dSphs and GCs lie in two very distinct parts of parameter space. We ran mass models for the GCs in identical ways to the dSphs (i.e. we allowed for an extended DM halo) and yet the data strongly rule out any significant presence of DM. In the end of the day, there are two critical components that makes a system a dSph instead of a GC: 1) being deeply embedded within a DM halo and 2) having a wide spread in metallicity.

A. JAN VAN MARLE: Would the upper limit on galaxy mass not depend on composition since a small number of massive stars has a much higher luminosity than a large number of red dwarfs?

J. WOLF: From Figure 3 there does not seem to be an upper limit to the mass of a galaxy. However, since we live in a finite physical universe, the cut-off will be determined by lookback time, amongst other factors. Compare the middle figure to the right figure with regard to the luminosity ceiling. This interesting feature is much more striking.

J. TOOMRE: How you can possibly have thousands of dwarf galaxies and only be finding tens of them?

J. WOLF: This simply has to do with surface brightness limits. Bullock *et al.* (2010) discuss this effect in detail, and show that low mass DM halos (which dominate in sheer number compared to more massive DM halos due to a steep mass function) have very low surface brightnesses.

E. ZWEIBEL: Are these dSphs in equilibrium?

J. WOLF: Signs for tidal disruption include an elongated photometric distribution, stellar streams, and a rise in the observed dispersion profile near the region where the stellar number density approaches the background density. A few of the faint dSphs show some signs of possible disruption, but most of the MW dSphs seem to be well-behaved.

Astrophysical Dynamics: From Stars to Galaxies
Proceedings IAU Symposium No. 271, 2010
N. Brummell, A.S. Brun, M. S. Miesch & Y. Ponty, eds.

© International Astronomical Union 2011
doi:10.1017/S1743921311017522

Galaxy Dynamics: Secular Evolution and Accretion

Francoise Combes

Observatoire de Paris, LERMA and CNRS, 61 Av. de l'Observatoire, F-75014 Paris, France
email: francoise.combes@obspm.fr

Abstract. Recent results are reviewed on galaxy dynamics, bar evolution, destruction and reformation, cold gas accretion, gas radial flows and AGN fueling, minor mergers. Some problems of galaxy evolution are discussed in particular, exchange of angular momentum, radial migration through resonant scattering, and consequences on abundance gradients, the frequency of bulgeless galaxies, and the relative role of secular evolution and hierarchical formation.

Keywords. galaxies: abundances, galaxies: bulges, galaxies: evolution, galaxies: formation, galaxies: kinematics and dynamics, galaxies: spiral, galaxies: structure

1. Disk formation and "viscosity"

There are at least three main scenarios invoked for galaxy formation: the first one, inspired by Eggen et al.(1962) is the monolithic collapse, where the initial gas clouds collapse gravitationally while forming quickly stars before the end of their collapse, so that they end up with a stellar spheroid. Disks must then form later on, through slower gas accretion. The second is the hierarchical scenario, where the first systems to form are disks (the star formation time-scale being longer than the collapse time). Then the interaction and merger of two disks lead to the formation of a spheroid. The third scenario, developped in this talk, considers that disks form first, as in the second scenario. However, disks then may evolve without any merger with other galaxies. Their own internal evolution could also produce a spheroid at the center, through secular evolution. This scenario assumes external gas accretion from filaments of the cosmic web (e.g. Kormendy & Kennicutt 2004).

Spontaneously, a disk evolves through its gravitational instability, producing non-axisymmetric features or waves, that will transfer angular momentum. This is equivalent to an effective kinematic viscosity (Lin & Pringle 1987a). The stability of the disk is ensured at small scales by the equivalent pressure, due to the disordered motions, or velocity dispersions. All scales smaller than the Jeans length $\lambda_J = \sigma t_{ff} = \sigma/(2\pi G\rho)^{1/2}$ are stabilised, where σ is the velocity dispersion, and ρ the density. At large scales, the disk is stabilised by the differential rotation, and the subsequent shear. All scales larger than the critical L_{crit} are stable, where $L_{crit} \sim G\Sigma/\kappa^2$, with Σ the disk surface density, and κ the epicyclic frequency. Scales larger than λ_J and lower than L_{crit} remain unstable, unless the velocity dispersion is increased until $\lambda_J = L_{crit}$: this condition is the Toomre criterion (Toomre 1964). The parameter $Q^2 = \lambda_J/L_{crit}$ must be larger than 1 for stability. From hydrostatic equilibrium in the z-direction, the disk thickness must be $h = h_r \min(1, Q)$, with $h_r = \sigma/\Omega$.

When instabilities occur, they transfer momentum on scale L_{crit}, with time scale Ω^{-1} (or dynamical time-scale); therefore a prescription for effective viscosity is $\nu \sim L_{crit}^2/\Omega^{-1} \sim Q^{-2}h_r^2\Omega$. In addition, it can be shown that, when the viscous time-scale is of the same order as the star-formation time-scale, i.e. $t_\nu \sim t_*$, then an exponential disk is formed (Lin & Pringle 1987b).

2. Bars and gas flows

Bars are the most frequent non-axisymmetries developped in galaxy disks. 75% of spiral galaxies are barred, when viewed in the near-infrared. They form easily in numerical simulations, even when $Q > 1$. Bars are very efficient to trasnfer angular momentum. They are robust and generate long-lived gravity torques, acting on the gas, to concentrate mass towards the center. The amount of gas driven towards the center can be quantified by observations. It can be shown that the sign of the torques change at each Lindblad resonance, in particular they are negative (with respect to the sense of rotation) inside corotation (CR), and positive outside. The amplitude of the torques depends on the bar strength, and on the phase shift between the gas response and the stellar potential. Inside CR, the gas leads the stars, and this corresponds to the characteristic morphology of dust lanes leading the bar, commonly observed in barred galaxies.

From the near-infrared image, representing the old stellar component, i.e. most of the mass in the visible disk, it is possible to deduce the gravitational potential of the galaxy. From the gas map observed either through HI or CO lines, it is then possible to compute the average torque exerted on the gas at each radius, assuming a stationary state at least for one rotation. These computations have shown that typically in a strongly barred galaxy, the gas may lose 30% of its angular momentum at each rotation (Garcia-Burillo *et al.* 2005). The stellar bar receives the angular momentum lost by the gas, and therefore weakens: the bar is a wave with negative momentum, with most the orbits sustaining the bar with high eccentricity. The absorption of angular momentum makes the orbits rounder and the bar weaker. This can lead to the destruction of the bar (e.g. Bournaud & Combes 2002).

The gas driven towards the center could fuel the supermassive black hole, present in every galaxy with bulge, and trigger nuclear activity (AGN). It is however difficult to find direct evidence of correlation between AGN and bars, since the evolution time-scales are widely different: the dynamical time-scale for the gas to flow to the center, and to weaken the bar is a few 10^8 yr, while the AGN active phase is a few 10^7 yr. To account for the large frequency of bars, in spite of their easy weakening and destruction, another mechanism has to re-form bars: this is done through external gas accretion, which replenishes the disk, and makes it unstable again to bar formation. First gas is stalled outside the outer resonance by the positive torques, when the bar is strong. When the bas is weakened, the external gas can then enter and fuel the disk, intermittently.

There is therefore a self-regulated cycle of bar formation and destruction: the bar forms in an unstable cold disk, rich in gas. The strong bar produces gas inflow, which weakens or destroys the bar. Gas accretion can then enter and re-juvenate the disk, and a new bar form. A few percent in mass of gas infall is enough to transform a bar in a lens (Friedli *et al.* 1994, Berentzen *et al.*1998, Bournaud *et al.* 2002, 2005). External gas accretion is essential to drive the secular evolution of galaxy disks, and to maintain spirals and bars frequent enough in galaxies. The observed bar frequency can be used to quantify the accretion rate (Block *et al.* 2002). The required accretion rate corresponds to the baryonic galaxy mass doubling in 7 Gyrs. Cosmological simulations have recently revealed the importance of cold gas inflow in filaments. The predicted rate of gas accretion is similar to what is required to maintain the observed bar frequency (Dekel & Birnboim 2006).

A consequence of angular momentum redistribution over the disk by the bar, is the formation of exponential disks, with radial breaks, and different slopes in the outer parts.

The break is expected at the outer Lindblad resonance (Pfenniger & Friedli 1991). When several patterns develop in the disk, there could be several breaks at differnt radii. Alternatively, breaks may also be the consequence of gas accumulation, due to external accretion, together with a threshold of star formation in gas surface density (Roskar *et al.* 2008). The galaxy disk then form inside out, and the break moves radially outwards. This is confirmed through numerical simulations, which show how the successive spirals and bars transfer some of the inner stars to the outer parts. Age and abundance gradients can then change suddenly in the outer parts, after the break limiting the location of new star formation. This can explain the frequent observations of these surprising reversals (for instance after a radius of 8kpc in M33, Williams *et al.* 2009).

3. Angular momentum transfers, radial migration

There is a large efficiency in the exchange of angular momentum, by resonant scattering at resonances, due to several successive spiral patterns, as shown by Sellwood & Binney (2002), leading to radial migration of stars and gas (cf Figure 1).

The principle is that, at corotation, it is possible to exchange angular momentum L almost without heating. Assuming a nearly steady spiral wave, at least for a few rotations, there is an energy invariant, the Jacobi integral in the Ω_p rotating frame:

$E_J = E - \Omega_p L,$

and the energy and angular momentum are related by $\Delta E = \Omega_p \Delta L$. To separate energy in the radial motions, defining the radial action J_R, it can be shown that:

$\Delta J_R = \frac{\Omega_p - \Omega}{\kappa} \Delta L.$

This shows that at corotation, where $\Omega_p = \Omega$, changes in L do not cause changes in J_R, i.e. no radial heating. Exchanges of L will be most efficient near CR, and the orbits which are almost circular will be preferentially scattered, as shown in Figure 1. This kind of change in L without heating is called churning by Sellwood & Binney (2002), while the change with increase of epicyclic amplitude, through heating, is called blurring, since it has specific signatures in the stellar orbits. Gas contributes to churning, while it is also radially driven inwards.

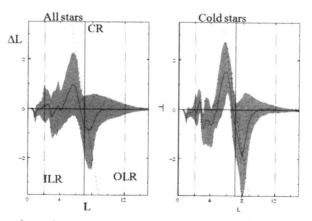

Figure 1. Exchange of angular momentum ΔL, as a function of L, for all stars (left) and only stars with small epicycles (right). The locations of Lindblad resonances ILR, CR and OLR are indicated by vertical lines. The resonance scattering occurs mainly at corotation, from Sellwood & Binney (2002).

In real galaxies, spiral waves are not steady, but transient, and develop with different pattern speeds, so the CR could span a wide range in radii. Radial migration could then involve most of the galaxy disk.

This has important consequences on the chemical evolution, and abundance gradients. The relation between metallicity and age is considerably scattered around a gross trend, as is observed (Shoenrich & Binney 2009). There is also a large scatter in the O/Fe versus Fe/H relation. Radial migration can also produce a thick disk, since stars in the inner disks have higher z-velocity dispersion, due to the higher restoring force of the disk. When migrating outwards, they will feel a smaller restoring force towards the disk, and they will flare. This could explain both the α-enriched and low metallicity property of the thick disk.

The transfer of angular momentum can be multiplied if several patterns exist, with resonances in common. Both bars and spirals can participate to this overlap of resonances. Minchev & Famaey (2010) have shown that due to the non-linear coupling, the region affected by the migration is widened, and the migration rate accelerated by a factor 3. Numerical simulations have confirmed the high migration rate, due to coupled patterns (Minchev et al. 2010). The exchange of angular momemtum has now maxima not only at corotation, but also at the OLR. The presence of the gas increases the rate of angular momentum exchange by about 20%. Metallicity gradients can flatten in less than 1 Gyr

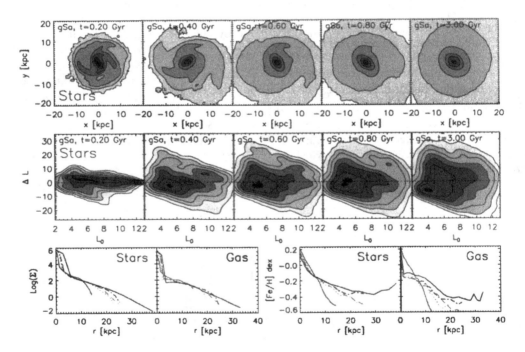

Figure 2. Results of a Tree-SPH simulation, studying the exchange of angular momentum due to resonance overlap of bar and spiral waves. **Top row:** Stellar disk density contours of a giant Sa galaxy in 5 snapshots. **Middle row:** Changes in angular momentum, ΔL as a function of the initial angular momentum, L_0. The locations of the bar CR and OLR are indicated by the dotted and solid lines, respectively. **Bottom row:** The evolution of the radial profiles of surface density (left) and metallicity (right) for the stellar and gaseous disks. The initial disk scale-lengths are indicated by the solid lines. The 5 time steps shown are as in the Top row, indicated by solid red, dotted orange, dashed green, dotted-dash blue and solid purple, respectively, from Minchev et al. (2010).

as shown in Figure 2. Again, the migration is the most important, for almost circular orbits. It can explain the absence of clear age metallicity relation, or age velocity relation.

4. Bulge formation

Several mechanisms are invoked to form bulges or spheroids. First, as suggested by Toomre & Toomre (1972), major mergers of spiral galaxies can form an elliptical, with a remaining disk, according to the relative alignment of their initial spins (e.g. Bendo & Barnes, 2000). In minor mergers, disks are more easily kept and progressively enrich the classical bulge. Secular evolution can also form spheroids: bars and their vertical resonance elevate stars in the center into what is called a pseudo-bulge. The latter is intermediate between a spheroid and a disk. Pseudo-bulges are observed more frequently in late-type galaxies. A third mechanism is provided by clumpy galaxies at high redshift. The massive clumps through dynamical friction, quickly spiral in, and form a bulge. Bulges in fact are too easily formed, and the main problem of the hierarchical scenario is to form bulgeless galaxies. These are observed with unexpected abundance, at $z = 0$.

As for mergers, it is possible that most ellipticals are the result of multiple minor mergers, instead of a 1:1 mass ratio remnant. Numerical simulations have shown that the issue is not the mass ratio of individual mergers, but the total mass accreted, if at least 30-40% of initial mass. A large number of successive minor mergers can form an elliptical galaxy, for instance 50 mergers of 50:1 mass ratio. Given the mass function of galaxies, multiple minor mergers are even more frequent than equal-mass major mergers (Bournaud, Jog Combes 2007a).

The formation of a box-peanut bulge, from disk stars which are in vertical resonance with the bar, i.e. when $\Omega - \nu_z/2 = \Omega_b$, where ν_z is the vertical oscillation frequency, has been known for a long time (Combes & Sanders 1981, Combes *et al.* 1990). During

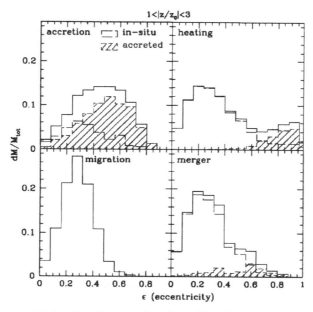

Figure 3. Can the eccentricity distribution of stellar orbits discriminate between the formation scenarios of the thick disk? Here the distributions are compared between four thick disc formation models (accretion, heating, migration, merger), for stars in the range 1.3 scaleheights and cylindrical distance $2 < R/R_d < 3$, from Sales *et al.* (2009).

secular evolution, the resonance can move radially, following the slowing down of the bar, and several peanuts can be formed, larger and larger (Martinez-Valpuesta *et al.* 2006). Even after the bar destruction, the central stellar component will keep its thickness, forming a pseudo-bulge. These spheroids, formed in numerical simulations, correspond perfectly to the kind of pseudo-bulges so frequently observed in late-type galaxies. They have characteristics intermediate between a classical bulge (or an elliptical) and a normal disk (Kormendy & Kennicutt 2004), their main properties are:

- a luminosity distribution μ (in magnitude), with a Sersic index $\mu \sim r^{1/n}$, with $n = 1 - 2$ (exponential disks have $n = 1$, elliptical galaxies $n = 4$ or larger)
- a flattening degree similar to disks, with typical box/peanut shapes, and blue colors
- their kinematics show more rotation support than classical bulges.

Concerning the early evolution of galaxies, their percentage of gas is so high that their disk is unstable and fragments in a few massive clumps. With stars forming in those clumps, this scenario explains the formation of the clumpy galaxies observed at high redshift, which appear as chain galaxies, when edge-on. Numerical simulations then show the rapid formation of exponential disks and bulges, through dynamical friction (Noguchi 1999, Bournaud *et al.* 2007b). The evolution time-scale might even be shorter than with spirals and bars, given the contrast of the structures.

Many spiral galaxies are observed with a thick disk, distinct from their younger thin disk, where contemporary stars are forming. This thick disk could trace some features of the galaxy past formation. At least 4 scenarios have been invoked for the thick disk formation:

- 1) Accretion and disruption of satellites (like in the stellar halo)
- 2) Disk heating due to minor merger
- 3) Radial migration, via resonant scattering
- 4) In-situ formation from thick gas disk (mergers, or clumpy galaxies).

Sales *et al.* (2009) recently proposed that orbit excentricity of stars could help to disentangle among these scenarios.

5. Conclusion

The gravitational stability of disks, the dynamics of spiral perturbations, and the feedback cycle generated may explain the formation of exponential disks, with efficient kinematic viscosity. Angular momentum exchange are also very efficient with bars, which gravity torques drive the gas towards the center, in a few dynamical times. In addition, resonant scattering by spiral waves, constantly re-generated with different pattern speed, are efficient to produce radial migration of stars and gas. When overlap of resonances occurs, between spirals, or between spiral and bars, the migration is strong and spans the whole disk.

If classical bulges and spheroids are commonly the results of mergers, secular evolution can also form the pseudo-bulges, through vertical resonance with the bar. Bulges in early and late-type galaxies could be formed by a combination of these two mechanisms. Presently, it is very difficult to explain the presence of a large number of bulge-less galaxies, the more so as clumpy galaxies formed by gas instabilities at high redshift will also form massive bulges through dynamical friction.

Several scenarios have been invoked for thick disk formation, and a detailed study of stellar orbits could help to discriminate among them.

References

Bendo G. & Barnes J.: 2000, *MNRAS* 316, 315

Berentzen, I., Heller, C. H., Shlosman, I., & Fricke, K. J.: 1998, *MNRAS* 300, 49

Block D., Bournaud F., Combes F. *et al.*: 2002 *A&A* 394, L35

Bournaud F. & Combes F.: 2002, *A&A* 392, 83

Bournaud F., Combes F., & Semelin B.: 2005, *MNRAS* 364, L18

Bournaud F., Jog C., & Combes F.: 2007a, *A&A* 476, 1179a

Bournaud F., Elmegreen, B. G., & Elmegreen, D. M.: 2007b *ApJ* 670, 237

Combes F. & Sanders, R. H.: 1981, *A&A* 96, 164

Combes F., Debbasch F., Friedli D., & Pfenniger D.: 1990, *A&A* 233, 82

Dekel A., & Birnboim Y.: 2006, *MNRAS* 368, 2

Eggen, O. J., Lynden-Bell, D., & Sandage, A. R.: 1962, *ApJ* 136, 748

Friedli, D., Benz, W., & Kennicutt, R.: 1994 *ApJ* 430, L105

Garcia-Burillo S., Combes, F., Schinnerer, E. *et al.*: 2005, *A&A* 441, 1011

Kormendy, J. & Kennicutt, R. C., 2004, *ARAA* 42, 603

Lin D. N. C. & Pringle J. E.: 1987a, *MNRAS* 225, 607

Lin D. N. C. & Pringle J. E.: 1987b, *ApJ* 320, L87

Martinez-Valpuesta, I., Shlosman, I., & Heller, C.: 2006, *ApJ* 637, 214

Minchev I. & Famaey B.: 2010, *MNRAS* in press, arXiv0911.1794

Minchev I., Famaey, B., Combes, F. *et al.*: 2010ib, in press arXiv1006.0484

Noguchi M.: 1999, *ApJ* 514, 77

Pfenniger D. & Friedli D.: 1991, *A&A* 252, 75

Roskar, R., Debattista, V. P., Stinson, G. S. *et al.*: 2008, *ApJ* 675, L65

Sales, L. V., Helmi, A., Abadi, M. G. *et al.*: 2009, *MNRAS* 400, L61

Shoenrich R. & Binney J.: 2009, *MNRAS* 399, 1145

Sellwood G. & Binney J.: 2002, *MNRAS* 336, 785

Toomre A.: 1964, *ApJ* 139, 1217

Toomre A. & Toomre J.: 1972, *ApJ* 178, 623

Williams B. F., Dalcanton, J. J., Dolphin, A. E. *et al.*: 2009, *ApJ* 695, L15

Discussion

J. TOOMRE: Are there ways to assess how much mass of gas is flowing in from the cosmic web to replenish a galaxy? What are the signatures of such replenishment of gas?

F. COMBES: It is difficult to find direct evidence, since the infalling gas is diffuse, too dilute to be detected (either as HI atomic gas at 21cm, ionised or molecular gas). Around our own galaxy, high velocity HI clouds have been observed for a long time, and with simple models giving their distance, they would correspond to an infall of a few solar masses per year. Around external galaxies, the search is difficult, by lack of sensitivity. It is possible that the ubiquitous warps in spiral galaxies come from external accretion. A stream of HI gas emission has been observed in the edge-on galaxy NGC 891 up to 25kpc distance.

But indirectly, it is possible to infer how much is required for the spiral maintenance in galaxies, and in particular the bar frequency in late-type spirals. From the observed bar frequency compared to numerical simulations, this gives a doubling of mass of a typical Sb in 7-10 Gyrs, which corresponds to the right order of magnitude of the cold gas accretion in cosmological simulations. Another indirect way is to quantify the frequency of bulgeless spiral galaxies today. If galaxies acquire their mass only through merging, then there should be massive bulges in all galaxies, and it is not what is observed. At least half of the spiral galaxies did not have any major merger in their life, nor a lot of minor mergers, but should have grown through gas accretion.

E. NTORMOUSI: (About cold flows forming disk galaxies) What is the density / metallicity / speed of these flows? Are they stable to hydrodynamical instabilities? Could they be star-forming?

F. COMBES: They are not dense enough to be star-forming. They are low metallicity and they would indeed be unstable to hydrodynamical instabilities.

J. WOLF: Have you examined what happens with 10 1:10 mergers instead at 50 1:50 mergers? Recent work suggests that 1:10 mergers are more common.

F. COMBES: Yes, we did explore all these mass ratios. Averaged over the geometrical parameters, 10 1:10 mergers have a remnant quite similar to the 50 1:50 remnant.

A. S. BRUN: I did not realize there was so much radial displacement of stars (or the sun). What is exactly the proper motion of stars?

F. COMBES: The velocity dispersion of stars is of the order of 30km/s in average, more for the older stars, and less for the youngest ones. However, the migration is done through spiral in (and out) of the orbits, due to exchange of angular momentum, and is equivalent to a radial velocity of 10km/s.

A. CURIR: How does the result depend on the concentration of the satellites, since more concentrated galaxies will not be disrupted by tidal forces before the final merger?

F. COMBES: We compared the results using satellites having Hernquist's profiles and NFW density profiles, and we did not find significative differences, even if the concentrations of these profiles is noticeable.

Astrophysical Dynamics: From Stars to Galaxies
Proceedings IAU Symposium No. 271, 2010
N. H. Brummell, A. S. Brun, M. S. Miesch & Y. Ponty, eds.
© International Astronomical Union 2011
doi:10.1017/S1743921311017534

KPG 390: A pair of trailing spirals.

P. Repetto[1], M. Rosado[1], R. Gabbasov[1] and I. Fuentes-Carrera[2]

[1]Instituto de Astronomía, Universidad Nacional Autonoma de México (UNAM),
Apdo. Postal 70-264, 04510, México, D.F., México.
email: prepetto@astroscu.unam.mx

[2]Department of Physics, Escuela Superior de Física y Matemáticas, IPN, U.P. Adolfo López
Mateos, C.P. 07738, Mexico city IPN, México.

Abstract. In this study we present scanning Fabry-Perot $H\alpha$ observations of the isolated interacting galaxy pair NGC 5278/79. We derived velocity fields, various kinematic parameters and rotation curves for both galaxies. These kinematical results together with the fact that dust lanes have been detected in both galaxies, as well as the analysis of surface brightness profiles along the minor axis, allowed us to determine univocally that both components of the interacting pair are trailing spirals. We have also estimated the mass of NGC 5278 fitting its rotation curve with a disk-halo component. We have tested three different types of halo (pseudo-isothermal, Hernquist and Navarro Frenk White) and we have obtained that the rotation curve can be fitted either with a pseudo-isothermal, an Hernquist halo or a Navarro Frenk White halo component, although in the first case the amount of dark matter required is about ten times smaller than for the other two halo distributions.

Keywords. Instrumentation: interferometers, methods: data analysis, techniques: image processing, techniques: interferometric, techniques: radial velocities, astrometry, galaxies: interactions, galaxies: kinematics and dynamics, galaxies: spiral, (cosmology): dark matter.

1. Introduction

Interactions and mergers of galaxies are common phenomena in the Universe. Isolated pairs of galaxies represent a relatively easy way to study interactions between galaxies because these systems, from a kinematical point of view, are simpler than associations and compact groups of galaxies, where so many galaxies participate in the interaction process, that it is difficult to discriminate the role of each galaxy in the interaction.

Obtaining kinematical information on interacting galaxies systems is useful to understand the effect the interacting process could have on each of the members of the pair (Marcelin *et al.* (1987), Amram *et al.* (2002), Fuentes-Carrera *et al.* (2007)). Rotation curves determination, for instance, is a very efficient tool to study the mass distribution in spiral galaxies. It allows to discover the significant discrepancy between the luminous mass and the gravitational mass that has led to the assumption of a large amount of dark matter in the Universe (Rubin *et al.* (1976), Bosma *et al.* (1977), Blais-Ouellette *et al.* (2001)) (among other authors). The decomposition of the rotation curve is made considering various mass components such as bulge, disk and dark matter halo (van Albada *et al.* (1985)).

In this work we study the system NGC5278/79 (Arp 239 KPG 390) belonging to a particular class of interacting pair of galaxies: the M51-type galaxies. According to Reshetnikov & Klimanov(2003), the two empirical criteria to classify an M51-type pair of galaxies are that the B-band luminosity ratio of the components (main/satellite) vary between 1/30 and 1/3, and that the projected distance of the satellite does not exceed two optical diameters of the main component. In the case of NGC5278/79 the B-band luminosity ratio is 0.30 and the projected separation is 16.8 kpc (optical diameter of the main component is 39.2 kpc).

For this particular system there are HST images (one of these images is showing in Fig. 1) (Windhorst *et al.* (2002)) showing dust lanes across the nuclei of both galaxy components that could help us determine which are the nearest sides of the galaxies and thus, to determine whether the spiral arms are leading or trailing, as well as to have an extensive view of the geometric conditions of the encounter in order to perform numerical simulations. It is important to recall that according to several studies (Byrd *et al.* (1989), Keel (1991)), leading spiral arms can only exist in interacting systems with retrograde encounters triggering the formation of m = 1 spiral arms (Athanassoula (1978), Thomasson *et al.* (1989)), conditions that, in principle, taking into account the morphology, could be fulfilled by NGC5278/79.

The aim of this study is to perform detailed kinematic and dynamic analysis of NGC 5278/79 using Hα kinematical data in order to study the mass distribution of this pair of galaxies and to determine the type of spiral arms (leading or trailing) in the galaxy members with the intention of reproducing both its morphology and kinematics with future numerical simulations that could shed more light on the interaction process.

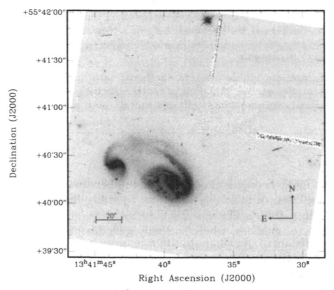

Figure 1. HST image at $\lambda = 8230$ Å with the F814W filter (Windhorst *et al.* (2002)).

2. Observations and data reduction

Observations of NGC 5278/79 (Arp 239, KPG 390) were done in 2002 July at the f/7.5 Cassegrain focus of the 2.1 m telescope at the Observatorio Astronómico Nacional in San Pedro Mártir (México), using the scanning Fabry-Perot interferometer PUMA (Rosado *et al.* (1995)). The data reduction and most of the analysis were done using the ADHOCw† software. For more information about the observations strategy, data reduction and analysis see Repetto *et al.* (2010)

† http://www.oamp.fr/ adhoc/ developped by J. Boulesteix.

3. Velocity field

The total velocity field of KPG 390 with each component of the pair and the bridge, is shown in Fig. 2 with Hα isophotes superposed. The outer isophote clearly shows a common envelope enclosing both galaxies.

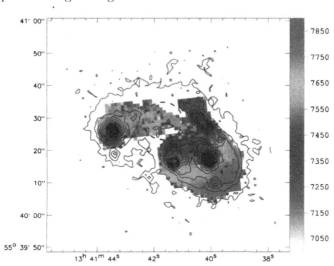

Figure 2. The full velocity field of KPG 390 (Arp 239) with overplotted Hα image isophotes. The isophotes are separated by a factor of 200 in arbitrary intensity units and the color scale shows heliocentric systemic velocity in km s^{-1}. The dashed circle indicates the bridge region.

In the disk of NGC 5278 the radial velocity values are in the range $7400 - 7860$ km s^{-1}. On the other hand, inside the disk region of NGC 5279 the radial velocity values are in the range $(7550 - 7650$ km s$^{-1})$. For NGC 5279 the mean radial velocity value is of ≈ 7600 km s^{-1}. The radial velocity values in the bright arm region of the primary galaxy (north side of NGC 5278) are in the range $7350 - 7480$ km s^{-1}. In this zone the velocity profiles are slightly broader than those in the disk of NGC 5278.

In the bridge region indicated by a dash circle in Fig. 2, the radial velocity profiles are double, or distorted, and the kinematics is more complicated. Such profiles shown in Fig. 3 (region inside the square) have a main radial velocity component and a faint secondary one. In this figure the north part shows profiles with lower signal–to–noise ratio, whereas in the south part the profiles have higher signal to noise ratio. The mean radial velocity of the brightest peak in the double profiles of the bridge region is ≈ 7600 km s^{-1}, which is close to the mean systemic velocity of both galaxies. The faintest velocity component (at $7200 - 7330$ km s^{-1}) could be associated with an extended gas outflow due to the interaction.

Summarizing, we conclude that there is a transference of material between the two components of the pair, indicatating the ongoing interacion process.

4. Rotation curves

The rotation curves were obtained from the corresponding velocity fields considering that the inner parts of these two galaxies are not strongly perturbed by the interaction process. This is true at least up to a certain radius. In the case of NGC 5278 this radius is ≈ 7 kpc ($\approx 14''$) and for NGC 5279 it is ≈ 6 kpc ($\approx 12''$). Thus, we can accurately

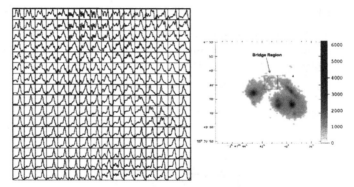

Figure 3. Hα radial velocity profiles showing the region of interaction between the two galaxies, superposed onto the Hα image of KPG 390. The profiles are normalized by the corresponding intensity in each pixel. The square indicates the region of double profiles displayed on the left. These are the original spectra.

determine the rotation curve of both galaxies considering a region of the velocity field within a sector of a specified angle inside these radii.

4.1. *NGC 5278*

The rotation curve of NGC 5278 was obtained with pixels in the velocity field within an angular sector of 20° around the galaxy major axis. The photometric center of this galaxy is the position of the brightest pixel in the continuum map. The physical coordinates of the photometric center are R.A.= $13^h 41^m 39.36^s$ and Dec.= $55°40'47.13''$. The kinematic center, derived as the position around the photometric center at which the scatter in rotation curve is minimized, is R.A.= $13^h 41^m 39.33^s$ and Dec.= $55°40'44.39''$.

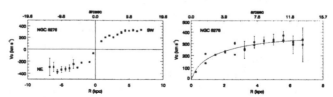

Figure 4. *Left:* Rotation curve of NGC 5278 for both sectors of the galaxy superposed: approaching (NE) and receding (SW) part. *Right:* Overplotted both parts of the rotation curve together with an exponential fit. The error bars are dispersion of values within the considered sector.

The kinematic parameters that give us the most symmetric, smooth, and low-scattered curve inside a radius of 12'' are P.A.= $(42 \pm 2)°$, $i = (42 \pm 2)°$, and $V_{syst} = (7627 \pm 10)$ km s^{-1}. The corresponding rotation curve is shown in Fig. 4.

4.2. *NGC 5279*

The brightest pixel of NGC 5279 in our continuum map has coordinates: R.A.= $13^h 41^m 44.240^s$ and Dec.= $55°41'1.45''$. The coordinates of the kinematical center are R.A.= $13^h 41^m 43.901^s$ and Dec.= $55°41'0.32''$.

The kinematic parameters that reduce significantly the asymmetry and scatter in the rotation curve were in this case: P.A.= $(141.5 \pm 1)°$, $i = (39.6 \pm 1)°$ and $V_{syst} = (7570 \pm 10)$ km s^{-1}. Due to the lack of data points it was impossible to minimize the scattering and asymmetries of the rotation curve. The asymmetry of the rotation curve did not allowed

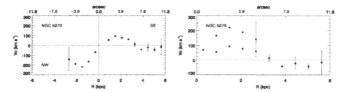

Figure 5. *Left:* Rotation curve of NGC 5279 for both sectors of the galaxy: approaching (NW) and receding (SE) side. *Right:* Overplotted both parts of the rotation curve. The error bars represent dispersion of velocity points within the considered sector.

us to perform a detailed analysis of the mass distribution of NGC 5279. The corresponding rotation curve is shown in Fig. 5.

5. Rotation curve decomposition of NGC 5278

We consider that NGC 5278 has two components that contribute to the rotation curve: an exponential disk and a massive dark matter (DM) halo. The disk was assumed to be thin and not truncated with an exponential density distribution (Freeman (1970)).

Photometric observations provide the surface brightness profiles from which we can obtain the central surface brightness in magnitude units and the disk scale lenght in kpc. In order to transform these observable parameters to mass density distribution, it is assumed that the M/L ratio is uniform and constant over the disk. In principle, the disk M/L could be known from photometric and spectroscopic observations of the disk which allow us to know the colors, or to perform a population synthesis analysis. In this case we do not have any detailed photometric study of KPG 390 so, we perform the rotation curve decomposition for NGC 5278 using the photometric data given by Mazzarella & Boroson (1993) and Paturel *et al.* (2003) as upper limits.

We tested three different types of DM halos: Hernquist halo Hernquist (1990), Navarro, Frenk & White halo (NFW) Navarro *et al.* (1996), and spherical pseudo-isothermal halo. To accomplish the rotation curve decomposition we fit the rough rotation curve data using an exponential function (Fig. 4). Then we use this fit to perform the disk-halo decomposition.

Since for this galaxy we do not have any robust restiction on the luminous mass distribution, we vary the disk scale radius $h \in [1.0 - 11.0]$ kpc and the mass to luminosity ratio $M/L_B \in [1.3 - 6.3]$. The halo mass and the halo scale length are in the range $[0 - 10^{13}]$ $M_\odot$ and $[0 - 20]$ kpc.

The rotation curve of NGC 5278 can be well fitted with spherical pseudo-isothermal halo, Hernquist halo and NFW halo (Fig. 6). The results obtained are summarized in Table 1.

6. NGC 5278/79: two trailing spirals

This kinematic study sheds light on the geometry of the galaxy encounter by determining the real orientation in the sky of the galaxy members, as well as the kind of spiral arms they possess. The latter point is not irrelevant in the case of interacting systems where a possibility of having leading spiral arms is open. Indeed, even if leading spiral arms in galaxies are a very uncommon phenomenon, the only examples where are found are interacting systems.

Following Sharp & Keel(1985) there is a criterion that determines if any particular spiral galaxy has trailing or leading arms. This criterion is based on three main clues

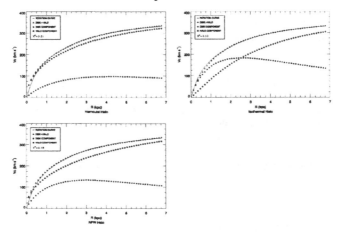

Figure 6. Disk-Halo decomposition of the rotation curve of NGC 5278.

Table 1. Mass determination from rotation curve decomposition.

Rotation curve mass	Pseudo-isothermal[a]	Hernquist	NFW
Disk component ($M_\odot$)	2.42×10^{10}	1.13×10^{10}	1.5×10^{10}
Disk lenght scale (kpc)	1.2	2.0	1.42
Disk M/L	2.8	1.3	1.6
Halo component ($M_\odot$)	1.9×10^{11}	2.1×10^{12}	6.3×10^{12}
Halo lenght scale (kpc)	2.8	17.7	16.8
Total Mass ($M_\odot$)	2.1×10^{11}	2.1×10^{12}	6.3×10^{12}
χ^2 [b]	0.10	0.21	0.18

[a] Maximum Disk
[b] Normalized χ^2 by 58 degrees of freedom.

(receding-approaching side, direction of spiral arms and the tilt of the galaxy, i.e., which side is closer to observer).

In our particular case, we have both the kinematic information in order to establish which side of the galaxy is receding and which side approaching, as well as very conspicuous morphological aspects such as well defined spiral arm patterns and the presence of dust lanes in both galaxy members running near the galaxy nuclei. This last issue will be used in what follows in two main ways: 1) we will suspect that the nearest side of the galaxy is the side hosting the dust lane and, 2) we will check it by getting an intensity profile of the galactic nucleus along the minor axis. In this kind of profiles, the nearest side is the steepest one (because of the presence of the dust lane).

In the case of NGC 5278, the receding radial velocities are in the south-western part, while the approaching radial velocities are at the north-eastern side. From Fig. 1 it is clear that the arms of NGC 5278 point in anti-clockwise direction and the dust lane is located at the concave side of the bulge, thus the northern side is the nearest. This fact is confirmed by the profile extracted along the kinematic minor axis of NGC 5278 (see Fig. 7). From these figures and the above criteria we have decided that NGC 5278 is a trailing spiral because the sense of rotation is opposite to the direction of the arms. We were able to apply similar arguments to NGC 5279. In this case the receding radial velocities are at the north-eastern side of the galaxy and the approaching radial velocities are at the south-western part. The arms of NGC 5279 point in clockwise direction and the nearest side is the southern side. As in the case of NGC 5278 this fact is confirmed by the profile extracted along the kinematic minor axis of NGC 5279 (see Fig. 7) Väisänen

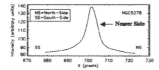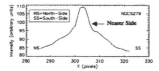

Figure 7. Intensity profiles along the kinematic minor axis of NGC 5278 and NGC 5279. The cross–section is captured in the Hubble image (Fig. 1). From the profile on left it is clear that the northern–side is the nearer in NGC 5278, because the profiles fall more abruptly than along the southern–side. In the case of NGC 5279 the southern–side is the nearer because the profiles fall more abruptly than along the northern–side, as one can see from the profile on right.

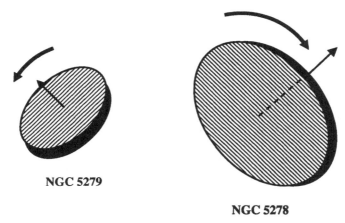

Figure 8. A possible spatial configuration of KPG 390.

et al. (2008). We can conclude that NGC 5279 is a trailing spiral also because the sense of rotation is opposite to the direction of the arms. A scheme of 3D orientation of KPG 390 derived from our kinematic analysis is shown in Fig. 8.

7. Conclusions

In this work we presented Fabry-Perot observations of the isolated pair of galaxies NGC 5278/79 (Arp 239, KPG 390) showing that for an interacting and asymmetric system it is important to have kinematic information of large portions of the galaxies participating in the interaction process. From the analysis of the velocity field, we obtain the rotation curves for NGC 5278 and NGC 5279. We perform the decomposition of the rotation curve for NGC 5278 and determine the content of dark matter of this galaxy, using different types of halos (pseudo-isothermal, Hernquist and NFW halo). We also study the kind of spiral arms of the two spirals of the pair and we determine that NGC 5278/79 are trailing spirals. We will use the kinematic information as a starting point in future numerical simulations of this pair.

Acknowledgements

We acknowledge DGAPA-UNAM grant: IN102309 and CONACYT grant: 40095-F.

References

Marcelin, M., Lecoarer, E., Boulesteix, J., Georgelin, Y., & Monnet, G. 1987, *A&A*, 179, 101

Amram, P., Mendes de Oliveira, C., Plana, H., Balkowski, C., Boulesteix, J. & Carignan, C. 2002, *Ap&SS*, 281, 389

Fuentes-Carrera, I., Rosado, M., Amram, P., Salo, H. & Laurikainen, E. 2007, *A&A*, 466, 847

Rubin, V. C., Peterson, C. J. & Ford, W. K., Jr. 1976, *BAAS*, 8, 297

Bosma, A., van der Hulst, J. M. & Sullivan, W. T., III 1977, *A&A*, 57, 373

Blais-Ouellette, S., Amram, P. & Carignan, C. 2001, *AJ*, 121, 1952

van Albada, T. S., Bahcall, J. N., Begeman, K. & Sancisi, R. 1985, *ApJ*, 295, 305

Reshetnikov, V. P. & Klimanov, S. A. 2003, *Astron. Lett.*, 29, 429

Windhorst, R. A., *et al.* 2002, *ApJS*, 143, 113

Byrd, G. G., Thomasson, M., Donner, K. J., Sundelius, B., Huang, T. Y. & Valtonen, M. J. 1989, *Celestial Mechanics*, 45, 31

Keel, W. C. 1991, *ApJ*, 375, L5

Athanassoula, E. 1978, *A&A*, 69, 395

Thomasson, M., Donner, K. J., Sundelius, B., Byrd, G. G., Huang, T.-Y. & Valtonen, M. J. 1989, *A&A*, 211, 25

Freeman, K. C. 1970, *ApJ*, 160, 811

Hernquist, L. 1990, *ApJ*, 356, 359

Navarro, J. F., Frenk, C. S. & White, S. D. M. 1996, *ApJ*, 462, 563

Sharp, N. A. & Keel, W. C. 1985, *AJ*, 90, 469

Väisänen, P., Ryder, S., Mattila, S. & Kotilainen, J. 2008, *ApJ*, 689, L37

Mazzarella, J. M. & Boroson, T. A. 1993, *ApJS*, 85, 27

Paturel, G., Petit, C., Prugniel, P., Theureau, G., Rousseau, J., Brouty, M., Dubois, P., & Cambrésy, L. 2003, *A&A*, 412, 45

Rosado, M., *et al.* 1995, *Rev. Mexicana AyA*, 3, 263

Repetto, P., Rosado, M., Gabbasov, R. & Fuentes-Carrera, I. 2010, *AJ*, 139, 1600

Discussion

WOLF: Why is there a factor of 3 discrepancy between total halo mass when adopting a NFW vs Hernquist profile? Also, how do you define the total halo mass for the NFW? What is your cutoff radius?

REPETTO: This discrepancy is due to the fact that we chose the solution with the minimum χ_2 value; however, there were other solutions with higher χ_2 but similar halo masses. The total halo mass for the NFW halo was defined following Navarro *et al.* (1996) as M_{200} and the cutoff radius was defined as R_{200}. In the case of NGC 5278 the cutoff radius was located at $\approx$234 kpc.

Astrophysical Dynamos: From Stars to Galaxies
Proceedings IAU Symposium No. 271, 2010
N.H. Brummell, A.S. Brun, M.S. Miesch Y. Ponty, eds.

© International Astronomical Union 2011
doi:10.1017/S1743921311017546

Magnetic Fields in Galaxies

Ellen G. Zweibel[1]

[1] Departments of Astronomy & Physics, & Center for Magnetic Self-Organization
University of Wisconsin-Madison
475 N. Charter St, Madison, Wisconsin 53706 USA
email: zweibel@astro.wisc.edu

Abstract. The origin and evolution of magnetic fields in the Universe is a cosmological problem. Although exotic mechanisms for magneotgenesis cannot be ruled out, galactic magnetic fields could have been seeded by magnetic fields from stars and accretion disks, and must be continuously regenerated due to the ongoing replacement of the interstellar medium. Unlike stellar dynamos, galactic dynamos operate in a multicomponent gas at low collisionality and high magnetic Prandtl number. Their background turbulence is highly compressible, the plasma $\beta \sim 1$, and there has been time for only a few large exponentiation times at large scale over cosmic time. Points of similarity include the importance of magnetic buoyancy, the large range of turbulent scales and tiny microscopic scales, and the coupling between the magnetic field and certain properties of the flow. Understanding the origin and maintenance of the large scale galactic magnetic field is the most challenging aspect of the problem.

Keywords. galaxies:magnetic fields, cosmology:miscellaneous,plasmas

1. Introduction

Most of the papers presented at this symposium are about the Sun and other stars, the topics on which Juri Toomre has concentrated for most of his career. This paper is about how magnetic fields in galaxies originated, and how they evolve. Because many aspects of this topic have recently been reviewed elsewhere in great detail (Widrow 2002, Kulsrud & Zweibel 2008, Beck 2009, Han 2009), I will focus on contrasting the stellar and galactic dynamo problems. I will write from a personal perspective, and will not attempt to give a comprehensive list of references. The paper is organized as follows: §2 briefly reviews existing theories of and observational constraints on cosmic magnetogenesis, §3 does the same for galactic magnetic fields, §4 discusses similarities and differences between the stellar and galactic dynamo problems, §5 discusses the likely ingredients of galactic dynamo theory, and §6 is a summary.

2. Cosmic Magnetogenesis

The origin of magnetic fields is a cosmological problem. It is widely accepted that magnetic fields were not produced in the Big Bang, but later. It is useful to classify theories of magnetogenesis as top-down and bottom-up. In top-down theories, the entire Universe is magnetized by a global process. Some of these theories are based on new and as yet unconfirmed physics. Others rely on conventional plasma processes, such as the Biermann Battery. Examples of top down theories based on conventional physics are magnetization by the Biermann battery in cosmological ionization fronts or cosmological shock fronts (Kulsrud *et al.* 1997, Gnedin *et al.* 2000). In bottom up theories, magnetic fields are generated in small objects and propagated to large scales by a combination of outflows, free expansion, and diffusion. Examples are magnetization of galaxies by magnetized stellar ejecta (Rees 1987), and magnetization of portions of the intergalactic medium by fossil radio lobes from AGN (Fulanetto & Loeb 2001, Kronberg *et al.* 2001).

The best evidence for a top down theory would be the discovery of a pervasive intergalactic magnetic field. For many years there were only rather high upper limits on the existence of such a field. One is an upper limit on the magnetic pressure at the time of Big Bang Nucleosynthesis; this corresponds to a field at the current epoch of about 10^{-7} G, which is about 1% - 10% of the fieldstrengths measured in galaxies. Limits on a *coherent* field come from the lack of evidence for an increase in Faraday rotation with redshift in some particular direction, and are 2-3 orders of magnitude smaller.

Recent high resolution observations of extragalactic, pointlike TeV γ-ray sources has made it feasible to search for much weaker magnetic fields. If the intergalactic medium is weakly magnetized, then $e^{\pm}$ pairs created by collisions between TeV and ambient photons will gyrate through a small angle before producing additional γ photons by inverse Compton scattering. This produces a γ-ray helo around the original source.

Recently Neronov & Vovk (2010) have set a lower limit of 3 10^{-16} G for large scale fields along the line of sight to two TeV blazar sources observed by the *Fermi* satellite, while Ando & Kusenko (2010) claim to have detected a field of 10^{-15}G, also from *Fermi*. As weak as these fields are, they exceed the minimum predictions for fields produced by the Biermann Battery, as we shall now see.

According to a simple variant of the battery mechanism, an electron pressure gradient which arises, e.g. due to ionizing radiation, must be balanced by an electric field

$$en_e E = -\nabla P_e. \tag{2.1}$$

From Faraday's Law, the rate of change in the magnetic flux Φ threading a surface moving with the fluid is then

$$\frac{d\Phi}{dt} = \frac{c}{e} \int_C \frac{\nabla P_e \cdot dl}{n_e}, \tag{2.2}$$

where C is the contour bounding the surface. If the contours of constant n_e are also contours of constant P_e, then the integrand in eqn. (2.2) can be written as an exact differential, and vanishes. But if n_e and P_e are not functions of one another - which for an ideal gas implies the that gradients of n_e and T_e are nonparallel - there is a net EMF around C, which allows creation of magnetic flux.

It can also be estimated from Faraday's Law that the ion gyrofrequency $\omega_{ci} \equiv ZeB/m_i c$ evolves according to

$$\omega_{ci} \sim \frac{v_i^2 t}{l_n l_T}, \tag{2.3}$$

where v_i is the ion thermal velocity, t is time, and l_n and l_T are the electron density and temperature lengthscales, respectively. In a disk geometry, where one of the l is of order the radius R and the other of order the scale height H, we estimate that the ratio of gyrofrequency to rotation frequency, ω_{ci}/Ω, satisfies $\omega_{ci}/\Omega \sim \Omega t(H/R)$. That is, R/H rotations must elapse before the ions are magetized. This argument, and the inverse dependence of ω_{ci} on l_n and l_T, makes it clear that the Biermann battery can make appreciable magnetic fields only in small objects, likes stars or accretion disks. Numerical calculations of the battery on cosmological scales find fieldstrengths on the order of 10^{-18} - 10^{-21}G. It should be mentioned that the simulations in question have rather coarse spatial resolution. Small scale structure produces stronger fields, in accord with eqn. (2.3). But it is interesting that the large scale intergalactic fields probed by γ-ray halo observations are 2-3 orders of magnitude larger than the fields produced by the cosmological version of the Biermann Battery. Possibly the Battery fields are amplified by strong, localized shear flows (Ryu *et al.* 2008), and are distributed intermittently along the line of sight. Possibly they were created by an entirely different process. Extending these pioneering observations to many more sources would put firmer constraints on the magnitude and coherence length of the field, and on its spatial distribution.

3. Galactic Magnetic Fields

Galactic magnetic fields are probed through several complementary techniques:

• *Diffuse nonthermal continuum:* Cosmic ray electrons gyrating in the interstellar magnetic field emit polarized synchrotron radiation. The orientation of the polarization depends on the orientation (but not the direction) of magnetic field projected on the plane of the sky. If the field were uniform the radiation would be about 70% polarized; lower polarization indicates unresolved or line of sight magnetic strucutre. The nonthermal continuum probes the orientation, strength, and angular dispersion of the plane of sky field, convolved with the cosmic ray electron spectrum.

• *Faraday rotation:* When linearly polarized radiation propagates along a magnetic field in a medium with free electrons, its plane of polarization rotates. Faraday rotation occurs in pulsars, extragalactic radio sources, and the diffuse nonthermal continuum. Faraday rotation is a probe of the strength and directedness of the line of sight field, convolved with the thermal electron spectrum.

• *Zeeman splitting:* Magnetic fields lift the degeneracy of magnetic sublevels in atoms and molecules. Unlike the Sun, where the full vector field is measured, only the longitudinal Zeeman effect has been observed in the interstellar medium. The longitudinal Zeeman effect effect is a probe of the strength and directedness of the line of sight field, convolved with the local density of atoms and molecules.

• *Polarization by aligned dust grains:* Aspherical interstellar dust grains are aligned by the interstellar magnetic field. Starlight is polarized by intervening dust, as is the thermal infrared emission from dust. Polarization probes the orientation of the plane of sky magnetic field in regions of moderate to high density.

Application of these techniques has led to a fairly consistent picture of the magnetic field within a few kiloparsecs of the Sun. The mean orientation is nearly azimuthal - the deviation is consistent with alignment along the spiral arms - and nearly parallel to the Galactic plane. The rms fieldstrength is about $5.5\mu G$, which puts the magnetic energy density near equipartition with the energy densities in random gas motion and in cosmic rays. The field has a nonzero average of about $1.6\mu G$ over the observed volume, with the remaining field in the form of fluctuations. The azimuthal field reverses sign at least once with Galactocentric radius (Brown *et al.* 2007), possibly more (Han 2009). At high Galactic latitudes, the azimuthal field is antisymmetric about the Galactic plane (so-called dipole symmetry); the parity of the field near the Galactic plane is less clear. As to the fluctuation spectrum, there is evidence for an outer scale of 50 - 100 parsecs, with structure down to subparsec scales. The field declines with distance from the Galactic plane, with a scale height of about 1.5 kpc, similar to that of the diffuse ionized gas, but larger than that of the atomic and molecular gas distributions. The fieldstrength increases toward the Galactic Center.

By the nature of the magnetic field probes described above, the fieldstrength is relatively well measured in the diffuse ionized and neutral components, and in the molecular component, but not well measured in the very hot, low density component, and in the halo. Within the density range over which the field is measured, there is no relationship between fieldstrength and gas density in the diffuse ionized and neutral gas. The field does increase with gas density in the denser, molecular gas, with $B \propto \rho^{1/2}$ giving an upper envelope. However, this relationship cannot be extrapolated to the diffuse gas; the mean density of giant molecular clouds is about 50 cm^{-3}, 50 times denser than the mean density of diffuse gas, but the mean B in these clouds is only about twice that measured in the diffuse gas, not 7 times larger, as $B \propto \rho^{1/2}$ would suggest (Crutcher *et al.* 2003).

The existence of a mean vertical magnetic field, or lack of it, is a strong constraint on dynamo models. Such a field would be difficult to remove, and hence could be primordial. Mean field dynamo theories can also produce vertical fields with either dipole symmetry (constant across the Galactic midplane) or quadrupole symmetry (reversing across the midplane). Searches for a mean vertical field in the solar neighborhood have had mixed results.Taylor *et al.* (2009) and Mao *et al.* (2010) found a north-directed field of 0.30 μG below the Galactic plane. The former study reports a south directed field of 0.14 μG above the Galactic plane while the latter finds no net field there.

There is a close relationship between star formation rate, magnetic fieldstrength, and energetic particle density. The far-infrared luminosity (essentially, the luminosity of young, massive stars) and radio synchrotron luminosity are observed to be linearly proportional over a wide range of galaxy luminosities and star formation rates (Bell 2003 and references therein). Since the rate of core collapse supernovae is correlated with the star formation rate, and supernovae are the dominant source of energy in the interstellar medium, this is evidence that magnetic field amplification, cosmic ray acceleration, and supernova remnants are linked.

Galaxies, including the Milky Way, have been magnetized throughout their history. In the Milky Way this is inferred from the presence of light elements detected in the atmospheres of some of the oldest halo stars. These elements are thought to have been produced by cosmic ray spallation reactions in the material from which these stars formed (Parizot & Drury 1999). Acceleration and confinement of cosmic rays requires a magnetic field, although it could have been several orders of magnitude weaker than the present Galactic field (Zweibel 2003). Coherent, microgauss magnetic fields have been measured in damped Lyα systems at redshift $z \sim 1$ (Oren & Wolfe 1995).

4. Stellar *vs* Galactic Dynamos

The solar cycle is the best evidence for a solar dynamo. Although the roles of the convection zone, the tachocline, and the radiative core are not yet fully understood, it is clear that differential rotation, poloidal circulation, and thermal convection, together with an outer boundary condition that permits the magnetic field to escape, act jointly to produce large scale toroidal and poloidal fields which reverse over a stable 22 year period.

The evidence for a Galactic dynamo is less clear. The timescales for cyclic behavior are far too long to be detected, and the variety of magnetic field morphologies observed in other galaxies has never been interpreted as evidence for cycles. Perhaps the best argument for dynamos in galaxies is the ongoing replacement of the interstellar medium. Large disk galaxies such as the Milky Way accrete gas through mergers with other galaxies, engulfing dwarf satellites, and infall of extragalactic gas, and eject gas via winds. Gas is added to the interstellar medium through stellar mass loss, and lost through star formation. These processes result in complete replacement of the interstellar medium within 10^9 yr, or about 4-5 rotation periods. Maintaining a coherent field of the observed strength and orientation as magnetized gas is added and deleted requires the continuous operation of dynamo processes.

Stellar and galactic dynamos operate in very different regimes, both macroscopically and microscopically. Normal stars are pressure supported, with little rotational support.

Convective turbulence in stars is highly subsonic, and slow even with respect to rotation. Therefore, magnetic fields reach equipartition with convection and differential rotation at large plasma β. Under these conditions most of the field is in pressure confined magnetic flux ropes which occupy a small fraction of the volume. Galactic disks are rotationally supported in the radial direction, with turbulent velocities that are supersonic but well below the rotational velocity. Therefore, when the magnetic field reaches equipartition with the turbulence, it is too strong for confinement in flux ropes, but must be volume filling. Thermal and dynamical gas pressure, cosmic ray pressure, and magnetic forces jointly determine the vertical structure of galactic disks.

The mean flows in the Sun - both the differential rotation and the circulation are quite well determined from surface measurements and from helioseismology. The turbulent convection in the solar interior is not directly observed, although its primary effect, the transport of heat, is well measured. It is highly subsonic except near the photosphere, and compressibility effects over most of the solar interior are weak. The mean flow in the Galaxy is differential rotation, which is well measured. Flow perpendicular to the Galactic plane, whether a wind or a poloidal circulation or "fountain flow" driven by supernovae, has been inferred, but not directly measured. The turbulence is probably driven more by supernovae than by *in situ* instabilities. A combination of density and velocity diagnostics show that the turbulence exists on scales from the thermal ion gyroradius to tens of parsecs. The large scale turbulence is highly supersonic, and compressibility effects are large.

Despite these differences, some dynamical features are common to both stars and galaxies. Magnetic buoyancy plays a role in both types of system. In stars it acts radially; in galactic disks it acts vertically. Buoyancy driven escape can limit the strength of the field. The twisting of rising magnetic loops by Coriolis forces produces a net helicity flux, which may promote the growth of a large scale field (Blackman & Field 2001). Both types of system are turbulent over a large range of scales. Neither stars nor galaxies are magnetically dominated, but magnetic fields affect the slower flows, and certainly the turbulence; therefore the dynamo is coupled to the gas dynamics.

At the microscopic level, stars and galaxies also differ substantially. The plasma in stellar interiors is highly collisional: particle collision times τ_c are much less than gyroperiods $2\pi/\omega_c$, so transport coefficients are isotropic with respect to the magnetic field. The ratio of magnetic to viscous diffusivity, or magnetic Prandtl number Pm, is much less than one: the kinetic energy spectrum extends to much smaller scales than the magnetic spectrum. The transport coefficients in stellar interiors are well understood at the level needed for dynamo theory: both plasma and radiation contribute to the transport, the relative amounts varying with electron degeneracy and stellar luminosity, and it is known how to calculate them. In the outer atmosphere, the physics is more complicated and less certain; partial ionization effects and kinetic effects, in particular, may affect the rate and onset conditions of magnetic reconnection, and thus magnetic flux escape.

Interstellar gas is much more complex than stellar plasma. Most interstellar gas is sufficiently collisionless ($\omega_c \tau_c \gg 1$) that plasma transport coefficients are highly anisotropic. At the low densities typical of interstellar gas, $Pm \gg 1$, so the magnetic spectrum extends far below the velocity spectrum. Interstellar gas has multiple components; most of it (by mass, but not by volume) is weakly ionized, and pervaded by cosmic rays. On large scales, the plasma and neutrals can be treated as a single, conducting fluid, but on small scales - typically less than a parsec, but much larger than the resistive scale - the plasma decouples from the neutrals. At small scales, the plasma pressure and Alfvén Mach number are low $\beta \ll 1$, $M_A \ll 1$, and the medium is magnetically controlled. Partial ionization promotes the formation of current sheets (Brandenburg & Zweibel 1994), which accelerates magnetic reconnection in neutral sheets (Heitsch & Zweibel 2003). Ion neutral drift in a turbulent medium leads to fast diffusion of the magnetic field through the neutrals (Heitsch *et al.* 2004) and also suppresses the growth of magnetic fields at scales below the ion-neutral decoupling scale (Zweibel & Heitsch 2008). Cosmic rays are coupled to the thermal gas by scattering from small scale magnetic fluctuations. On scales larger than their mean free path (typically a few pc), they behave as a fluid, and are a source of buoyancy. The effect of cosmic ray buoyancy on large scale dynamos was first proposed by Parker (1992). On small scales, cosmic rays amplify helical Alfvénic fluctuations, which if the cosmic ray flux is sufficently large can lead to significant magnetic field amplification (Zweibel 2003, Bell 2004). The effects of these additional components on galactic dynamos is not yet fully understood.

Again, despite their differences, stars and galaxies have common features. In both, the microscopic transport coefficients imply that the velocity and magnetic spectra extend to scales many orders of magnitude less than the global scales. This creates conceptual difficulties for the generation of large scale magnetic fields and makes direct numerical simulation of the full system all but impossible.

5. The Ingredients of Galactic Dynamo Theory

Certain features of the Galactic magnetic field seem easily explained. The overall strength of the field is consistent with the expectation of equipartition between turbulent kinetic and magnetic energies. Its strength and scale height are also consistent with stability arguments: if too large a fraction of the weight of the interstellar medium is magnetically supported, it is unstably stratified (Parker 1966). The predominantly azimuthal orientation of the field is naturally produced by the strong differential rotation of the Galactic disk.

None of these arguments explain the existence of the large scale field, which is coherent on scales of at least several kiloparsecs, possibly more. This scale is one to two orders of magnitude larger than the largest forcing scale in the interstellar medium: the scale of superbubbles, formed by the winds and supernova explosions of clusters of massive stars. This coherent field must be sustained over the $\sim 10^9$yr timescale on which it is disrupted by stellar ejecta, star formation, infall of material, and possibly a wind.

There are two current models, and an emerging third model, for the mean field. The primordial model (Howard & Kulsrud 1997) assumes the field was present at the time the Galactic disk formed, and inclined at an angle to it. The vertical magnetic flux associated with this field is almost impossible to remove, and is a source of azimuthal field through windup by differential rotation. In the absence of diffusion, the windup amplifies the field linearly with time and produces reversals with radius on a scale $\Delta r \sim R/N$, where R is the radius of the disk and N is the number of times the disk has rotated. The windup is counteracted by diffusion, which HK take to be ion-neutral drift perpendicular to the Galactic plane. One difficulty with the model is the presence of the thick ionized gas layer, which envelopes the neutral hydrogen disk and blocks ion-neutral drift. The second difficulty is that no vertical field which runs through the Galactic plane has yet been detected, and it is even possible that there is a vertical field which reverses across the Galactic plane, contrary to the model. However, the primordial model is appealing in its simplicity, and some aspects of the scenario it describes may be relevant to galaxies.

The second model is the mean field dynamo, which was first applied to the Galaxy by Parker 1971. Ferrière (Ferrière 1993a, Ferrière 1993b) has calculated the α and β tensors when the small scale induction and diffusion are produced by expanding shells associated with supernova remnants and superbubbles. Hanasz & Lesch 1993 and Kowal *et al.* (2006)

calculated α for Parker instabilities driven by cosmic ray buoyancy. Calculations of mean field dynamo modes, some incorporating prescriptions for nonlinear saturation, have been calculated, for example, by Ferrière & Schmitt (2000). Unstable modes with both dipole and quadrupole symmetries are, with growth times typically in the 10^8 - 10^9 year range. These models make clear predictions, and have been used to interpret observations. However, they encounter the objections that plague mean field dynamo theory in any system with large magnetic Reynolds number Rm: the small scale fields are predicted to grow much faster than the mean field (Kulsrud & Anderson 1992. The existence of dominant magnetic power on the resistive scale - about $R_\odot$ - is firmly precluded by observations.

A third type of model is direct numerical simulation of the gas dynamics in galaxies (e.g. Kulesza-Zydik *et al.* 2010). Such work can include a variety of interstellar processes, and make no assumptions about mean field theory. Like mean field models, direct numerical simulations can be compared to observations of galaxies. However, they are of necessity restricted to large scales, and therefore do not address the problem of dominance of the small scales.

Although we lack a complete explanation for the properties of the Galactic magnetic field, we can make some educated guesses as to what the ingredients of such a theory would be:

• *Seed magnetic field*: As discussed in §, possible sources include a "primordial" field generated by an exotic physical process in the early Universe, a large scale intergalactic field generated by the Biermann Battery operating in shock and/or ionization fronts, or a field from stellar ejecta magnetized by battery and dynamo processes. The resulting magnetic field is not completely independent of initial conditions, as the net vertical magnetic flux through the galaxy should be preserved. Since coherent μG magnetic fields seem to have existed by the time the Universe was 1/10 its present age, the seed field must be large enough to grow to the observed level in 10^9 yr. For example, if $B = B_i e^{t/\tau}$ and $\tau = 10^8$yr, we must have $B_i \geqslant 5 \times 10^{-11}$G. Magnetic field estimates from stellar ejecta satisfy this constraint, fields produced in cosmological fronts do not, unless they are pre-amplified by flows in the intergalactic medium. Even very weak magnetic fields can be amplified by plasma mechanisms. For example, Krolik & Zweibel (2006) showed that accretion disks are unstable to magnetorotational instabilities even when the ions are unmagnetized. However, the wavelength of the fastest growing modes scale inversely with B. In collisional disks, the modes are therefore subject to strong damping, but in collisionless disks they can survive. Thus, the battery to MRI scenario appears to work, at least in hot, low density disks. Other amplification mechanisms may exist in cold, dense disks, such as protoplanetary disks (Tan & Blackman 2004). Once the battery fields are amplified, the next step is to eject them, in supernova explosions, jets, or winds. Then they may remain in the ambient medium as confined structures, or they may diffuse throughout the volume. This seems to be a plausible mechanism for magnetizing the interstellar medium. For example, if we assume a turbulent lengthscale of 10 pc and a characteristic velocity of 10 km s^{-1}, the turbulent diffusivity D_t of the interstellar medium is $D_t \sim 3 \times 10^{25}$ cm^2 s^{-1}, the rms displacement of magnetized material is 300 pc in 10^9 yr. Empirically, the apparently uniform chemical composition of interstellar gas in the solar neighborhood is good evidence that the interstellar medium is well mixed.

• *Energy source*: Galactic gas dynamics are determined by gravity, which is due primarily to dark matter and stars, and by energy from massive stars, in the form of radiation, winds, and supernova explosions. The correlation between far-IR luminosity and synchrotron luminosity is evidence that energy input from massive stars is important in driving the dynamo. Stellar energy drives turbulence, while cosmic ray acceleration at shock fronts amplifies small scale magnetic fields.

- *Fieldline stretching:* Fieldine stretching is fundamental to any dynamo. In galactic dynamos, fieldlines are stretched by differential rotation, small scale turbulence, expansion of bubbles, and possibly a galactic wind or fountain flow. Accretion of intergalactic gas clouds or dwarf galaxies, and mergers with larger galaxies, generate large tidal forces, which may help to drive a dynamo. and by tidal stretching in mergers. Turbulence seems to exist over a large range in scales, and because $Pm \gg 1$, the fieldlines develop structure on scales far below the viscous cutoff for the velocity scale.

- *Fast diffusive transport:* Stellar ejecta is undermagnetized in comparison with the interstellar medium; infalling intergalactic material may also be undermagnetized. Magnetic fields must diffuse rapidly into this material, possibly by ion - neutral drift, acting jointly with turbulence, or by fast magnetic reconnection. In models with buoyancy driven escape, the field must separate itself from the matter, by reconnection or diffusion. The problem of fast diffusive magnetic field transport is related to the problem of thermal and chemical mixing of interstellar gas.

- *Fast magnetic reconnection:* Dynamos require magnetic reconnection to irreversibly change magnetic topology. Since reconnection typically operates at small scales, it can also suppress the growth of small scale fields. Classical reconnection rates in the interstellar medium are very low, but a number of mechanisms have been proposed to accelerate the reconnection rate (Zweibel & Yamada 2009). Among these are current sheet formation by ion-neutral drift (Brandenburg & Zweibel 1994), externally driven MHD turbulence in the reconnection layer (Lazarian & Vishniac 1999), and self-disruption of the reconnection layer by the plasmoid instability (Loureiro *et al.* 2007), possibly accompanied by kinetic effects. Rapid progress is being made in this area.

6. Summary

This paper is a short review of the galactic dynamo problem, with comparison to the stellar dynamo problem. The best evidence that dynamos operate in galaxies is the ongoing replacement of their interstellar gas. The greatest challenge in galactic dynamo theory is explaining the existence of a coherent magnetic field, comparable in magnitude to the random field, over scales much larger than the forcing scale, Such fields have existed at least since the Universe was about 1/10 its present age.

The seed magnetic fields for the dynamo early in the history of the Galaxy are more likely to have been produced by many small sources than by a cosmological scale Biermann Battery process. The fields produced by the latter cannot be amplified to μG levels in 1/10 the age of the Universe. Intriguing recent evidence for an intergalactic field 2-3 orders of magnitude stronger than predicted by the Battery mechanism may hint at a true intergalactic dynamo, or at a magnetogenesis mechanism based on exotic physics.

Galactic dynamos differ from stellar dynamos. The interstellar medium is relatively collisionless, much of its mass is electrically neutral, it is pervaded by cosmic rays, and its turbulence is supersonic. Unlike stars, which accommodate many magnetic cycles over their lifetimes, galactic gas is replaced in 4-5 rotation periods, and galaxies themselves are no older than ~ 50 rotation periods.

Galactic dynamo theory needs a breakthrough. It may come through basic plasma physics, such as improved understanding of the small scale dynamo in a medium with anisotropic viscosity, or of magnetic reconnection under interstellar conditions. It may come as astrophysical processes such as gas dynamics during galaxy mergers and impacts of intergalactic or galactic halo clouds on galaxy disks are more accurately modeled and better understood. New observations may modify our characterization of the galactic dynamo problem, and the basic constraints. Each of these aspects is fascinating on its own.

I met Juri Toomre at a solar physics meeting in 1978. He was as friendly, inquisitive, canny, witty, and bald then as he is now, in 2010. We were colleagues from 1981, when I became a faculty member at the University of Colorado, until 2002, when I left for the University of Wisconsin. I learned an enormous amount from him about science itself, and about being a scientist. He created many opportunities for me and was very helpful to me in my own career and in being effective as both Department Chair and JILA Chair. He probably helped me also in ways that I will never know. This meeting has been a wonderful way to witness Juri's enormous impact on the lives of other many other scientists, and on science.

I am happy to acknowledge the support of the National Science Foundation through grants PHY0821899 and AST0907837.

References

Ando, S. & Kusenko, A. 2010 *ApJ*, 722, L39

Beck, R. 2009 *Ast. & Sp. Sci. Trans*, 5, 43

Bell, E. F. 2003 *ApJ*, 586, 794

Bell, A. R. 2004 *MNRAS*, 353, 550

Blackman, E. G. & Field, G. B 2001 *Phys. Plasmas*, 8, 2407

Brandenburg, A. & Zweibel, E. G. 1994 *ApJ* 427, L91

Brown, J. C., Haverkorn, M., Gaensler, B. M., Taylor, A. R., Bizunok, N. M., Dickey, J. M., & Green, A. J. 2007 *ApJ*, 663, 258

Crutcher, R. M., Heiles, C., & Troland, T 2003 *Turbulence & Magnetic Fields in Astrophysics, eds. E. Falgarone & T. Passot, Lecture Notes in Physics*, 614, 155

Ferrière, K. M. 1993 *ApJ*, 404, 162

Ferrire, K. M. 1993 *ApJ*, 409, 248

Ferrière, K. M. & Schmitt, D. 2000 *A&A* 358, 125

Furlanetto, S. R. & Loeb, A. 2001 *ApJ*, 556, 619

Gnedin, N. Y., Ferrara, A., & Zweibel, E. G. 2000 *ApJ*, 505, 516

Han, J. L. 2009 *Cosmic Magnetic Fields: Proceedings of IAU Symposium 259*, 455

Hanasz, M. & Lesch, H. 1993 *A&A*, 278, 561

Heitsch, F. & Zweibel, E. G. 2003 *ApJ* 583, 229

Heitsch, F., Zweibel, E. G., Slyz, A., & Devriendt, J. E. G. 2004 *ApJ*, 603, 175

Howard, A. M. & Kulsrud, R. M. 1997 *ApJ*, 483, 648

Kowal, G., Otmianowska-Mazur, K., & Hanasz, M. 2006 *A&A*, 445, 915

Krolik, J.H. & Zweibel, E. G. 2006 *ApJ*, 644, 651

Kronberg, P. P., Dufton, Q. W., Li, H., & Colgate, S. A 2001 *ApJ*, 560, 178

Kulesza-Zydik,B., Kulpa-Dybel, K., Otmianowska-Mazur, K., Soida, M., & Urbanik, M. 2010 *A&A*, 522, 61

Kulsrud, R. M. & Anderson, S. W. 1992 *ApJ*, 396, 606

Kulsrud, R. M., Cen R., Ostriker, J. P., & Ryu, D. 1997 *ApJ*, 480, 481

Kulsrud, R. M. & Zweibel, E. G. 2008 *Rep. Prog. Phys*, 71, 046901

Lazarian, A. & Vishniac, E. T. 1999 *ApJ*, 517, 700

Loureiro, N. F., Schekochihin, A., & Cowley, S. C 2007 *Phys. Plasmas*, 14, 100703

Mao, S. A., Gaensler, B. M., Haverkorn, M., Zweibel, E. G., Madsen, G. J., McClure-Griffiths, N. M., Shukurov, A., & Kronberg, P. P. 2010 *ApJ*, 714, 1170

Neronov, A. & Vovk, I 2010 *Science*, 328, 73

Oren, A. L. & Wolfe, A. M. 1995 *ApJ*, 425, 624

Parizot, E. & Drury, L. 1999 *A&A*, 349, 673

Parker, E. N. 1966 *ApJ*, 145, 811

Parker, E. N. 1971 *ApJ*, 166, 295
Parker, E. N. 1992 *ApJ*, 401, 137
Rees, M. J. 1987 *QJRAS*, 28, 197
Ryu, D., Kang, H., Cho, J., & Das, S. 2008 *Science*, 320, 909
Tan, J. C. & Blackman, E. G. 2004 *ApJ*, 603, 401
Taylor, A. R., Stil, J. M., Sunstrum, C. 2009 *ApJ*, 702, 1230
Widrow, L. M. 2002 *Rev. Mod. Phys.*, 74, 775
Zweibel, E. G. 2003 *ApJ*, 587, 625
Zweibel, E. G. & Heitsch, F. 2008 *ApJ*, 684, 373
Zweibel, E. G. & Yamada, M 2009 *ARAA* 47, 291

Discussion

STEVE TOBIAS: There are some theories that have the kinematic field generated on the resistive scale followed by a scale by scale saturation. Is this feasible given the timescale?

ELLEN ZWEIBEL: The turnover times of the largest eddies are probably of order 10^6 yr, which is quite short compared to the replacement time for the interstellar medium. The field is observed to be coherent on scales 10 - 30 times larger than the largest eddies. I'm not sure how to estimate the time required for amplification on the global scale.

AXEL BRANDENBURG: The growth rate of the MRI may not be relevant because the MRI relies on the existence of an existing background field, which still needs to be amplified by a dynamo. I would guess that this mechanism is only when there are no other sources of turbulence, e.g. in the outskirts of the galaxy? In any case I do not think this MRI dynamo would be more efficient than the turbulent dynamo. Also, it would only produce large scale fields if the usual ingredients (stratification, rotation) are present.

ELLEN ZWEIBEL: Krolik and I contend that the Battery fields in accretion disks become unstable to the MRI when they are still very weak. To the extent that the MRI acts as a dynamo, a large scale field will then be amplified. Of course, there are other possible sources of turbulence, e.g. gravitational instabilities in protostellar disks, where the MRI is likely to be damped by collisional processes.

ANNICK POUQUET: Is equipartition between rms magnetic field and turbulent gas pressure exact, or is there a winner?

ELLEN ZWEIBEL: The observations refer only to averages, so I don't think it's possible to to say.

KEITH MOFFATT: Turbulence in the interstellar medium is characterized by very large magnetic Prandtl number. Can you estimate the Kolmogorov scale of the turbulence on which the most efficient stretching and intensification of the magnetic field takes place? I discussed this problem as early as 1962:JFM (magnetic eddies...)

ELLEN ZWEIBEL: If one uses the parallel viscosity coefficient and assumes a fully ionized gas with $T = 10^4$K, $n = 1$, and an outer scale of 30 parsec, then the Kolmogorov scale is about 10^{14} cm, or 7 AU - very small.

Astrophysical Dynamics - from Stars to Galaxies
Proceedings IAU Symposium No. 271, 2010
N. H. Brummell, A. S. Brun, M. S. Miesch & Y. Ponty, eds.
© International Astronomical Union 2011
doi:10.1017/S1743921311017558

The dual nature of the Milky Way stellar halo

Anna Curir[1], Giuseppe Murante[1], Eva Poglio [2], Álvaro Villalobos[3]

[1]INAF- Astronomical Observatory of Torino, strada Osservatorio 20, 10025 Pino Torinese
(Torino) Italy
email: curir@oato.inaf.it

[2]Dept. of Physics, Turin University, Italy
email: epoglio@studenti.ph.unito.it

[3]INAF - Astronomical Observatory of Trieste,via Tiepolo 11, Trieste, Italy
email: villalobos@oats.inaf.it

Abstract. The theory of the Milky Way formation, in the framework of the ΛCDM model, predicts galactic stellar halos to be built from multiple accretion events starting from the first structure to collapse in the Universe.

Evidences in the past few decades have indicated that the Galactic halo consists of two overlapping structural components, an inner and an outer halo. We provide a set of numerical N-body simulations aimed to study the formation of the outer Milky Way (MW) stellar halo through accretion events between a (bulgeless) MW-like system and a satellite galaxy. After these minor mergers take place, in several orbital configurations, we analyze the signal left by satellite stars in the rotation velocity distribution. The aim is to explore the orbital conditions of the mergers where a signal of retrograde rotation in the outer part of the halo can be obtained, in order to give a possible explanation of the observed rotational properties of the MW stellar halo.

Our results show that the dynamical friction has a fundamental role in assembling the final velocity distributions originated by different orbits and that retrograde satellites moving on low inclination orbits deposit more stars in the outer halo regions and therefore can produce the counter-rotating behavior observed in the outer MW halo.

Keywords. Galaxy: formation, kinematics and dynamics, halo

1. Introduction

Evidences in the past few decades indicated that the MW halo has a more complex structure than a single component. Recently Carollo *et al* (2007), analyzing a large sample of calibration stars from the Sloan Digital Sky Survey (SDSS), confirmed the existence of two broadly overlapping structural components -an inner and an outer halo- that exhibit different spatial density profiles, stellar orbits and stellar metalicities. In particular the inner halo shows a net prograde rotation around the center of the Galaxy. The outer halo, instead, shows a clear retrograde net rotation.

The theory of galaxy formation in a Λ Cold Dark Matter (ΛCDM) Universe predicts galactic stellar halos to be built from multiple accretion events starting from the first structures to collapse in the Universe. Halos, composed of both dark and baryonic matter, grow by merging with other halos. While the gas from mergers and accretions loses its energy through cooling and settles into a disk, the non dissipative material (accreted stars and dark matter (DM)) forms a halo around the Galaxy. We have many likely evidences of past accretion into the MW, therefore in this study we adopt the scenario in which the MW halo has been formed by subhalos accretion. We then focus on a possible origin of a retrograde outer stellar halo, defined as the set of stars orbiting around the Galaxy at large distance from the disk plane, namely $z > 15$ kpc. It is expected that such halo is made of stars stripped from merging or disrupted satellites of low mass.

With the aim to determine if, and in what condition, we can obtain a signal of retrograde rotation in the stellar distribution in the outer halo, we simulate minor mergers of a satellite halo with a main halo of Galactic mass (Murante *et al.* 2010). We used controlled and simplified numerical experiments instead of cosmological simulations, in order to focus on the dynamics of the outer halo, disentangling the effect of the pure gravitational dynamics from the complex gas physics that must be considered when performing self-consistent simulations of galaxy formation. Our simulations, are nevertheless cosmological motivated by the use of halos with characteristics in accord with cosmological simulations. After the satellite completes its merging/disruption process, we identify stars at distance larger than 15 kpc from the disk plane and analyze their rotational velocities in order to determine if an excess of counter-rotating signal in the outer halo can be produced by couples of mergers having orbits with identical inclination and opposite initial rotation. We also investigate the possible influence of the spin parameter of the primary halo, the numerical resolution and the distribution of the stellar component, on the results.

2. The Milky Way Stellar Halo

A spiral galaxy like the Milky Way has three basic components in its visible matter: the bulge,the disk and the stellar halo. The galactic disk is indeed surrounded by a spheroid halo of old stars and globular clusters, of which 90% lie within 30 kpc, suggesting a stellar halo diameter of 60 kpc.

The observed rotation of the stars and gas clouds indicates that the visible matter is surrounded by a DM halo containing the major portion of the total galaxy mass and extending very far beyond the visible matter. Some indirect means suggest that the DM halo may extend as far as 170 kpc/h from the center of the galaxy.

According to Carollo *et al.* (2007), the inner-halo component of the MW dominates the population of halo stars found at distance up to 10-15 kpc from the Galactic Center. An outer-halo component dominates in the region beyond 15-20 kpc. Inner-halo stars possess generally high orbital eccentricities, and exhibit a modest prograde rotation (between 0 and 50 km s^{-1}) around the center of the Galaxy. The distribution of metallicities for stars in the inner halo peaks at $[Fe/H] = -1.6$. Outer-halo stars cover a wide range of orbital eccentricities, and exhibit a clear retrograde net rotation (between -40 and -70 km s^{-1}) about the center of the Galaxy. The metallicity distribution function of the outer halo peaks at lower metallicity than that of the inner halo, around $[Fe/H] = -2.2$.

These properties indicate that the individual halo components probably formed in fundamentally different ways through successive dissipational (inner halo) and dissipationless (outer halo) mergers and tidal destruction of proto-Galactic clumps.

In our work we focus on the outer halo. To this purpose, we performed a set of numerical N-body simulations for studying the formation of the outer MW's halo through accretion events. After simulating minor mergers of a satellite with a DM main halo containing a stellar disk, we analyze the signal left by satellite stars in the rotational velocities distribution. Above all, we explore the orbital conditions starting from which a signal of retrograde rotation in the outer part of the halo can be obtained, in order to give a possible explanation of the rotational properties of the MW stellar halo. As we will discuss, the dynamical friction has a fundamental role in assembling the final velocity distributions originated by different orbits. Therefore, we will describe this effect in the next section.

3. Dynamical Friction

A particle of mass M_s moving through a homogeneous background of individually much lighter particles with an isotropic velocity distribution suffers a drag force (Chandrasekar, 1943):

$$F_d = -\frac{4\pi G^2 M_s^2 \rho_f(<v_s)ln\Lambda}{v_s^2} \tag{3.1}$$

where v_s is the speed of the satellite with respect to the mean velocity of the field and $\rho(<v_s)$ is the total density of the field particles with speeds less than v_s, Λ is the Coulomb Logarithm (Binney and Tremaine, 1987).

It is expected that the outer halo is made of stars stripped from merging or disrupted satellites of low mass. It is very important to note that, from equation (3.1), we expect that, the higher is v_s, the weaker is the dynamical friction force. Retrograde satellites have higher v_s with respect to prograde one, since in the first case velocity of the satellite is opposite to that of the disk and to rotational velocities of the main halo particles. As a consequence, prograde orbits decay faster. Particle stripped from the satellites will remain on the orbit on which the satellite was when they were stripped. How we can see from equation (3.1) the dynamical friction depends upon the satellite mass, so a single star does not feel the effect of this force, due to its to much small mass, and consequently conserves its energy remaining on the same orbit.

Another important gravitational effect, which happens during mergers, is the tidal disruption of satellites. While the tidal disruption is most important near the center of the main halo, where the gravitational potential is stronger, the dynamical friction is exerted both by the main halo DM particles and by the disk star particles.

4. The simulations

We perform two kinds of simulations. In both simulations, we use a primary DM halo with a NFW radial density profile (Navarro *et al* 1997), containing a stellar disk. What changes is the satellite: in the first set of simulations we realize a DM+bulge satellite configuration, in which the DM halo contains a stellar bulge, having an Hernquist (1990) radial density profile; in the second set the satellite is an halo with a NFW profile, in which we assume that the DM in the inner region traces the stars.

We simulated minor prograde mergers, in which a satellite co-rotates with the disk spin, and retrograde ones, with a counter-rotating satellite. Stars in the outer halo can be stripped from accreted DM satellites, hosting dwarf galaxies, during their orbits before the final merger. Minor mergers are the most probable sources for these kind of stars, firstly because they do not destroy the galactic disk and can thus happen in all evolutionary phases of the life of a disk galaxy. Moreover, a smaller satellite suffers less dynamical friction, and can deposit stars at larger distances from the galactic center, for a longer period of time.

We run all our simulation using the public parallel Treecode GADGET2 (Springel, 2005).

Into our halo, we embed a truncated stellar disk supported by rotation, having an exponential surface density law. The disk is in gravitational equilibrium with the DM halo.

An analytical mass model for spherical galaxies and bulge has been proposed by Hernquist (1990). This model is one of the most successful analytic model for these stellar systems. It as a $\rho \propto r^{-4}$ profile in the outer parts and a central density cusp of strength $\rho \propto r^{-1}$.

In the first set of simulations, in which we used a DM+bulge satellite configuration, we performed a suite of numerical experiments only in high resolution. In the second one, in which the satellite is a DM halo with a NFW profile, we run a suite of numerical experiments, varying the force and mass resolution of both the primary and the secondary.

We choose our minor mergers with a mass ratio $M_{prim}/M_{sat} \approx 40$, similar to the estimated mass ratio of the Large Magellanic Cloud (LMC) to the MW halo.

All the physical parameter of the main DM halo, of the DM satellite, of the DM+bulge satellite and of the main disk are listed in Tables 1, 2, 3 and 4.

Halo	M_{200}	R_{200}	C_{200}	R_{trunc}	N	ϵ	M_{DM}
Main-HR	10^{12}	165	7.25	1300	10^6	0.5	10^6
Main-LR	10^{12}	165	7.25	1300	10^5	1.0	10^7
Satellite-HR	$2.4 \cdot 10^{10}$	47	8.54	400	10^5	0.5	$2.4 \cdot 10^5$
Satellite-LR	$2.4 \cdot 10^{10}$	47	8.54	400	10^4	1.0	$2.4 \cdot 10^6$

Table 1. Properties of main and satellite halos (NFW profile).Column 1: Halo. Column 2: Virial mass in $(M_\odot/h)$; here we refer all our virial quantities to an overdensity of 200 times the mean cosmic density. Column 3: Virial radius in (kpc/h). Column 4: NFW concentration parameter $(c_{200} = r_{200}/r_s)$. Column 5: Truncation radius. Column 6: Softening length in (kpc/h). Column 7: Mass of DM particle in $(M_\odot/h)$.

Halo	M_{200}	R_{200}	C_{200}	ϵ	a	M_b
Satellite-stars	$2.4 \cdot 10^{10}$	47	8.54	0.25	0.709	$2.4 \cdot 10^9$

Table 2. Properties of the satellite halo (Hernquist profile). Column 1: Halo. Column 2: Virial mass in $(M_\odot/h)$. Column 3: Virial radius in (kpc/h). Column 4: NFW concentration parameter. Column 5: Softening length in kpc/h. Column 6: Hernquist scale radius in kpc/h. Column 7: Mass of the stellar bulge in $M_\odot/h$. In the case of the Hernquist profile, the secondary halo has an exponential cut-off in density.

Halo	N_{DM}	N_*	M_{DM}	M_*
Satellite-stars	$1.1 \cdot 10^5$	10^5	$1.95 \cdot 10^5$	$2.38 \cdot 10^4$

Table 3. Properties of the satellite halo (Hernquist profile). Column 1: Halo. Column 2: DM particles number inside the virial radius. Column 3: Star particles number inside the virial radius. Column 4: Mass of DM particle in $(M_\odot/h)$. Column 5: Mass of star particle in $(M_\odot/h)$.

We put our satellite on a prograde orbit, in which the satellite co-rotates with respect to the disk spin, and a retrograde (counter-rotating) one. We choose two orbits used in Read *et al.* (2008) for studying the thickening of the disk due to the same kind of minor merger: a low-inclination one, with a 10 degree angle with the disk plane, and a high inclination one with a 60 degree angle.

Initially, the center of the primary halo stays in the origin of our coordinate system. The satellite coordinates and its components of the velocity are listed in Table 5.

Disk	M_*	r_{disk}	r_0	N	ϵ	Q
Disk-HR	$5.7 \cdot 10^{10}$	20	4	10^6	0.5	2
Disk-LR	$5.7 \cdot 10^{10}$	20	4	10^5	1.0	2

Table 4. Properties of the disk. Column 1: Disk. Column 2: Disk stellar mass in $(M_\odot/h)$. Column 3: Disk truncation radius in (kpc/h). Column 4: Disk scale radius in (kpc/h). Column 5: Particles number inside the disk radius. Column 6: Softening length in (kpc/h). Column 7: Toomre parameter.

	Orbit	x	y	z	v_x	v_y	v_z
Pro1	Prograde-1 $(10°)$	80.0	0.27	-15.2	-6.3	62.5	0.35
Ret1	Retrograde-1 $(10°)$	80.0	0.27	-15.2	-6.3	-62.5	0.35
Pro2	Prograde-2 $(10°)$	29.5	0.27	-5.2	-6.3	89.3	0.35
Ret2	Retrograde-2 $(10°)$	29.5	0.27	-5.2	-6.3	-89.3	0.35
Pro	Prograde $(60°)$	15.0	0.12	-26.0	-1.2	80.1	2.0
Ret	Retrograde $(60°)$	15.0	0.12	-26.0	-1.2	-80.1	2.0

Table 5. In the first three columns, are listed the initial satellite coordinates in kpc, while in the last three the components of velocity in km/s. At low inclination, we simulate mergers with two different departures, one further from the disk center (1) and one nearer to the disk center (2).

In our full suite of numerical experiments, we vary the force resolution and the DM angular momentum of both the primary and the secondary. In our first set of simulation, our secondary halo always has a spin parameter $\lambda = 0$, where we define

$$\lambda = \frac{J}{\sqrt{2}MVR} \tag{4.1}$$

whit M being the mass inside a radius R and $V = GM/R$ the circular velocity, as in Bullock et al.(2001).

We use simulations with a primary halo with a spin parameter $\lambda = 1$ (A case) and $\lambda = 0$ (B case). We assign the angular momentum to DM particles using a rigid body rotation profile. In our main halo, the angular momentum of DM particles is always aligned with that of stellar disk.

In our second set of simulations, we repeated our experiments using a Bullock's profile instead of a rigid rotation one for the DM rotation velocity (only in the case in which the secondary halo has a DM+bulge configuration). Bullock et al.(2001) found indeed that in cosmological simulations the specific angular momentum j follows a power law: $j(r) \propto r^\alpha$ with α roughly distributed over the halos like a Gaussian. In this case, we set the spin parameter of the DM halo to $\lambda = 0.06$, a value suggested by cosmological simulations.

We note that the average rotational speed of haloes having a Bullock profile with $\lambda = 0.06$ and a rigid rotation profile with $\lambda = 1$ are of the same order of magnitude in the inner part, where the disk resides.

In the DM-only simulations the satellite is a DM halo with a NFW radial density profile. In this set of simulations, we follow the evolution of the inner region of the satellite, in particular a sphere of 8 kpc radius, centered on the center of mass of the satellite, and we assume that the DM particles in this inner regions trace the stars. The choice of such a value is due to the fact that, the gravitational mass contained within 8 kpc radius, is comparable with the gravitational mass of the LMC. Besides, since the NFW profile tends to r^{-3} at $r > r_s$, in good approximation one can consider that the majority halo mass is contained in the inner regions. We run our DM-only numerical experiments, varying the mass and force resolution in both the primary and the secondary: the case of LR has 10^4 particles within the virial radius of the DM satellite, whereas in the case of HR we have 10^5 particles within the virial radius of the DM satellite.

The time unit of the code is $T_* = 0.98$ Gyrs. We run all our simulations for $T = 4.725\,T_*$, corresponding to ~ 16 dynamical time of the main halo. We define a merger to be completed when the z coordinate of the center of mass of all satellite stars remains within 2 kpc from the initial disk plane.

The simulations were carried out at CASPUR, with CPU time assigned with the "Standard HPC grant 2009", and at the CINECA, Bologna, with CPU time assigned under an INAF/CINECA grants.

5. Distribution of v_Φ

In order to evaluate the v_Φ we rotate the coordinate system so that the disk lies in the XY plane. In Figure 1, we show histograms of the rotation velocity obtained in the

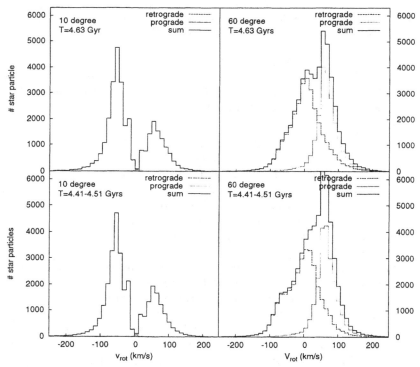

Figure 1. Histograms of rotation velocities for star particles in the outer halo, at the end of the set A of merger simulations. Upper panels show the histograms at the final time, lower panel show the same histograms, averaged over five simulation snapshots. In red (dashed line), we plot the histogram obtained for our retrograde orbits; in green (dotted line), those for our prograde orbits. In black (solid line) we show the sum of the two. Left column is for the low inclination orbits, right column for the high inclination ones. We used 50 velocity class, equispaced, between -300 and 300 km/s in all histograms.

four simulations of the set A: the satellite is either co-rotating or counter-rotating with respect to the disk, and the orbit has either a low or a high inclination with respect to the disk plane. In the upper panels, we show histograms at the final time of our simulations; in the lower ones, we performed an average over five consecutive snapshots, which cover a period of time of 100 Myr, to get rid of possible sampling effect on the particle orbits which can rise using a particular instant of time. We show both the distributions of stars from single simulations, and the one resulting from taking together star particles from prograde and retrograde orbits having the same initial inclination. The latter gives us a hint on the possible origin of the rotation signal in the case a similar number of prograde and retrograde minor accretion events happen during the formation history of a galaxy, under the hypothesis that their inclination is the same. It is immediately clear from the Figure1 that our pair of low inclination orbits show an excess of counter-rotating stars in the outer halo. It remains to be determined if the counter-rotating signal is caused by the interaction with the stellar disk of the main halo, or with the DM of the halo itself.

Figure 2 shows the results of the same analysis performed on our set A, but in the case in which no halo spin is present (set B). Here we only show the results for our last snapshot; as in Figure 1, averaging over five snapshots makes no appreciable difference. Also in our set B, low inclination retrograde orbits produce more counter-rotating star

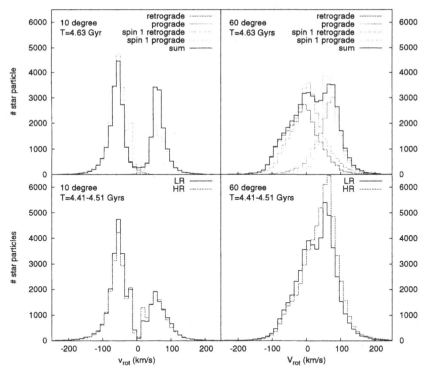

Figure 2. Upper row: histograms of rotation velocities for star particles in the outer halo, at the end of the set B of merger simulations, in which the main halo has no spin. In red (dashed line), we plot the histogram obtained for our retrograde orbits; in green (dotted line), those for our prograde orbits. In black (solid line) we show the sum of the two. Left panel is for the low inclination orbits, right panel for the high inclination ones. We also show the histogram for the case of spin 1 (dashed-dotted lines). Lower row: histogram of rotation velocities for star particles in the outer halo, for our set A, at the basic resolution ("LR", black continuous line) and at a ten times better mass resolution ("HR", red dashed line).

particle than co-rotating stars produced by prograde orbits. Again, this is not the case for high inclination orbits. From Figure 2, it is clear that the excess of counter-rotating star is anyway smaller: and this is due to the fact that *more* co-rotating stars are produced by prograde orbit in our no spin case than in spin 1 case.

Therefore, both disk rotation and halo spin contribute to the slowing-down of prograde orbits and to the consequent smaller amount of high-energy star particle stripped from satellites that can reach the outer halo. However, disk rotations appear to be the main driver of such an effect.

As far as the resolution comparison is concerned, in the lower row of Figure 2 we note that our result does not depend on the mass resolution of our satellite halo. Even using 10 times more particles in the secondary, only our low inclination couple of mergers shows an excess of counter-rotating stars in the outer halo.

6. Conclusions

Our main results are the following:

• low inclination mergers do produce an excess of counter-rotating satellite stellar particles in the outer halo, independently on the spin of the DM;

• high inclination orbits do not seem to produce a significant counter-rotating signal;

• the fraction of counter-rotating to co-rotating satellite stars in the outer stellar halo is higher if the DM has spin.

In the ΛCDM model, there is no reason for expecting an excess of retrograde mergers (Sales *et al.* 2008). Thus, if such an excess was needed to produce a counter-rotating signal in the outer stellar halo, we should require a peculiar accretion history for the MW. The main result of this work shows that such a fine-tuning in the MW history is not needed. Even if, statistically, the number of prograde and retrograde minor merger is the same, still a counter-rotating signal can arise *if such mergers predominantly happen along low inclination orbits*. Since matter accretion, in a CDM dominated Universe, mainly occurs along filaments, this will be the case *if the galaxy disk is co-planar* to the (majority of) filaments. The disk-filament alignment issue is still debated (Brunino *et al.* 2007): from our results, we expect that if the galactic disk were perpendicular to the main accretion streams, no counter-rotating signal should be observed in the outer halo star distribution.

References

Carollo, D., Beers, T., C., Sun Lee, Y., Chiba, M., E. Norris, J. E., Wilhelm, R., Sivarani, T., Marsteller, B., Munn, J., A., Bailer-Jones, C., A., L., Re Fiorentin, P. & York, D., G. 2007, *Nature*, 450, 1020

Murante, G., Poglio, E., Curir, A. & Villalobos, A. 2010, *ApJ* (Letters), 716, L115

Chandrasekar, S. 1943, *ApJ*, 97, 255

Binney and Tremaine 1987, in: *Galactic Dynamics* (Princeton, NJ: Princeton University Press), p. 427

Navarro, J. F., Frenk, C. S. & White, S. D. M. 1997, *ApJ*, 190, 493

Hernquist, L. 1990, *ApJ*, 356, 359

Springel, V. 2005, *ApJ*, 364, 1105

Read, J. I., Lake, G., Agertz, O. & Debattista, V. P. 2008, *MNRAS* 389, 1041

Bullock, J., Dekel, A., Kolatt, T., Kravtsov, A.,V., Klypin, A. & Porciani, C. and Primack, J. 2001, *ApJ*, 555, 240

Sales, L., V., Helmi, A., Starkenburg, E., Morrison, H., L., Engle, E., Harding, P., Mateo, M., Olszewski, E., W. & Sivarani, T. 2008, *MNRAS* 389, 1391

Brunino, R., Trujillo, I., Pearce, F. R. & Thomas, P. A. 2007, *MNRAS* 375, 184

Astrophysical Dynamics: From Stars to Galaxies
Proceedings IAU Symposium No. 271, 2010
N. H. Brummell, A. S. Brun, M.S. Miesch & Y. Ponty, eds.

© International Astronomical Union 2011
doi:10.1017/S174392131101756X

Large Dynamic Range Simulations of Galaxies Hosting Supermassive Black Holes

Robyn Levine[1]

[1]Canadian Institute for Theoretical Astrophysics,
60 St. George St., Toronto, ON, M6G 1A6, Canada
email: levine@cita.utoronto.ca

Abstract. The co-evolution of supermassive black holes (SMBHs) and their host galaxies is a rich problem, spanning a large-dynamic range and depending on many physical processes. Simulating the transport of gas and angular momentum from super-galactic scales all the way down to the outer edge of the black hole's accretion disk requires sophisticated numerical techniques with extensive treatment of baryonic physics. We use a hydrodynamic adaptive mesh refinement simulation to follow the growth and evolution of a typical disk galaxy hosting an SMBH, in a cosmological context (covering a dynamical range of 10 million!). We have adopted a piecemeal approach, focusing our attention on the gas dynamics in the central few hundred parsecs of the simulated galaxy (with boundary conditions provided by the larger cosmological simulation), and beginning with a simplified picture (no mergers or feedback). In this scenario, we find that the circumnuclear disk remains marginally stable against catastrophic fragmentation, allowing stochastic fueling of gas into the vicinity of the SMBH. I will discuss the successes and the limitations of these simulations, and their future direction.

Keywords. galaxies: evolution, galaxies: high-redshift, galaxies: nuclei

1. Introduction

Observations suggest that most galaxies host massive compact objects in their centers, likely to be supermassive black holes (SMBHs) with masses ranging from $10^6 \, M_\odot$ to $10^9 \, M_\odot$ (e.g., Kormendy & Richstone 1995; Magorrian et al. 1998). The ubiquity of SMBHs at the centers of galaxies includes our own galaxy, where the use of adaptive optics has constrained the mass of the Milky Way's black hole to be $\approx 4.5 \times 10^6 \, M_\odot$ (Ghez *et al.* 2008).

Observations of galaxies hosting SMBHs indicate a relationship between black hole mass and various properties of the host galaxies, such as the spheroid mass (Magorrian *et al.* 1998) and the velocity dispersion of stars in the bulge (Ferrarese & Merritt 2000; Gebhardt *et al.* 2000; Tremaine *et al.* 2002). In order for such correlations to arise, there must be some processes linking the growth of SMBHs (objects with a sphere of influence of order tens of parsecs) to the evolution of their host galaxies (governed by interactions on much larger, cosmological scales).

Active galactic nuclei (AGNs) undoubtedly are an important piece of the puzzle of galaxy and SMBH co-evolution. They output an enormous amount of energy into the host galaxy through both radiative and mechanical feedback, which can potentially influence galaxy evolution on scales ranging from that of the interstellar medium (ISM) to the intracluster medium. Since AGN are driven by accretion onto SMBHs, it is essential to understand black hole fueling. On galactic scales, fueling is determined by both the secular evolution of the host galaxy and by outside influences from the galaxy's environment (such as mergers with neighboring galaxies). Fueling may occur continuously over the course of some large-scale dynamical instability in the galaxy, or it may be an intermittent process, dependent entirely on the dynamics on small scales. Intermittent accretion episodes consisting of in-falling clouds of gas with randomly oriented angular momentum vectors may contribute to the spin-down of SMBHs, thus lowering their radiative efficiency and allowing them to grow faster (e.g., King & Pringle 2007; King, Pringle, & Hofmann 2008). Modeling the many physical processes relevant for the growth of SMBHs and their host galaxies is a challenging numerical feat, spanning a large range of scales.

In an attempt to meet the numerical challenge,we use cosmological adaptive mesh refinement (AMR) simulations with a large dynamic range to study the transport of gas and angular momentum through the circumnuclear disk of a SMBH host galaxy over time. Here we present some results of these simulations, which are presented in greater detail in our previous work (Levine *et al.* 2008; Levine 2008; Levine, Gnedin, & Hamilton 2010).

2. Simulations of Black Hole-Galaxy Co-Evolution

Different Approaches. There are a range of numerical tools that are individually suited for addressing different pieces of the problem of SMBH-Galaxy co-evolution. Cosmological simulations follow the evolution of large scale structure and are well suited for following sequences of events occuring slowly over cosmic time (such as following merger histories, etc.). Often cosmological simulations implement sub-grid models to describe processes occurring on scales smaller than the resolution of the simulations (e.g. gas cooling, star formation, and feedback processes). Large-scale simulations often cannot follow the circumnuclear regions of galaxies with high enough resolution to describe the accretion onto the SMBH in detail, and must make approximations using the properties of the galaxies on scales that are resolved. A common practice is to estimate the Bondi accretion rate to describe the growth of the black hole (Bondi 1952). Such simulations can be sued to study the effects of merger-driven fueling and AGN feedback on SMBH growth and galaxy morphology and demographics (Sijacki et al. 2007; Di Matteo *et al.* 2008; Colberg & Di Matteo 2008; Croft et al. 2009).

Small-scale simulations focus their attention on individual galaxies or even smaller scales inside them (e.g. accretion disks, star formation, or the ISM). High-resolution simulations of subgalactic scale disks, like those found is SMBH host galaxies, have found the development of a turbulent, multi-phase ISM (e.g., Wada 2001; Escala 2007; Wada & Norman 2007). If the galactic disk is also self-gravitating, then this implies that accretion onto a central SMBH in such an environment is not well modeled by the Bondi accretion rate, as is often approximated in larger-scale simulations.

Adaptive Mesh Refinement and the Zoom-In Technique. Following both the complex baryonic physics of the ISM and star formation, and the cosmological evolution of a galaxy simultaneously, requires the use of adaptive simulation techniques. We address the issue of SMBH fueling using state-of-the-art cosmological simulations which are ideally suited to meet the numerical challenges presented above. The simulations use the AMR technique to self-consistently model the gas dynamics in a single galaxy at high resolution (subparsec resolution in the center of the galaxy). A large dynamic range (> 10 million), achievable with the AMR technique, allows us to bridge cosmological scales to scales relevant for molecular cloud formation (the birthplace of stars) and AGN fueling. It is a complex task to implement mergers, feedback, and secular evolution in large, cosmological simulations all at once. Our approach is to split the problem into pieces to be addressed one at a time, ultimately building a more realistic simulation. After studying the physics in this basic model galaxy, we can begin to include physical processes that are directly relevant to the problem of SMBH growth in the context of galaxy evolution, such as AGN feedback.

We have performed simulations using the Adaptive Refinement Tree (ART) code (Kravtsov, Klypin, & Khokhlov 1997; Kravtsov 1999; Kravtsov, Klypin, & Hoffman). The code follows gas hydrodynamics on an adaptive mesh, and includes dark matter and stellar particles (with stars forming at an observationally motivated rate), as well as gas cooling by heavy elements and dust, using rates tabulated from the CLOUDY code (Ferland *et al.* 1998). Additionally, radiative transfer and feedback and enrichment from stars are included in the pre-zoomed, cosmological portion of the simulation.

Beginning with a cosmological simulation with a maximum resolution of $\approx 50\,\mathrm{pc}$ proper at $z = 4$, the resolution is slowly increased one refinement level at a time in a region centered on a typical disk galaxy (which will evolve into a galaxy about the size of the Milky Way at $z = 0$). The simulated galaxy reaches a quasi-stationary state on each level before the simulation refines to the next level. The final maximum resolution is $\approx 0.03\,\mathrm{pc}$, corresponding to 20 levels of refinement. After reaching the maximum resolution, a fraction of the gas in the center of the galaxy is replaced with a black hole particle of equal mass and momentum. For simplicity, the mass of the black hole particle does not change over the course of any of our current zoom-in simulations. After the introduction of the black hole particle, the simulation continues to evolve with the maximum resolution for several hundred thousand years, allowing us to follow the evolution of the circumnuclear region of the galaxy at high-resolution. The zoom-in technique is also applied at different redshifts of the cosmological simulation, and for different black hole masses.

3. Results

Mass Accretion Rate. As the simulation evolves at high resolution, the circumnuclear disk becomes globally unstable, driving super-sonic turbulence. The turbulence supports the gas disk against catastrophic fragmentation into star forming clumps, maintaining marginal stability in the disk. The complex dynamical state of the circumnuclear disk results in transient structures forming on a range of scales. This behavior causes the gas mass interior to any radius within the disk to vary significantly on short timescales. Figure 1 shows the amplitude of fluctuations in the gas mass interior to four different radii throughout the circumnuclear region (different panels) over time for three different redshift simulations (different lines). The Figure gives an indication of the turbulent nature of the gas in the circumnuclear disk of the galaxy, and suggests that a characteristic accretion rate onto the SMBH is not straightforwardly determined by any one individual snapshot of the circumnuclear region. Instead, the accretion rate fluctuates randomly on short time scales because of the complicated dynamics of the circumnuclear disk. This result is qualitatively similar to the findings of Hopkins & Quataert (2010), who use a different simulation technique (involving increasingly smaller re-simulations of the central disk) to reach high resolution in their SPH galaxy simulations.

Angular Momentum. Several authors have discussed the likelihood that gas which accretes onto a black hole can influence its spin (e.g., King & Pringle 2007; Volonteri, Sikora, & Lasota 2007; Berti & Volonteri 2008; King *et al.* 2008). Whether accreting gas increases or decreases the spin of the black hole (ultimately determining its accretion efficiency and potentially the mode of feedback it produces) depends on the mass and angular momentum of the accreting material. Our simulations model gas on parsec scales, enabling us to resolve the direction of the angular momentum vector of gas (measuring position and velocity relative to that of the black hole particle) as it evolves over time.

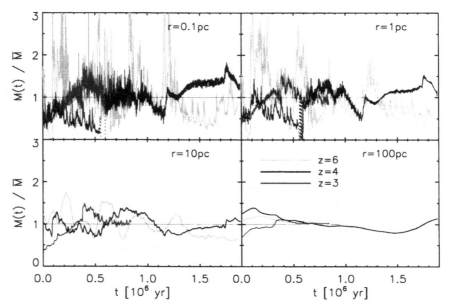

Figure 1. Amplitude of gas mass fluctuations at redshifts 3, 4, and 6 for 4 different radii (measured with respect to the location of the black hole particle). The thin solid line shows $M = \overline{M}$ for comparison.

Figure 2 shows a map projection of the direction of the normalized angular momentum vector, $\hat{\mathbf{L}}\mathbf{L}/L$, as it evolves at 1 and 100 pc from the black hole particle (for our $z = 4$ simulation). The maps are oriented so that the angular momentum vector of the disk on kiloparsec scales lies in the center. The disk is slightly warped in the circumnuclear region, so that the axis of the disk is oriented at an angle to the large-scale disk.

At 1 pc (bottom), the direction of the angular momentum vector starts out at approximately the same orientation as kiloparsec scales, but shows slightly more scatter. The map shows two sudden changes in the direction of the angular momentum vector by 100 degrees each. The first incident, at $t \sim 0.55$ Myr, occurs as a clump of gas with mass comparable to that of the black hole falls into the center. The scatter in the measurement of the angular momentum vector near this time (visible at 100 pc as well) corresponds to a temporary displacement of the black hole particle as the clump reaches the center. The second incident, at $t \sim 1.2$ Myr, is the result of gravitational interaction with a massive clump of gas which develops at > 100 pc and continues to move toward the center of the disk at late times. Similar angular momentum flips occur over the course of several other runs with varying black hole masses.

At 100 pc (top), the angular momentum vector slowly changes direction, showing little scatter over the course of the simulation. The slow change in direction at 100 pc may correspond to the increased warping of the disc as the simulation progresses. The behavior of the angular momentum vector, particularly on small scales in the simulations, consistently shows that the gas ultimately delivered to the SMBH may have varying angular momentum, which can change as massive clumps of gas develop and move through the disk.

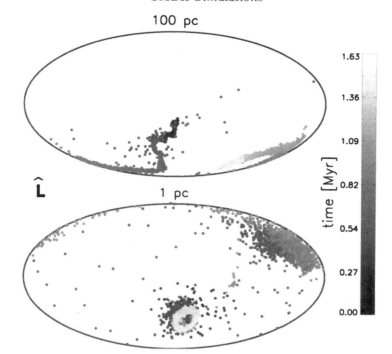

Figure 2. Map showing the direction of the normalized angular momentum vector, $\mathbf{L}/L$, as it changes over time in the $z = 4$ simulation. Grey-scale corresponds to the elapsed time since the introduction of the black hole particle. The maps are aligned so that the mean rotation axis of the galactic disk on a scale of 1 kpc sits in the center.

4. Conclusions

With a high-resolution cosmological simulation of a galaxy hosting an SMBH, we have modeled a simplified picture of accretion through a circumnuclear disk (without feedback from the black hole and in the absence of any major mergers). We found the development of a turbulent, self-gravitating, and globally unstable disk in the central kiloparsec. Accretion through this disk is stochastic, and not well modeled by the usual prescriptions employed in coarser resolution simulations. As clumps develop in the disk, they can potentially influence the angular momentum of the accretion disk on small scales, and subsequently the spin of the black hole.

A detailed analysis of accretion and angular momentum of gas requires accurate modeling of an array of physical processes that are important for galaxy evolution. In particular, resolving star forming regions and accurately modeling ISM physics can potentially change the dynamics of the circumnuclear disk. In new and improved simulations with the ART code, we plan to add additional physics important for studying black hole and galaxy co-evolution (such as feedback from black holes and black hole mergers) in order to better understand accretion onto black holes and the influence they have on their environment.

References

Berti, E. & Volonteri, M. 2008, *ApJ*, 684, 822
Bondi, H. 1952, *MNRAS*, 112, 195

Colberg, J. & Di Matteo, T. 2008, *MNRAS*, 387, 1163

Croft, R. A. C., Di Matteo, T., Springel, V., & Hernquist, L. 2009, *MNRAS*, 400, 43

Di Matteo, T., Colberg, J., Springel, V., Hernquist, L., & Sijacki, D. 2008, *ApJ*, 676, 33

Escala, A. 2007, *ApJ*, 671, 1264

Ferland *et al.* 1998, *PASP*, 110, 761

Ferrarese, L. & Merritt, D. 2000, *ApJL*, 539, L9

Gebhardt, K. *et al.* 2000, *ApJL*, 539, L13

Ghez, A. M. *et al.* 2008, *ApJ*, 689, 1044

Hopkins, P. F. & Quataert, E. 2010, *MNRAS*, pub. online 21 July, 2010

King, A. R., Pringle, J. E., & Hofmann, J. A. 2008, *MNRAS*, 385, 1621

King, A. R. & Pringle, J. E. 2007, *MNRAS*, 377, L25

Kormendy, J. & Richstone, D. 1995, *ARAA*, 33, 581

Kravtsov, A. V., Klypin, A. A., & Khokhlov, A. M. 1997, *ApJS*, 111, 73

Kravtsov, A. V. 1999, Ph.D. Thesis, New Mexico State University

Kravtsov, A. V., Klypin, A. A., & Hoffman, Y. 2002, *ApJ*, 571, 563

Levine, R., Gnedin, N. Y., Hamilton, A. J. S., & Kravtsov, A. V. 2008, *ApJ*, 678, 154

Levine, R. 2008, *Simulating the growth of a disk galaxy and its supermassive black hole in a cosmological context*, Ph.D. thesis, University of Colorado at Boulder

Levine, R., Gnedin, N. Y., & Hamilton, A. J. S. 2010, *ApJ*, 716, 1386

Magorrian, J. *et al.* 1998, *AJ*, 115, 2285

Sijacki, D., Springel, V., Di Matteo, T., & Hernquist, L. 2007, *MNRAS*, 380, 877

Tremaine, S. *et al.* 2002, *ApJ*, 574, 740

Volonteri, M., Sikora, M., & Lasota, J. P. 2007, *ApJ*, 667, 704

Wada, K. 2001, *ApJL*, 559, L41

Wada, K. & Norman, C. A. 2001, *ApJ*, 547, 172

Discussion

VASIL: Is the Zeldovich approximation era over once the solutions to burger equation form shocks?

LEVINE: Yes, because once shocks develop, the temperature of the gas rises, introducing additional pressure so that the motion of gas particles is no longer described by the Zeldovich approximation. The Zeldovich approximation era precedes the formation of proto-galaxies, and we apply the zoom-in technique well after this point.

BRUN: Angular momentum transport depends strongly on the turbulence level. How confident are you that your simulations are "converged" in a hydrodynamical sense? What is the Reynolds number of your simulation?

LEVINE: We are convinced that the level of turbulence we measure at our resolution limit (and the implied turbulent viscosity) is consistent with the dynamics of the gas that we see. We've done a couple different things– measuring the turbulent viscosity by modeling the evolution of passive scalars, varying our resolution and the refinement technique, and all of these things produce consistent results. However, even though our results are self-consistent, there could be additional important effects that we fail to model because they're beneath our resolution limit. Our Reynolds number is probably enormous because the circumnuclear disk is very centrally concentrated (so that the velocities are quite high), but it's possible that fragmentation on small scales would lead to star formation, which would change the structure of the disk (and subsequently the dynamics).

COMBES: Your gaseous accretion disk appears very clumpy and unstable around the black hole; however it should be dominated by the mass of the black hole. Shouldn't it be only slightly self-gravitating?

LEVINE: Actually, the mass of the disk is enormous compared to the mass of the black hole, which we arbitrarily assigned to be $3 \times 10^7 \, M_\odot$ (in our fiducial simulation). So the disk is highly self-gravitating and the black hole doesn't influence the dynamics at all. The disk in these simulations has a very high gas fraction, but in future simulations, done with a newer version of the code, the gas fraction of the galaxy will be lower, so the black hole could potentially play a bigger role.

NTORMOUSI: How do you compare your black hole accretion model to Bondi-Hoyle accretion model usually employed in cosmological simulations?

LEVINE: Since the gas disk is so massive, it is self-gravitating and really doesn't resemble the Bondi regime at all. We attempted to estimate the Bondi accretion rate by using modified prescriptions accounting for turbulence and vorticity, but the rates even they suggest are enormous and don't at all resemble the behavior we observe. One of our goals though is to understand how the high-resolution simulations map onto lower resolution, cosmological simulations, where a Bondi prescription seems to give reasonable results (on average). That's a work in progress.

Astrophysical Dynamics: From Stars to Galaxies
Proceedings IAU Symposium No. 271, 2010
N. H. Brummell, A. S. Brun, M. S. Miesch & Y. Ponty, eds.

© International Astronomical Union 2011
doi:10.1017/S1743921311017571

Star formation in galaxy mergers: ISM turbulence, dense gas excess, and scaling relations for disks and starbusts

Frédéric Bournaud, Leila C. Powell, Damien Chapon and Romain Teyssier

CEA, IRFU, SAp – CEA Saclay, F-91191 Gif-sur-Yvette, France.
email: frederic.bournaud@cea.fr

Abstract. Galaxy interactions and mergers play a significant, but still debated and poorly understood role in the star formation history of galaxies. Numerical and theoretical models cannot yet explain the main properties of merger-induced starbursts, including their intensity and their spatial extent. Usually, the mechanism invoked in merger-induced starbursts is a global inflow of gas towards the central kpc, resulting in a nuclear starburst. We show here, using high-resolution AMR simulations and comparing to observations of the gas component in mergers, that the triggering of starbursts also results from increased ISM turbulence and velocity dispersions in interacting systems. This forms cold gas that are denser and more massive than in quiescent disk galaxies. The fraction of dense cold gas largely increases, modifying the global density distribution of these systems, and efficient star formation results. Because the starbursting activity is not just from a global compacting of the gas to higher average surface densities, but also from higher turbulence and fragmentation into massive and dense clouds, merging systems can enter a different regime of star formation compared to quiescent disk galaxies. This is in quantitative agreement with recent observations suggesting that disk galaxies and starbursting systems are not the low-activity end and high-activity end of a single regime, but actually follow different scaling relations for their star formation.

Keywords. galaxies: formation, galaxies: mergers, galaxies : star formation

1. Introduction: Galaxy mergers as triggers of star formation

Evidence that interactions and mergers can strongly trigger the star formation activity of galaxies has been observed for more than two decades (e.g., Sanders *et al.* 1988). Nevertheless, both the underlying mechanisms the role of mergers in the cosmic budget of star formation remain largely unknown. Mergers have long appeared to potentially dominate the star formation history of the Universe. Starbursting galaxies with specific star formation rates much above the average, such as Ultraluminous Infrared Galaxies (ULIRGs), are almost exclusively major interactions and mergers – in the nearby Universe (see Duc *et al.* 1997), not necessarily at high redshift. The decrease in the cosmic density of star formation between $z = 1$ and $z = 0$ (Le Floc'h *et al.* 2005) may follow, and might even result from, the decrease in the galaxy merger rate. Other studies based on disturbed UV morphologies and/or kinematics also suggested high merger fractions among star forming galaxies at $z \sim 1$ or even below (Hammer *et al.* 2005).

However, disturbed morphologies and kinematics can also arise internally, without interactions, especially at high redshift when disk galaxies are wildly unstable, clumpy and irregular (Elmegreen *et al.* 2007; Förster Schreiber *et al.* 2009). Merger-induced ULIRG-like starbursts may also be less important in the cosmic budget than the more moderate but more numerous objects with internally sustained star-formation, as is possibly the case for most high-redshift LIRGs (e.g., Daddi *et al.* 2010a). Recent studies aimed at accurately distinguishing the signatures of mergers from internal evolution actually suggest that major interactions and mergers account only for a small fraction of the cosmic star formation history (Jogee *et al.* 2009; Robaina *et al.* 2009). Moreover, observational estimates appear to be quite dependent on the chosen tracers of star formation. For instance, mergers are probably a more important trigger of dust-obscured star formation than of general star formation, and thus the fraction of mergers among objects with high infrared-traced star formation rate can be higher (as discussed by Robaina *et al.* 2009).

Numerical simulations are then required to understand the mechanisms leading to starbursting activity in mergers, but also to further probe the contribution of mergers to cosmic star formation since observational estimates remain uncertain. Section 2 reviews the standard knowledge on merger-induced star formation, which is mostly based on "subgrid" modeling: all steps from the formation of cold/dense gas clouds to the formation of actual stars remain unresolved, and described with arbitrary recipes. This standard understanding cannot account for some general observationed features that we briefly review. Section 3 presents a new generation of high-resolution models in which the first steps of galactic-scale star formation, namely ISM turbulence and cold/dense gas cloud formation, are explicitly resolved using high-resolution codes. Based on this, we provide a substantially different explanation for interaction-triggered star formation and show that it could better account for recent observations of disk galaxies and starbursting mergers.

2. Standard models versus observations

2.1. *Merger-induced gas flows*

In an interacting galaxy pair, the gas content of one galaxy undergoes tidal forces from the companion galaxy, but this is actually not the main direct driver of the gas response. The gas distribution becomes non-axisymmetric, because of the asymmetry in the gravitational field itself induced by the companion. This results in gravitational torquing of the gas. - a thorough description of the mechanism can be found in Bournaud (2010, Section 2 and Figure 1). Gas initially inside the corotation radius (typically a radius of a few kpc) undergoes negative gravity torques and flows inwards in a more and more concentrated central component. Gas outside the corotation, i.e. initially in the outer disk, undergoes positive gravity torques and gains angular momentum, forming long tidal tails.

2.2. *Nuclear starbursts*

Gravitational torquing in the inner disk increases the gas concentration in the central kiloparsec or so. Any model for star formation will then predict an increase of the star formation rate (global Schmidt-Kennicutt law, models based on cloud-cloud collisions, etc). The result is thus a centralized or nuclear starburst. As the driving process is gravitational torquing, early restricted three-body models could already describe the effect (Toomre & Toomre 1972). Later models have added extra physical ingredients leading to more accurate predictions on the star formation activity: self-gravity (Barnes & Hernquist 1991), hydrodynamics and feedback processes (Mihos & Hernquist 1996; Cox *et al.* 2008), etc.

A large library of SPH simulations of galaxy mergers, in which the driving process is mostly the one presented above (tidal torquing of gaseous disk) was performed and analyzed by (Di Matteo *et al.* 2007, 2008). This study highlighted various statistical properties of merger-induced starbursts. In particular, it showed that some specific cases can lead to very strong starbursts with star formation rate (SFRs) are increased by factors of 10–100 or more, but that on average the enhancement of the SFR in a random galaxy collision is only a factor of a few (3–4 being the median factor) and only lasts 200-400 Myr. These results were confirmed with code comparisons, and found to be independent of the adopted sub-grid model for star formation (Di Matteo *et al.* 2008). Models including an external tidal field to simulate the effect of a large group or cluster found that the merger-induced starbursts could be somewhat more efficient in such cases – but the SFR increase remains in general below a factor of 10 (Martig & Bournaud 2008).

2.3. *Theoretical predictions versus observations*

The intensity of merger-induced starbursts Numerical simulations reproducing the interaction-induced inflow of gas and resulting nuclear starbursts can sometimes trigger very strong starbursts, but in general the SFR enhancement peaks at 3–4 times the sum of the SFRs of the two pre-merger galaxies. This factor of 3–4 seems in good agreement with the most recent observational estimates (Jogee *et al.* 2009; Robaina *et al.* 2009, e.g.). In fact, there is substantial disagreement: the factor 3–4 in simulation samples is the peak amplification of the SFR in equal-mass mergers. In observations, it is the average factor found at random (observed) instant of interactions, and in mergers that are "major" ones but not strictly equal-mass ones. Given that typical duration of a merger is at least twice longer than the starburst activity in the models, and that unequal-mass mergers make substantially weaker starbursts (Cox *et al.* 2008), one would need a peak SFR enhancement factor of about 10–15 (as measured in simulations) to match the average enhancement of 3–4 found in observations. There is thus a substantial mismatch between the starbursting activity predicted by existing samples of galaxy mergers, and that observed in the real Universe – although the observational estimates remain debated and may depend on redshift.

The spatial extent of merger-induced starbursts

A disagreement between these 'standard' models/theories and observations of mergers is also found in the geometry and spatial extent of star formation. The very strongest merger-induced starbursts (ULIRGs) are centrally-concentrated, but there are many examples of significant merger-induced starbursts that are spatially extended. A well-known example are the Antennae galaxies, where the burst of star formation proceeds, for a large fraction, in a few big star-forming clumps in extended disks and in the bridge between the two galaxies (Wang *et al.* 2004). Another well-known example is the IC2163/NGC2207 pair, which has an extended starburst in big gas clouds with remarkably high gas velocity dispersions (Elmegreen *et al.* 1995). There are many other examples of extended star formation in merging systems (Cullen *et al.* 2006; Smith *et al.* 2008).

Quantitative comparisons of the extent of star formation in observations to that predicted by "standard" models have shown a significant disagreement Barnes (2004); Chien & Barnes (2010). These authors also suggested that a sub-grid model of shock-induced star formation may better account for the spatial extent of merger-induced star formation (see also Saitoh *et al.* 2009).

The standard mechanism for merger-induced star formation, as reproduced in low-resolution simulations, certainly takes place in real mergers – signs of nuclear merger-driven starbursts are a plenty. But it seems impossible to explain the typical intensity of merger-induced starbursts and their often relatively larger spatial extend, based on these "standard" models. Some relatively old observations may actually hold the key to correctly understanding merger-induced star formation:

2.4. *A key observation: ISM turbulence in interacting galaxies*

The cold gas component of interacting galaxies has high velocity dispersions, which can reach a few tens of km s^{-1}. This has been noticed in the 90s in at least two well-resolved interacting and merging pairs (Irwin 1994; Elmegreen *et al.* 1995). Many interacting pairs and mergers in Green *et al.* (2010) also have large velocity dispersions (in the ionized gas). More generally, massive star clusters in merging systems suggest the Jeans mass is high, which is indicative of high velocity dispersions. A typical number could be a factor 4 of increase for the cold gas velocity dispersion in equal-mass mergers. An interesting question will be understand whether these dispersions result from the tidal interaction and are a trigger star formation, or whether they just result from the starburst and associated feedback effects.

Traditional SPH simulations model a relatively warm gas for the ISM, because the limited spatial resolution translates into a minimal temperature under which gas cooling should not be modeled (it would generate artificial instabilities, Truelove *et al.* 1997). Modeling gas cooling substantially below 10^4 K requires "hydrodynamic resolutions"† better than 100 pc. Cooling down to 100 K and below can be modeled only at resolutions of a few pc. The vast majority of existing merger simulations hence have a sound speed of at least 10 km/s and cannot explicitly treat the supersonic turbulence that characterizes most of the mass in the real ISM (Burkert 2006). Turbulent speeds in nearby disk galaxies are of 5-10 km/s, for sound speed of the order of 1-2 km/s in molecular clouds, i.e. turbulent Mach numbers up to a few. These are even higher in high-redshift disks (Förster Schreiber *et al.* 2009), and in mergers (references above), but not necessarily for the same reason.

Increased ISM turbulence in galaxy mergers is thus absent from the modeling used in most hydrodynamic simulations to date. Some particle-based models have nevertheless been successful in reproducing these increased gas dispersions (Elmegreen *et al.* 1993; Struck 1997; Bournaud *et al.* 2008), indicating that it is a consequence of the tidal interaction which triggers non-circular motions, rather than a consequence of starbursts and feedback. It should then arise spontaneously in hydrodynamic models, if these a capable of modeling gas below 10^{3-4}K.

3. Merger-induced starbursts with resolved ISM turbulence and cluster star formation

3.1. *High-resolution AMR simulations*

Adaptive Mesh Refinement (AMR) codes allow hydrodynamic calculations to be performed at very high resolution on adaptive-resolution grids. The resolution is not high everywhere, but the general philosophy is to keep the Jeans length permanently resolved until the smallest cell size is reached. That is, the critical process in the collapse of dense star-forming clouds, namely the Jeans instability (or Toomre instability in a rotating disk) is constantly resolved up to a typical scale given by the smallest cell size, or a small multiple of it.

AMR simulations of whole galaxies have recently reached resolutions of a few pc for disk galaxies (Agertz *et al.* 2009b; Tasker & Tan 2009), and even 0.8 pc lastly (Bournaud *et al.* 2010). Such techniques have been first employed to model ISM dynamics and star formation in galaxy mergers by Kim *et al.* (2009); Teyssier *et al.* (2010).

We here study the properties of star formation in a sample of a few AMR simulations of 1:1 mergers of Milky Way-type spiral galaxies, performed with a resolution of 4.5 pc

---

† i.e., the average smoothing length in SPH codes, or the smallest cell size in AMR codes

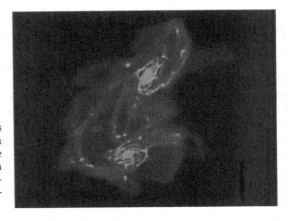

Figure 1. Surface density of the cold gas component in an AMR merger simulation at 4.5 pc resolution. Dense gas clouds are more numerous and more massive than in similar models of isolated disk galaxies, owing to strong non-circular motions in interacting systems.

and a barotropic cooling model down to ~50 K, technically similar to the isolated disk simulation described in (Bournaud *et al.* 2010). Star formation takes place above a fixed density threshold and is modeled with a local Schmidt law, i.e. the local star formation rate density in each grid cell is $\rho_{SFR} = \epsilon_{ff}\rho/t_{ff} \propto \rho^{1.5}$ where ρ is the local gas density and t_{ff} the gravitational free-fall time, and the efficiency ϵ_{ff} is a fixed parameter. Supernova feedback is included. Further details and results for whole sample will be presented in Powell *et al.* (in preparation). An individual model of this type (but at lower resolution and without feedback), matching the morphology and star formation properties of the Antennae galaxies, was presented in Teyssier *et al.* 2010.

3.2. *Starburst properties*

In the models, the pre-merger isolated spiral galaxies spontaneously develop ISM turbulence at a about 10 km s^{-1} under the effect of gravitational instabilities (and/or feedback), and most star formation takes place in dense complexes of dense gas along spiral arms. In some sense, the large-scale star formation process is not entirely sub-grid anymore in these simulations, as the first steps of star formation, namely are the development of ISM turbulence and the formation of dense molecular gas clouds in this turbulent ISM, are explicitly captured – the subsequent steps of star formation at smaller scales, inside the densest parts of these cold clouds, remain sub-grid.

A merger simulation is shown in Figure 1. The mass-weighted average of the gas velocity dispersion reaches ~ 30 km/s. This strong turbulence is consistent with the observations reviewed above. It induces numerous local shocks that increase the local gas density, which in turn triggers the collapse of gas into cold clouds. Also, gas clouds become more massive and denser than in the pre-merger spiral galaxies. The fraction of gas that is dense-enough and cold-enough to form star increases, and the timescale for star formation in these dense gas entities (the gravitational free-fall time) becomes shorter. As a result, the total SFR becomes several times higher than it was in the pre-interaction pair of galaxies. The standard process of merger-induced gas inflow towards the central kpc or so is also present, but the timescale is substantially longer, so this process dominates the triggering the star formation by enhancing the global gas density only in the late stages of the merger (Fig. 1).

An example of star formation distribution is shown in Figure 2. Two consequences of modeling a cold turbulent ISM in merging galaxies are: (1) the peak intensity of the starburst can become stronger (as shown by Teyssier *et al.* 2010 for the Antennae) although this is not a systematical effect, and (2) the spatial extent of star formation in the starbursting phase is larger. The radius containing 50% of the star formation rate

Figure 2. Instantaneous star formation in an AMR merger model. Note the central nuclear starburst component from global gas inflows, but also the extended starburst component in massive clusters throughout the system. Dense knots of star formation along tidal tails are reminiscent of "beads on a string" star formation, as observed (e.g., Smith *et al.* 2008).

(half-SFR radius) can more than double. This is because increased ISM turbulence is present throughout the disk, and triggers, through locally convergent flows and shocks, the collapse of efficiently star-forming clouds even at large distances from the nuclei. At least quantitatively, these results put the models in better agreement with observations. In these models, the increased ISM turbulence is also obtained without feedback, showing that it is not a consequence of the starbursting activity, but is rather driven by the tidal forces in the interaction.

3.3. *Dense gas fractions and star formation scaling relations*

Density PDFs and galactic-scaled star formation

The connection between ISM turbulence, dense gas phases and star formation is best understood by examining the probability distribution function of the local gas density (density PDF). In its mass-weighted (resp. volume-weighted) version, the density PDF represents the mass (volume) fraction of the ISM in bins of density. We here use mass-weighted versions.

In an system at equilibrium (e.g. isolated disk galaxy), supersonic turbulence generates a log-normal density PDF (Wada *et al.* 2002). Only small mass fractions of the whole ISM are at found number densities below ~ 1 cm^{-3} or in very high density regions above, say, 10^4 cm^{-3}, most mass being at ten to a few hundreds of atoms per cm^3 (typically observed as HI or CO-traced molecular gas). The width of the log-normal PDF depends on various factors, but at first order the main dependence is on the turbulent Mach number (Krumholz & Thompson 2007). A more turbulent ISM will have a larger density spread in its density PDF, because the turbulent flows will sometimes diverge and create low-density holes, and sometimes converge (or even shock) into very dense structures that can further collapse gravitationally. Note that log-normal density PDFs are observed, at least at the scale of molecular complexes (Lombardi *et al.* 2010).

Our simulations of Milky Way-like disk galaxies do have quasi log-normal PDFs (see example on Figure 3). The PDFs are not exactly log-normal because the radial density gradient in the disk induces some deviations (the PDF is expected to be exactly log-normal only at fixed average density). Also, the spatial resolution limit converts into a density limit at which the density PDF is truncated, and which corresponds to the smallest/densest entities that can be resolved. The maximal density resolution in the

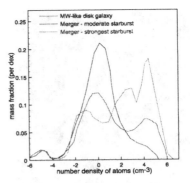

Figure 3. Gas density PDFs (see text for details) of an isolated disk galaxy and two starbursting mergers, in our sample of high-resolution AMR simulations.

present merger models is around 10^6 cm^{-3}, but parsec-scale resolution can capture even higher densities (as in Bournaud *et al.* 2010).

The density PDF is a useful tool to describe the star-forming activity of a galaxy at a given gas content. For a simplified description, one can consider that star formation takes place only in the densest gas phases (i. e. above some density threshold), and that in these dense regions the local star formation rate follows, for instance, a fixed efficiency per free-fall time: $\rho_{SFR} = \epsilon_{ff}\rho_{gas}/t_{ff} \propto \rho_{gas}^{1.5}$ (see detailed theory in Elmegreen 2002 and Krumholz & Thompson 2007). Even if the local rate of star formation follows a different prescription than this purely density-dependent model (which is physically motivated by the gravitational collapse timescale), then the first step remains that star formation proceeds only in the densest gas phases. Thus, the fraction of dense gas remains a key parameter for the global star formation activity of galaxies.

Dense gas excess and starbursts in mergers

Representative density PDFs are shown in our models on Figure 3 for an isolated spiral galaxy, a moderately starbursting merger (SFR increased by a factor ~ 3 compared to the two pre-merger galaxies taken separately), and a stronger merger-induced starburst (factor ~ 10).

The PDFs of merging galaxies have a larger width than those isolated disk galaxies, as expected from the higher turbulent speeds that result from the tidal interaction. These PDFs are not necessarily log-normal†, but they clearly show a substantial excess of dense gas in starbursting mergers. Such density PDFs naturally imply high SFRs, idenpendently from the local SFR prescription, since the fraction of efficiently star-forming gas is high.

High fractions of dense gas in mergers were already proposed by (Juneau *et al.* 2009), based on detailed post-processing of merger simulations aimed at re-constructing dense molecular gas phases not resolved in these simulations. Here we obtain a qualitatively similar conclusion using simulations that explicitly resolve turbulent motions, local shocks and small-scale instabilities in cold ISM phases. This excess of dense gas should have signatures in molecular line ratios (as also observed by Juneau *et al.* 2009). If, in a rough approach, we assume that low-J CO lines are excited for densities of 100 cm^{-3} and above, and HCN lines for densities of 10^4 cm^{-3} and above, then the HCN/CO line ratios could be up to 5-10 times higher in the starbursting phases of major galaxy mergers. Simulations with an somewhat higher resolution would actually be desirable to accurately quantify the emission of dense molecular tracers.

Star formation scaling relations

† presumably because the turbulent motions do not follow a quasi-isotropic cascade and/or some of the involved non-circular motions are not really cascading turbulence

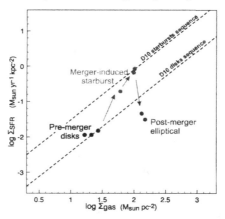

Figure 4. Kennicutt-Schmidt diagram showing the evolution of a major merger simulation in the $(\Sigma_{gas}; \Sigma_{SFR})$ plane (see text for details). The two dashed lines are the two star formation "laws" for disks and starbursts proposed by Daddi *et al.* (2010b). The pre-merger isolated disks evolve on the "disk sequence". The interaction and merger trigger a starburst not just through global gas inflows increasing the global averaged gas density, but also by increasing ISM turbulence and forming an excess of dense gas in massive cold clouds. The system then evolves towards the "starburst" mode as observed, while the increased fractions of dense gas could independently suggest a high excitation of molecular lines. The post-merger early-type galaxy settles back on a quiescent or even deficient mode.

The interpretation of merger-induced starbursts proposed from our high-resolution models is that it is not just a global gas inflow that increases the average gas density and increases the SFR, but also that there are strong non-circular motions, high turbulent velocity dispersions, causing many small-scale convergent flows and local shocks, that in turn initiate the collapse of dense star-forming clouds with high Jeans masses. The former "standard" process does take place, but the later can be equally important especially in the early phases of mergers.

We here note Σ_{gas} the average gas surface density of a galaxy. This is the quantity that observers would typically derive from the total gas mass and half-light radius, or similar quantities. The second mechanism above is a way to increase the SFR of a system, and its SFR surface density Σ_{SFR}, without necessarily increasing its average Σ_{gas}. Actually in our merger models Σ_{gas} does increase (as there are global merger-induced gas inflows), but Σ_{SFR} increases in larger proportions (as the starburst is not just from the global merger-induced inflow but also from the exacerbated fragmentation of high-dispersion gas). Going back to the density PDFs shown previously, one can note that the fraction of very dense gas (say, in the $\sim 10^{4-6}$ cm^{-3} range) can increase by a factor of 10–20 in mergers while the average surface density Σ_{gas} increases by a factor 3–5 (see also on Figure 4). As a consequence, the Σ_{SFR} activity of these systems is unexpectedly high compared to their average surface density Σ_{gas}.

Figure 4 shows the evolution of a system throughout a merger simulation in the $(\Sigma_{gas}; \Sigma_{SFR})$ plane. While our pre-merger spiral galaxy models lie on the standard Kennicutt relation, starbursting mergers have high $\Sigma_{SFR}/\Sigma_{gas}$ ratios. This is in agreement with observational suggestions that quiescent disks and starbursting mergers do not follow the same scaling relations for star formation, but could actually display two different star formation "laws" (Daddi *et al.* 2010b; Genzel *et al.* 2010). The offset between the disk and merger sequences proposed by Daddi *et al.* (2010b) is quantitatively recovered in our simulations (Figure 4). Post-starburst, post-merger systems lie back on the quiescent sequence, or even somewhat below it: these systems contain some dense gas which is somewhat stabilized by the stellar spheroid. This is another example of a "morphological quenching" effect in early-type galaxies (Martig *et al.* 2009), and the location of our post-merger early-type galaxies in the $(\Sigma_{gas}; \Sigma_{SFR})$ diagram may be consistent with observations of nearby ellipticals (Crocker *et al.* 2010, but see Fabello *et al.* 2010).

The proposal that disks and mergers follow two different regimes of star formation by Daddi *et al.* (2010b) and Genzel *et al.* (2010) relies for a part (but not entirely) on the assumption that different CO luminosity-to-molecular gas mass conversion factors apply in quiescent disks and starbursting mergers. Interestingly, our simulations recover the two regimes of star formation without any assumption on such conversion factors since gas masses are directly known. But at the same time, excess of dense gas found in these merger models suggests that the excitation of CO lines would naturally be higher in mergers/starbursting phases (although this needs to be quantified in the models), which would mean that the assumption of different conversion factors by Daddi *et al.* and Genzel *et al.* could be physically justified. High molecular gas excitation in Sub-Millimeter Galaxies (SMGs, Tacconi *et al.* 1998) could then naturally result if these are starbursting major mergers with high gas surface densities and a clumpy turbulent ISM (e.g. Narayanan *et al.* 2010; Bournaud *et al.* 2011).

4. Summary

The results presented here were based mostly on low-redshift merger simulations. A recent set of high-redshift merger simulations with AMR is presented in Bournaud *et al.* (2011), and shows similar increase in the ISM velocity dispersions and clumpiness in mergers, with extended starbursts and high $\Sigma_{SFR}/\Sigma_{gas}$ ratios.

Galaxy interactions and mergers play a significant but still debated and poorly understood role in the star formation history of galaxies. Numerical and theoretical models have significant difficulties in accounting for the properties of merger-induced starbursts, including their intensity and their spatial extent. Usually, the mechanism invoked to explain the triggering of star formation by mergers is a global inflow of gas towards the central kpc, resulting in a nuclear starburst. We show here, using high-resolution AMR simulations and comparing to observations of the gas component in mergers, that the triggering of starbursts also results from increased ISM turbulence and velocity dispersions in interacting systems. This results in the formation and collapse of dense and massive gas clouds in the regions of convergent flows and local shocks, these clouds being denser, more massive and/or more numerous than in quiescent disk galaxies. The fraction of dense cold gas largely increases, modifying the global density distribution of these systems, and efficient star formation results. Because the starbursting activity is not just from a global compacting of the gas to high average surface densities, but also from higher turbulence and fragmentation into massive and dense clouds, merging systems can enter a different regime of star formation compared to quiescent disk galaxies. This is in quantitative agreement with recent observations suggesting that disk galaxies and starbursting systems are not the low activity end and high activity end of a single regime, but actually follow different scaling relations for their star formation.

Acknowledgements

FB acknowledges numerous discussions on gas dynamics and star formation in galaxy mergers with Bruce Elmegreen, Pierre-Alain Duc, Emanuele Daddi and many others, and is grateful to the organizers of the symposium for the exciting discussions along a wide range of astrophysical problems.

References

Agertz, O., *et al.* 2009a, *MNRAS*, 392, 294

Barnes, J. E. & Hernquist, L. E. 1991, *ApJL*, 370, L65

Barnes, J. E. 2004, *MNRAS*, 350, 798

Bournaud, F., Duc, P.-A., & Emsellem, E. 2008, *MNRAS*, 389, L8

Bournaud, F. 2010, ASP Conference Series, 423, 177, arXiv:0909.1812v2

Bournaud, F., Elmegreen, B. G., Teyssier, R., Block, D. L., & Puerari, I. 2010, *MNRAS* 409, 1088

Bournaud, F., *et al. ApJ* submitted, arXiv:1006.4782

Burkert, A. 2006, *Comptes Rendus Physique*, 7, 433

Chien, L.-H. & Barnes, J. E. 2010, *MNRAS*, 407, 43

Cox, T. J., Jonsson, P., Somerville, R. S., Primack, J. R., & Dekel, A. 2008, *MNRAS*, 384, 386

Crocker, A. F., Bureau, M., Young, L. M., & Combes, F. 2010, submitted to *MNRAS*

Cullen, H., Alexander, P., & Clemens, M. 2006, *MNRAS*, 366, 49

Di Matteo, P., Combes, F., Melchior, A.-L., & Semelin, B. 2007, *A&A*, 468, 61

Di Matteo, P., Bournaud, F., Martig, M., Combes, F., Melchior, A.-L., & Semelin, B. 2008, *A&A*, 492, 31

Daddi, E., *et al.* 2010a, *ApJ*, 713, 686

Daddi, E., *et al.* 2010b, *ApJL*, 714, L118

Duc, P.-A., Brinks, E., Wink, J. E., & Mirabel, I. F. 1997, *A&A*, 326, 537

Elmegreen, B. G., Kaufman, M., & Thomasson, M. 1993, *ApJ*, 412, 90

Elmegreen, D. M., Kaufman, M., Brinks, E., Elmegreen, B. G., & Sundin, M. 1995, *ApJ*, 453, 100

Elmegreen, B. G. 2002, *ApJ*, 577, 206

Elmegreen, D. M., Elmegreen, B. G., Ravindranath, S., & Coe, D. A. 2007, *ApJ*, 658, 763

Le Floc'h, E., *et al.* 2005, *ApJ*, 632, 169

Fabello, S., *et al.* 2010, *MNRAS* in press, arXiv:1009.4309

Förster Schreiber, N. M., *et al.* 2009, *ApJ*, 706, 1364

Genzel, R., *et al.* 2010, *MNRAS*, 407, 2091

Green, A. W., *et al.* 2010, *Nature*, 467, 684

Hammer, F., Flores, H., Elbaz, D., Zheng, X. Z., Liang, Y. C., & Cesarsky, C. 2005, *A&A*, 430, 115

Irwin, J. A. 1994, *ApJ*, 429, 618

Jogee, S., *et al.* 2009, *ApJ*, 697, 1971

Juneau, S., Narayanan, D. T., Moustakas, J., Shirley, Y. L., Bussmann, R. S., Kennicutt, R. C., & Vanden Bout, P. A. 2009, *ApJ*, 707, 1217

Kim, J.-h., Wise, J. H., & Abel, T. 2009, *ApJL*, 694, L123

Krumholz, M. R. & Thompson, T. A. 2007, *ApJ*, 669, 289

Lombardi, M., Lada, C. J., & Alves, J. 2010, *A&A*, 512, A67

Martig, M. & Bournaud, F. 2008, *MNRAS*, 385, L38

Martig, M., Bournaud, F., Teyssier, R., & Dekel, A. 2009, *ApJ*, 707, 250

Mihos, J. C. & Hernquist, L. 1996, *ApJ*, 464, 641

Narayanan, D., *et al.* 2010, *MNRAS*, 401, 1613

Robaina, A. R., *et al.* 2009, *ApJ*, 704, 324

Saitoh, T. R. *et al.* 2009, *PASJ*, 61, 481

Sanders, D. B., *et al.* 1988, *ApJ*, 325, 74

Smith, B. J., *et al.* 2008, *AJ*, 135, 2406

Struck, C. 1997, *ApJS*, 113, 269

Tacconi, L. J., *et al.* 2008, *ApJ*, 680, 246

Tasker, E. J. & Tan, J. C. 2009, *ApJ*, 700, 358

Teyssier, R., Chapon, D., & Bournaud, F. 2010, *ApJL* 720, 149

Toomre, A. & Toomre, J. 1972, *ApJ*, 178, 623

Truelove, J. K., *et al.* 1997, *ApJL*, 489, L179

Wada, K., Meurer, G., & Norman, C. A. 2002, *ApJ*, 577, 197

Wang, Z., *et al.* 2004, *ApJS*, 154, 193

Astrophysical Dynamics: From Stars to Galaxies
Proceedings IAU Symposium No. 271, 2010
N. H. Brummell, A. S. Brun, M. S. Miesch & Y. Ponty, eds.
© International Astronomical Union 2011
doi:10.1017/S1743921311017583

The interstellar magnetic field near the Galactic center

Katia Ferrière[1]

[1]Laboratoire d'Astrophysique de Toulouse-Tarbes, Université de Toulouse, CNRS,
14 avenue Edouard Belin, F-31400 Toulouse, France
email: `ferriere@ast.obs-mip.fr`

Abstract. We review the observational knowledge that has built up over the past 25 years on the interstellar magnetic field within ~ 150 pc of the Galactic center. We also provide a critical discussion of the main observational findings and comment on their possible theoretical interpretations. To conclude, we propose a coherent view of the interstellar magnetic field near the Galactic center, which accounts at best for the vast body of observations.

Keywords. ISM: magnetic fields - ISM: general - ISM: structure - (ISM:) cosmic rays - Galaxy: center - Galaxies: magnetic fields

---

1. Introduction

The interstellar medium (ISM) near the Galactic center differs significantly from the ISM in the rest of the Galaxy. The ordinary matter (made of gas and small amounts of dust) tends to be denser, warmer, and more metal-rich, while the magnetic field reaches higher values and has a more poloidal geometry.

The physical characteristics and the spatial distribution of the interstellar gas in the innermost 3 kpc of the Galaxy were reviewed by Ferrière, Gillard, & Jean (2007). To summarize, the interstellar gas can be found in molecular, atomic, and ionized forms. The molecular gas is globally ten times more abundant than the atomic gas, and together these two neutral components enclose roughly the same amount of mass ($\sim 10^8\ M_\odot$) as the ionized component. Spatially, the molecular gas tends to concentrate in the so-called central molecular zone (CMZ), a thin sheet parallel to the Galactic plane, which, in projection on the sky, extends out to a radius $r \sim 250$ pc at positive longitudes and $r \sim 150$ pc at negative longitudes, and has a FWHM thickness ~ 30 pc. The CMZ itself contains a ring-like feature with mean radius ~ 180 pc, now known as the 180-pc molecular ring, and, deeper inside, a population of dense molecular clouds. Outside the CMZ, the molecular gas is confined to a significantly tilted disk, extending in projection out to $r \sim 1.3$ kpc on each side of the Galactic center and having a FWHM thickness ~ 70 pc. The spatial distribution of the atomic gas is arguably similar to that of the molecular gas, with this difference that the atomic layer is about three times thicker than the molecular layer. In addition, it is likely that the tilted disk encloses proportionally more atomic gas than the CMZ. Finally, the ionized gas is not confined to either the CMZ or the tilted disk; it appears to fill the entire Galactic bulge and to merge smoothly with the ionized gas present in the Galactic disk.

In the present paper, we focus on the interstellar magnetic field in a much smaller region around the Galactic center, which extends only ~ 300 pc along the Galactic plane and ~ 150 pc in the perpendicular direction. Radially, this small region is entirely contained within the CMZ. Outside the CMZ, observational data are scanty, so that not much can be said about the magnetic field properties.

In Section 2, we present the observational picture that emerged at the end of the 1980s, and in Section 3, we discuss the main problems inherent in this picture. In Section 4, we describe some new developments arising from a variety of recent observations. Finally, in Section 5, we try to piece everything together into a coherent picture. We also propose theoretical explanations for the existing observations, and we examine how the magnetic field near the Galactic center might connect with the magnetic field in the rest of the Galaxy.

Throughout the paper, the magnetic field vector is denoted by $\vec{B}$, its line-of-sight component by $B_{\parallel}$, the field strength by B, and the Alfvén speed by V_{A}. Besides, l and b denote Galactic longitude and latitude, respectively, r and z denote Galactocentric radial and vertical (i.e., perpendicular to the Galactic plane) coordinates, respectively, and the Sun is assumed to lie at $r_{\odot} = 8.5$ kpc. At this distance from the Galactic center, an angle of $1°$ corresponds to a length of 150 pc.

2. The Old Picture

The first observational clues to the direction and strength of the interstellar magnetic field near the Galactic center date back to the 1980s, when radio-astronomers (starting with Yusef-Zadeh, Morris, & Chance 1984 and Liszt 1985) discovered systems of radio continuum filaments running nearly perpendicular to the Galactic plane. As summarized by Morris (1996), these filaments are typically a few to a few tens of parsecs long and a fraction of a parsec wide, they appear straight or mildly curved all along their length, and their radio continuum emission is linearly polarized and characterized by a spectral index consistent with synchrotron radiation, hence the denomination of nonthermal radio filaments (NRFs or NTFs).

The long and thin shape of NRFs strongly suggests that they follow magnetic field lines. This suggested alignment is confirmed by the measured radio polarization angles (corrected for Faraday rotation), which indicate that, in the plane of the sky, the magnetic field inside NRFs is indeed oriented along their long axes (e.g., Tsuboi *et al.* 1985, 1986; Reich 1994; Lang *et al.* 1999a). From this, it has naturally been concluded that the interstellar magnetic field near the Galactic center is approximately vertical, at least close to the Galactic plane. Farther from the plane, NRFs tend to lean somewhat outwards, consistent with the interstellar magnetic field having an overall poloidal geometry (Morris 1990).

The magnetic field strength inside NRFs is much more uncertain and controversial. The equipartition/minimum-energy field strength is typically $B_{\mathrm{eq}} \sim (50 - 200)\ \mu\mathrm{G}$ (Anantharamaiah *et al.* 1991; LaRosa *et al.* 2004, and references therein). However, there appears to be no particular reason why NRFs would actually be in an equipartition/minimum-energy state with cosmic rays. In fact, their apparent rigidity and organized structure suggest instead that they are magnetically dominated (Anantharamaiah *et al.* 1991; Lang, Morris, & Echevarria 1999b), so that their actual field strength is probably higher than the equipartition/minimum-energy value.

A completely independent estimate of the magnetic field strength inside NRFs relies on a simple dynamical argument, originally proposed by Yusef-Zadeh & Morris (1987). According to these authors, the fact that NRFs remain nearly straight, even as they pass through the layer of molecular clouds, suggests that their magnetic pressure, P_{mag}, is higher than the cloud ram pressure, P_{ram}. For a presumably conservative value of P_{ram}, this condition is equivalent to $B \gtrsim 1$ mG inside NRFs.

Going one step further, Morris (1990) argued that NRFs must be pressure-confined. He estimated that the ambient gas pressure, which is dominated by the thermal pressure of the very hot ($T \sim 10^8$ K) plasma, is too low by a factor ~ 30 to confine NRFs, and he concluded that NRFs must be confined by magnetic pressure. In other words, the mG magnetic field inferred to exist inside NRFs must also prevail outside.

These considerations led to the notion that a pervasive mG magnetic field, with an overall poloidal geometry, fills the region containing NRFs – NRFs were initially observed out to $\gtrsim 70$ pc of the Galactic center (Morris 1990), but they are now detected out to $\gtrsim 150$ pc (LaRosa et al. 2004; Yusef-Zadeh, Hewitt, & Cotton 2004). In this view, NRFs would simply be magnetic flux tubes which happen to be illuminated by the injection of synchrotron-radiating electrons (Morris 1990).

3. Problems with a Pervasive mG Magnetic Field

The dynamical argument leading to $B \gtrsim 1$ mG inside NRFs presents several shortcomings: (1) Not all the NRFs remain nearly straight. Some clearly display severe distortions (e.g., the so-called Snake; Gray et al. 1995), while others could have deformations that escape detection from Earth because of projection effects. (2) Although most NRFs pass through the layer of molecular clouds, it is likely that only a fraction of them are actually colliding with clouds. (3) Even for truly colliding NRFs, the condition $P_{\mathrm{mag}} \gtrsim P_{\mathrm{ram}}$ is probably too stringent. Unless they collide at more than one location along their length, a more appropriate condition would be $V_A \gg v_{\mathrm{cloud}}$ (Chandran 2001) or, equivalently, $B \gg 10$ μG.

The pressure-balance argument leading to $B \sim 1$ mG in the general ISM raises even more serious objections: (1) The thermal pressure of the very hot plasma, P_{hot}, might be significantly higher than estimated by Morris (1990). According to the X-ray spectroscopic results of Koyama et al. (1996), P_{hot} could be comparable to the pressure of a 1 mG magnetic field, and hence high enough to confine NRFs. (2) NRFs could also be confined by magnetic tension forces (Lesch & Reich 1992; Uchida & Guesten 1995), as suggested by the helical fields detected in or around some NRFs (e.g., Yusef-Zadeh & Morris 1987; Gray et al. 1995). (3) More fundamentally, NRFs do not need to be confined at all; they could very well be transient or dynamic structures out of mechanical balance with their surroundings. For instance, Boldyrev & Yusef-Zadeh (2006) argued that turbulence in the Galactic center region naturally leads to a highly intermittent magnetic field distribution, with strongly magnetized filaments arising in an otherwise weak-field background, and they suggested that the turbulent filaments could correspond to the observed NRFs. Alternatively, Shore & LaRosa (1999) proposed that NRFs are the long and thin magnetic wakes produced by a weakly magnetized Galactic wind impinging on molecular clouds near the Galactic center.

Aside from the potential flaws in the dynamical and pressure-balance arguments, a magnetic field as strong as 1 mG implies synchrotron lifetimes that might be too short to explain certain observations (see, e.g., Yusef-Zadeh 2003; Morris 2007). At the radio frequencies (typically 1.5 GHz and 5 GHz) of most NRF observations, the synchrotron lifetimes in a 1 mG field are only $\sim 2 \times 10^4$ yr (2.7×10^4 yr at 1.5 GHz and 1.5×10^4 yr at 5 GHz). If the synchrotron-radiating electrons are injected somewhere along the NRFs (for instance, at the site of interaction with a molecular cloud) and if they stream away along the NRFs at about the Alfvén speed, then, over their lifetimes, they travel distances $\sim (10 - 60)$ pc, which might be too short to account for the length of the longest NRFs. At the lower radio frequencies (74 MHz and 330 MHz) of the diffuse nonthermal emission detected by LaRosa et al. (2005) (see Section 4), the synchrotron lifetimes in a 1 mG field are somewhat longer, but still only $\sim 10^5$ yr (1.2×10^5 yr at 74 MHz and 0.6×10^5 yr at 330 MHz), which the authors considered to be "shorter than any plausible replenishment timescale".

Finally, a 1 mG magnetic field filling the $\sim (300 \text{ pc})^2 \times 150$ pc region over which the NRF phenomenon is observed (LaRosa *et al.* 2004) encloses $\sim 10^{55}$ ergs in magnetic energy. This huge amount of magnetic energy corresponds to the energy released by $\sim 10^4$ supernova explosions. It is roughly comparable to the kinetic energy associated with Galactic rotation in the CMZ, while being significantly larger than the kinetic energy associated with turbulent motions in the same region. It is also larger than, or comparable to, the thermal energy of the very hot plasma present in the Galactic center region. What the origin of such a huge magnetic energy could be remains unclear.

4. New Developments

The picture of a poloidal mG magnetic field pervading the central 300 pc of the Galaxy is further challenged by a variety of more recent observations, which we now review.

Yusef-Zadeh *et al.* (2004) put together a catalog of all the (well-established + likely candidate) NRFs detected at 1.5 GHz, and they presented a schematic diagram of their spatial distribution in the plane of the sky. Their diagram conveys a sense that the vast majority of NRFs are indeed nearly straight (with the notable exception of the Snake) and that they tend to align with the vertical. In reality, however, only the longer NRFs strictly follow this tendency; the shorter NRFs exhibit a broad range of orientations, with only a loose trend toward the vertical.

Lower-frequency (74 MHz and 330 MHz) radio imaging of the Galactic center region by LaRosa *et al.* (2005) revealed a $6° \times 2°$ source of diffuse nonthermal (presumably synchrotron) emission. The inferred minimum-energy field strength (on spatial scales $\gtrsim$ 5 pc) is $\simeq (6 \ \mu\text{G}) \ (k/f)^{2/7}$, where k is the cosmic-ray proton-to-electron energy ratio and f the filling factor of the synchrotron-emitting gas. Both parameters are quite uncertain, but we may reasonably assume that k lies somewhere between the two canonical values 1 (generally adopted in the vicinity of powerful cosmic-ray sources) and 100 (generally adopted in the Galactic disk), while $f \gtrsim 0.01$. The minimum-energy field strength must then lie in the range $\simeq (6 - 80) \ \mu\text{G}$.

More recently, Crocker *et al.* (2010) found that the spectrum of the diffuse radio emission from the Galactic center region exhibits a downward break at $\sim$ 1.7 GHz, which can be explained by the synchrotron-radiating electrons undergoing a transition from bremsstrahlung to synchrotron cooling. The measured break frequency imposes a relationship between magnetic field strength (which governs synchrotron cooling) and gas density (which governs bremsstrahlung cooling), and this relationship makes it possible, for any given value of B, to model the cooled electron distribution as well as its γ-ray (inverse-Compton + bremsstrahlung) emission. Requiring that the latter do not exceed the 300 MeV γ-ray emission measured by EGRET then leads to the constraint $B \gtrsim$ 50 μG.

$* \ * \ * \ * \ *$

Far-infrared/submillimeter polarization studies of dust thermal emission enable one to probe the direction (in the plane of the sky) of the interstellar magnetic field inside dense molecular clouds. As a general rule, far-infrared polarimetry applies to the warmer parts of molecular clouds, while submillimeter polarimetry applies to their colder parts.

Davidson (1996), who reviewed the existing far-infrared polarization measurements toward dense regions located within $\sim$ 100 pc of the Galactic center, noted that the measured magnetic field direction is generally roughly parallel to the Galactic plane. She argued that this field direction could be explained by the dense gas moving relative to the surrounding diffuse gas and either distorting the local poloidal field or dragging its own distorted field from another Galactic location.

The magnetic field direction inferred from far-infrared polarimetry is largely backed up by submillimeter polarization observations. In particular, the 450 μm polarization map of Novak et al. (2003), which covers a 170 pc × 30 pc area around the Galactic center, confirms that the magnetic field threading molecular clouds is, on the whole, approximately parallel to the Galactic plane. To reconcile the horizontal field measured in molecular clouds with the poloidal field traced by NRFs, Novak et al. (2003) proposed that the large-scale magnetic field near the Galactic center is predominantly poloidal in the diffuse ISM and predominantly toroidal in dense regions along the Galactic plane, where it was sheared out in the azimuthal direction by the differential rotation of the dense gas.

The conclusions of Novak et al. (2003) were refined by Chuss et al. (2003), who found that the measured magnetic field direction depends in fact on the molecular gas density, being generally parallel to the Galactic plane in high-density regions and generally perpendicular to it in low-density regions. According to their preferred interpretation, the large-scale magnetic field near the Galactic center was initially poloidal everywhere, but in dense molecular clouds, where the gravitational energy density exceeds the magnetic energy density, it became sheared out into a toroidal field by the clouds' motions. In the framework of this scenario, Chuss et al. (2003) estimated a characteristic field strength ~ 3 mG inside molecular clouds, by assuming that clouds where the field is half-way between toroidal and poloidal are those where gravitational and magnetic energy densities are equal.

Near-infrared polarization observations of starlight absorption by dust also offer a promising tool to trace the magnetic field direction (again in the plane of the sky) in dense regions near the Galactic center. The recent 50 pc × 50 pc near-infrared polarization map of Nishiyama et al. (2009) exhibits a strong tendency for the polarization vectors to align with the Galactic plane, in good agreement with the results of far-infrared/submillimeter polarimetry. However, this map shows no hint of a potential correlation between field direction and gas density.

* * * * *

Zeeman splitting of radio (atomic or molecular) spectral lines offers a direct means of measuring the line-of-sight component of the magnetic field, $B_{\parallel}$, in dense, neutral regions.† So far, unfortunately, Zeeman splitting measurements have only yielded mixed results.

For the circumnuclear disk, the ~ 10 pc sized innermost molecular region, Killeen, Lo, & Crutcher (1992) derived $B_{\parallel} \sim -2$ mG both in the southern part (firm detection) and in the northern part (marginal detection). A little later, Marshall, Lasenby, & Yusef-Zadeh (1995) obtained only an upper limit ~ 0.5 mG in each of the northern and southern parts, whereas Plante, Lo, & Crutcher (1995) reported 7 detections (1 positive and 6 negative values of $B_{\parallel}$) ranging between -4.7 mG and $+1.9$ mG toward the northern part. These disparate Zeeman results are not necessarily contradictory; they can be reconciled if $B_{\parallel}$ varies substantially (especially if $B_{\parallel}$ changes sign) across the circumnuclear disk.

Farther away from the Galactic center, Crutcher et al. (1996) measured values of $B_{\parallel}$ ranging between $\simeq -0.1$ mG and -0.8 mG toward the Main and North cores of Sgr B2.

---

† By convention in the Zeeman splitting community, a positive (negative) value of $B_{\parallel}$ corresponds to a magnetic field pointing away from (toward) the observer.

In contrast, Uchida & Guesten (1995) reported only non-detections, with 3σ upper limits $\sim (0.1 - 1)$ mG, toward 13 selected positions within a few degrees of the Galactic center (including Sgr B2).

The mixture of positive detections, with $|B_\parallel| \sim (0.1 - 1)$ mG, and non-detections, with $|B_\parallel| \lesssim (0.1 - 1)$ mG, outside the circumnuclear disk can probably be attributed partly to possible dilution of the Zeeman signal by averaging over the observed area and along the line of sight and partly to genuine differences in the local values of $B_\parallel$. Genuine differences in $B_\parallel$ are expected if the magnetic field inside molecular clouds is roughly horizontal, as indicated by far-infrared/submillimeter polarization observations, and has random directions in the Galactic plane. Zeeman results then suggest that the total field strength inside molecular clouds is roughly ~ 1 mG, with a possible range from a few 0.1 mG to a few mG. This broad range encompasses the characteristic field strength ~ 3 mG estimated from submillimeter polarimetry (Chuss *et al.* 2003).

* * * * *

Faraday rotation measures (RMs), for their part, provide information on $B_\parallel$ in the diffuse ionized medium.† Both positive and negative RMs have been obtained toward the Galactic center, with typical values ranging from a few hundred to a few thousand rad m^{-2} (e.g., Tsuboi *et al.* 1985; Yusef-Zadeh & Morris 1987; Gray *et al.* 1995; Lang *et al.* 1999a, 1999b). In principle, if the line-of-sight depth of the Faraday-rotating screen and the free-electron number density within it are known, $B_\parallel$ can be inferred from the measured RMs. In practice, though, both parameters are difficult to estimate, partly because the Faraday-rotating screen itself is often hard to locate. Nevertheless, with the parameter values adopted in the existing RM studies, the measured RMs generally translate into $B_\parallel \sim \pm$ a few μG (e.g., Tsuboi *et al.* 1985; Yusef-Zadeh & Morris 1987; Gray *et al.* 1995).

The values of $|B_\parallel|$ inferred from RM measurements are compatible with both a dynamically dominant magnetic field ~ 1 mG and a minimum-energy field $\simeq (6 - 80)$ μG. It is, however, noteworthy that if the large-scale magnetic field in the diffuse ISM near the Galactic center has a poloidal geometry, with only a small component along the line of sight, the low end of the minimum-energy range might be difficult to reconcile with $|B_\parallel| \sim$ a few μG.

Examination of the measured RM signs has led to contradictory conclusions. Novak *et al.* (2003), who collected all the available RMs toward NRFs within $1°$ of the Galactic center, brought to light a definite pattern in the sign of RM, such that RM > 0 in the quadrants $(l > 0, b > 0)$ and $(l < 0, b < 0)$ and RM < 0 in the quadrants $(l > 0, b < 0)$ and $(l < 0, b > 0)$. This pattern, they explained, could result from azimuthal shearing by the Galactic differential rotation of an initially vertical magnetic field pointing north (dense molecular clouds, which are confined close to the Galactic plane, tend to rotate faster than the diffuse gas).

In contrast, Roy, Rao, & Subrahmanyan (2005), who derived the RMs of 60 background extragalactic sources through a $12° \times 4°$ window centered on the Galactic center, obtained mostly positive values, with no evidence for a sign reversal with latitude or with longitude. Roy, Rao, & Subrahmanyan (2008) pointed out that this predominance of positive RMs is consistent with either the large-scale Galactic magnetic field having a bisymmetric spiral configuration or the magnetic field near the Galactic center being oriented along the Galactic bar.

---

† The convention for the sign of $B_\parallel$ in the Faraday rotation community is opposite to that adopted in the Zeeman splitting community. Here, a positive (negative) value of $B_\parallel$ corresponds to a magnetic field pointing toward (away from) the observer.

5. Conclusions

Based on the critical observational overview presented in the preceding sections, we (tentatively) propose that the interstellar magnetic field near the Galactic center has the following properties:

(1) In the diffuse intercloud medium, the magnetic field is approximately poloidal on average. There exist a number of localized filamentary structures (the so-called NRFs), wherein the field is almost certainly above equipartition with cosmic rays [$B_{eq} \sim (50 - 200)\ \mu G$] and could, in some cases, be as strong as $B \gtrsim 1$ mG. Outside NRFs, the field is almost certainly weaker, although not necessarily by a huge factor, as there is some evidence pointing to $B \gtrsim 50\ \mu G$; one may not rule out the possibility that such a field could be in equipartition with cosmic rays [$B_{eq} \simeq (6 - 80)\ \mu G$].

(2) In dense molecular clouds, the magnetic field is approximately horizontal. The field strength is probably quite high, with typical values ranging between a few 0.1 mG and a few mG.

The approximately poloidal direction of the magnetic field in the diffuse intercloud medium can be explained by various scenarios, including inward advection from the Galactic disk (Sofue & Fujimoto 1987; Chandran, Cowley, & Morris 2000), outflows from the Galactic nucleus (Sofue 1984; Chevalier 1992), and a local dynamo. The reason why the magnetic field within dense molecular clouds is approximately horizontal can be understood if the poloidal intercloud field was sheared out in a horizontal direction by the cloud bulk motions (from differential rotation or from turbulence) with respect to the diffuse intercloud medium (Novak et al. 2003; Chuss et al. 2003) or by the forces (of compressive or tidal nature) that created and/or shaped the clouds (Morris & Serabyn 1996, and references therein). Alternatively, it is possible that the field within molecular clouds became decoupled from the intercloud field (Morris & Serabyn 1996; Morris 2007), for instance, as a result of cloud rotation.

If the magnetic field near the Galactic center is approximately poloidal in the diffuse intercloud medium, it is most likely antisymmetric.[†] This inferred antisymmetry is fully supported by the set of RMs collected by Novak et al. (2003), which displays a clear sign reversal across the midplane (see Section 4). RMs also indicate that the magnetic field runs counterclockwise above the midplane and clockwise below it. If this configuration results from azimuthal shearing by the Galactic differential rotation, then the vertical field must be pointing north ($B_z > 0$). Interestingly, these directions of the azimuthal and vertical field components coincide with those thought to prevail in the inner Galactic halo (Han, Manchester, & Qiao 1999; Mao et al. 2010).

Hence, the magnetic field near the Galactic center could possibly be a natural continuation of the magnetic field in the inner halo. The combined field would evidently be antisymmetric. However, its poloidal component would not be a pure dipole (as often assumed in the cosmic-ray propagation community), otherwise B_z would have opposite signs along the rotation axis and at the position of the Sun. Instead, it seems more likely that the combined Galactic-center + inner-halo field would have a poloidal component pointing everywhere north, together with a (possibly strong) azimuthal component produced by Galactic shear.

---

† A magnetic field is said to be symmetric (antisymmetric) in z, or, equivalently, quadrupolar (dipolar), if its horizontal component is an even (odd) function of z and its vertical component an odd (even) function of z.

References

Anantharamaiah, K. R., Pedlar, A., Ekers, R. D., & Goss, W. M. 1991, *MNRAS*, 249, 262
Boldyrev, S. & Yusef-Zadeh, F. 2006, *ApJ*, 637, L101
Chandran, B. D. G. 2001, *ApJ*, 562, 737
Chandran, B. D. G., Cowley, S. C., & Morris, M. 2000, *ApJ*, 528, 723
Chevalier, R. A. 1992, *ApJ*, 397, L39
Chuss, D. T., Davidson, J. A., Dotson, J. L., *et al.* 2003, *ApJ*, 599, 1116
Crocker, R. M., Jones, D. I., Melia, F., Ott, J., & Protheroe, R. J. 2010, *Nature*, 463, 65
Crutcher, R., Roberts, D. A., Mehringer, D. M., & Troland, T. H. 1996, *ApJ*, 462, L79
Davidson, J. A. 1996, in ASP Conf. Ser., 97, Polarimetry of the Interstellar Medium, ed. W. G. Roberge & D. C. B. Whittet (San Francisco: ASP), 504
Ferrière, K., Gillard, W., & Jean, P. 2007, *A&A*, 467, 611
Gray, A. D., Nicholls, J., Ekers, R. D., & Cram, L. E. 1995, *ApJ*, 448, 164
Han, J. L., Manchester, R. N., & Qiao, G. J. 1999, *MNRAS*, 306, 371
Killeen, N. E. B., Lo, K. Y., & Crutcher, R. 1992, *ApJ*, 385, 585
Koyama, K., Maeda, Y., Sonobe, T., *et al.* 1996, *PASJ*, 48, 249
Lang, C. C., Anantharamaiah, K. R., Kassim, N. E., & Lazio, T. J. W. 1999a, *ApJ*, 521, L41
Lang, C. C., Morris, M., & Echevarria, L. 1999b, *ApJ*, 526, 727
LaRosa, T. N., Brogan, C. L., Shore, S. N., *et al.* 2005, *ApJ*, 626, L23
LaRosa, T. N., Nord, M. E., Lazio, T. J. W., & Kassim, N. E. 2004, *ApJ*, 607, 302
Lesch, H. & Reich, W. 1992, *A&A*, 264, 493
Liszt, H. S. 1985, *ApJ*, 293, L65
Mao, S. A., Gaensler, B. M., Haverkorn, M., Zweibel, E. G., Madsen, G. J., McClure-Griffiths, N. M., Shukurov, A., & Kronberg, P. P. 2010, *ApJ*, 714, 1170
Marshall, J., Lasenby, A. N., & Yusef-Zadeh, F. 1995, *MNRAS*, 274, 519
Morris, M. 1990, in IAU Symp., 140, Galactic and Intergalactic Magnetic Fields, ed. R. Beck, R. Wielebinski & P. P. Kronberg (Dordrecht: Kluwer), 361
Morris, M. 1996, in IAU Symp., 169, Unsolved Problems of the Milky Way, ed. L. Blitz & P. J. Teuben (Dordrecht: Kluwer), 247
Morris, M. 2007, ArXiv Astrophysics e-prints
Morris, M. & Serabyn, E. 1996, *ARAA*, 34, 645
Morris, M., Uchida, K., & Do, T. 2006, *Nature*, 440, 308
Nishiyama, S., Tamura, M., Hatano, H., *et al.* 2009, *ApJ*, 690, 1648
Novak, G., Chuss, D. T., Renbarger, T., *et al.* 2003, *ApJ*, 583, L83
Plante, R. L., Lo, K. Y., & Crutcher, R. 1995, *ApJ*, 445, L113
Reich, W. 1994, in NATO ASI, 445, The Nuclei of Normal Galaxies: Lessons from the Galactic Center, ed. R. Genzel & A. Harris (Dordrecht: Kluwer), 55
Roy, S., Rao, A. P., & Subrahmanyan, R. 2005, *MNRAS*, 360, 1305
Roy, S., Rao, A. P., & Subrahmanyan, R. 2008, *A&A*, 478, 435
Serabyn, E. & Morris, M. 1994, *ApJ*, 424, L91
Shore, S. N. & LaRosa, T. N. 1999, *ApJ*, 521, 587
Sofue, Y. 1984, *PASJ*, 36, 539
Sofue, Y. & Fujimoto, M. 1987, *PASJ*, 39, 843
Tsuboi, M., Inoue, M., Handa, T., Tabara, H., & Kato, T. 1985, *PASJ*, 37, 359
Tsuboi, M., Inoue, M., Handa, T., *et al.* 1986, *AJ*, 92, 818
Uchida, K. I. & Guesten, R. 1995, *A&A*, 298, 473
Yusef-Zadeh, F. 2003, *ApJ*, 598, 325
Yusef-Zadeh, F., Hewitt, J. W., & Cotton, W. 2004, *ApJS*, 155, 421
Yusef-Zadeh, F. & Morris, M. 1987, *ApJ*, 322, 721
Yusef-Zadeh, F., Morris, M., & Chance, D. 1984, *Nature*, 310, 557

Discussion

ZWEIBEL: Is the cosmic-ray density in the NRFs higher than ambient?

FERRIÈRE: We don't know for sure. The cosmic-ray density cannot be measured directly. It can be constrained, to some extent, by the measured synchrotron intensity, but the constraints depend on the assumed magnetic field strength. If the magnetic field is roughly uniform throughout the medium (as argued by Morris 1990), then the relativistic-electron density, and presumably also the cosmic-ray density, must be higher in NRFs. In this case, NRFs can be regarded as magnetic flux tubes illuminated by a local injection of relativistic electrons. In contrast, if the magnetic field is stronger inside NRFs (as suggested by all the problems inherent in the uniform-field picture; see Section 3), which most likely supposes that NRFs are compressed flux tubes, then two antagonistic effects may come into play. On the one hand, when an NRF forms by compression, the attached cosmic rays are compressed together with the field lines and their density increases. On the other hand, once cosmic rays find themselves inside a high-B NRF, they stream away along field lines (at about the Alfvén speed) and escape into the halo faster than cosmic rays in the ambient medium, so that their density decreases relative to the ambient cosmic-ray density. In addition, relativistic electrons cool off (via synchrotron radiation) more rapidly than in the ambient medium.

TOOMRE: Could you expand upon what may be the origin of the lengthy filaments near the center of our Galaxy, which has long been described as the "violent center of our Galaxy" (aside from the supermassive black hole Sgr A*)?

FERRIÈRE: Different scenarios have been put forward to explain the origin of the lengthy radio filaments (the so-called NRFs) near the Galactic center. A first possibility is that NRFs result from a local injection of relativistic electrons. For instance, when a molecular clump moves across the ambient magnetic field lines, magnetic reconnection may occur at its leading edge and lead to particle acceleration (Serabyn & Morris 1994). Another possibility is that NRFs result from a local compression or shearing of magnetic field lines. Here, two possible candidates are turbulence, which, in the Galactic center region, is expected to produce strongly magnetized filaments (Boldyrev & Yusef-Zadeh 2006), and a Galactic wind impinging on molecular clouds, which could generate long magnetic wakes behind the clouds (Shore & LaRosa 1999).

BRANDENBURG: Is anything known about the orientation of the helical structures and their position relative to the midplane (above/below)? Also, the slightly different directions of the filaments might be explicable by them having different distances to the observer.

FERRIÈRE: Observational evidence has been found for the existence of helical magnetic fields winding about some of the radio filaments, including the filaments of the Radio Arc (e.g., Yusef-Zadeh & Morris 1987; Gray et al. 1995; Morris, Uchida, & Do 2006). In none of these cases is the evidence very strong, and the putative helical fields are not perceptible over more than two or three wavelengths. The orientation of the helical structures and their position relative to the midplane can be determined in individual cases, but there are not enough cases to infer a general trend of orientation *versus* position above/below the midplane.

Regarding the second part of the question, I think the slightly different directions of the NRFs is more likely explained by turbulence.

Astrophysical Dynamics: From Stars to Galaxies
Proceedings IAU Symposium No. 271, 2010
N.H. Brummell, A.S. Brun, M.S. Miesch & Y. Ponty, eds.

© International Astronomical Union 2011
doi:10.1017/S1743921311017595

Simulations of stellar convection, pulsation and semiconvection

Herbert J. Muthsam[1], Friedrich Kupka[1], Eva Mundprecht[1], Florian Zaussinger[2,1], Hannes Grimm-Strele[1] & Natalie Happenhofer [1]

[1]Faculty of Mathematics, University of Vienna
Nordbergstrasse 15, A-1090 Vienna, Austria
email: `herbert.muthsam@univie.ac.at`

[2]Max Planck Institute for Astrophysics,
Karl Schwarzschildstrasse 1, D-85741 Garching, Germany

Abstract. We report on modelling in stellar astrophysics with the ANTARES code. First, we describe properties of turbulence in solar granulation as seen in high-resolution calculations. Then, we turn to the first 2D model of pulsation-convection interaction in a cepheid. We discuss properties of the outer and the HeII ionization zone. Thirdly, we report on our work regarding models of semiconvection in the context of stellar physics.

Keywords. Sun: granulation, stars: oscillations, stars: variables: Cepheids, stars: interiors

1. Introduction

In this paper we present applications of the ANTARES code to various problems in stellar physics. ANTARES is a radiation-hydrodynamics code. Its core functionality has recently been described by Muthsam *et al.* (2010). Using this core functionality, we first discuss properties of turbulence in *solar granulation*, studied in extremely high resolution in 3D.

We have in the meanwhile extended the core functionality of ANTARES, and the applications discussed in the next two chapters make use of these extensions. So, we have included a spherical and optionally radially moving grid for the study of the interaction of radial stellar pulsation with convection. We present a first description of the pulsation-convection interaction of a 2D model of a *cepheid* with realistic equation of state and grey radiative transfer. For the importance of studying the pulsation-convection interactions consult, for example, the reviews Buchler (1997) and Buchler (2009). Consider that since the pioneering work due to Christy, Cox and Kippenhahn in the 1960's improvement in nonlinear stellar pulsation modelling has basically been due to better microphysics and to some extent to improvement in numerics. Convection has, however, (practically) always been included by a mixing-length type approach where already the basic form of the equations may be prone to some doubt and where additionally a number of parameters needs to be set in a way which can barely considered to be really convincing. Problems probably stemming from such shortcomings are discussed, e.g., in Buchler (1997). For a recent discussion of a specific uncertainty which arises from traditional convection modelling in pulsating stars (namely, excitation of double mode pulsations or lack thereof) see Smolec & Moskalik (2009).

Subsequently, we turn our attention towards modelling of *semiconvection* in the context of stellar physics. Although it is textbook knowledge that a faithful description of semiconvection is highly important for properly dealing with the late stages of stellar evolution, but little work has been undertaken with regard to multidimensional modelling in the stellar context. The only major numerical studies are due to Merry-field (1995) and Biello (2001). See also the results due to Bascoul (2007). We present here our first 2D results discussing, in particular, the properties of diffusive interfaces. These results mainly refer to Boussinesq models, although we report also on compress-ible models and discuss their connection with the Boussinesq case. For enabling a more extensive investigation of compressible models it is necessary to implement a method which allows large time-steps, restricted by the macroscopic rather than the sound ve-locity. In this connection, an implementation of the method of Kwatra *et al.* (2009) is in an advanced stage of development.

2. High-resolution solar granulation simulations

In order to investigate the turbulent state of solar granulation we have performed hydrodynamic simulations with extremely high resolution. While they are basically sim-ilar to the ones described in Muthsam *et al.* (2010), they differ in that the highest resolution is achieved by making use of *two* grid refinement zones (stacked within each

Figure 1. Solar granulation at two instances of time. Horizontal extent of the domain shown: 1 Mm. The grey surface is an isosurface for $T = 6000$ K. The vortex tubes are made visible by plotting isosurface for a (large) value of the modulus of the gradient of a suitably normalized pressure. A comparison of the upper and lower figure shows the effect of increased degree of turbulence when applying higher resolution.

other) instead of one, and the resolution in the smaller one is considerably finer than in the old run. In addition, the old run referred to an exploding granule, whereas we concentrate here, in the highest resolution, on a strong downdraft surrounded by normal granules in order to figure out to what extent our previous results apply here as well.

More specifically, the basic computational domain spans now $3.68 \times 6 \times 6$ Mm^3 (the first coordinate is the vertical) with a cell size of $10.8 \times 22.2 \times 22.2$ km^3. The first grid refinement zone has an extension of $1.9 \times 2.5 \times 2.8$ Mm^3 and a grid spacing of $5.4 \times 7.4 \times 7.4$ km^3. The second and therefore finest grid refinement zone contains the downdraft mentioned above and parts of the surrounding granules. It has an extent of $1 \times 1.2 \times 1.2$ Mm^3 with a cell size of $2.7 \times 3.7 \times 3.7$ km^3.

Using the high-resolution calculations, we want to discuss the role of resolution in the representation of the turbulent state of the solar granulation layer. In earlier simulations of an exploding granule Muthsam *et al.* (2010) a host of vortex tubes of small diameter has shown up, located preferentially near the granular lanes and the downdrafts. The questions popping up are now whether a similar large number of vortex tube shows up even in more normal granulation and whether these previous calculations were fully resolved; they used a grid spacing of $7.1 \times 9.8 \times 9.8$ km^3, i.e. worse than even our present *first* refinement.

In order to answer the question at what resolution the turbulent field of solar granulation is (possibly) fully resolved, we compare in figure 1 two snapshots of our highest resolution subdomain. The first figure represents the state shortly after we have turned on the second grid refinement and thus still have the effective resolution due to the first refinement, whereas the next snapshot is taken 153 seconds later. Due to increased resolution, the number and intensity of the vortex tubes has drastically increased. A closer study of the properties of this special sort of turbulence encountered mainly in granular downdrafts is under way.

3. Cepheid pulsation and convection: a 2D model

Initial condition and computational domain. We model a cepheid with an effective temperature of 5125K, luminosity $L = 912.797L_\odot$, mass $M = 5.0M_\odot$, hydrogen content $X = 0.7$ and metallicity $Z = 0.01$. Our computational domain reaches from 4000 K to 320.000 K, thus the outer 42% of the star are computed.

This domain is equipped with a polar grid. In radial direction $N_x = 510$ grid points plus 4 ghost cells at each boundary are used. The r-range covers $r \in [r_{top}, r_{bot}]$, where r_{bot} is fixed at 15.5 Gm and r_{top} varies with time from about 26.2 Gm to 27.4 Gm. The grid is stretched in radial direction by a factor q. The mesh sizes are $\Delta r_{i+1} = q \Delta r_i$ varying from $\Delta r_0 = 0.046$ Mm at the top to $\Delta r_{N_x-1} = 12$ Mm at the bottom. In angular direction there are 800 grid points, the distance $r_i \Delta \varphi$ between two adjacent gridpoints in the H-ionisation zone is 5.8 Mm and 5.6 Mm in the He-ionisation zone. The corresponding aspect ratios are $1 : 2.5$ and $1 : 0.6$.

Fluxes and pulsation in the HeII-ionisation zone. The work integral can be decomposed into an average part $PdV_0 = \partial_r u_0 (p/\rho)$ and a perturbational part defined as $PdV_{pert} = \partial_r u_r'(p/\rho)$. Here, $u_0 = \bar{I}_r/\bar{\rho}$ and $u_r' = u_r - u_0$. $\vec{I} = (I_r, I_\varphi)$ is the momentum vector and $\vec{u} = (u_r, u_\varphi)$ the velocity. $\bar{I}_r$ denotes the horizontal average of the radial momentum component.

While PdV_{pert} varies locally from $-6.e8$ to $+5.e8$ in the HeII convection zone and is greater than $PdV_0 \in [-8.e7, +5.e7]$, the horizontal average is considerably smaller, $\overline{PdV_{pert}} \in [-9.e6, +7.e6]$, compared to $\overline{PdV_0} \approx PdV_0$.

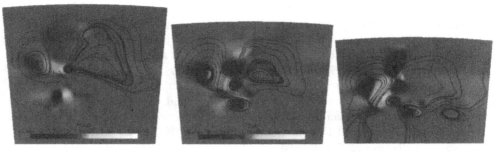

Figure 2. Convective flux in the course of contraction (HeII convection zone)

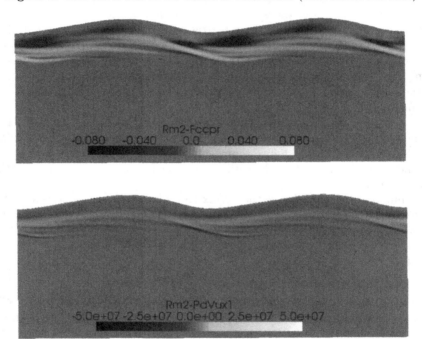

Figure 3. Horizontal averages of convective flux (upper figure) and work integral (lower figure)

The averaged convective flux varies from -3% to $+8\%$ of the input energy flux at the lower boundary, the mean value during one period being $+4\%$. Locally values of up to $1.4e11$ corresponding to 30% can be reached. The averaged kinetic flux varies from -3% to $+0.2\%$ the mean value is -2%. The data are from 12 consecutive periods and the same flux pattern can be observed in each though the extent may vary. In the expanded state one observes two convection centers: a new one forming at the top and an older one which is the remainder of the downwards moving and slowly disapearing plume of the previous period. At contraction the two centers are very close to each other. These centers of convective activity lead also to the two (occasionally more) stripes situated atop of each other and visible, for example, in the quantities depicted in figure 3. The centers of activity can also directly be seen in figure 2.

Remarks on the H-ionisation zone. The upper boundary of our computational domain is situated at an optical deph of a few thousandths. Figure 4 shows a part (in $\phi-$direction) of the upper zone. Color coding is for momentum densities (bright resp. red is downwards). The convective motions are very vigorous. We have observed Mach

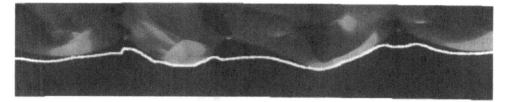

Figure 4. Radial momentum in H-ionisation zone. The bright line denotes the optical depth $\tau = 100$

numbers for inflow of up to 4 and for outflow of up to 3. Remarkably, the lower boundary of the downdrafts often coincides the line of maximum temperature gradient (just above the bright line in figure 4. These results could only be achieved with grid refinement. The refined grid spacing is one third of the coarse grid spacing in the radial and one fourth in the angular direction.

4. Numerical simulation of semiconvection

When the convective transport does not depend on the temperature gradient only but also on an additional scalar field like salt (in the ocean) or helium (in stars), then *double-diffusive convection* can occur. This name implies that two diffusion coefficients (κ_T and κ_S/κ_{He}) influence the fluids motion. In the oceans double-diffusive convection can form *salt-fingers*, while the well-known Latte Macchiato layers are a result of another double-diffusive phenomenon, which in astrophysics is called *semi-convection*.

In the evolution of stars semi-convection plays an important role. When cold hydrogen-rich matter is stratified above hot helium-rich matter, double-diffusive layering can occur under certain circumstances (Ledoux stable, $\nabla_\mu > \nabla - \nabla_{ad}$). The main question, which arises is the influence of the double-diffusive mixing processes on the evolution of high mass stars ($M > 15 M_\odot$) around main sequence turnoff.

A stable layered structure in semi-convection zones (cf. Latte Macchiato), separated by diffusive boundary layers, is assumed (Huppert & Moore (1976), Spruit (1992)). In particular, the mass and heat transport through these diffusive boundaries are of major interest for explaining mixing time scales and the life span of the entire semi-convective region. There are several unknowns when describing single layers. Especially, the thickness of a single layer as function of thermal and saline Rayleigh number (Ra_T, Ra_S) and the superadiabatic gradient is not well understood. A comprehensive hydrodynamical model, based on Spruit (1992) and Muthsam *et al.* (1999) (see also Muthsam *et al.* (1995)) was derived for the double diffusive convection case to tackle these uncertainties.

In Zaussinger & Spruit (2010) numerical simulations in 2D of semi-convection, based on compressible and incompressible formulations, have been performed for a wide initial parameter range of the Prandtl number $\sigma = \nu/\kappa_T$, the Lewis number $\tau = \kappa_S/\kappa_T$, the modified Rayleigh number $Ra_* = Ra_T \sigma$ and the stability parameter $R_\rho = Ra_S/Ra_T$. It is proven there that simulations done in the Boussinesq approximation are directly comparable with fully compressible fluids as long as the layer heights are small enough ($H_P \gg H$). The aim of this study was the verification and extent of validity of existing power laws in the form $Nu_T = \alpha Ra_T^\beta$, where Nu_T is the thermal Nusselt number. An extrapolation to the stellar parameter space was done subsequently.

In Zaussinger (2010) the parameter dependence of the fluxes is analysed. Single and double layer simulations were at the focus of that study. While the diffusion associated dimensionless numbers (σ, τ) can be compared directly between compressible and incompressible fluids, other physical quantities like the fluxes (F_T, $F_{S.He}$) or the Rayleigh numbers (Ra_T, Ra_S) need a special treatment. The thermal Nusselt number, which compares the diffusive heat flux to the total heat flux, has to be corrected by the adiabatic stratification flux F_{ad}, which is by definition not present in incompressible fluids modelled by the Boussinesq approximation. Neglecting this difference would lead to wrong comparisons in simulations with flat temperature gradients.

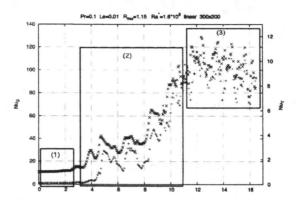

Figure 5. Nusselt numbers as function of time. The formation of a semi-convective zone (3) is the consequence of a diffusive phase (1) and a oscillatory phase (2). For the specified parameter space the convective rolls are established within 12 turnover cycles.

Linear stability analysis shows that semi-convection in an oscillatory instability. This could be observed in all our simulations. Figure 5 illustrates the development of a semi-convection cell in terms of the Nusselt number when starting from a linear stratification. Nusselt numbers in the range $1 < Nu_T < 5$ indicate very diffusive regimes, where the total heat flux is dominated by conduction. High values correspond to more convective regimes. A diffusive phase (1) is followed by an oscillatory phase (2). The fully evolved semi-convective role (3) leads to the expected (Spruit (1992)) step like structure in helium and temperature at the upper and lower boundaries. With a suitable choice of parameters, a preassigned step in the middle of the domain is found to be stable for many eddy turnover times.

While adoption of linear stratification as initial condition leads to very long simulation times, a 'direct initial step stratification' is considered. It has been shown that both approaches lead to the same results. Strong damping mechanisms in the parameter space $R_\rho > 2$ and $Le > 10^{-2}$ lead to very long simulation times. By starting from a step like stratification these could be reduced significantly and a formerly inaccessible parameter range came into reach.

The verification of theoretical and experimental power laws under consideration for the stellar parameter space was the main focus of this study. It turned out that several existing 'fitting formulas' (Castaing *et al.* (1989), Spruit (1992), Niemela *et al.* (2000)) describe an upper limit. Even the flux relation,

$$Nu_S = \tau^{1/2} Nu_T \qquad (4.1)$$

could be verified for semi-convective stable layers (see figure 6). Equipped with these results an extension to the stellar parameter regime was considered. The main problem of existing 1D stellar evolution codes is the lack of information about to the semi-convective mixing efficiency. The superadiabatic gradient $\nabla - \nabla_{ad}$ is not known in advance and has to be estimated. The semi-convective zone in a stellar model is limited in extent and consequently the evolution of the star on small time-scales does not depend much on the way it is calculated. But the additional diffusion in helium could be important for later evolutionary stages. This leads to the assumption that the heat flux could be considered as known rather than the real temperature gradient ∇. Based on Spruit (1992) an implicit function was derived, relating the temperature gradient, the Rayleigh number, the fluxes and the the vertical extent of the semi-convective zone to each other.

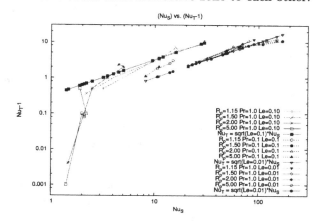

Figure 6. Nu_S versus $NuT - 1$ plot. Black dots represent the theoretical solutions. Each line stands for a different Ra_*. Along each line the stability parameter decreases. The red lines denote convective sub-critial simulations, where $Ra_T < Ra_{crit}$.

Acknowledgements This research has been supported by grants P18224, P20762, P20973 and P21742 of the Austrian Science Foundation and by grant KU 1954/3 within SPP 1276 of the Deutsche Forschungsgemeinschaft. Calculations were performed within DEISA (project SOLEX), at the Leibniz Computer Center, Munich (project SOLAR-SURF), at computers of the Max Planck Society and at the VSC cluster, Vienna. We are grateful to G. Houdek, Vienna, for providing us with starting models for cepheids.

References

Bascoul, G. P. 2007, IAU Symposium, 239, 317

Biello, J. A. 2001, Ph.D. Thesis

Buchler, J. R. 1997, Variables Stars and the Astrophysical Returns of the Microlensing Surveys, 181

Buchler, J. R. 2009, American Institute of Physics Conference Series, 1170, 51

Castaing, B., Gunaratne, G., Kadanoff, L., Libchaber, A. & Heslot, F. 1989, *Journal of Fluid Mechanics*, 204, 1

Huppert, H. E. & Moore, D. R. 1976, *Journal of Fluid Mechanics*, 78, 82

Kupka, F., Ballot, J. & Muthsam, H. J. 2009, *Communications in Asteroseismology*, 160, 30

Kwatra, N., Su, J., Grétarsson, J. T. & Fedkiw, R. 2009, *Journal of Computational Physics*, 228, 4146

Merryfield, W. J. 1995, *ApJ*, 444, 318

Mundprecht, E. 2010, PhD thesis, University of Vienna

Muthsam, H. J., Göb, W., Kupka, F., Liebich, W. & Zoechling, J. 1995, *A&A*, 293, 127

Muthsam, H. J., Göb, W., Kupka, F. & Liebich, W. 1999, *NewAst*, 4, 405

Muthsam, H. J., Kupka, F., Löw-Baselli, B., Obertscheider, C., Langer, M. & Lenz, P. 2010, *NewAst*, 15, 460

Niemela, J. J., Skrbek, L., Sreenivasan, K. R. & Donnelly, R. J. 2000, APS Meeting Abstracts, 2

Smolec, R. & Moskalik, P. 2009, American Institute of Physics Conference Series, 1170, 73

Spruit, H. C. 1992, *A&A*, 253, 131
Zaussinger, F. 2010, PhD thesis, University of Vienna
Zaussinger, F. & Spruit, H. 2010 in preparation for *A&A*

Discussion

TRAMPEDACH: Your simulation of semi-convection looked 2-D. Is that correct? Will it be extended to 3-D?

MUTHSAM: Yes, the present models are 2D. We will turn to 3D in the forseeable future.

GOUGH: The stated purpose of the final solution you presented was to determine whether the interface between turbulent convection layer is purely diffusive, but you did not tell us whether that is the case. Please would you tell us whether the transport of heat and helium across the interface which you showed us, which maintains its integrity despite being modulatory is purely diffusive or whether there is significant (small scale) convective transport?

MUTHSAM: You obviously refer to the movie with the pre-assigned step in the middle of the domain. The interface itself looks quite stable. Only on isolated spots (upstreams or downstreams) these streams can penetrate it somewhat.

POUQUET: Do you see any advantage to use a hybrid (MPI/OpenMP) parallelization method?

MUTHSAM: For really large runs you have to use MPI anyway for lack of shared memory. If you have, on one node, multiple cores you save messages when using them in the OpenMP rather than in the MPI mode.

Astrophysical Dynamics: From Galaxies to Stars
Proceedings IAU Symposium No. 271, 2010
N. H. Brummell, A. S. Brun, M. S. Miesch & Y. Ponty, eds.

© International Astronomical Union 2011
doi:10.1017/S1743921311017601

Magnetic Fields in Molecular Clouds

Paolo Padoan[1] Tuomas Lunttila[2], Mika Juvela[2], Åke Nordlund[3], David Collins[4], Alexei Kritsuk[4], Michael Normal[4], and Sergey Ustyugov[5]

[1]ICREA & ICC, University of Barcelona, Marti i Franquès 1, E-08028 Barcelona, Spain
email: ppadoan@icc.ub.edu

[2]Department of Physics, University of Helsinki, Helsinki, Finland

[3]Niels Bohr Institute, University of Copenhagen, Denmark

[4]CASS / Department of Physics, University of California, San Diego, USA

[5]Keldysh Institute of Applied Mathematics, Russian Academy of Sciences, Moscow, Russia

Abstract. Supersonic magneto-hydrodynamic (MHD) turbulence in molecular clouds (MCs) plays an important role in the process of star formation. The effect of the turbulence on the cloud fragmentation process depends on the magnetic field strength. In this work we discuss the idea that the turbulence is super-Alfvénic, at least with respect to the cloud mean magnetic field. We argue that MCs are likely to be born super-Alfvénic. We then support this scenario based on a recent simulation of the large-scale warm interstellar medium turbulence. Using small-scale isothermal MHD turbulence simulation, we also show that MCs may remain super-Alfvénic even with respect to their rms magnetic field strength, amplified by the turbulence. Finally, we briefly discuss the comparison with the observations, suggesting that super-Alfvénic turbulence successfully reproduces the Zeeman measurements of the magnetic field strength in dense MC clouds.

Keywords. ISM: kinematics and dynamics, MHD, stars: formation, turbulence

1. Star Formation and Supersonic Turbulence

What are the physical processes that determine the mass distribution and the formation rate of stars? We know that stars originate from the gravitational collapse of prestellar cores, but gravity alone cannot determine the wide range of stellar masses, nor the slow rate of star formation. Stellar masses span the approximate range 0.01-100 $m_\odot$ and the most numerous stars have a mass $\ll 1$ $m_\odot$, while the gravitational instability sets a characteristic Jeans mass of 1-10 $m_\odot$ in molecular clouds (MCs). Star-forming gas is converted into stars at a rate of approximately 2% per free-fall time (Krumholz & Tan 2007), while gravity alone would cause the collapse of all gas in one free-fall time.

The observed random velocities in MCs carry a kinetic energy comparable to the cloud gravitational energy, thus turbulence and self-gravity are of comparable importance on scales of several parsecs. Assuming uniform density, the gravitational energy scales as L^2, while the turbulent energy scales as $L^{0.4-0.5}$, based on observations (Larson 1981; Heyer & Brunt 2004; Padoan *et al.* 2006) and simulations (Kritsuk *et al.* 2007; Kritsuk *et al.* 2009). On the average, the turbulent energy must therefore exceed the gravitational energy within a MC, increasingly so towards smaller scales. As shown by Rosolowsky *et al.* (2008), the virial parameter (ratio of turbulent to gravitational energies) in molecular clouds is almost everywhere much larger than unity, except in some of the densest regions. The turbulence is thus able to regulate the star formation rate to a level much lower than that set by gravity alone (Krumholz & McKee 2005, Padoan & Nordlund 2010, in preparation).

The supersonic turbulent flows in MCs result in a complex network of interacting shocks responsible for the observed filamentary structure. Such flows can naturally assemble dense cores spanning the mass range of stars, suggesting that the turbulence can be directly responsible for the origin of the stellar mass distribution (Padoan & Nordlund 2002; Padoan *et al.* 2007). Although individual prestellar cores eventually collapse into stars due to their own gravity, the initial conditions for the gravitational collapse, specifically the total mass brought into the collapsing region, may be determined by the turbulent flow with little cooperation from the local gravitational force.

Having recognized the importance of turbulence in the fragmentation of MCs, we must establish its nature with respect to the magnetic field strength. How strong is the magnetic field in MCs?

2. The Magnetic Field Strength in MCs

The idea that MCs are magnetically supported against their gravitational collapse is reviewed in Shu *et al.* (1987). In that scenario, the observed random velocities correspond to MHD waves, or perturbations of a strong mean field. Gravitationally bound prestellar cores are initially subcritical and contract because of ambipolar drift until they become supercritical and collapse. Padoan & Nordlund (1997, 1999) investigated the possibility that the mean magnetic field in molecular clouds is weak and the observed turbulence is super-Alfvénic. By comparing results of two simulations, one with a weak field and the other with a strong field, with observational data, they showed that the super-Alfvénic case reproduced the observations better. Further results in support of the super-Alfvénic scenario were presented in Padoan *et al.* (2004) and, more recently, by Lunttila *et al.* (2008) and Lunttila *et al.* (2009), based on simulated Zeeman measurements.

In the following sections, we first address the super-Alfvénic nature of the turbulence in MCs in the context of their formation process. We then present results of large-scale multiphase MHD simulations of driven turbulence to illustrate the origin of the weak mean magnetic field in MCs. Finally, we discuss results of small-scale isothermal MHD turbulence simulations to show that MC turbulence may remain super-Alfvénic even with respect to their rms magnetic field, amplified by the turbulence.

3. Why Are MCs Born Super-Alfvénic?

Skipping a detailed discussion of the different processes that may contribute to the formation of MCs, and entirely neglecting the issue of turning atomic gas into molecules, one can at least say that MCs must be the result of large scale compressions of the warm interstellar medium (WISM). When such compressions reach the pressure threshold of the thermal instability, the compressed gas rapidly cools and compresses further to a characteristic mean density of MCs. This process may be driven, for example, by the evolution of supernova remnants. Irrespective of the specific driving mechanism, we can characterize the large-scale turbulence of the WISM based on its rms sonic and Alfvénic Mach numbers, M_s and M_a. It is generally believed that the large-scale turbulence in the WISM is transonic and trans-Alfvénic, meaning $M_s \sim 1$ and $M_a \sim 1$. This WISM turbulence regime is the fundamental reason why MCs are born super-Alfvénic.

Because of the transonic nature of the WISM turbulence, the large-scale velocity field can occasionally cause compressions strong enough to bring large regions above the thermal instability threshold (this is more likely within spiral arms, where the mean gas

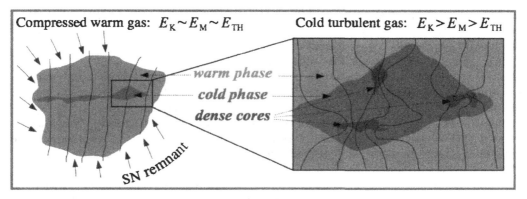

Figure 1. Schematic scenario of the formation of super-Alfvénic MCs.

density has been increased by a spiral arm shock). As the gas is further compressed thanks to its cooling, the magnetic field cannot be compressed because of the initially trans-Alfvénic nature of the flow. In other words, the mean magnetic field is strong enough that the initial compression is forced to be primarily along the magnetic field. Assuming that turbulent velocities are not significantly decreased, the characteristic increase in density, $\rho_{\text{cold}} \sim 100\rho_{\text{warm}}$, results in a comparable increase in turbulent kinetic energy, $E_{\text{K,cold}} \sim 100E_{\text{K,warm}}$, or a corresponding drop in the rms Alfvén velocity, $V_{\text{a,cold}} \sim V_{\text{a,warm}}/10$. As a consequence, the turbulence in the rapidly cooling gas must be initially super-Alfvénic with respect to the mean magnetic field. Compression and stretching in this super-Alfvénic flow can then locally amplify the magnetic field, but the mean field (averaged over the whole MC) cannot change much, because the large-scale compressive flow was initially directed to be primarily along the mean magnetic field direction. Because of the reduced temperature, the turbulence in the cold gas is also supersonic, so dense cores with enhanced magnetic field strength are naturally formed by shocks in the turbulent flow. This sequence of events is schematically depicted in Figure 1.

The prediction of a weak mean magnetic field in MCs is further supported by a numerical simulation discussed in the next section. Despite the low mean field, the rms value of B (or the magnetic energy) within MCs is expected to grow due to compression and stretching in the super-Alfvénic flow, and possibly also due to the action of a turbulent dynamo. The evolution of the rms B is investigated in the section after the next one.

4. The Mean B: Large-Scale Multiphase Simulations

In order to test the validity of our general argument about the super-Alfvénic nature of MCs with respect to their mean magnetic field, we have developed a simulation on a numerical mesh of 512^3 computational cells, reproducing global properties of the WISM turbulence. The simulation assumes a mean density of $n_0 = 5$ cm^{-3}, and drives the turbulence to the rms Mach number values of $M_s \approx 1.8$ and $M_a \approx 0.6$ with respect to the warm phase. The turbulent pressure fluctuations keep 49% of the gas mass and 6% of its volume in the cold phase, while 5% of the mass and 24% of the volume remains in the warm phase. The rest of the gas is found at intermediate temperatures that would be considered thermally unstable in the absence of turbulent pressure. The mean magnetic field strength in the simulation is $B_0 = 3.2$ μG, with an rms value of the same order. The parameters of this simulation may characterize a slightly over-dense region of the ISM within a spiral arm, on a scale of approximately 200 pc. Details of the simulation will be presented in Kritsuk *et al.* (2010, in preparation). Here we only report some preliminary results on the magnetic field strength.

The left panel of Figure 2 shows the projected density from a snapshot of the simulation. If the box size is assumed to be $L = 200$ pc, filamentary dense regions of length up to approximately 40 pc are common. These are regions of cold gas with a mean density of approximately 100 cm^{-3} and mass up to several thousands solar masses, characteristic values for MCs. We collect a sample of these cold clouds by selecting spatially connected regions within isodensity contours corresponding to $n > 30$ cm^{-3}. We compute the volume-averaged magnetic field strength within each cloud, and plot its distribution on the right panel of Figure 2, as a black histogram. The distribution of the mean magnetic field strength in the clouds is found to peak at 5 μG, less than a factor of two larger than the mean field of the simulation, and covers the approximate range of 0.5-10 μG. This result confirms that approximately transonic and trans-Alfénic turbulence in a gas with the cooling properties of the WISM naturally generates dense regions of cold gas with global properties characteristic of MCs and mean magnetic field strength only a little larger than the large scale mean value in WISM.

The right panel of Figure 2 shows also the histograms of B for all computational cells with $n < 2$ cm^{-3} and with $n > 30$ cm^{-3}. Values much larger than the mean can be found in both histograms, but particularly so in the case of the dense and colder gas. That is because, as explained in the previous section, the turbulence is super-Alfvénic and supersonic in the cold gas, so large enhancements of both B and n are expected within the cold clouds. The magnetic field distribution in the cold gas shows a very extended exponential tail, as in isothermal simulations of super-Alfvénic turbulence (Padoan & Nordlund 1999). However, the average field within each cloud remains rather weak (see black histogram), as argued in the previous section.

Although the turbulence within the largest clouds in the simulation is super-Alfvénic, their value of M_a is under-estimated due to the limited numerical resolution. A much larger resolution is needed to resolve well the internal cloud turbulence. This should not affect the derived mean magnetic field strength in the clouds, but it is certainly a significant numerical limitation with respect to the evolution of the cloud rms B. In order to study the evolution of the rms B within regions of cold gas, we focus in the next section on simulations of isothermal super-Alfvénic turbulence.

5. The rms B: Small-Scale Isothermal Simulations

If their super-Alfvénic turbulence generates a dynamo, MCs with a very weak mean magnetic field may in principle have their magnetic energy amplified to equipartition with the turbulent kinetic energy. If that were achieved during their lifetime, MCs that are born (and remain) super-Alfvénic with respect to their mean magnetic field, may evolve to become trans-Alfvénic at least with respect to their rms magnetic field. To investigate the saturated value of the rms magnetic field strength we have run a set of isothermal MHD turbulence simulations with rms Mach numbers $M_s \approx 10$ and $M_{a,0} \approx 30$, 10, and 3, where $M_{a,0}$ is the rms Alfvénic Mach number with respect to the Alfvén velocity defined with the mean magnetic field, B_0, and the mean gas density, n_0. The turbulence is driven by a random force within the wavenumber range $1 \leqslant k \leqslant 2$ ($k = 1$ corresponds to the computational box size) for several dynamical times, starting from uniform magnetic and density fields. The runs are repeated at three different numerical resolutions, on mesh sizes of 256^3, 512^3, and 1024^3 computational cells. Details about the PPML code and the simulations can be found in Ustyugov *et al.* (2009) and in Kritsuk *et al.* (2009a,b).

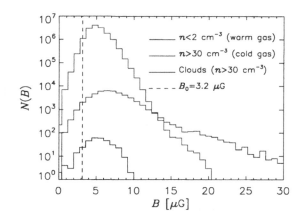

Figure 2. *Left:* Projected density field from the multiphase turbulence simulation representing a region of approximately 200 pc (the grey scale is linearly proportional to the column density). *Right:* Probability distribution of the magnetic field strength in the warm gas (red histogram), the cold gas (blue histogram), and averaged inside MCs (black histogram). The vertical dashed line corresponds to the mean magnetic field in the simulation, $B_0 = 3.2$ μG, representing the mean magnetic field strength in the Galactic disk. The characteristic mean field in MCs is approximately 5 μG, very close to the mean Galactic value.

The non-dimensional parameters of the runs are listed in Table 1, where β_0 is the ratio of gas to magnetic pressure, $\beta_0 = 2(M_{a,0}/M_s)^2$, based on the mean magnetic field and the mean density. The two last columns of Table 1 give the saturated rms values of the Alfvénic Mach number and of the ratio of gas to magnetic pressure, M_a and β, after the rms field has been amplified to saturation. Based on those values of M_a, one can see that only the simulation with the strongest mean magnetic field becomes trans-Alfvénic with respect to the rms magnetic field.

The time evolution of magnetic and turbulent kinetic energy is plotted in Figure 3, with the energy shown with linear and logarithmic scales on the left and right panels respectively. The magnetic energy is amplified and reaches saturation after approximately three dynamical times. The amplification can be entirely accounted for by a combination of compression and stretching events. There is no evidence of a real turbulent dynamo, meaning constructive twisting and folding events that could further amplify the magnetic field. In the simulation with the weakest mean magnetic field there is a very slow field amplification during most of the evolution. Even if that were due to a turbulent dynamo, the growth rate would be extremely low and the effect of the turbulent dynamo irrelevant within a MC lifetime. Only the simulation with the strongest mean magnetic field reaches equipartition of kinetic and magnetic energy, while in the other two runs the magnetic energy remains approximately 4 and 10 times smaller than the turbulent kinetic energy.

The saturation level of the rms magnetic field strength could depend on numerical resolution. By performing a numerical convergence test, we have verified that at a resolution of 1024^3 computational cells the saturation level of the magnetic energy is almost converged. Dynamo action may also depend on the magnetic Prandtl number, P_m, giving the ratio of viscosity and resistivity, $P_m = \nu/\eta$. In numerical simulations where the resistivity and viscosity are not explicitly included, the effective value of P_m is of order unity. In real MCs it is many orders of magnitude larger. The turbulence regime with a value of P_m as large as in MCs cannot be tested numerically. However, previous numerical work by Haugen *et al.* (2004) suggests that the growth rate of the turbulent dynamo in the

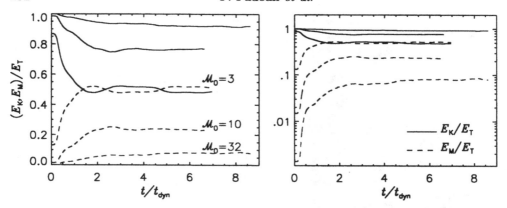

Figure 3. *Left:* Time evolution of magnetic and kinetic energies (normalized to the total energy), in the three 512^3 runs with different values of B_0. The energy scale is linear *Right:* Same as the left panel, but with a logarithmic energy scale to show better the magnetic energy amplification in the run with the weakest mean magnetic field.

Based on B_0 and n_0:			Saturated values:	
M_s $M_{a,0}$		β_0	M_a	β
10 31.6		20.0	9.2	0.11
10 10.0		2.0	3.0	0.03
10 3.2		0.2	1.1	0.01

Table 1. Non-dimensional parameters of the isothermal MHD turbulence simulations.

supersonic regime may be significantly reduced. It is possible that when gas elements of a supersonic turbulent flow dissipate their energy in shocks, their ability to twist and fold constructively is largely reduced compared to the case of fluid elements following the vortical motions of an incompressible turbulent flow.

Taking our numerical results at face value, we would conclude that MCs may remain super-Alfvénic also with respect to their rms magnetic field strength over their whole lifetime. Only MCs that are born with a relatively strong mean magnetic field may have a chance to reach equipartition of kinetic and magnetic energies. However, equipartition is always rapidly reached locally. The magnetic pressure in the postshock regions has to balance the gas ram pressure, which is the same in three runs, because $M_s \approx 10$ in all of them. The largest values of B, typically found in the densest regions, are thus weakly dependent on the mean or rms magnetic field strength.

5.1. $B - n$ correlation in super-Alfénic turbulence

In super-Alfvénic turbulence, the local magnetic field strength correlates with the gas density, as long as the turbulence remains super-Alfvénic with respect to both the mean and the rms magnetic field strength. Figure 4 shows scatter plots of magnetic versus gas pressure in the case of the run with the largest value of B_0, where the magnetic energy has reached equipartition with the turbulent kinetic energy (left panel), and in the case of the lowest value of B_0, which is still super-Alfvénic also with respect to the rms magnetic field strength (right panel). In the strong B_0 case there is almost no $B - n$ correlation. This would be even more true in the case of a trans-Alfvénic run with respect to the mean magnetic field (Padoan & Nordlund 1999). In the weak B_0 case, despite the large scatter, there is a clear trend of increasing B with increasing n. The

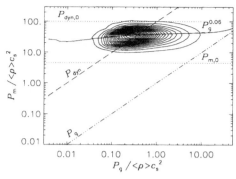 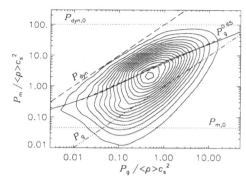

Figure 4. *Left:* Scatter plot of magnetic versus gas pressures, normalized to the mean gas pressure, for the case with the strongest mean magnetic field strength, $M_{a,0} = 3$. The top dotted line shows the characteristic value of the dynamic pressure, defined with the mean density and the global rms velocity. The bottom dotted line corresponds to the value of the mean magnetic pressure. The diagonal dashed line is the dynamic pressure defined with the global rms velocity and the local density. The diagonal dashed-dotted line is the local gas pressure. The solid line shows the mean magnetic pressure as a function of gas pressure. The power law fit gives $P_m \propto P_g^{0.06}$ *Right:* Same as on the left panel, but for the simulation with $M_{a,0} = 32$. The mean magnetic pressure increases with gas pressure following approximately a power law, $P_m \propto P_g^{0.65}$.

mean value of magnetic pressure grows with gas pressure approximately as a power law, $P_m \propto P_g^{0.65}$, corresponding to $B \propto n^{0.32}$. The upper envelope of the scatter plot shows that, at each density, the largest values of magnetic pressure scale approximately as the dynamic pressure defined with the global rms velocity and the local density. While the magnetic pressure is almost always in excess of the gas pressure, it almost never exceeds the dynamic pressure defined above, showing that the magnetic field plays primarily a passive role in the dynamics. In the strong B_0 case, instead, the magnetic pressure often exceeds the dynamic pressure.

6. Comparison with Observations

Padoan & Nordlund (1999) discussed several observational tests of numerical simulations of supersonic MHD turbulence that could be used to constrain the mean magnetic field. Based on those tests, they suggested that turbulence in MCs is super-Alfvénic with respect to their mean magnetic field on scales of few to several parsecs. This super-Alfvénic model of star-forming regions was recently used to generate simulated measurements of the Zeeman effect on 18 cm OH lines (Lunttila *et al.* 2008). It was shown that a super-Alfvénic turbulence simulation with the characteristic size, density, and velocity dispersion of star-forming regions could produce dense cores with the same $|B_{los}|$-N relation as observed cores. Lunttila *et al.* (2008) also computed the relative mass-to-flux ratio $\mathcal{R}_\mu$, defined as the mass-to-flux ratio of a core divided by that of its envelope, following the observational procedure proposed by Crutcher *et al.* (2009). They found a large scatter in the value of $\mathcal{R}_\mu$, and an average value of $\mathcal{R}_\mu < 1$, in contrast to the ambipolar-drift model of core formation, where the mean magnetic field is stronger and only $\mathcal{R}_\mu > 1$ is allowed. The observational results of Crutcher *et al.* (2009) confirmed $\mathcal{R}_\mu < 1$ in observed cores, as predicted by Lunttila *et al.* (2008) for the super-Alfvénic model.

Here we present further evidence that the same super-Alfvénic simulation compares well with the observational data, summarizing the main results of Lunttila *et al.* (2009).

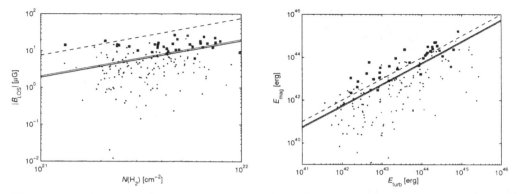

Figure 5. *Left:* Inferred line-of-sight magnetic field as a function of H_2 column density. Black symbols show the results from our simulation, red symbols are cores from Troland & Crutcher (2008). Squares show 3σ detections, and dots are non-detections. The black solid line shows the mean mass-to-flux ratio of the simulated cores, and the red solid line the mean mass-to-flux ratio calculated from Troland & Crutcher (2008) observations. The dashed line is the critical mass-to-flux ratio with no geometrical correction, $\lambda = 1$. *Right:* Magnetic energy versus turbulent kinetic energy for the same cores and with the same symbols as in the left panel. The dashed line corresponds to the energy ratio $\beta_{turb} = E_{turb}/E_{mag} = 1$. The solid black line shows the mean energy ratio for the simulated cores, and the red solid line for Troland & Crutcher (2008) observations.

Lunttila *et al.* (2009) used simulated OH Zeeman measurements to compute the mass-to-flux ratio relative to the critical one, λ, and the ratio of turbulent to magnetic energies, β_{turb}, in molecular cores selected from simulated maps. They followed closely the observational procedure of Troland & Crutcher (2008), and found mean values of λ and β_{turb} in good agreement with the observational results.

The work of Lunttila *et al.* (2009) is based on a simulation run on a mesh of 1000^3 zones with the Stagger Code Padoan *et al.* (2007), with periodic boundary conditions, isothermal equation of state, random forcing in Fourier space at wavenumbers $1 \leqslant k \leqslant 2$ ($k = 1$ corresponds to the computational box size), uniform initial density and magnetic field, and random initial velocity field with power only at wavenumbers $1 \leqslant k \leqslant 2$. The rms sonic Mach number is $\mathcal{M}_s = \sigma_{v,3D}/c_s = 8.91$. The initial and evolved rms Alfvénic Mach numbers are $M_{a,0} = 29.7$ and $M_a = 2.8$ respectively.

For the computation of synthetic Zeeman spectra the data cube was scaled to physical units, assuming $L = 9$ pc, $\langle n(H_2) \rangle = 67$ cm^{-3} (typical for that scale in the sample of Falgarone *et al.* (1992)), and $T_{kin} = 10$ K. A constant fractional OH abundance of $[OH]/[H] = 4.0 \times 10^{-8}$ was assumed, following Crutcher (1979). Radiative transfer calculations were performed to compute simulated Zeeman measurements of the 1665 and 1667 MHz OH lines observed by Troland & Crutcher (2008). The synthetic observations were made with a 3 arcmin (fwhm) beam, corresponding to the angular resolution of the Arecibo telescope, and using a channel separation of 0.05 km s^{-1}.

In order to compare the simulations with the observations of Troland & Crutcher (2008), dense cores were selected from the OH emission position-position-velocity data cubes using the clumpfind algorithm of Williams *et al.* (1994). Before applying the clumpfind routine, the data cubes were resampled to an angular resolution of ~ 1.2 arcmin (approximately Nyquist sampled), and uncorrelated Gaussian noise with rms of 0.08 K was added to simulate observational noise.

Core physical parameters obtained from the simulated observations agree well with those in the sample of Troland & Crutcher (2008). In particular, the mean values of the mass-to-flux ratio relative to the critical one, λ, and of the ratio of turbulent to magnetic energies, β_{turb} are $\langle\lambda\rangle_{sim} = 3.9$ and $\langle\beta_{turb}\rangle_{sim} = 1.8$, almost identical to those from the observed core sample, $\langle\lambda\rangle_{obs} = 3.8$ and $\langle\beta_{turb}\rangle_{obs} = 1.9$. As shown in Figure 5, also the scatter around these mean values found in the observations is reproduced well by the simulated cores. Even using only detections, both observed and simulated cores appear to be supercritical, while their turbulent kinetic energy is of the order of their magnetic energy, but with a significant scatter.

If the MC turbulence is indeed super-Alfvénic, the scatter found around the mean values of λ and β_{turb} is only partly due to the random orientation of the magnetic field with respect to the line of sight (the Zeeman measurements are only sensitive to the line-of-sight component of the magnetic field). Part of the scatter originates from intrinsic variations of the magnetic field strength from core to core. Such intrinsic variations of magnetic field strength are not expected in the traditional picture of MCs were the mean magnetic field is strong (Shu *et al.* 1987).

7. Conclusions

We have argued that MCs are born super-Alfvénic with respect to their mean magnetic field, because of the trans-Alfvénic nature of the turbulence in the WISM. Using a multiphase turbulence simulation meant to represent an overdense region (spiral arm) of the WISM turbulence on a scale of order 200 pc, we have shown that large regions of cold dense gas are formed by transonic or mildly supersonic large-scale turbulent compressions, with physical properties (mean density, size, and temperature) characteristic of MCs. We have shown that the mean field within these clouds is on the average around 5 μG, only slightly larger than the assumed mean field of 3.2 μG (with a comparable rms). This simulation confirms our general argument about the super-Alfvénic nature of clouds formed out of trans-Alfvénic turbulence in a gas with the cooling properties of the WISM. If large scale turbulence, driven for example by the evolution of SN remnants, plays an important role in their formation, MCs should be born with a rather weak mean magnetic field, as suggested by Padoan & Nordlund (1997, 1999). The actual process of MC formation is of course much more complex, especially with regard to the formation of molecular species.

We have then addressed the question of the time evolution of the rms magnetic field strength, or magnetic energy. Using a set of isothermal simulations of supersonic and super-Alfvénic turbulence, we have shown that magnetic energy can reach equipartition with the turbulent kinetic energy only in the case where the mean field is already not too far from equipartition. If a turbulent dynamo operates at all, beyond the saturated field level reached in the simulation, its growth rate may be too small to be relevant within a MC lifetime. We conclude that a significant fraction of MCs may remain super-Alfvénic also with respect to their rms magnetic field strength.

Despite the small value of their mean magnetic field strength, super-Alfvénic MCs are expected to naturally generate regions of stronger magnetic field, particularly where dense cores are formed. When the gas density is enhanced by turbulent shocks, the magnetic field is also enhanced due to the combined effects of the compression of the field component perpendicular to the shock direction and of the field stretching by the shear flow in the postshock layers. We have shown that the properties of dense MC cores formed in this way in the simulations are consistent with those of real cores observed by Troland & Crutcher (2008). The mean values of the mass-to-flux ratio relative to the critical one, λ, and of the ratio of turbulent to magnetic energies, β_{turb}, are $\langle\lambda\rangle_{sim} = 3.9$ and $\langle\beta_{turb}\rangle_{sim} = 1.8$, almost identical to those from the observational sample, $\langle\lambda\rangle_{obs} = 3.8$

and $\langle \beta_{\text{turb}} \rangle_{\text{obs}} = 1.9$. The relative mass-to-flux ratio (core to envelope) was predicted to be less than unity in super-Alfvénic turbulence, at least for cores where the magnetic field could be detected, in contrast to the basic ambipolar drift model of core formation, where that ratio was predicted to be larger than unity. Recent observations by Crutcher *et al.* (2009) have confirmed this prediction of the super-Alfvénic model.

Acknowledgements

This work was supported in part by the National Science Foundation under Grants AST0808184, and AST0908740. Computer simulations utilized NSF TeraGrid resources provided by NICS and TACC through allocation MCA07S014 as well as by NCCS through the DOE INCITE allocation AST015.

References

M. R. Krumholz & J. C. Tan, *ApJ* **654**, 304–315 (2007).

R. B. Larson, *MNRAS* **194**, 809–826 (1981).

M. H. Heyer & C. M. Brunt, *ApJL* **615**, L45–L48 (2004).

P. Padoan, M. Juvela, A. Kritsuk & M. L. Norman, *ApJL* **653**, L125–L128 (2006).

A. G. Kritsuk, M. L. Norman, P. Padoan & R. Wagner, *ApJ* **665**, 416–431 (2007).

A. G. Kritsuk, S. D. Ustyugov, M. L. Norman & P. Padoan, *Journal of Physics Conference Series* **180**, 012020 (2009).

E. W. Rosolowsky, J. E. Pineda, J. Kauffmann & A. A. Goodman, *ApJ* **679**, 1338–1351 (2008).

M. R. Krumholz & C. F. McKee, *ApJ* **630**, 250–268 (2005).

P. Padoan & A. Nordlund, *ArXiv e-prints* (2009), `arXiv:astro-ph/0907.0248`.

P. Padoan & Å. Nordlund, *ApJ* **576**, 870–879 (2002).

P. Padoan, Å. Nordlund, A. G. Kritsuk, M. L. Norman & P. S. Li, *ApJ* **661**, 972–981 (2007).

F. H. Shu, F. C. Adams & S. Lizano, *ARA&A* **25**, 23 (1987).

P. Padoan & Å. Nordlund, *ArXiv e-prints* (1997) `arXiv:astro-ph/9706176`.

P. Padoan & Å. Nordlund, *ApJ* **526**, 279 (1999).

P. Padoan, R. Jimenez, M. Juvela & Å. Nordlund, *ApJL* **604**, L49–L52 (2004).

T. Lunttila, P. Padoan, M. Juvela & Å. Nordlund, *ApJL* **686**, L91–L94 (2008).

T. Lunttila, P. Padoan, M. Juvela & Å. Nordlund, *ApJL* **702**, L37–L41 (2009).

S. D. Ustyugov, M. V. Popov, A. G. Kritsuk & M. L. Norman, *Journal of Computational Physics* **228**, 7614–7633 (2009).

A. G. Kritsuk, S. D. Ustyugov, M. L. Norman & P. Padoan, ASPC, 406, 15 (2009)

N. E. L. Haugen, A. Brandenburg, & A. J. Mee, *MNRAS* **353**, 947–952 (2004).

T. H. Troland & R. M. Crutcher, *ApJ* **680**, 457–465 (2008).

E. Falgarone, J. L. Puget & M. Pérault, *A&A* **257**, 715 (1992).

R. M. Crutcher, *ApJ* **234**, 881–890 (1979).

J. P. Williams, E. J. De Geus & L. Blitz, *ApJ* **428**, 693 (1995).

R. M. Crutcher, N. Hakobian & T. H. Troland, *ApJ* **692**, 844–855 (2009).

Astrophysical Dynamics: From Stars to Galaxies
Proceedings IAU Symposium No. 271, 2010
N. H. Brummell, A. S. Brun, M. S. Miesch & Y. Ponty, eds.
© International Astronomical Union 2011
doi:10.1017/S1743921311017613

The influence of stratification upon small-scale convectively-driven dynamos

Paul J. Bushby[1], Michael R. E. Proctor[2] and Nigel O. Weiss[2]

[1]School of Mathematics and Statistics, Newcastle University,
Newcastle upon Tyne, NE1 7RU, U.K.
email: paul.bushby@ncl.ac.uk

[2]Department of Applied Mathematics and Theoretical Physics (DAMTP),
University of Cambridge, Cambridge CB3 0WA, U.K.
email: mrep@cam.ac.uk (MREP) and now@cam.ac.uk (NOW)

Abstract. In the quiet Sun, convective motions form a characteristic granular pattern, with broad upflows enclosed by a network of narrow downflows. Magnetic fields tend to accumulate in the intergranular lanes, forming localised flux concentrations. One of the most plausible explanations for the appearance of these quiet Sun magnetic features is that they are generated and maintained by dynamo action resulting from the local convective motions at the surface of the Sun. Motivated by this idea, we describe high resolution numerical simulations of nonlinear dynamo action in a (fully) compressible, non-rotating layer of electrically-conducting fluid. The dynamo properties depend crucially upon various aspects of the fluid. For example, the magnetic Reynolds number (Rm) determines the initial growth rate of the magnetic energy, as well as the final saturation level of the dynamo in the nonlinear regime. We focus particularly upon the ways in which the Rm-dependence of the dynamo is influenced by the level of stratification within the domain. Our results can be related, in a qualitative sense, to solar observations.

Keywords. Convection, magnetic fields, (magnetohydrodynamics:) MHD, Sun: magnetic fields, Sun: photosphere

---

1. Introduction

High resolution observations of the solar surface have greatly enhanced our theoretical understanding of the ways in which magnetic fields interact with turbulent convection in an electrically-conducting fluid. The time-dependent, near-surface, convective motions in the quiet Sun form a characteristic granular pattern at the solar photosphere (see, for example, Stix 2004). The dark intergranular lanes, which correspond to the convective downflows, contain mixed-polarity concentrations of vertical magnetic flux. These localised magnetic features show up as bright points in G-band images of the solar surface (see, for example, Sánchez Almeida *et al.* 2010). Peak magnetic field strengths in these regions are typically in excess of a kilogauss (see, for example, de Wijn *et al.* 2009, and references therein), which implies that the magnetic energy density of these features is about an order of magnitude larger than the mean kinetic energy density of the surrounding granular convection

It is plausible that the convective motions that are observed at the solar surface are themselves responsible for the generation of quiet Sun magnetic fields. A small-scale dynamo of this type would proceed independently of the large-scale dynamo processes that are responsible for driving the solar cycle. As demonstrated by Cattaneo (1999), standard Boussinesq convection in an electrically-conducting fluid (in the absence of rotation or shear) can drive a small-scale dynamo. More recent calculations have demonstrated that dynamo action is also possible in fully compressible convection (Abbett

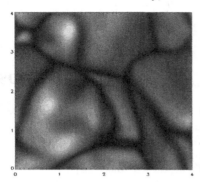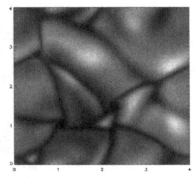

Figure 1. The temperature distribution for hydrodynamic convection in a horizontal plane just below the upper surface of the computational domain. Bright contours correspond to warmer fluid. Left: Moderately-stratified convection ($\theta = 3$). Right: Highly-stratified convection ($\theta = 10$).

2007; Vögler & Schüssler 2007; Käpylä, Korpi & Brandenburg 2008; Bushby, Proctor & Weiss 2010; Brummell, Tobias & Cattaneo 2010). However, there are still some aspects of convectively-driven dynamos that have not yet been addressed. The aim of this short paper is to investigate some of the ways in which the level of stratification within the fluid influences the efficiency of a convectively-driven dynamo.

2. An idealised model

We consider the behaviour of a compressible, electrically-conducting, monatomic gas, in the presence of a magnetic field. In this idealised model, we adopt constant values for the viscosity μ, the thermal conductivity K, the magnetic diffusivity η, and the magnetic permeability μ_0. The constant specific heat capacities (c_P and c_V) satisfy $c_P/c_V = 5/3$, whilst the gas constant is defined by $R_* = c_P - c_V$. Choosing a (non-rotating) Cartesian frame of reference in which the z-axis points vertically downwards, the gas occupies the region $0 \leqslant x \leqslant 4d$, $0 \leqslant y \leqslant 4d$, $0 \leqslant z \leqslant d$. The constant gravitational acceleration is given by $\mathbf{g} = g\hat{\mathbf{z}}$. The state of this system at position $\mathbf{x}$ and time t is defined by the density $\rho(\mathbf{x}, t)$, the temperature $T(\mathbf{x}, t)$, the velocity $\mathbf{u}(\mathbf{x}, t)$ and the magnetic field $\mathbf{B}(\mathbf{x}, t)$. The gas pressure, $P(\mathbf{x}, t)$, is given by $P = R_*\rho T$. The fluid variables satisfy periodic boundary conditions in each horizontal direction. The upper and lower bounding surfaces are held at fixed, uniform temperatures, with $T = T_0$ at $z = 0$ and $T = T_0 + \Delta T$ at $z = d$ (where a positive value of ΔT implies that the layer is heated from below). These boundaries are also assumed to be impermeable and stress-free in this idealised model. Vertical field boundary conditions are imposed on $\mathbf{B}$ at $z = 0$ and $z = d$. The evolution of this system is governed by the standard equations of (non-ideal) compressible magnetohydrodynamics (see, for example, Bushby *et al.* 2008).

In the absence of any magnetic fields, the governing equations have a hydrostatic solution, corresponding to a polytropic layer, in which $T = T_0(1 + \theta z/d)$ and $\rho = \rho_0(1 + \theta z/d)^m$. Here, the parameter $\theta = \Delta T/T_0$ is a measure of the thermal stratification of the layer, whilst $m = (gd/R_*\Delta T) - 1$ defines the polytropic index. For an unmagnetised monatomic gas, this polytropic equilibrium is unstable to convective perturbations provided that $m < 3/2$. The evolution of any convective perturbation depends crucially upon the other parameters in the system. We define the Prandtl number to be $\sigma = \mu c_P/K$, whilst the dimensionless thermal diffusivity is defined to be $\kappa = K/d\rho_0 c_p (R_*T_0)^{1/2}$. The parameter $\zeta_0 = \eta c_P \rho_0/K$ gives the ratio of the magnetic diffusivity to the thermal diffusivity at the upper surface of the domain. Two other relevant parameters for convection are the (mid-layer) Reynolds number,

$$Re = \frac{\rho_{mid} U_{rms} d}{\mu}, \tag{2.1}$$

(where U_{rms} is the rms velocity of the convection and ρ_{mid} is the mid-layer density of the unperturbed polytrope) whilst

$$Rm = \frac{U_{rms} d}{\eta}, \tag{2.2}$$

corresponds to the magnetic Reynolds number of the flow.

By carrying out three-dimensional numerical simulations (see, for example, Bushby *et al.* 2008 for numerical details), we investigate the dynamo properties of convection in two different polytropic layers, one moderately-stratified, the other highly-stratified. In both cases, $m = 1$ and $\sigma = 1$. In the moderately-stratified case, we set $\theta = 3$, which implies that the density and temperature both vary by a factor of 4 across the depth of the unperturbed layer. In the highly-stratified layer, we set $\theta = 10$, which means that the initial density and temperature profiles both vary by a factor of 11. Finally, $\kappa = 0.0245$ for the highly-stratified case and $\kappa = 0.00548$ for the moderately-stratified layer. These parameters have been chosen so that both calculations produce hydrodynamic convection with $Re \approx 150$. This is illustrated in Figure 1, which shows the temperature distribution (for each case) in a horizontal plane just below the upper surface of the computational domain. Note that the "granular" pattern is similar in both simulations, despite the differing levels of stratification. Some aspects of dynamo action in the highly-stratified case were described in a previous paper (Bushby, Proctor & Weiss 2010).

Once statistically-steady hydrodynamic convection has developed, we insert a weak vertical magnetic field into the flow. This has a simple cosine dependence upon x and y in order to ensure that there is no net flux across the computational domain. The fate of this field depends crucially upon the magnetic Reynolds number. If everything else is fixed, the value of Rm depends solely upon the value of ζ_0, with $Rm \propto (1/\zeta_0)$. Hence for a given hydrodynamic flow, different values of Rm can be investigated simply by changing the value of ζ_0. Since Re is fixed, varying Rm is equivalent to varying the magnetic Prandtl number (which is given by $Pm = Rm/Re$).

3. Numerical results

3.1. *The kinematic regime*

Provided that the energy of the initial magnetic field is very much smaller than the kinetic energy of the flow, the Lorentz force does not play a significant dynamical role during the early stages of the field evolution. During this kinematic phase, the magnetic energy fluctuates in time, but (on average) either grows or decays exponentially, depending upon the magnitude of the magnetic Reynolds number, Rm. Dynamo growth can only occur if Rm is large enough such that the inductive effects due to the fluid motions are strong enough to outweigh the dissipative effects of magnetic diffusion. Of course, exponential growth cannot proceed indefinitely. Eventually, the dynamo-generated magnetic field will become strong enough to exert a significant Lorentz force upon the flow, which forces the dynamo to saturate in the nonlinear regime.

Initially, however, we focus on the kinematic regime, indefinitely extending this phase of evolution by "switching off" the magnetic terms in the momentum and heat equations.

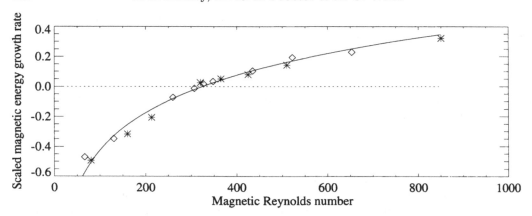

Figure 2. Scaled kinematic growth rates for the magnetic energy, as a function of the magnetic Reynolds number. Stars correspond to the moderately-stratified case, diamonds correspond to highly-stratified convection. The solid line is the curve $\Gamma = 0.36 \log_e (Rm/325)$.

A prolonged kinematic phase allows us to average growth rates over a long period of time, thus obtaining more accurate values for these quantities. In a previous paper (Bushby et al. 2010), it was shown that dynamo action is possible in the highly-stratified case provided that the magnetic Reynolds number exceeds a critical value of $Rm_{crit} \approx 325$. Furthermore, the results from that paper suggested that the kinematic growth rates (denoted here by Γ) had a logarithmic dependence upon Rm. More precisely, expressing the growth rates in terms an inverse (isothermal) acoustic travel time at the top of the computational domain, $(R_* T_0)^{1/2}/d$, the measured growth rates are a good fit to the following curve, $\Gamma = 0.21 \log_e (Rm/325)$. Although this logarithmic fit is rather empirical, a similar Rm-dependence has been found in a previous analytic study (Rogachevskii & Kleeorin 1997). In order to assess the influence that stratification has upon the kinematic regime, we repeated these calculations for our moderately-stratified convective layer. Again we found a critical magnetic Reynolds of $Rm_{crit} \approx 325$, whilst $\Gamma = 0.12 \log_e (Rm/325)$ gives a good fit to the Rm-dependence of the dynamo growth rates in moderately-stratified convection.

From these results, it is tempting to draw the conclusion that dynamo action is more efficient in highly stratified convection, with a higher kinematic growth rate for a given value of Rm. In fact, some care is needed here. The acoustic travel time that has been used to scale these growth rates corresponds to the time taken for a sound wave to cross a horizontal distance d across the upper surface of the domain. Clearly the surface surface sound speed is more representative of a "typical" sound speed in the moderately-stratified case than it is in the highly stratified simulation. So, in order to make a fairer comparison between the two cases, we scale the growth rates by the convective turnover time, d/U_{rms}. The results of this rescaling, for both stratifications, are shown in Figure 2. It is apparent from Figure 2 that all the data points lie close to the same best fit curve, $\Gamma = 0.36 \log_e (Rm/325)$. This suggests that the rescaled growth rates are effectively independent of the level of stratification in the domain. In other words, it is purely the convective turnover time that is responsible for setting the growth rate rather than any of the topological aspects of the flow that are associated with the compressibility.

3.2. Nonlinear results

The process of flux expulsion leads to the formation of vertical magnetic flux concentrations in the downflow regions at the edges of the granular convective cells. Whilst the field

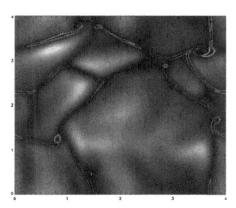
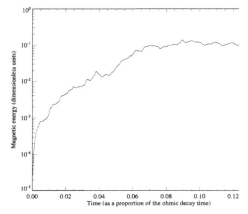

Figure 3. A nonlinear dynamo calculation, in moderately-stratified convection, for $Rm \approx 480$. Left: Contours of constant temperature, overlaid with contours of the vertical component of the magnetic field. Solid (dashed) contours correspond to positive (negative) values of B_z. Right: A log-linear plot showing the magnetic energy (in dimensionless units) as a function of time (expressed in terms of ohmic decay times).

is weak, this is a kinematic process, with no magnetic feedback upon the flow. In fully nonlinear dynamo simulations, the high magnetic pressure that is associated with these magnetic regions causes them to become partially evacuated, particularly in the upper layers of the domain. Amongst other things, this leads to an increase in the local Alfvén speed, as well as a reduction in the timescale that is associated with thermal diffusion. Both of these factors lead to a reduction in the time-step that must be taken in order to ensure stability of the (explicit) numerical scheme. Preliminary nonlinear results for highly-stratified convection were described in Bushby et al. (2010). Even for a relatively modest values of Rm ($Rm \approx 350$, which is less than 10% above the critical value for dynamo action), the partial evacuation that occurs is extremely significant. In such cases, the corresponding reduction in the time-step is so severe that it is not possible to carry out the calculation on a reasonable time-scale without (e.g.) imposing an artificial lower limit on the minimum density within the domain.

Partial evacuation is also a feature of nonlinear dynamo action in moderately-stratified convection. However, because the effects of compressibility are reduced (for example, the peak mach number at the surface is approximately 0.6, as opposed to 1.0 in the highly-stratified case), the effect is less dramatic. So, in this case, it is possible to carry out these simulations without having to deal with overly restrictive time-step constraints. A calculation of dynamo action in moderately-stratified convection, for $Rm \approx 480$, is illustrated in Figure 3. The field geometry is similar to that of the kinematic regime, with mixed polarity concentrations of vertical magnetic flux accumulating in the convective downflows. The minimum gas density within these magnetic regions is highly time-dependent, but a typical minimum value would be (approximately) 25% of the mean density of the non-magnetic surroundings. This dynamo has saturated in the nonlinear regime, reaching a state in which the total magnetic energy is approximately 5% of the total kinetic energy in the domain. The saturation level of such a dynamo clearly depends upon the magnetic Reynolds number, with larger values of Rm expected to produce more efficient dynamos. Early results for an ongoing calculation at $Rm \approx 800$ certainly support this idea. For a lower magnetic Reynolds number of $Rm \approx 360$, the magnetic energy saturates at approximately $3 - 4\%$ of the kinetic energy. For such a marginal dynamo ($Rm \approx 360$ is only 10% above Rm_{crit}), the magnetic energy exhibits significant fluctuations, so it is

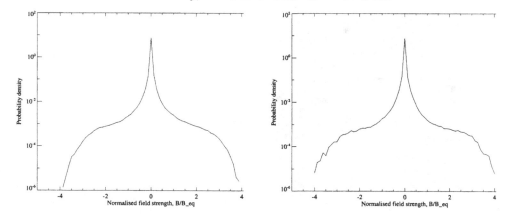

Figure 4. Probability density functions for the vertical component of the magnetic field (scaled by the equipartition value of B_z) in a nonlinear dynamo that is driven by moderately-stratified convection. Left: $Rm \approx 360$. Right: $Rm \approx 480$.

necessary to average over a significant fraction of an ohmic decay time in order to obtain a reliable estimate for the saturation level in this case. As described in Bushby *et al.* (2010), a highly-stratified calculation at $Rm \approx 350$ produces a dynamo in which the magnetic energy appears to be saturating at a similar level. However, due to fluctuations in the magnetic energy, it is unclear whether or not the dynamo has truly saturated in this highly-stratified case, so we will not attempt to draw any conclusions on the basis of this particular comparison.

Figure 4 shows probability density functions (PDFs) for the vertical magnetic field component at the upper surface, in moderately-stratified convection, at $Rm \approx 360$ and $Rm \approx 480$. In these PDFs, the strength of the magnetic field has been scaled by the value of B_z that would be in energy equipartition with the local convective motions at the surface of the domain. Both PDFs are consistent with a stretched exponential distribution, centred at $B_z = 0$. The peak fields that are obtained are roughly four times the equipartition value. This can be related to the quiet Sun, where the observed kilogauss-strength magnetic features exceed the equipartition field strength by a similar multiplicative factor. The processes which lead to the formation of such strong magnetic fields are discussed elsewhere (see, for example, Bushby *et al.* 2008), so will not be discussed in any detail here. The only thing that we note is that these super-equipartition magnetic features can only be in near-pressure balance with their non-magnetic surroundings if they are partially-evacuated.

4. Conclusions

These results clearly demonstrate that compressible convection in an electrically-conducting fluid can drive a small-scale dynamo, provided that the magnetic Reynolds number is large enough. For the two models of convection that are considered in this paper, which both have a Reynolds number of $Re \approx 150$, dynamo action is possible provided that Rm exceeds a critical value of $Rm_{crit} \approx 325$. In the kinematic regime, the growth rate of the magnetic energy appears to have a logarithmic dependence upon Rm. Furthermore, if the growth rates are scaled in terms of the convective turnover time, the result seems to be independent of the level of stratification, with all (scaled) growth rates appearing to lie close to the same curve, $\Gamma = 0.36 \log_e (Rm/325)$. Hence, although the flow structure will clearly be sensitive to the level of stratification within the domain, it appears to be solely the convective turnover time that is responsible for determining the growth rate of any dynamo. As in previous calculations (see, for example, Cattaneo 1999; Vögler & Schüssler 2007) the growth rate at large values of Rm is of the same order as the convective turnover time. There is no evidence from these kinematic results that high levels of stratification hinder the operation of convectively-driven dynamos. Therefore, the lack of dynamo action in the calculations of Stein, Bercik & Nordlund (2003) is probably not associated with the vertical asymmetry of the convective motions. In this context, it should be noted that the calculations of Stein *et al.* (2003) have an open (rather than an impermeable) lower boundary that allows magnetic flux to leave the domain. However, the penetrative dynamo calculations of Brummell *et al.* (2010) strongly suggest that an open lower boundary condition should not prohibit dynamo action. So, perhaps the lack of dynamo action in the calculations of Stein et al. (2003) was simply a consequence of Rm being too small.

In the nonlinear regime, the high magnetic pressure leads to the partial evacuation of the upper regions of the vertical magnetic flux concentrations. With our existing explicit code, this leads to numerical time-step constraints that rapidly become prohibitive in the case of highly-stratified convection. Only for a marginally excited dynamo, has it been possible to carry out a nonlinear dynamo calculation in the highly-stratified case, and even there it was necessary to impose an artificial lower bound on the minimum density within the computational domain. An implicit code would be needed to address this parameter regime in a more satisfactory manner. For moderately-stratified convection, on the other hand, it has been possible to carry out nonlinear dynamo simulations. At moderate values of Rm, the dynamo saturates in the nonlinear regime, reaching a state in which the total magnetic energy is a few percent of the total kinetic energy. At the surface of the computational domain, the partially-evacuated magnetic features achieve super-equipartition field strengths, as observed in the quiet Sun.

References

Abbett, W. P. 2007, *ApJ*, 665, 1469

Brummell, N. H., Tobias, S. M., & Cattaneo, F. 2010, *Geophys. Astrophys. Fluid Dyn.*, in press

Bushby, P. J., Houghton, S. M., Proctor, M. R. E., & Weiss, N. O. 2008, *MNRAS*, 387, 698

Bushby, P. J., Proctor, M. R. E., & Weiss, N. O. 2010, in: N.V. Pogorelov, E. Audit, G.P. Zank (eds.), *Numerical Modeling of Space Plasma Flows: ASTRONUM-2009*, ASP Conference Series Vol. 429 (San Francisco: ASP), p. 181

Cattaneo, F. 1999, *ApJ*, 515, L39

de Wijn, A. G., Stenflo, J. O., Solanki, S. K., & Tsuneta, S. 2009, *Space Sci. Revs*, 144, 275

Käpylä, P. J., Korpi, M. J., & Brandenburg, A. 2008, *ApJ*, 491, 353

Rogachevskii, I. & Kleeorin, N. 1997, *Phys. Rev. E*, 56, 417

Sánchez Almeida, J., Bonet, J. A., Viticchié, B., & Del Moro, D. 2010, *ApJ*, 715, L26

Stein, R. F., Bercik, D., & Nordlund, Å. 2003, in A. A. Pevtsov & H. Uitenbroek (eds.), *Current Theoretical Models and Future High Resolution Solar Observations: Preparing for ATST*, ASP Conference Series Vol. 286 (San Francisco: ASP), p. 121

Stix, M. 2004, *The Sun: An Introduction* (Berlin: Springer)

Vögler, A. & Schüssler, M. 2007, *A&A*, 465, L43

Discussion

E. ZWEIBEL: Convective collapse seems to be playing an important role in making the strongest fields (as a dynamo – yesterday's talk).

P. BUSHBY: Yes, we do observe an intensification process of this type, although we see more of a convective adjustment rather than a well-defined convective collapse instability.

Certainly, without the associated reduction in the internal gas pressure, it would be impossible to generate the observed super-equipartition fields at the upper surface of the domain.

G. VASIL: What type of numerical method did you use to solve the equations?

P. BUSHBY: Horizontal derivatives are evaluated in Fourier space, whilst fourth-order finite differences are used to calculate the vertical derivatives. An explicit third-order Adams-Bashforth scheme is used to time-step the equations.

S. TOBIAS: You were very diplomatic about avoiding the argument as to whether putting more scale heights into a dynamo calculation "switches off" the dynamo. Do you have a comment on this?

P. BUSHBY: From these calculations, there is no evidence to suggest that an increased level of stratification will inhibit the dynamo.

D. HUGHES: Would one expect very different levels of saturation for the compressible dynamo as compared to the Boussinesq model? What determines the peak fields in the compressible case?

P. BUSHBY: That is a difficult question to answer at the moment, as we have not yet reached high enough values of Rm to make a direct comparison with the Boussinesq calculation of Cattaneo (1999). Certainly it is feasible that the effects of magnetic pressure will have some influence upon the saturation level of the dynamo in the compressible case. Where significant partial evacuation occurs in the compressible case, the local magnetic pressure within a vertical flux concentration is comparable to the external gas pressure. Although the convective motions also play a role in the confinement of the flux concentrations, this total pressure balance sets an (approximate) upper limit on the peak magnetic fields that can be generated in the upper layers of the domain.

N. BRUMMELL: Raising Rm having fixed other parameters implies changing Pm. Comments?

P. BUSHBY: Yes, that's right. For $Re \approx 150$, a marginal dynamo has a magnetic Prandtl number of $Pm \approx 2$. At fixed Re, $Pm \propto Rm$, so wherever dynamo action is found in these simulations, $Pm > 2$.

A. BRANDENBURG: For a small-scale dynamo one expects the Kazantsev scaling, where magnetic energy increases with wavenumber, k, like $k^{3/2}$, reaching a peak at the resistive scale k_η. The growth rate should then scale with the turnover rate at that scale, so it should scale with Rm like $Rm^{1/2}$ (see Haugen et. al 2004, PRE; Käpylä et. al 2008).

P. BUSHBY: Although the logarithmic fit to the growth rate curve is empirical, it seems to be fairly convincing, with a large number of data points spread over a significant range of values for Rm. The scalings that you refer to both seem to cover a small number of data points over a narrow range of magnetic Reynolds numbers. It's not clear to me that we should expect to see a Kazantsev-like scaling for dynamo action in compressible convection. Furthermore, I believe that you need $Pm \ll 1$ in order to justify this $Rm^{1/2}$ scaling. This is not the case here.

Astrophysical Dynamics: From Galaxies to Stars
Proceedings IAU Symposium No. 271, 2010
N. H. Brummell, A. S. Brun, M. S. Miesch & Y. Ponty, eds.

© International Astronomical Union 2011
doi:10.1017/S1743921311017625

Time-dependent Turbulence in Stars

W. David Arnett[1] and Casey Meakin[1]

[1]Steward Observatory, University of Arizona,
Tucson AZ 85721, USA
email: wdarnett@gmail.com

Abstract. Three-dimensional (3D) hydrodynamic simulations of shell oxygen burning by Meakin & Arnett (2007b) exhibit bursty, recurrent fluctuations in turbulent kinetic energy. These are shown to be due to a global instability in the convective region, which has been suppressed in simulations of stellar evolution which use mixing-length theory (MLT). Quantitatively similar behavior occurs in the model of a convective roll (cell) of Lorenz (1963), which is known to have a strange attractor that gives rise to random fluctuations in time. An extension of the Lorenz model, which includes Kolmogorov damping and nuclear burning, is shown to exhibit bursty, recurrent fluctuations like those seen in the 3D simulations. A simple model of a convective layer (composed of multiple Lorenz cells) gives luminosity fluctuations which are suggestive of irregular variables (red giants and supergiants, see Schwarzschild (1975). Details and additional discussion may be found in Arnett & Meakin (2011).

Apparent inconsistencies between Arnett, Meakin, & Young (2009) and Nordlund, Stein, & Asplund (2009) on the nature of convective driving have been resolved, and are discussed.

Keywords. turbulence, irregular variables, convection

1. Introduction

Three-dimensional fluid dynamic simulations of turbulent convection in an oxygen-burning shell of a presupernova star show bursty fluctuations which are not seen in one-dimensional stellar evolutionary calculations (which use various versions of mixing-length theory, MLT, Böhm-Vitense (1958)).

Our particular example is a set of simulations of oxygen burning in a shell of a star of $23M_\odot$ (Meakin & Arnett (2007b), Meakin & Arnett (2010). This is of astronomical interest in its own right as a model for a supernova progenitor, but also happens to represent a relatively simple and computationally efficient case, and has general implications for the convection process in all stars.

Three-dimensional hydrodynamic simulations of shell oxygen burning exhibit bursty, recurrent fluctuations in turbulent kinetic energy (see Fig. 1, left panel). These simulations show a damping, and eventual cessation, of turbulent motion if we artificially turn off the nuclear burning (Arnett, Meakin, & Young (2009)). Further investigation by Meakin & Arnett (2010) shows that nearly identical pulsations are obtained with a volumetric energy generation rate which is constant in time, so that *the cause of the pulsation is independent of any temperature or composition dependence in the oxygen burning rate.* Heating is necessary to drive the convection; even with this time-independent rate of heating, we still get pulses in the turbulent kinetic energy. Such behavior is fundamentally different from traditional nuclear-energized pulsations dealt with in the literature (e.g., the ε-mechanism, Ledoux (1941), Ledoux (1958), Unno, *et al.* (1989)), and is a consequence of time-dependent turbulent convection (it might be called a "τ-mechanism", with τ standing for turbulence).

2. A Controversy Resolved.

This section was written with the aid of extensive email discussions with A. Nordlund and R. Stein. Arnett, Meakin, & Young (2009), analyzing the 3D simulations of Meakin & Arnett (2007b), found that buoyancy driving is balanced by viscous (Kolmogorov) dissipation; the remaining terms (mostly the kinetic energy flux escaping the convection zone) were less than 6% of the total. Nordlund, Stein, & Asplund (2009) found that the buoyancy terms exactly cancelled, and that the viscous dissipation was balanced by average gas pressure work. Which is correct? Upon careful reanalysis, both are, as we shall show.

2.1. The Convective Mach Number.

There are two limiting cases for convective flow, depending upon the convective Mach number $\mathcal{M}_{conv}$ (the ratio of the fluid speed to the local sound speed). The case $\mathcal{M}_{conv} \ll 1$ corresponds to "incompressible" flow. For turbulent motion, the pressure perturbation P' is related to the convective Mach number by $P'/P \sim \rho u_{rms}^2/P \sim \mathcal{M}_{conv}^2$, and must be small. Sound waves outstrip fluid motion, so that pressure differences quickly become small, except possibly for a static background stratification. Most of the historical research on convection (e.g., the Bènard problem, Chandrasekhar (1961), Landau & Lifshitz (1959)) is done in this limit (the Bousinesq approximation, Chandrasekhar (1961)).

Let $\langle a \rangle$ denote the average of any variable a over a spherical shell (lagrangian stellar coordinate). The density perturbation ρ' and the velocity perturbation u' are both first order in $\mathcal{M}_{conv}$, while the pressure terms (both gas and turbulent pressure) and the average velocity $\langle u \rangle$ are of second order. At low $\mathcal{M}_{conv}$ the first order terms dominate, but as $\mathcal{M}_{conv}$ rises, the second and higher order terms become important, changing the physical behavior of the system. $\mathcal{M}_{conv}$ does not increase indefinitely for a quasi-static system; as it approaches unity, kinetic energy approaches internal energy in magnitude, and the system becomes gravitationally unbound, allowing no quasi-static solution. There is a narrow range, $0.1 \leqslant \mathcal{M}_{conv} \leqslant 1$, in which this interesting transition occurs. Except in dynamic situations, convection in stellar interiors satisfies $\mathcal{M}_{conv} \leqslant 0.1$, for which the Boussinesq limit is a reasonable approximation. Near stellar surfaces of cool stars, $\mathcal{M}_{conv} \geqslant 0.1$, and the Boussinesq limit is no longer accurate.

2.2. Convective Driving.

We are interested in the rate of transfer of energy into different forms, so we consider various terms for power per unit volume, or energy per unit volume per unit time. In order to fix the terminology, let us define some terms:

- buoyancy power is $W_B = \langle \rho' \mathbf{u}' \cdot \mathbf{g} \rangle$,
- gas pressure power is $W_P = \langle -\mathbf{u} \cdot \nabla P \rangle$,
- net rate of work done by gravity is $W_G = \langle \rho \mathbf{u} \cdot \mathbf{g} \rangle$,
- net rate of work done by gravity on the mean flow is $W_{Gm} = \rho_0 \mathbf{u_0} \cdot \mathbf{g}$, and
- net rate of gas pressure work on the mean flow is $W_{Pm} = -\mathbf{u_0} \cdot \nabla P_0$.

For a quasi-steady state, $\langle \partial \rho / \partial t \rangle = 0$ so that $\langle \nabla \cdot \rho \mathbf{u} \rangle = 0$, which requires $0 = \langle \rho u_z \rangle = \langle \rho \rangle \langle u_z \rangle + \langle \rho' u_z' \rangle$. This is true for all values of $\mathcal{M}_{conv}$. This implies that the total work done by gravity vanishes, but can be split into a (positive) work done on the convective motions, called buoyancy power, and a (negative) work done on the mean flow, that is, $W_G = W_B + W_{Gm} = 0$. Ignoring boundary and wave effects for the moment, the gas pressure power is balanced by viscous (Kolmogorov) dissipation, $W_P = \varepsilon_K$, see Meakin & Arnett (2007b).

2.3. Rotational Flow.

Historically W_B has been used in the Boussinesq approximation as the driving term for convection (Chandrasekhar (1961)). Using the Cowling approximation we may move the gravity outside the average, so that $W_B = g_z \langle \rho' u_z' \rangle = -g_z \langle \rho \rangle \langle u_z \rangle$. In the Boussinesq limit, the equation of hydrostatic equilibrium holds, $dP/dz = -\rho g_z$, and the rate of work done on the fluid passing through the pressure gradient is

$$\langle -\mathbf{u} \cdot \nabla P \rangle \rightarrow g_z \langle \rho u_z \rangle = W_B. \tag{2.1}$$

The mathematical result is the same as if the buoyancy power $\mathcal{B}$ alone were used, and provides consistency with the literature (e.g., Chandrasekhar (1961), Landau & Lifshitz (1959), Davidson (2004)). Thus, the general result that, integrated over a convection zone, the buoyancy work is balanced by the viscous dissipation, is still valid in the incompressible limit (Chandrasekhar (1961)). The same result is found for turbulent flow by Arnett, Meakin, & Young (2009): buoyancy power is balanced by Kolmogorov dissipation.

For the simulations of Meakin & Arnett (2007b), $\mathcal{M}_{conv} \leqslant 0.03$ so that the pressure fluctuations are small ($P'/P \leqslant 0.001$), clearly in the incompressible regime. In this limit Arnett, Meakin, & Young (2009) show that $\langle \rho' u' / \rho \rangle \propto \langle T' u' / T \rangle$, so that the the convective velocity field is directly related to the enthalpy flux. This connection is ignored in MLT, but is important for stellar evolution because *it removes the freedom to adjust* the MLT parameter α.

The flow is accelerated by a torque in the horizontal plane, and tends to be divergence free (solenoidal, or "rotational") because of mass conservation. The velocities are small enough so that, to the same level of approximation, ram pressure may be ignored, and the background stratification is hydrostatic.

2.4. Divergent Flow.

The "compressible" limit is $\mathcal{M}_{conv} \simeq 1$, for which $P'/P \sim 1$. Shock formation is the most startling change in the flow character. Sound wave generation increases rapidly as $\mathcal{M}_{conv} \rightarrow 1$ (Landau & Lifshitz (1959), §75). The flow becomes diverging, or "irrotational" (consider the extreme limit of a point explosion which is pure divergence). Even in the case of convective flow, ram pressure levitation begins to become important (Stein & Nordlund (1998)). Generally the flow begins to cause structural change and becomes even more complex. In this case the simple picture of pure buoyancy driving begins to break down; increasingly more work is done to generate diverging flows as $\mathcal{M}_{conv}$ increases.

If the convective region does not have an overall divergence (i.e., is in a quasi-steady state), the excess pressure fluctuations (beyond those implied by the buoyancy flux) must generate waves (both sound waves and gravity waves), which may propagate beyond the convective region. The ram pressure from the flow will cause the quasi-steady convection zone to expand as $\mathcal{M}_{conv}$ increases. Because of turbulence, episodes of vigorous local pulsation may occur. Further increase in $\mathcal{M}_{conv}$ may lead to vigorous global pulsation and even explosion ($\mathcal{M}_{conv} \sim 1$ implies the kinetic energy is of order the internal energy). Notice the transition from rotational flow toward diverging flow as $\mathcal{M}_{conv}$ increases.

2.5. Relevance to Astrophysics

Which of the $\mathcal{M}_{conv}$ limits is relevant for astrophysics? Both are. *Almost all* the matter in stellar convection zones, during *almost all* evolution, is in the limit of incompressible flow. For the Sun, the region for which $\mathcal{M}_{conv} \geqslant 0.1$ (so that the pressure fluctuations are greater than 0.01), contains only about 10^{-7} of a solar mass. However all stars with cool surfaces have a superadiabatic region in which the convective Mach numbers rise

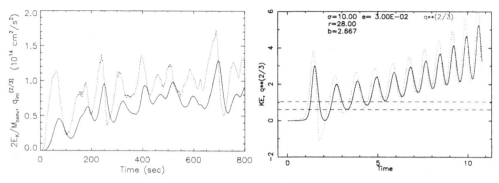

Figure 1. *Left.* Turbulent velocity squared $u_t^2 = 2E_{turb}/M_{CZ}$ (solid), and the corresponding buoyancy flux (dotted, plotted in the same units) $q_{int}^{2/3}$ in the convection zone, versus time. The kinetic energy lags the buoyant flux by roughly 20 seconds. *Right.* Initial Behavior of an extended Lorenz Model of Convection, for $\sigma = 10$, $r = 28$, $b = 8/3$, and $e = 3.0 \times 10^{-2}$. The parameters are similar to those inferred for the oxygen shell (see Arnett & Meakin (2011)). There is an increase in kinetic energy and in the amplitude of the pulses, somewhat like the left panel. The primary difference seems to be that the Lorenz model uses a single mode while the simulation is multi-mode.

to $\mathcal{M}_{conv} \sim 0.1$ or more, and *these regions are the ones observed.* Quasi-static oxygen burning, one of the most vigorous nuclear processes to drive convection, has $\mathcal{M}_{conv} \ll 1$, and this is true for other thermonuclear processes as well; but they evolve to violent and explosive events for which $\mathcal{M}_{conv} \sim 1$. The exceptions to $\mathcal{M}_{conv} \ll 1$ are important: (1) explosions, such as supernovae and novae, (2) vigorous thermonuclear flashes, (3) vigorous pulsations, especially radial ones, and (4) the sub-photospheric layers of stellar surface convection zones, which are strongly non-adiabatic, to name a few. Consequently the transition (near $\mathcal{M}_{conv} \sim 0.3$, implying pressure perturbations of order 0.1) between these two limiting cases is emerging as an important problem in astrophysics.

We have made some theoretical progress in understanding the $\mathcal{M}_{conv} \ll 1$ case, and will now follow Eddington's advice to "push a theory until it breaks," to see what happens.

3. Turbulence and the Lorenz Model

Lorenz (1963) devised a simple model (three degrees of freedom) of a convective cell which captured the seeds of chaos in terms of the Lorenz strange attractor, which is part of the foundation of the study of instabilities in nonlinear systems (Cvitanović (1989), Gleick (1987), Thompson & Steward (1986)). The model is an extreme truncation of the fluid dynamics equations, reducing a system of seven variables and about 10^7 grid points (Fig. 1 (left)) to one of amplitudes of three variables (Fig. 1 (right)). This reduces the degrees of freedom from $\sim 7 \times 10^7$ to merely 3. To attain this simplification, only a single mode, of low Mach number flow, was examined in two dimensions ("a convective roll"). The three variables are the speed of the convective roll, the vertical temperature fluctuation, and the horizontal temperature fluctuation. For sufficiently large Reynolds number (essentially luminosity in excess of that which can be carried by radiative diffusion), the flow becomes chaotic. In MLT, the vertical and horizontal temperature fluctuations are assumed identical, and this two variable model is not chaotic (Arnett & Meakin (2011)).

We examine the conjecture that the fluctuating behavior of the Lorenz model is a simple version of that seem in our 3D compressible, multi-mode convection simulations. Arnett & Meakin (2011) make the simple generalization of the Lorenz model to include nuclear heating. In Fig. 1 is a comparison of the fluctuating behavior of kinetic energy

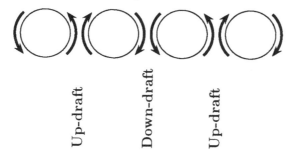

Figure 2. The Lorenz Model extended: Convection in a shell composed of cells. Notice the alternation of the sign of rotation. This may be thought of as a cross sectional view of infinitely long cylindrical rolls, or of a set of toroidal cells, with pairwise alternating vorticity (see Chandrasekhar (1961), §16). Each cell may exhibit independent fluctuations in time and space.

for the 3D simulation (left panel) and the generalized Lorenz model (right panel). The behavior is similar, with the largest differences due to the fact that the 3D simulation exhibits many modes, while the Lorenz model has only one by definition.

This suggests that the fluctuations seen in the 3D simulation are related to the onset of chaotic behavior in the much simpler Lorenz model. Analytically, the equations of 3D fluid flow used in the simulations may be directly simplified to obtain the Lorenz equations (see Arnett & Meakin (2011)), which strengthens the suggestion.

4. Convective Cells

Schwarzschild (1975) suggested, based upon MLT, that convection was dominated by cells, whose size was determined by the size of a local pressure scale height. Simulations suggest that this is qualitatively true; mass conservation in a stratified medium implies cells of a density scale height in size (see Nordlund, Stein, & Asplund (2009)). Suppose that we imagine the convective zone to be made of cells, each of which we approximate by a Lorenz model? Fig. 2 illustrates a cross-section segment of part of such a planar convective layer. We do not imagine that the cells are fixed as in a crystalline structure, but that the cells are unstable, forming and decaying in a dynamic way, with average properties approximated by independent Lorenz models. Each Lorenz cell has a fluctuation in energy flux it carries, and this is identified with a fluctuation in luminosity of that horizontal patch of the convective layer. For simplicity we ignore the fluctuations of cells in underlying layers, and in the same spirit, the fact that stellar photospheres often exhibit convective Mach numbers which are not small.

Schwarzschild (1975) noted that for the sun, the local pressure scale height was relatively small, implying that the size of a convective cell was also small relative to the radius, similar to the size of a granule. Such a large number of cells ($\sim 10^6$) would average out fluctuations over a stellar disk. For red supergiants, the scale heights approach the radius in size, so that a few cells would be sufficient to cover the surface, and the fluctuations in luminosity would be more obvious.

5. Irregular Variables

Combining Schwarzschild's idea of convective cells with Lorenz's model of a cell allows us to construct a simple model of a stellar convective surface, which contains an element of the fluctuating nature of turbulence, although it may to be a crude approximation

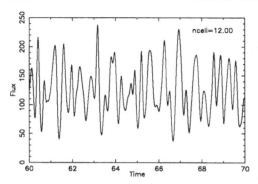 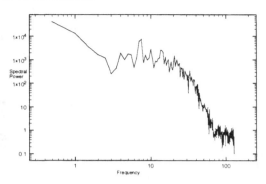

Figure 3. (left) Fluctuations of Luminosity in Convective Layer of 12 Cells of Random Phase, for $\sigma = 10$, $r = 28$, and $b = 8/3$. The dimensionless flux (luminosity) is shown for a convective layer with 12 visible Lorenz cells. The luminosity variations are large and seemingly chaotic, suggestive of irregular variables and Betelgeuse in particular. (right) Power specta of the luminosity. There is no sharp peak, but a broad distribution of power, as would be expected from a chaotic source.

in that it has no realistic photospheric layer. Fig. 3 shows the luminosity fluctuations implied by such a model (left panel). The amplitude of the luminosity fluctuations are simply those of the classical Lorenz model. They are larger than but comparable to those found in 3D simulations of Betelgeuse (α Orionis) by Chiavassa, *et al.* (2010), and observed in red supergiants (Kiss, Szabo, & Bedding (2006)). This qualitative agreement for such a simple model is suggestive.

The right panel in Fig. 3 shows the spectral power density of the luminosity fluctuations shown in the left panel. There are a few broad, marginal peaks, and a lot of broad-band noise. The frequency is normalized to the time interval shown in the left panel. In a real star, this broad-band noise would resonate with normal modes of pulsation, driving them to finite amplitude and thereby giving additional sharp peaks to the spectrum.

Joel Stebbins (pioneer of photoelectric astronomy) monitored the brightness of Betelgeuse (α Orionis) from 1917 to 1931, and concluded that "there is no law or order in the rapid changes of Betelgeuse" (Goldberg (1984)), which seems apt for Fig. 3 as well. More modern observations (see Kiss, Szabo, & Bedding (2006)) show a strong broad-band noise component in the photometric variability. The irregular fluctuations of the light curve are aperiodic, and resemble a series of outbursts. This should be no surprise; the 3D equations have embedded in them the strange attractor of Lorenz. It will be interesting to apply these ideas to the analysis of solar-like oscillations in other stars (Kjeldsen & Bedding (1995)).

6. Summary

By quantitatively examining the implications of 3D simulations of turbulent convection, we are led to an identification of similarities in convective cells and the Lorenz strange attractor. Applying this connection in a simple model implies a broad-band component of the power spectrum of luminosity in stars with convective surfaces, much like that observed by Kiss, Szabo, & Bedding (2006).

However, there is no free lunch. Application of these ideas to stellar surfaces implies their use in regimes in which the convective Mach number $\mathcal{M}_{conv}$ is not much less than unity. Will the compressible effects make qualitative changes in this picture? We shall see. Fortunately the surface of the Sun is a nearby example on which we can test our theoretical ideas, and these ideas will have wide ramifications in astrophysics.

Acknowledgements

This work was supported in part by by NSF Grant 0708871 and NASA Grant NNX08AH19G at the University of Arizona. We are grateful for helpful discussions with Aake Nordlund and Bob Stein, and thank the Symposium organizers for allowing us to participate. Happy birthday Juri!

References

Arnett, W. D., Meakin, C., & Young, P. A., 2009, *ApJ*, 690, 1715
Arnett, W. D., Meakin, C., & Young, P. A., 2010, *ApJ*, 710, 1619
Arnett, W. D. & Meakin, D., 2011, submitted.
Böhm-Vitense, E., 1958, *Zeit. für Ap.*, 46, 108
Chandrasekhar, S. 1961, *Hydrodynamic and Hydromagnetic Instability*, Oxford University Press, London
Chiavassa, A., Haubois, X., Young, J. S., Plez, B., Josselin, E., Perrin, G., & Freytag, B., 2010, *A&A*, in press
Cvitanović, P., *Universality in Chaos*, Adam Hilger, Bristol and New York
Davidson, P. A., 2004, *Turbulence*, Oxford University Press, Oxford
Gleick, J., 1987, *Chaos: Making a New Science*, Penguin Books, New York
Goldberg, L., 1984, *Publ. Astron. Soc. Pacific*, 96, 366
Kiss, L. L., Szabo, Gy. M., & Bedding, T. R., 2006, *MNRAS*, 372, 1721
Kjeldsen, H. & Bedding, T. R., 1995, *A&A*, 293, 87
Landau, L. D. & Lifshitz, E. M. 1959, *Fluid Mechanics*, Pergamon Press, London
Ledoux, P., 1941, *ApJ*, 94, 537
Ledoux, P. & Walraven, Th., 1958, in *Handbuch der Physik*, 51, ed. S. Flugge, (Springer-Verlag, Berlin), p. 353
Lorenz, E. N., 1963, *Journal of Atmospheric Sciences*, 20, 130
Meakin, C. & Arnett, D., 2007b, *ApJ*, 667, 448
Meakin, C. & Arnett, D., 2010 *ApJ*, submitted
Nordlund, A., Stein, R., & Asplund, M., 2009, *Living Reviews in Solar Physics*, 6, 2
     {http://www.livingreviews.org/lrsp-2009-2}
Schwarzschild, M., 1975, *ApJ*, 195, 137
Stein, R. F. & Nordlund, A., 1998, *ApJ*, 499, 914
Thompson, J. M. T. & Stewart, H. B., 1986, *Nonlinear Dynamics and Chaos*, John Wiley and Sons, New York
Unno, W., Osaki, Y., Ando, H., Saio, H., & Shibahashi, H., 1989, *Nonradial Oscillations of Stars*, 2nd. ed., University of Tokyo Press, Tokyo

Discussion

N. WEISS: A word of caution about the meaning of the Lorenz equations: chaos is ubiquitous, and Lorenz's realisation is this was original and important – but one should not rely on his equations to describe a specific physical system.

W. D. ARNETT: I agree. We started with the simplest possibility, which we knew contained chaotic behavior.

The Lorenz model is only a 2D and incompressible model, and should be generalized to 3D, and compressible flow if possible. Hopefully, analysis of 3D simulations for a variety of conditions will illuminate this issue.

R. COLLET: 1. In your analysis you derive a value of the MLT parameter α that is four times as large as the typical value used in stellar evolution calculations. This result is essentially based on simulations of one massive star, if I understand correctly. How can you generalise this result to other stellar masses, e.g solar?

2. Does this result apply to stellar interiors or to surface convection as well?

W. D. ARNETT: Let me clarify. We claim that the α parameter is not adjustable, but an eigenvalue determined by the stellar structure (depth of the convective zone).

Our original compressible 3D simulations (Meakin & Arnett (2007b)) showed that the *turbulent dissipation length (essentially the "mixing length") was simply equal to the depth of the convection zone* (2 pressure scale heights for that stellar model). Based on this and results from many 3D simulations by others, we conjectured that this relation was a general result, but found that it seemed limited on the high side by a value of roughly 4 pressure scale heights. We have since verified the relation with 3D simulations of shells from 0.5 to 4 pressure scale heights in depth for low Mach-number flows (stellar interiors). The solar convection zone is 20 pressure scale heights deep, so we used 4 as the limiting value, which is $4/1.65 \approx 2.4$ times the standard value of α for the same solar atmospheric model.

We believe this limit may be a general result for low Mach-number flows. For the outer, sub-photospheric layers in a stellar convective zone, the convective Mach-number rises, and pressure perturbations become increasingly important. We do not yet have a more general theory applicable to this small but important region. As a first step we simply see what happens if we ignore the effects of compressibility.

R. TRAPENDACH: How realistic in your simulation is the surface superadiabatic gradient when compared to 3-D calculations. How to you model those layers?

W. D. ARNETT: They are not realistic enough. We use the radiative diffusion approximation, but have not directly simulated an atmosphere. However, we have focused here on the possible implications of chaotic behavior (turbulence) for stellar variability, with the simplest model we know.

We are attempting to develop a general *theoretical understanding* of stellar turbulence. It may be that our computational abilities now run ahead of our ability to assimilate what we compute. How can we develop a mathematical (rather than numerical) understanding? So far, we have a promising quantitative solution for low Mach-number convective zones (stellar interiors).

Stellar atmospheres and immediate sub-surface regions have convective Mach-numbers which increase toward unity. In such cases, compressible fluid effects can no longer be neglected (shock, g- and p-mode wave generation, dynamic expansion and contraction, to mention some of the most important).

These effects also occur in other interesting astrophysical situations: explosions of novae and supernovae, thermonuclear ignition flashes, thin-shell flashes, and vigorous radial pulsations, for example.

Many groups, including yours, have already solved this problem *numerically* for stellar atmospheres. We should join forces in the quest of a theoretical model capable of reproducing the general features of turbulent convection, which is not restricted to small Mach-number flow. We have a developing theory and many of you have numerical solutions; we would be happy to work with you.

Astrophysical Dynamics: from Stars to Galaxies
Proceedings IAU Symposium No. 271, 2010
N. H. Brummell, A. S. Brun, M. S. Miesch & Y. Ponty, eds.

© International Astronomical Union 2011
doi:10.1017/S1743921311017637

Dynamo coefficients
from the Tayler instability

Rainer Arlt and Günther Rüdiger

Astrophysikalisches Institut Potsdam,
An der Sternwarte 16, D-14482 Potsdam, Germany
email: rarlt@aip.de, gruediger@aip.de

Abstract. Current-driven instabilities in stellar radiation zones, to which we refer as Tayler instabilities, can lead to complex nonlinear evolutions. It is of fundamental interest whether magnetically driven turbulence can lead to dynamo action in these radiative zones. We investigate initial-value simulations in a 3D spherical shell including differential rotation. The Tayler instability is connected with a very weak kinetic helicity, stronger current helicity, and a positive $\alpha_{\phi\phi}$ in the northern hemisphere. The amplitudes are small compared to the effect of the tangential cylinder producing an eddy with negative kinetic helicity and negative $\alpha_{\phi\phi}$ in the northern hemisphere. The $\alpha_{\phi\phi}$ from the Tayler instability reaches about 1% of the rms velocity.

Keywords. Keyword1, keyword2, keyword3, etc.

1. Introduction

Stellar radiation zones are often hosting magnetic fields. The majority of solar dynamo models include the presence of strong toroidal magnetic fields in the tachocline, the transition from differential to uniform rotation below the convection zone. Also a fair fraction of intermediate-mass stars shows strong magnetic fields hosted by their radiative envelopes and are called magnetic Ap stars.

These magnetic fields may become unstable because they are connected with currents (Vandakurov 1972; Tayler 1973). We will use the term Tayler instability for this class of current-driven instabilities. Rotation and differential rotation alter the stability limits of the Tayler instability.

The limiting magnetic fields strengths have been determined under various conditions and in various contexts in a series of studies, e.g., Pitts & Tayler (1985), Gilman & Fox (1997), Cally (2000), Dikpati *et al.* (2004), Braithwaite (2006b), Arlt *et al.* (2007), Rüdiger & Kitchatinov (2010). The linear stability of magnetic fields is fairly well understood. Non-axisymmetric, large-scale modes are typically the consequence of the Tayler instability.

The nonlinear development of the instability may lead to turbulence as well as enhanced diffusivity and angular-momentum transport in radiative stellar zones. Another issue is the existence of a sustained dynamo, if the magnetic field which becomes unstable has a source of replenishment (Spruit 2002). While the large-scale non-axisymmetric unstable mode will only be able to produce a non-axisymmetric magnetic field, a turbulent state may imply the generation of a substantial axisymmetric part. The non-axisymmetry is then hidden in the turbulent field as to comply with Cowling's theorem. Mixed results have been obtained in attempts to show sustained dynamo action in simulations (e.g. Braithwaite 2006a; Brun & Zahn 2006; Gellert *et al.* 2008).

Our paper deals with the possible dynamo-effect from the Tayler instability by measuring the mean-field dynamo coefficients during the nonlinear evolution of the system.

Zahn *et al.* (2007) pointed out that a turbulent electromotive force (EMF) is the only way of regenerating the large-scale magnetic field from the non-axisymmetric instability. Measuring the EMF appears to be a suitable approach, even though we do not have an energy source in the system which would be necessary for sustained generation of magnetic fields from a continuously excited instability.

2. Numerical setup

We consider a spherical shell extending from the normalized radii $r_{\rm i} = 0.5$ to $r_{\rm o} = 1$. The colatitude θ and the azimuth ϕ are covered in their full extent. The spectral spherical MHD code in Boussinesq approximation by Hollerbach (2000) is employed for the simulations. In this study, the buoyancy term and the equation for temperature fluctuations are dropped for the sake of simplicity. The remaining, normalized equations are

$$\frac{\partial \boldsymbol{u}}{\partial t} = -(\boldsymbol{u} \cdot \nabla)\boldsymbol{u} + (\nabla \times \boldsymbol{B}) \times \boldsymbol{B}$$
$$-\nabla p + \mathrm{Pm}\triangle \boldsymbol{u}, \tag{2.1}$$

$$\frac{\partial \boldsymbol{B}}{\partial t} = \nabla \times (\boldsymbol{u} \times \boldsymbol{B}) + \triangle \boldsymbol{B}, \tag{2.2}$$

where $\boldsymbol{u}$ and $\boldsymbol{B}$ are the velocity and magnetic fields and p is the pressure. Additionally, the relations $\nabla \cdot \boldsymbol{u} = 0$ and $\nabla \cdot \boldsymbol{b} = 0$ hold. The equations are normalized by the magnetic diffusivity η, whence the magnetic Prandtl number $\mathrm{Pm} = \nu/\eta$ in the Navier-Stokes equation, where ν is the kinematic viscosity. The magnetic permeability and the density are set to unity and do not appear in this system of equations.

The initial velocity profile is that of a differential rotation with the angular velocity Ω depending on the axis distance, $s = r \sin\theta$,

$$\Omega(s) = \frac{\mathrm{Rm}}{\sqrt{1 + s^q}}, \tag{2.3}$$

where we start with $\mathrm{Rm} = 20\,000$ and $q = 2$. The initial magnetic field is purely poloidal and confined to the computational domain. The parameter Rm is the second dimensionless parameter entering the system in the initial conditions:

$$\mathrm{Rm} = \frac{R^2 \Omega_*}{\eta}, \tag{2.4}$$

where Ω_* is the angular velocity of the star and R is its radius. By comparing the Alfvén velocity of the magnetic field with the velocity of the fluid, one can convert the dimensionless magnetic fields of the simulations into physical units,

$$B_{\rm phys} = \sqrt{\mu\rho}\,\Omega_* R \frac{B}{\mathrm{Rm}} \tag{2.5}$$

where μ is the permeability and ρ is the bulk density of the fluid, now in physical units.

The spectral truncations for these simulations were typically at 40 Chebyshev polynomials for the radial decomposition and 60 Legendre polynomials for the latitudinal decomposition. For each Legendre degree l, the spherical harmonics were running from $m = -l$ to $m = l$ accordingly.

3. Results

The simulations are initially entirely axisymmetric, but apply a non-axisymmetric perturbation after a given time t_{pert}. As long as no perturbation has been applied, the differential rotation winds up toroidal magnetic fields from the initial, purely poloidal one. Lorentz forces, however, start to act on the differential rotation and reduce it, while the total angular momentum of the spherical shell is preserved in the simulations. This scenario alone leads to uniformly rotating radiative zones.

During the amplification of toroidal fields, the stability properties of the magnetic field changes. The Tayler instability is suppressed by rotation, and non-axisymmetric instabilities are even further suppressed if differential rotation is present. Now, the differential rotation is gradually reduced while toroidal fields grow – the system will turn supercritical for the Tayler instability quite suddenly after some time. This is the time when the non-axisymmetric perturbation is injected into the system. In the case described here, $t_{\text{pert}} = 0.003$ diffusion times.

The instability now develops a fairly complex pattern with higher modes being excited through nonlinear coupling, and some energy dropping back into the $m = 0$ mode as well. The latitudinal structure exhibits a very steep energy spectrum of $E_l \sim l^{-3.7}$ at $t - t_{\text{pert}} = 0.001$ diffusion times. The system does not develop a fully turbulent state in these simulations. The azimuthal spectrum is even steeper with $E_m \sim m^{-6.7}$.

Here, we are interested in the question whether the averaged electromotive force can be expressed in terms of an (axisymmetric) α-effect. In other words, can the Tayler instability – through nonlinearities – act like a mean-field dynamo.

Assuming that the turbulent electromotive force **EMF** only depends on the mean magnetic field $\overline{\boldsymbol{B}}$ and its first spatial derivatives, one can write in general

$$\mathbf{EMF} = \quad \boldsymbol{\alpha}\overline{\boldsymbol{B}} + \boldsymbol{\gamma} \times \overline{\boldsymbol{B}} - \beta(\nabla \times \overline{\boldsymbol{B}}) - \boldsymbol{\delta} \times (\nabla \times \overline{\boldsymbol{B}}) - \boldsymbol{\kappa}(\nabla\overline{\boldsymbol{B}})^{(\text{sym})}, \qquad (3.1)$$

with the symmetric $\boldsymbol{\alpha}$- and $\boldsymbol{\beta}$-tensors, the vectors $\boldsymbol{\gamma}$ and $\boldsymbol{\delta}$, and the third-rank tensor $\boldsymbol{\kappa}$ acting on the symmetric part of the tensor gradient of $\overline{\boldsymbol{B}}$.

We employ the test-field method developed by Schrinner et al. (2007) to measure the components of the above mentioned tensors. While running the actual simulation, an additional set of 27 test equations is integrated delivering the various components of the mean-field tensors and vectors simultaneously as functions of the meridional location and time. The spatial distribution of the components of the symmetric part of the $\boldsymbol{\alpha}$-tensor, averaged over 0.0005 diffusion times or 1.6 rotation periods, is shown in Fig. 1.

The strongest effect comes from the inner cylinder which tends to produce an eddy of helical motion giving rise to an α-effect which is not primarily caused by the Tayler instability. The actual Tayler-α is the positive $\alpha_{\phi\phi}$ (alpha_pp) in the bulk of the northern hemisphere at $r > 0.6$. Note that the component $\alpha_{\phi\phi}$ – the one which generates the poloidal field from the toroidal one – is the smallest among the diagonal components of the tensor. The rms velocity fluctuations in the computational domain are around $0.005R\Omega_*$. The $\alpha_{\phi\phi}$ in the Tayler unstable region of the domain is roughly 1% of this rms value or $5 \cdot 10^{-5}R\Omega_*$. The tangent-cylinder effect delivers an $\alpha_{\phi\phi}$ of about 7% of the rms velocity.

The velocity fluctuations are mostly horizontal, even in this unstratified setup. Radial velocity fluctuations are about five times smaller than horizontal motions. The positive $\alpha_{\phi\phi}$ which we think results from the Tayler instability is associated with a positive current helicity and very weak, negative kinetic helicity. This is certainly an indication for the magnetic nature of the instability and the raising turbulence which is different from convection. However, this actually means the test-field method will probably underesti-

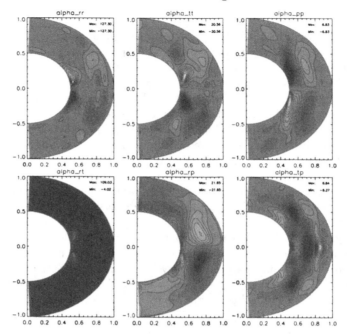

Figure 1. Distributions of α_{rr}, $\alpha_{\theta\theta}$, and $\alpha_{\phi\phi}$ in the upper row, and $\alpha_{r\theta}$, $\alpha_{r\phi}$, and $\alpha_{\theta\phi}$ in the lower row, averaged from a period which is 0.0005 to 0.0010 diffusion times after the perturbation was injected. Light areas (solid lines) represent positive α_{ij} while dark areas (dashed lines) represent negative values.

mate the values of $\boldsymbol{\alpha}$. The magnetic-field fluctuations are by a factor of 3–4 larger than the velocity fluctuations, so $\alpha_{\phi\phi}$ could also be about 3–4% of the rms velocity instead of only 1%.

4. Outlook

While the simulations show that the Tayler instability may act like a mean-field dynamo, the α is very small. There is no energy source in these simulations, sustained dynamo action is thus not possible. Future simulations with some sort of energy source will tell whether a continuous dynamo is possible and feasible for stellar radiation zones. Just as for convection-driven dynamos, the question also matters of how long it takes for the large-scale magnetic field to grow. If this happens on a resistive time-scale, one again has to be puzzled with the slow generation of large-scale magnetic fields from a small-scale dynamo, as it was seen in forced and convective turbulence (see Brandenburg & Subramanian 2005 for a review, and e.g. Käpylä *et al.* 2010 for a solution).

References

Arlt, R., Sule, A., & Rüdiger, G. 2007, *A&A*, 461, 295
Brandenburg, A. & Subramanian, K. 2005, *Physics Reports*, 417, 1
Braithwaite, J. 2006a, *A&A*, 449, 451
Braithwaite, J. 2006b, *A&A*, 453, 687
Brun, A. S. & Zahn, J.-P. 2006 *A&A*, 457, 665
Cally, P. S. 2000 *Sol. Phys.*, 194, 189
Dikpati, M., Cally, P. S., & Gilman, P. A. 2004, *ApJ*, 610, 597

Gellert, M., Rüdiger, G., & Elstner, D. 2008, *A&A*, 479, L33

Gilman, P. A. & Fox, P. A. 1997, *ApJ*, 484, 439

Hollerbach, R. 2000, *Int. J. Num. Meth. Fluids*, 32, 773

Käpylä, P. J., Korpi, M. J., & Brandenburg, A. 2010, *A&A*, 518, A22

Pitts, E. & Tayler, R. J. 1985, *MNRAS*, 216, 139

Rüdiger, G. & Kitchatinov, L. L. 2010, *Geophys. Astrophys. Fluid Dyn.*, in press

Schrinner, M., Rädler, K.-H., Schmitt, D., Rheinhardt, M., & Christensen, U. R. 2007, *Geophys. Astrophys. Fluid Dyn.*, 101, 81

Spruit, H. 2002, *A&A*, 381, 923

Tayler, R. J. 1973, *MNRAS*, 161, 365

Vandakurov, Yu. V. 1972, *SvA*, 16, 265

Zahn, J.-P., Brun, A. S., & Mathis, S. 2007, *A&A*, 474, 145

Discussion

DONATI: Ap stars are unlikely no more dynamo-type fields as this would imply some kind of positive correlation between field strength and rotation rate that we don't see. It would also be difficult to explain why only 9–10% of massive stars are magnetic. However, this exotic dynamo may be responsible for the weak field cutoff of the Ap star histogram. And for the weak fields recently in normal A, B and O stars.

ARLT: The measurements of the dynamo coefficients were not particularly addressing Ap stars, but of general nature. The simulations indicate that dynamo action without a sharp inner boundary or with strongly dominating toroidal fields (like in the solar tachocline) will be very weak.

BRUN: Do you find a genuine dynamo action in your simulation of stellar radiative interior of a massive star, or just a transitory growth of field due to Tayler's instability? If dynamo action is present, at which magnetic Reynolds number is the onset?

ARLT: The simulations lack an energy source, so are not capable of showing a genuine dynamo. All we could do at this stage is measure the mean-field coefficients generated by the nonlinear evolution of the instability. Differential rotation and fields all decay at large times.

Astrophysical Dynamics: From Stars to Galaxies.
Proceedings IAU Symposium No. 271, 2010
N. H. Brummell, A. S. Brun, M.S. Miesch & Y. Ponty, eds.
© International Astronomical Union 2011
doi:10.1017/S1743921311017649

The Evolution of a Double Diffusive Magnetic Buoyancy Instability

Lara J. Silvers[1], Geoffrey M. Vasil[2], Nicholas H. Brummell[3] and Michael R. E. Proctor[4]

[1] Centre for Mathematical Science, City University London, Northampton Square, London, EC1V 0HB, U. K.
email: lara.silvers.1@city.ac.uk

[2] Canadian Institute for Theoretical Astrophysics, 60 St. George Street, Toronto, ON M5S 3H8, Canada
email: vasil@cita.utoronto.ca

[3] Department of Applied Mathematics & Statistics, University of California, Santa Cruz, CA 95064, U.S.A.
email: brummell@soe.ucsc.edu

[4] Department of Applied Mathematics and Theoretical Physics, University of Cambridge, Wilberforce Road, Cambridge, CB3 0WA, U. K.
email: mrep@damtp.cam.ac.uk

Abstract. Recently, Silvers *et al.* (2009b), using numerical simulations, confirmed the existence of a double diffusive magnetic buoyancy instability of a layer of horizontal magnetic field produced by the interaction of a shear velocity field with a weak vertical field. Here, we demonstrate the longer term nonlinear evolution of such an instability in the simulations. We find that a quasi two-dimensional interchange instability rides (or "surfs") on the growing shear-induced background downstream field gradients. The region of activity expands since three-dimensional perturbations remain unstable in the wake of this upward-moving activity front, and so the three-dimensional nature becomes more noticeable with time.

Keywords. instabilities, MHD, Sun: magnetic fields

1. Introduction

Although at this time there is no general consensus on all the elements necessary for the operation of a solar dynamo that creates the large-scale field that we observe as solar active regions, one ingredient that is generally considered to play an important role in all the proposed scenarios (see e.g. Parker (1993), Babcock (1961), Leighton (1969)) is magnetic buoyancy. It is widely believed that the strong shear in the solar tachocline can stretch existing poloidal magnetic field into strong toroidal magnetic field, and then magnetic buoyancy instabilities of the toroidal field create compact toroidal magnetic structures that rise into the overlaying convection zone to either be re-processed there or to emerge at the solar surface as active regions.

It is thus of considerable interest to seek to understand the formation and evolution of such magnetic structures. Until fairly recently, such investigations have principally focussed on the evolution of pre-conceived, idealised buoyant structures (e.g. Parker (1955), Fan *et al.* (1998), Emonet & Moreno-Insertis (1998), Hughes, Falle, & Joarder (1998), Hughes & Falle (1998), Wissink *et al.* (2000), Fan *et al.* (2003), Abbett *et al.* (2004), Jouve & Brun (2009)). Recently attention turned to the problem of the self-consistent generation of such structures as well as their evolution, in particular using the dynamics that are expected to be available in the solar tachocline, i.e. strong velocity shear

218

(Brummell, Cline & Cattaneo (2002), Cline, Brummell & Cattaneo (2003a), Cline, Brummell & Cattaneo (2003b), Cattaneo, Brummell & Cline (2006), Vasil & Brummell (2008), Vasil & Brummell (2009), Silvers, Bushby & Proctor (2009a), Silvers *et al.* (2009b)). For different configurations of shear and seed poloidal magnetic field, the creation of buoyantly-unstable magnetic layers that generate rising compact magnetic structures can be possible.

However, as was demonstrated in Vasil & Brummell (2008) and Vasil & Brummell (2009), for the case of an initial vertical (radial) poloidal field, the mere existence of a strong velocity shear does not guarantee the formation of a buoyant structures, as is often assumed in discussions of the large-scale solar dynamo. The instability will only occur for certain values of the parameters of the problem. Of particular importance are the strength and geometry of the shear flow as related to the background stratification via a Richardson number. In Vasil & Brummell (2008), it was shown that sufficiently strong shear forcing could lead to the generation of buoyant magnetic structures. However, the forcings required were extreme and unphysical for the context of the solar tachocline. Whilst attempting an analytical understanding of this problem, Vasil & Brummell (2009) outlined the possibility that the stabilizing effect of stratification could be mitigated by a strong difference between the stabilizing thermal diffusion time and the destabilizing magnetic diffusion time, as had been noted earlier Hughes (1985) for the simpler problem of the instability of a steady stratified horizontal magnetic field. When the ratio of these diffusivities $\zeta = \eta/\kappa$ (where η is the magnetic diffusivity, and κ is the thermal diffusivity) is small, instability may occur for more reasonable shear rates, via a double-diffusive instability mechanism.

This possibility was explored using numerical simulations in Silvers *et al.* (2009b). This paper provided clear evidence for the existence of the double diffusive magnetic buoyancy in the configuration of Vasil & Brummell (2008) for physically-realizable forcings, and showed that the onset of this instability is quasi two-dimensional in nature, in agreement with the previous linear stability calculations of Tobias & Hughes (2004). However, the purpose of Silvers *et al.* (2009b) was to establish the existence of the instability. Here, we wish to examine the subsequent non-linear evolution of such an instability. After substantial further computation, we present the results of the evolution of Case 1 from Silvers *et al.* (2009b) until the point where interactions with the boundaries of the computational domain invalidate further exploration.

2. Mathematical Modeling

The basic set up for this model is similar to that of several previous studies of magnetic buoyancy in a fully compressible plane layer (e.g. Vasil & Brummell (2008), Silvers, Bushby & Proctor (2009a)) and exactly the same as that described in Silvers *et al.* (2009b). The standard equations for compressible fluid dynamics in dimensionless form (see e.g. Matthews, Proctor & Weiss (1995)) are augmented with a forcing term that ensures the maintenance of a desired target shear flow against viscous decay in the absence of magnetic effects. The chosen target shear flow here (as in Silvers *et al.* (2009b)) is of the form $\mathbf{u} = (U_z, 0, 0)$ with $U(z)$ a hyperbolic tangent function, selected to be representative of the smooth radial shear transition believed to exist in the solar tachocline. Initially, we impose a weak uniform vertical magnetic field and examine the action of the shear on this seed field.

The problem is defined by a number of dimensionless quantities that appear in the governing equations and here we once again set these to be the values for Case 1 in

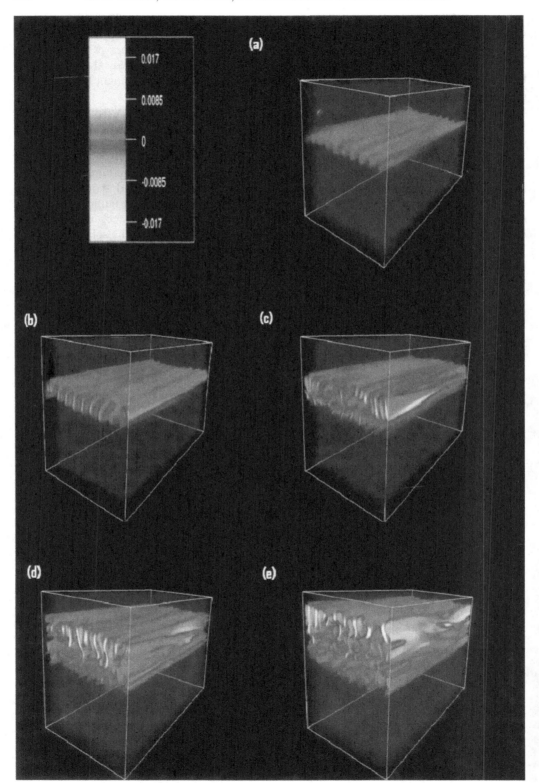

Figure 1. Volume rendered images of the vertical component of velocity at: (a) $t = 112.73$;
(b) $t = 135.63$; (c) $t = 158.53$; (d) $t = 181.43$; (e) $t = 204.33$.

Silvers *et al.* (2009b). That is, $\zeta = 5.0 \times 10^{-4}$, $\sigma = 2.5 \times 10^{-4}$, $C_K = 5 \times 10^{-4}$, $\alpha = 1.25 \times 10^{-5}$, $\theta = 5.0$, $m = 1.6$, with aspect ratio for the computational domain 2:1:1 so that the domain is elongated in the direction of the shear flow. The computation is performed by standard pseudospectral methods and was run on over 200 cores of a machine composed of 2GHz dual-core Opteron 270 processors for a few months.

3. Results

In order to investigate the full nonlinear evolution of a double diffusive magnetic buoyancy instability we evolve Case 1 from Silvers *et al.* (2009b) until there is significant interaction with the upper boundary and the computation has to be halted. This occurs at $t = 220$ in our dimensionless units.

Three-Dimensionality: Figure 1 shows volume renderings of the vertical component of velocity at a series of times that range from a short while after the onset of the instability to shortly before the run was terminated. Existence of a vertical velocity shows evidence of a mode of motion that is not driven by the forcing but which must spontaneously arise as an instability. As described in Silvers *et al.* (2009b), initially the instability appears as a wavelike perturbation with high frequency in the cross-stream y direction, with some but little variation in the x direction. These quasi two-dimensional rolls of motion aligned in the downstream direction are suggestive of the fastest growing modes that might be expected from the linear stability of a simple plane layer of magnetic field in a velocity shear (Tobias & Hughes (2004)). As time advances, the region of instability increases in size, extending upwards towards the top of the computational domain. It appears as though the strongest motions are occurring towards the top of this expanding region, but that action continues behind even after the peak of activity has passed on.

It also appears as though motions become more three dimensional as time progresses. To check this out, we examine the quantities shown in Figures 2. Initially, there is only a vertical magnetic field present. The action of the forced velocity profile is to draw out this initial field to induce a horizontal component in the region of shear. Eventually the gradients induced are sufficient for the double-diffusion magnetic buoyancy instability to occur. During the stretching phase, there are only two components of the magnetic field, the initial B_z and the induced B_x. When instability occurs, perturbations to all components of **B** may be present. If the instability is two-dimensional, we expect to see a signature of the perturbation to B_x in the y direction with little perturbation in the x direction. If three-dimensional motions appear, we expect to see a signature of the perturbation to B_x in the x direction too. We therefore examine $(\partial B_x/\partial y)^2$ and $(\partial B_x/\partial x)^2$ as measures of the existence of two- and three-dimensional perturbations. From Figure 2a, clearly the strongest two-dimensional perturbation grow with time, and appear at the top of the expanding region of activity, as was seen visually in Figure 1. Two-dimensional perturbations exist in the whole region, but are twice as strong near the top. From Figure 2b, three-dimensional perturbations appear to exist more homogeneously throughout the activity region. These perturbations exist for all times, but their measure is at least two orders of magnitude weaker than that of the two-dimensional perturbations. However, the relative strength of the three-dimensional measure does appear to increase approximately four-fold over time, whilst remaining substantially weaker overall, confirming our visual inspection of Figure 1.

The Depth of the Unstable Region: The point where the maximum activity is obtained moves towards the upper boundary. This appears to be a consequence of the underlying evolution of the induced B_x field. The shear induction of B_x produces an

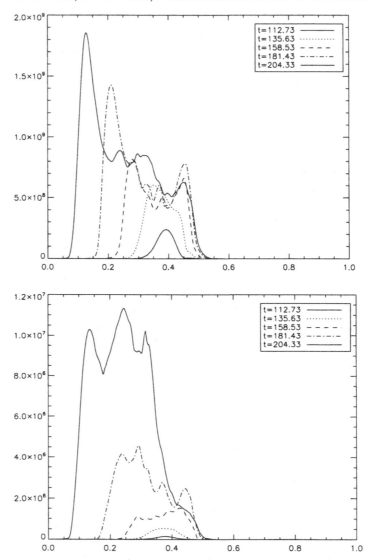

Figure 2. (a. *Top panel*) $(\partial B_x/\partial y)^2/\alpha$ horizontally-averaged as a function of depth at various times, showing the evolution of two-dimensional disturbances. (b. *Bottom panel*) $(\partial B_x/\partial x)^2/\alpha$ horizontally-averaged as a function of depth at various times, showing the evolution of three-dimensional disturbances.

increasing peak of B_x which leads to a rising "front" in the vertical gradient of B_x, on which the instability appears to "surf". This evolving peak can be seen as the short dot-dashed line in panels of Figure 3. Instability continues behind this front but appears to be strongest where the front is passing through. Eventually this front hits the upper boundary and at this point we can no longer continue the simulation. The problem under consideration therefore has the character of a dynamic bifurcation, where the background state that determines the linear instability is evolving with time. The timescale for the development of the basic horizontal field structure here appears to be comparable with the growth rate of the y-dependent disturbances, somewhat complicat-

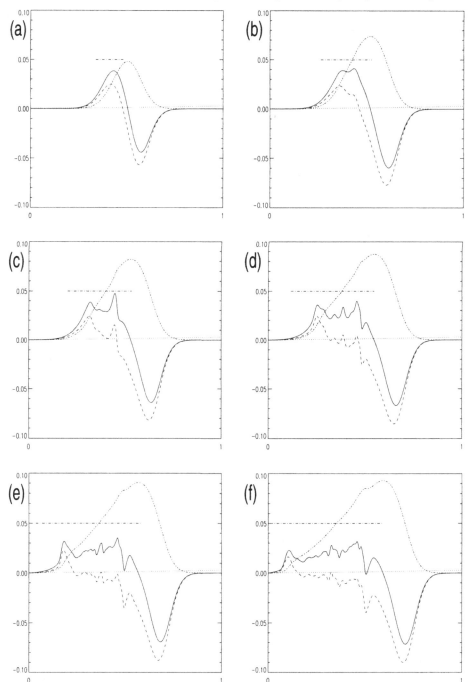

Figure 3. Images showing the depth dependency of the horizontal averages of the following quantities: $\alpha B_x \frac{dB_x}{dz}$ (solid line), $\alpha \rho B_x \frac{d}{dz}(\frac{B_x}{\rho})$ (long-dashed line) and $-\zeta \frac{1}{\rho^{-\gamma}} \frac{d}{dz} \rho^{1-\gamma} T$ (dotted line) at a series of times. The times for each panel are: (a) $t = 71.53$; (b) $t = 112.73$; (c) $t = 135.63$; (d) $t = 158.53$; (e) $t = 181.43$; (f) $t = 204.33$. These figures also show the region where $|w|$ (averaged in $x - y$ is greater than 1×10^{-7} (dash-dot thick horizontal lines) where the instability could be considered to be occurring in the simulation, and the horizontal average of $5\alpha B_x^2$ (short--dot-dashed line; scaling for clarity) to show the position of the growing background magnetic layer on which the instability "surfs".

ing matters. Nonetheless it is to be expected that the disturbance will grow principally in regions where the background state is locally unstable. From linear stability arguments (Newcomb (1961), Tobias & Hughes (2004)), we expect the fastest growing modes to have high cross-stream wavemnumber (k_y) and a downstream wavenumber (k_x) that is as long as possible – quasi interchange modes. We therefore here compare the extent of the instability found in our simulations with both the local two-dimensional inter-change and three-dimensional criteria which would hold for a non-evolving stationary field configuration. These can be written, in a simplified form as

$$\alpha B_x \frac{dB_x}{dz} > -\zeta \frac{1}{\rho^{-\gamma}} \frac{d}{dz} \rho^{1-\gamma} T \quad \text{(3D)}$$

$$\alpha \rho B_x \frac{d}{dz}\left(\frac{B_x}{\rho}\right) > -\zeta \frac{1}{\rho^{-\gamma}} \frac{d}{dz} \rho^{1-\gamma} T \quad \text{(interchange)}$$

(see e.g. Hughes (2007) for further details). Note that the right hand side of both in-equalities is the same. Figure 3 shows the regions where these inequalities are satisfied, namely where the lines corresponding to relevant field gradient criteria (left hand sides above) lie above the dotted line giving the stabilizing entropy gradient (right hand sides above). Initially, there is considerable overlap between the regions were both the two and three dimensional instability criteria are met, and as expected, the three-dimensional instability is slightly preferred. However, at later times we see that the region where the two dimensional instability criterion is met is confined to a very small region at the top of the expanding region. By contrast the region where the three dimensional instability criterion is satisfied occupies almost half the depth of the box at late times. The figure also shows the region where $|w|$(averaged in $x - y$) is significant (formally greater than 1×10^{-7}) to indicate the vertical extent of activity at each time. It can be seen that this expands reasonably in line with the extending region of (static) three-dimensional instability, and occupies at any time basically the whole region where there is significant positive gradient in B_x^2 since the stabilising entropy gradient is so low.

4. Discussion

From this simple analysis, one might hesitantly conclude that our initial visual im-pression is correct. Initially, the instability does appear to be dominated by quasi two-dimensional modes, although formally both (stationary not dynamic) instability criteria (for interchange and three-dimensional modes) are satisfied. The variation in y of the in-stability is on a rather shorter scale than that appearing for larger values of the diffusivity ratio, as shown in the results of Silvers, Bushby & Proctor (2009a). This short transverse scale is reminiscent of the salt finger instability that occurs in the oceans when warm salty water overlies cold fresh water. As the simulation progresses, the evolving region of the background field that provides the magnetic gradients to drive the instability dynam-ically creates a shifting and adjusting region of local instability. The strongest instability occurs at the top of this region, and appears to still have a strong quasi two-dimensional component, whereas the motions in the wake of this front appear to be more likely dom-inated by a variety of unstable three-dimensional modes as the pure interchange modes become damped. Overall, the region of instability broadens.

Considerably more work is required in order to understand fully the evolution of this instability. Taller boxes or domains that adjust to allow the rising front of activity are required for a complete picture of the long time evolution and saturation of this process.

The efficiency of this process in its transport of magnetic field is of primary interest in the solar context and is worthy of further study.

5. Acknowledgements

LJS wishes to thank the International Astronomy Union and the Royal Astronomical Society for the award of travel grants that enabled her to participate in this meeting. NHB acknowledges NASA grant NNX07AL74G. These computations were carried out at the UKMHD facility at St. Andrews University, supported by the UK STFC.

References

Abbett, W. P., Fisher, G. H., Fan, Y., & Bercik, D. J. 2004, *ApJ*, 612, 557
Babcock, H. W. 1961, *ApJ*, 131, 572.
Brummell, N. H., Cline, K. S., & Cattaneo, F. 2002, *MNRAS*, 329, L73
Cattaneo, F., Brummell, N. H., & Cline, K. S. 2006, *MNRAS*, 365, 727
Cline, K. S., Brummell, N. H., & Cattaneo, F. 2003, *ApJ*, 588, 630
Cline, K. S., Brummell, N. H., & Cattaneo, F. 2003, *ApJ*, 599, 1449
Emonet, T. & Moreno-Insertis F. 1998, *ApJ*, 492, 80
Fan, Y., Zweibel, E. G., & Lantz, S. R. 1998, *ApJ*, 493, 480
Fan, Y. Abbett, W. P. & Fisher, G. H. 2003, *ApJ*, 582, 1206
Hughes, D. W., 1985, *Geophys. Astrophys. Fluid Dynamics*, 32, 273
Hughes, D. W. 2007, in The Solar Tachocline, eds. D. W. Hughes, R. Rosner, & N. O. Weiss
      (Cambridge: Cambridge University Press) 11
Hughes, D. W., Falle, S. A. E.. G., & Joarder, P. 1998, *MNRAS*, 298, 433
Hughes, D. W. & Falle, S. A. E.. G. 1998, *ApJ*, 509, L57
Jouve, L. & Brun, A. S. 2009, *ApJ*, 701, 2, 1300.
Leighton, R. B. 1969, *ApJ*, 156, 1.
Matthews P. C., Proctor M. R. E., & Weiss N. O. 1995, *JFM*, 305, 281.
Newcomb, W. A. 1961, *Phys. Fluids*, 4, 391.
Parker, E. N. 1955, *ApJ*, 121, 491
Parker, E. N. 1993, *ApJ*, 408, 707
Silvers, L. J., Bushby P. J., & Proctor, M. R. E. 2009a, *MNRAS*, 400, 1, 337.
Silvers, L. J., Vasil G. M., Brummell N. H., & Proctor, M. R. E. 2009b, *MNRAS*, 400, 1, 337.
Tobias, S. M. & Hughes, D. W. 2004, *ApJ*, 603, 785.
Vasil, G. M. & Brummell, N. H. 2008, *ApJ*, 686, 709
Vasil G. M. & Brummell N. H. 2009, *ApJ*, 690, 783.
Wissink, J. G., Matthews, P. C., Hughes, D. W., & Proctor, M. R. E.. 2000, *ApJ*, 536, 982

Discussion

ROGERS: The R_I numbers isn't strictly large in the solar tachocline. So have you tried to run simulations in which N is varying rapidly, as in the solar tachocline?

SILVERS: So far we have only looked at this one case which is mildly stratified. It would indeed be interesting to look at cases with different, possibly more relevant, stratifications.

HUGHES: I believe that the time scale for stretching out the magnetic field is important. Where does it appear in the criterion that you displayed?

SILVERS: I feel that that is a question that is probably best answered by Nic or Geoff as it is from their earlier paper together.

BRUMMELL: The Richardson number criterion of Vasil & Brummell is basically a bound derived by examining the maximum value of the destabilizing gradients that can be built before the the induction of B_x by the shear halts. It is therefore a criterion that equates values at a particular time, and therefore the timescale drops out explicitly. This time is essentially the Alfven travel time on the original vertical field across (half of) the localized layer. Further details can be found in Vasil & Brummell (2009).

Astrophysical Dynamics: From Galaxies to Stars
Proceedings IAU Symposium No. 271, 2010
N. H. Brummell, A. S. Brun, M. S. Miesch & Y. Ponty, eds.
© International Astronomical Union 2011
doi:10.1017/S1743921311017650

3D Magnetic Reconnection

Clare E. Parnell[1], Rhona C. Maclean[1], Andrew L. Haynes[1] and Klaus Galsgaard[2]

[1]School of Mathematics & Statistics, University of St Andrews, North Haugh, St Andrews,
Fife, KY16 9SS, UK
email: clare@mcs.st-and.ac.uk
[2]Niels Bohr Institute, Julie Maries vej 30, 2100 Copenhagen 0, Denmark

Abstract. Magnetic reconnection is an important process that is prevalent in a wide range of astrophysical bodies. It is the mechanism that permits magnetic fields to relax to a lower energy state through the global restructuring of the magnetic field and is thus associated with a range of dynamic phenomena such as solar flares and CMEs. The characteristics of three-dimensional reconnection are reviewed revealing how much more diverse it is than reconnection in two dimensions. For instance, three-dimensional reconnection can occur both in the vicinity of null points, as well as in the absence of them. It occurs continuously and continually throughout a diffusion volume, as opposed to at a single point, as it does in two dimensions. This means that in three-dimensions field lines do not reconnect in pairs of lines making the visualisation and interpretation of three-dimensional reconnection difficult.

By considering particular numerical 3D magnetohydrodynamic models of reconnection, we consider how magnetic reconnection can lead to complex magnetic topologies and current sheet formation. Indeed, it has been found that even simple interactions, such as the emergence of a flux tube, can naturally give rise to 'turbulent-like' reconnection regions.

Keywords. magnetic fields, (magnetohydrodynamics:) MHD

1. Introduction

Magnetic fields pervade pretty much all the objects in not only our solar system, but throughout the Universe. The strength and scales of complexity of the magnetic fields vary depending on the objects they are associated with. For instance, the magnetic fields emanating from the planets in our solar system are essentially dipolar in nature and are relatively weak compared to the magnetic field strengths that can be found on the Sun. Although at large distances the Sun's magnetic field may be considered dipolar in nature, a closer look reveals that the Sun's surface is threaded by a patchwork of features with fluxes ranging over many orders of magnitude through which magnetic fields are directed into, or out from, the Sun (Parnell *et al.* 2009). This magnetic patchwork is not static, but highly dynamic (Hagenaar *et al.* 2003; Thornton & Parnell 2010) and results in a very complex and dynamic evolution of the magnetic field and plasma throughout the solar atmosphere. One key type of behaviour that results from this dynamic complexity is the fundamental plasma physics process of magnetic reconnection. Magnetic reconnection is not unique to the solar atmosphere, but also plays a key role in a wide range of astrophysical phenomena such as, the heating of stellar coronae, the acceleration of stellar winds and astrophysical jets, the generation of magnetic fields via dynamo mechanisms and the creation of aurora and substorms in planetary magnetospheres.

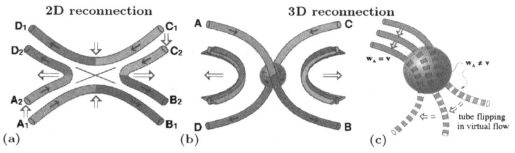

Figure 1. Illustrations highlighting the characteristics of (a) 2D and (b,c) 3D reconnection where the thick tubes represent flux tubes with arrows indicating the direction of the field, the block arrows represent the direction of the outflowing plasma and the purple shaded spheres are the diffusion volume in 3D with the arrows indicating the direction of plasma flow on its surface. (a) 2D reconnection at an X-type null point in which a pair of flux tubes $A_1 B_1$ and $C_1 D_1$ reconnect to form a new pair of flux tubes $A_2 D_2$ and $C_2 B_2$. (b) 3D reconnection in which a pair of flux tubes AB and CD reconnect, but they do not form a new pair of flux tubes. (c) In 3D thin flux tubes reconnecting in a diffusion region will, on one side of the diffusion volume, appear to be moving slowly, but on the other side will appear to be moving incredibly fast. In reality the plasma on this side is moving just as slowly as it is on the other side, and this 'virtual flow' is simply a consequence of fieldlines changing connectivity within the diffusion region. Images derived from figures 2, 6 and 7 of Priest *et al.* (2003) and taken from Cargill *et al.* (2010).

2. Three-Dimensional Reconnection

Magnetic reconnection enables a magnetic field to globally restructure by locally changing the mapping of field lines. This process has a number of important consequences for the plasma as it converts free magnetic energy into three different types of energy. Local Joule heating at the reconnection site raises the internal energy of the plasma, bulk acceleration of the plasma from the reconnection site by the magnetic tension force of the newly formed field lines can produce large kinetic energies and finally the large electric fields found at the reconnection site accelerate particles throughout the diffusion region volume to high velocities.

Over the past fifty years the main focus of researchers has been on the two-dimensional (2D) aspects of magnetic reconnection, since this permits significant simplifications to be made to analytical and numerical problems (see, for example Priest & Forbes 2000; Biskamp 2000, for a review). However, it is now known three-dimensional (3D) reconnection has different characteristics to 2D reconnection and is a much richer and more varied process (Schindler *et al.* 1988; Hesse & Schindler 1988; Hesse 1995; Hornig & Priest 2003) (Fig. 1). Indeed, even a slight departure from an exactly 2D configuration leads to considerably different behaviour. Below the characteristics of 2D and 3D reconnection are compared and contrasted, before the wide variety of locations where 3D reconnection can occur are discussed.

2.1. *Characteristics*

In 3D, reconnection occurs in a range of locations that can, but do not have to be, associated with null points (points at which all components of the magnetic field are zero), unlike in 2D where reconnection can only occur at X-type null points (Fig. 1a). Reconnection in 3D occurs in a finite volume, known as a diffusion region, within which the plasma and the field lines become 'unfrozen', i.e., the plasma elements can move independently to the field lines (Fig. 1b). In this diffusion volume the field lines continually and continuously diffuse through plasma and, as long as some portion of a field line is passing through the diffusion region volume, then it will reconnect with other field lines (Fig. 1c). Due to this behaviour it is not possible, in general, to find pairs of field lines that, after reconnection, match to form two new pairs of field lines, as occurs in 2D reconnection (c.f., Figs. 1a and 1b). Instead, it is only possible to find two surfaces (or volumes) of field lines that reconnect to form two new surfaces (or volumes). A consequence of re-connection throughout a finite volume is that the field line mappings are continuous, as opposed to discontinuous as they are in 2D reconnection.

2.2. *Where can it occur?*

A necessary and sufficient condition for 3D reconnection is that there exists a region where the ideal magnetohydrodynamic (MHD) assumption breaks down, i.e., a diffusion region through which

$$\int_{fl} E_{||} dl \neq 0 \, ,$$

where *fl* is the field line path and $E_{||}$ is the component of the electric field parallel to the field line (Schindler *et al.* 1988; Hesse & Schindler 1988; Hornig & Priest 2003). From the dot product of Ohm's law in MHD with the magnetic field, **B**,

$$\mathbf{E} \cdot \mathbf{B} + (\mathbf{v} \times \mathbf{B}) \cdot \mathbf{B} = \mathbf{j} \cdot \mathbf{B} / \sigma, \quad \Longrightarrow \quad E_{||} = j_{||} / \sigma \, ,$$

where **v** is the plasma velocity, **j** is the electric current and σ is the electrical conductivity of the plasma. Hence, the presence of electric currents are essential for 3D reconnection, just as they are for 2D reconnection, but in 3D it is the parallel component of current that plays the crucial role. In 3D, strong accumulations of current and current layers, can arise in a wide variety of locations and are not just associated with magnetic nulls, as they are 2D.

The locations for current layer formation in 3D may be divided into those that are associated with *topological* features and those associated with *geometrical* features. Quasi separatrix layers (QSLs) are an example of a geometric feature about which reconnection can occur (Priest & Démoulin 1995; Démoulin *et al.* 1996; Titov *et al.* 2003; Aulanier *et al.* 2006; Titov 2007; Titov *et al.* 2009). QSLs are regions, usually long and narrow, identified on a plane in a magnetic domain which is threaded by field lines whose foot-points significantly diverge at one end. Naturally if two field lines which start off running along a similar path end up in very different places, as do QSL field lines, then these lines will be associated with electric currents. If the divergence of the field is dramatic then the associated currents may be significant and reconnection (termed either QSL or slip-running reconnection) may result (Aulanier *et al.* 2006). Many papers have been written where observed phenomena, such as flares, bright points and CMEs, have been explained using QSL reconnection (Aulanier *et al.* 2005; Aulanier *et al.* 2006; Aulanier et al. 2007; Masson *et al.* 2009; Pariat *et al.* 2009; Baker *et al.* 2009).

Currents also accumulate when magnetic flux tubes are twisted (Browning *et al.* 2008; Hood *et al.* 2009) or braided (Parker 1991; Galsgaard & Nordlund 1996; Longbottom *et al.* 1998; Wilmot-Smith *et al.* 2009, 2010; Pontin *et al.* 2011). Neither twisting nor braiding has to be excessive for strong currents to form, as is shown in the braiding experiments conducted by (Wilmot-Smith *et al.* 2010; Pontin *et al.* 2011) (Fig. 2). Their experiment consists of magnetic field that runs in the same direction, i.e., out from the bottom and into the top of the numerical box. The field is braided (Fig. 2a) and the initial force-free field involving this braid is associated with a large-scale current (Fig. 2b). However, as this force-free system is allowed to resistively relax, it first collapses to form intense

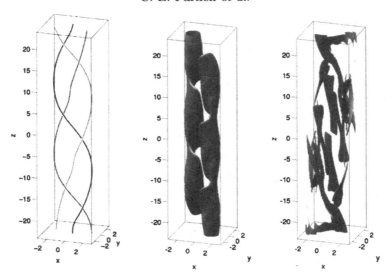

Figure 2. (a) Three sample magnetic field lines showing the force-free braiding structure of the initial magnetic field in the experiment of Pontin *et al.* (2011). (b) and (c) show isosurfaces of current at the start of the resistive relaxation and part way through once reconnection has started and fragmentation has broken up the current sheet providing rapid, long duration and widespread heating. Images taken from Pontin *et al.* (2011).

currents at which reconnection occurs leading to a cascade process in which the original large-scale homogeneous current fragments to smaller and smaller scales (Figs. 2c). This process is associated with rapid reconnection that occurs at multiple small-scale intense current accumulations throughout the domain. It is not simple for the magnetic field in this experiment to untangle itself and, since the plasma is not clever enough to work out the fastest way to untangle itself with the minimum amount of reconnection, magnetic flux is found to reconnect multiple times. This process of multiple reconnection was first observed by Parnell *et al.* (2009) and will be discussed in more detail in Section 3.3. The consequence of this type of behaviour is widespread reconnection throughout the flux tube, that lasts a long time. Hence, this 3D reconnection process can release a lot of energy in the whole of the flux tube over many hours and, hence, it may well be an important heating mechanism within the closed magnetic loop structures that fill the solar atmosphere.

3D magnetic null points are one type of topological feature which are prone to collapse to form a current layer, just like 2D nulls are. From a positive (negative) 3D null point (Fig. 3a) there are a set of field lines that extend out of (or into) the null forming what is known as a fan surface and a pair of field lines that extend into (out of) the null forming a curve known as the spine (e.g., Fukao *et al.* 1975; Lau & Finn 1990; Parnell *et al.* 1996). How the magnetic null is perturbed determines the nature of the collapse and the resulting current layer formed (Rickard & Titov 1996; Galsgaard & Nordlund 1997; Pontin *et al.* 2004; Pontin & Craig 2005; Pontin *et al.* 2005; Pontin *et al.* 2007; Priest & Pontin 2009). For instance, a rotational disturbance in planes perpendicular to the spine result in accumulations of current around the spine and/or fan. Two types of reconnection are found to be associated with these sorts of disturbances, namely, torsional-spine reconnection which occurs in response to a rotational disturbance of the fan plane, and torsional-fan reconnection which occurs in response to a rotational disturbance of the spine (Priest & Pontin 2009). However, the most common type of null-point reconnec-

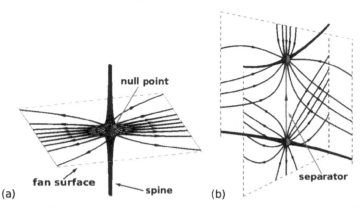

Figure 3. Magnetic field structure of (a) a positive 3D potential null point and (b) a separator formed by the intersection of two separatrix surfaces. Image taken from Pontin (2011).

tion, is spine-fan reconnection, which occurs as a result of any shearing motion in which the angle between the spine and fan are altered leading to a collapse of the spine and fan creating a current layer lying along both structures,(Pontin *et al.* 2007; Priest & Pontin 2009).

3D null points are not the only topological feature at which reconnection can occur. Fan surfaces from 3D nulls extend far out from the nulls themselves separating the magnetic field from topologically distinct field regions. Thus they are more generally known as separatrix surfaces and are bounded by either the edge of the domain investigated or by spine field lines from other nulls. If two separatrix surfaces intersect, special field lines called separators arise (Fig 3b). This manifestation of a separator is stable and resides at the intersection of four topologically distinct flux domains. This means they are in many ways the 3D equivalent of a 2D null point, although, of course, the magnetic field is only zero at the ends of the separators not along its length. It also means that reconnection at separators has global consequences and can lead to global restructuring of the magnetic field. In the following section, we focus on separator reconnection, in particular, we present examples showing how common separators are (Section 3.1), we discuss the nature of separator reconnection (Section 3.2) and consider the consequences of multiple separator reconnection (Section 3.3).

3. Separator Reconnection

3.1. *Examples of Separator Reconnection*

Determining the magnetic topology of a complex magnetic field is not trivial. Haynes & Parnell (2010) have recently published a method that can find the nulls, spines, separatrix surfaces and separators of magnetic fields that are known numerically on a discrete grid of points, or are known everywhere analytically. This method has been used successfully in a number of cases as discussed below.

Parnell *et al.* (2010b) analysed the magnetic topology of a flux tube emerging into an overlying coronal magnetic field from a 3D resistive MHD experiment (Archontis et al. 2005; Galsgaard *et al.* 2007). Although initially there were no null points in the region, when the flux tube rose up and started interacting with the overlying coronal magnetic field two clusters of magnetic nulls formed on either side of the emerged tube (Maclean *et al.* 2009). The clusters contain around 10-20 nulls most of which are short

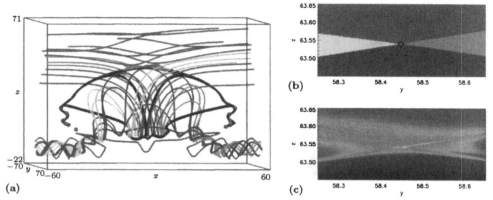

Figure 4. (a) Magnetic field structure at snapshot $109c_s$ during the emergence of a flux tube into an overlying coronal magnetic field. The positive/negative null points (red/blue spheres), separators (thick black lines), other fieldlines for context (overlying - red, flux tube - blue, flux tube to overlying - yellow and overlying to flux tube - green) and the strong regions of $E_\parallel$ (30% of maximum - cyan isosurfaces) are shown. (b) Connectivity map in the plane $x = 0$ about the top arched inter-cluster separator showing flux domains coloured according to the connectivity of the fieldlines that thread them. The separator (black diamond) is located at the junction of four flux domains. (c) Contour plot of integrated $E_\parallel$ along fieldlines threading the same region. The top arched separator threads the plane at the location of the maximum integrated $E_\parallel$ indicating its importance for reconnection. Images taken from Parnell *et al.* (2010b).

lived, but a couple in each cluster are long lived and last throughout the duration of the interaction between the emerging and overlying flux regions. Parnell *et al.* (2010b) found that inside each cluster the nulls are connected by separators which form a chain of nulls. Between the two clusters there is one, or more usually many, separators that link the null clusters. These intercluster separators connect just one (or occasionally two) nulls from each cluster. These nulls are the long lived ones. Figs. 4a and 5a show a few magnetic field lines, all the separators and nulls at two times during the emergence of the tube and its interaction with the overlying field. In Fig. 4a, taken at time $109c_s$, there are 3 intercluster separators and 19 separators within the null clusters. Only one of the intercluster separators lies solely in the corona and threads the isosurface of $\mathbf{E}_\parallel$ which indicates the main reconnection site. The other two intercluster separators initially rise up into the corona before dropping down under the emerging flux tube. This snapshot is taken during a relatively simple phase of the interaction. Fig. 5a is taken at an earlier time ($t = 86c_s$) during the most intense and dynamic phase of the reconnection. In this snapshot there are a total of 229 separators: 214 intercluster separators, which lie in the corona and all thread the large intense region of $\mathbf{E}_\parallel$ (reconnection site), and 15 separators inside the null clusters. Figure 6a shows a view from above of the separators colour coded with $\mathbf{E}_\parallel$ in this snapshot clearly highlighting the complex tangled mess formed by the intercluster separators.

All the separators reside at the intersection of the same four connectivities of flux. The two original flux domains (flux tube and overlying coronal field) and the two new flux domains which are created after reconnection (flux tube to overlying and overlying to flux tube). However, the separators do not all lie at the boundary between the same four flux domains. This is because flux of one connectivity may be divided into many topologically distinct flux domains, as explained by Parnell *et al.* (2008). The connectivity maps (Figs. 5b,c,d) show cuts through the mass of separators seen at $t = 86c_s$ and reveal the large number of distinct flux domains that have the same connectivity (same

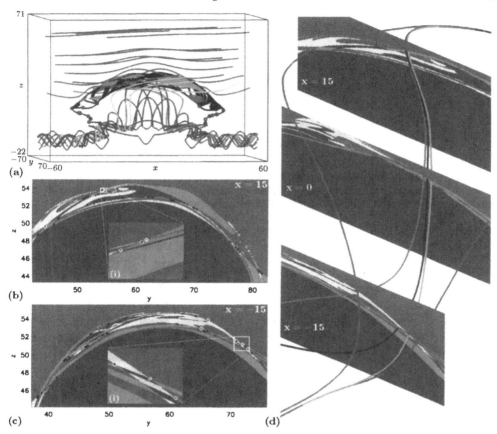

Figure 5. (a) Magnetic field structure at time $86c_s$ with the same features, indicated using the same nomenclature, as Figure 4. Connectivity maps in the (b) $x = 15$ and (c) $x = -15$ planes showing flux domains coloured according to field line connectivity. The separators (coloured and small black diamonds) are all located at the junctions of four flux domains. (d) Five separators are drawn threading connectivity maps plotted in 3 planes showing that separators may start out along similar paths before they diverge to follow very different paths. The coloured diamonds on (b) and (c) correspond to the coloured separators in (d). (b(i) and c(i)) Connectivity maps for the white boxed regions in (b) and (c), respectively. Images taken from Parnell *et al.* (2010b).

colour). Fig. 4b shows the connectivity map about the upper intercluster separator at $t = 109c_s$. One curious behaviour of separators is that they often start out along very similar paths before diverging at they move away from a null point at the end of the separators (Fig. 5d). This adds further to the difficulty of finding separators.

Furthermore, in all cases considered the isosurfaces of strong $\mathbf{E}_{\parallel}$ are always threaded by separators, although a separator does not always have to thread a region of high $\mathbf{E}_{\parallel}$ (see for example the isosurfaces in Figs. 4a and 5a). Fig. 6a shows a view from above of all the separators at $t = 86c_s$ colour coded according to the $\mathbf{E}_{\parallel}$ along them. The red regions indicate strong $\mathbf{E}_{\parallel}$ and thus these are likely to be the sites of strong reconnection. Fig. 6b shows a plot of the amount of reconnection (integral of the $\mathbf{E}_{\parallel}$) along all the separators shown in both Figs 4a and 5a against separator length. The intercluster separators are the long separators and the majority of them all show a significant amount of reconnection. The separators that are contained purely within the null clusters have no

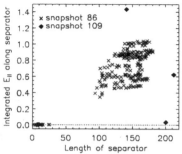

Figure 6. (a) View from above of the separators at time $86c_s$, colour coded according to the amount of $\mathbf{E}_{\parallel}$ along them (where red indicates strong $\mathbf{E}_{\parallel}$). (b) A plot of integrated $\mathbf{E}_{\parallel}$ along a separator against separator length. Images taken from Parnell *et al.* (2010b).

associated reconnection. This suggests that some, but not all, separators are important for reconnection.

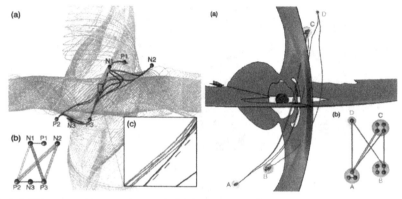

Figure 7. (a) 3D magnetic topology of the interaction, below the photosphere, between two emerging twisted flux tubes showing the null points (red and blue spheres), separators (thick lines), field lines of different connectivities (thin lines of different colours). The inset shows a close up of the separators. (b) Separators on the dayside of the magnetosphere for a Northwards interplanetary magnetic field of 45 degrees. Null points (red and blue spheres) and separators (thick lines) and pressure (filled contours). Examples taken from Haynes & Parnell (2010).

Fig. 7 shows sample snapshots of the magnetic topology from two other numerical MHD models. In each case examined so far the nulls have been found in clusters (as noted would occur by Albright 1999) and many separators have been found. It seems that, in general, pairs of nulls in the null clusters are linked by single separators and nulls with intercluster separators (separators linking the null clusters) are multiply connected nulls (i.e., they have many separators, as discussed in Parnell *et al.* 2008).

3.2. *Nature of Separator Reconnection*

A key question to answer is how does separator reconnection actually occur and why are some separators associated with reconnection whilst others are not?

In particular, Parnell *et al.* (2010a) focussed on answering the following key questions: Where are the enhanced regions of $\mathbf{E}_{\parallel}$ along separators (i.e., where are the diffusion/reconnection regions)? How do these reconnection sites vary temporally and spatially along separators? What is the nature of the magnetic and velocity fields in the vicinity of a separator? The answers to these questions revealed that, locally, separator

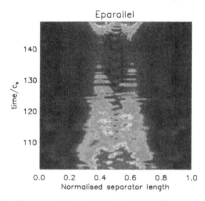
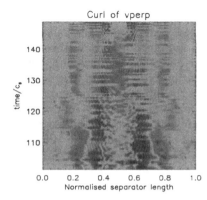

Figure 8. Temporal evolution of (a) $\mathbf{E}_\parallel$ and (b) the curl of the velocity perpendicular to the longest lived intercluster separator from the flux emergence experiment investigated by (Parnell *et al.* 2010b). The x-axis is the normalised length of the separator and the y-axis is time in units of the sound travel time across the box (c_s). In (a) red represents strong $\mathbf{E}_\parallel$ and in (b) blue/red represent negative/positive $\nabla \times \mathbf{v}_\perp$ (i.e. oppositely rotating plasma).

reconnection behaves very much like non-null reconnection as described by (Schindler *et al.* 1988; Hesse & Schindler 1988; Hesse 1995; Hornig & Priest 2003).

Parnell *et al.* (2010a) noted the following characteristics of separator reconnection. Here, though, we do not show the figures from their paper to illustrate the behaviour, but present new results determined from analysing the longest lived separator found in the flux emergence experiment considered by Parnell *et al.* (2010b). These results verify the characteristics found earlier by Parnell *et al.* (2010a) providing further support for their results. Regions of enhanced $\mathbf{E}_\parallel$ are found to occur along the lengths of the separators, as opposed to at their end null points (Fig. 6a). This means that in separator reconnection the field lines do not in general reconnect at the nulls, but instead they reconnect within the vicinity of the middle of the separator. Moreover, the extent, strength and location of the diffusion regions change in time along the length of the separator (Fig. 8a). In fact, multiple diffusion regions seem to be quite common along individual separators (Figs. 6a and 8a).

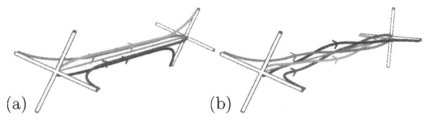

Figure 9. Cartoons showing the 3D global magnetic topology about (a) a separator with a hyperbolic local 3D field structure and (b) a separator with an elliptic local 3D magnetic field structure (equivalent to separator (a) twisted by $3\pi/2$). Each figure includes a separator (green), three field lines lying in the separatrix surface of the near null (blue, cyan, straight blue edge from near null) and three field lines lying in the separatrix surface of the far null (pink, orange, straight orange edge from far null), a spine from the near null (straight orange edge) and a spine from the far null (straight blue edge). Taken from Parnell *et al.* (2010a).

About a separator the magnetic field typically runs approximately parallel to the separator, but clearly, if there is a strong $\mathbf{E}_\parallel$ about the separator, there must also be anti-parallel field perpendicular to it. Parnell *et al.* (2009) considered the magnetic field perpendicular to the separator to investigate this component and found that this field does not have to be hyperbolic in nature, but can be elliptic (see Fig 9). The latter can result from a separator that has been twisted (Fig 9) or sheared.

In 2D reconnection, the magnetic field is carried into, and out of, the reconnection site by a stagnation type flow. In 3D reconnection scenarios, the magnetic field is obviously carried into the diffusion region and is also ejected out from the diffusion region (generally with high speed), however the inflow and outflow regions do not necessarily lie in a plane, and so the flow is not necessarily stagnation like. Additionally, a counter rotating plasma flow is found on either side of the diffusion region (Fig 8b). Such a flow is one of the main signatures of non-null reconnection Hornig & Priest (2003). The key reason that some separators are associated with reconnection, but others are not is because, in addition to a favourable magnetic field configuration for reconnection there must also be favourable plasma flows that drive the reconnection into these sites. If the flows about a separator are not driving magnetic flux in towards the separator then reconnection will not occur at that separator. However, note that in the flux emergence experiment of Parnell *et al.* (2010b) all the separators were created spontaneously at the onset of reconnection and did not exist before hand, so all were at sometime associated with reconnection, albeit in some cases only a very small amount.

3.3. *Multiple Separator Reconnection*

As already mentioned (and illustrated), multiple separators commonly arise in numerical 3D MHD experiments. This means that multiple reconnection sites also naturally arise (Haynes *et al.* 2007; Parnell *et al.* 2009; Wilmot-Smith *et al.* 2010; Pontin *et al.* 2011). Furthermore, in 3D, reconnection of current accumulations tends to cause the current to fragment into many smaller current layers where further reconnection occurs (Haynes *et al.* 2007; Hood *et al.* 2009; Wilmot-Smith *et al.* 2010; Parnell *et al.* 2010b; Pontin *et al.* 2011). That is, in 3D, reconnection often leads to a cascade in scales and, hence, turbulence.

One way to understand this type of behaviour is to realise that the plasma is not clever. It does not know the fastest way to untangle the magnetic field to release free magnetic energy and relax to a lower energy state. Instead, the plasma flows tend to squash the magnetic field together, accumulating currents, forming diffusion regions in which reconnection occurs. These reconnection sites squirt out newly reconnected field lines which result in further squashing of the field and current sheets forming off from the ends of the original diffusion regions. At these sites, more reconnection occurs and more current sheets are formed, etc. This occurs because the first reconnection of field lines is, in general, unlikely to be the most optimum for untangling the field. So the newly reconnected flux finds itself still tangled and must reconnect again and again before it is untangled. This type of behaviour was first noticed by Haynes *et al.* (2007) and was explained by Parnell *et al.* (2009). However, it has now been found in many different experiments (e.g., Haynes *et al.* 2007; Dorelli & Bhattacharjee 2008; Hood *et al.* 2009; Wilmot-Smith *et al.* 2010; Pontin *et al.* 2011). This behaviour was originally described by Parnell *et al.* (2009), who named it recursive reconnection, however, a better name is probably multiple reconnection.

In cases of multiple reconnection, the same flux may be reconnected multiple times, as the case suggests. Parnell *et al.* (2009) found that in their experiment the flux was reconnected 1.8 times more than it would have been if the flux was only reconnected once and the fastest untangling had occurred. This behaviour has also be found in other experiments. For instance, Pontin *et al.* (2011) studied how much reconnection occurred in the braiding experiment of Wilmot-Smith *et al.* (2010). They found that the flux in their experiment must have reconnected 1.6 times implying that the many reconnection sites that they found were multiply reconnecting the magnetic field.

The multiple reconnection of magnetic field at many different diffusion regions has some interesting consequences for the energetic behaviour of the plasma. In particular, the energy release (i) is wide spread as it occurs at multiple sites, (ii) occurs for longer as the magnetic field does not take the simplest and fastest path to untangle and (iii), in cases of driven reconnection, more energy is released than in the potential case. This happens because in the cases of multiple reconnection more Poynting flux can be injected into the system and hence more energy can be released.

4. Discussion

In this review, the complexities of 3D magnetic fields and 3D magnetic reconnection have been highlighted. The main differences between 2D and 3D reconnection have been discussed. Although 3D reconnection can occur in a wide range of locations the actual nature of the reconnection is very similar in all the cases of non-null reconnection, including separator reconnection. Although, separators are special field lines that link two null points, the null points themselves play no real role in separator reconnection and hence the reconnection at them is of non-null type.

A series of examples have been shown of globally complex 3D magnetic topologies that arise in a range of astrophysical magnetic fields. In all of these examples a multitude of separators are found and each reconnection site (region of strong $\mathbf{E}_{||}$) within the models has, so far, always been found to be threaded by one or more separators. In all the resistive MHD experiments discussed the magnetic field is found to become locally very complex upon the initiation of magnetic reconnection. That is to say, following the onset of reconnection, macroscopic current regions have a tendency to fragment into a multiscale array of current layers at each of which reconnection occurs. This turbulent like behaviour enables the process of 3D reconnection to be widespread and longer lasting (due to the multiple reconnection of flux) than one might imagine. This therefore, means that reconnection is a very good candidate for heating solar and stellar coronae. By managing to reconnect flux at multiple sites over a large area a lot of flux may be processed in a short space of time and hence reconnection can also release sufficient free energy rapidly enough to power a solar flare.

References

Albright, B. J. 1999, *Phys. Plasmas*, 6, 4222

Archontis, V., Moreno-Insertis, F., Galsgaard, K., & Hood, A. W. 2005, *Astrophys. J.*, 635, 1299

Aulanier, G., Golub, L., DeLuca, E. E., *et al.* 2007, *Science*, 318, 1588

Aulanier, G., Pariat, E., & Démoulin, P. 2005, "*Astron. Astrophys.*", 444, 961

Aulanier, G., Pariat, E., Démoulin, P., & Devore, C. R. 2006, *Solar Phys.*, 238, 347

Baker, D., van Driel-Gesztelyi, L., Mandrini, C. H., Démoulin, P., & Murray, M. J. 2009, "*Astrophys. J.*", 705, 926

Biskamp, D. 2000, Magnetic Reconnection in Plasmas (Cambridge, UK: Cambridge University Press)

Browning, P. K., Gerrard, C., Hood, A. W., Kevis, R., & van der Linden, R. A. M. 2008, *Astron. Astrophys.*, 485, 837

Cargill, P., Parnell, C., Browning, P., de Moortel, I., & Hood, A. 2010, *Astronomy and Geophysics*, 51, 030000

Démoulin, P., Henoux, J. C., Priest, E. R., & Mandrini, C. H. 1996, *Astron. Astrophys.*, 308, 643

Dorelli, J. C. Bhattacharjee, A. 2008, *Physics of Plasmas*, 15, 056504

Fukao, S., Ugai, M., & Tsuda, T. 1975, *Report Ionosphere Space Research Japan*, 29, 133

Galsgaard, K., Archontis, V., Moreno-Insertis, F., & Hood, A. W. 2007, *Astrophys. J.*, 666, 516

Galsgaard, K. Nordlund, Å. 1996, *J. Geophys. Res.*, 101, 13445

Galsgaard, K. Nordlund, Å. 1997, *J. Geophys. Res.*, 102, 231

Hagenaar, H. J., Schrijver, C. J., & Title, A. M. 2003, *Astrophys. J.*, 584, 1107

Haynes, A. L. Parnell, C. E. 2010, *Physics of Plasmas*, 17, 092903

Haynes, A. L., Parnell, C. E., Galsgaard, K., & Priest, E. R. 2007, *Royal Society of London Proceedings Series A*, 463, 1097

Hesse, M. 1995, in Reviews in Modern Astronomy, ed. G. Klare, *Reviews in Modern Astronomy*, 8, 323

Hesse, M. Schindler, K. 1988, *J. Geophys. Res.*, 93, 5559

Hood, A. W., Browning, P. K., & van der Linden, R. A. M. 2009, *Astron. Astrophys.*, in press

Hornig, G. Priest, E. R. 2003, *Phys. Plasma*, 10, 2712

Lau, Y. Finn, J. M. 1990, *Astrophys. J.*, 350, 672

Longbottom, A. W., Rickard, G. J., Craig, I. J. D., & Sneyd, A. D. 1998, *Astrophys. J.*, 500, 471

Maclean, R. C., Parnell, C. E., & Galsgaard, K. 2009, *Solar Phys.*, 260, 299

Masson, S., Pariat, E., Aulanier, G., & Schrijver, C. J. 2009, *Astrophys. J.*, 700, 559

Pariat, E., Masson, S., & Aulanier, G. 2009, "*Astrophys. J.*", 701, 1911

Parker, E. N. 1991, in Mechanisms of Chromospheric and Coronal Heating, ed. P. Ulmschneider, E. R. Priest, & R. Rosner, 615

Parnell, C. E., DeForest, C. E., Hagenaar, H. J., *et al.* 2009, *Astrophys. J.*, 698, 75

Parnell, C. E., Deforest, C. E., Hagenaar, H. J., Lamb, D. A., & Welsch, B. T. 2008, in First Results From Hinode, eds. S. A. Matthews, J. M. Davis, & L. K. Harra, *Astronomical Society of the Pacific Conference Series*, 397, 31

Parnell, C. E., Haynes, A. L., Galsgaard, K. 2010a, *J. Geophys. Res. (Space Physics)*, 115, 2102

Parnell, C. E., Maclean, R. C., Haynes, A. L. 2010b, *Astrophys. J. Letts.*, 725, L214

Parnell, C. E., Smith, J. M., Neukirch, T., & Priest, E. R. 1996, *Physics of Plasmas*, 3, 759

Pontin, D. I. 2011, "*Adv. Space Res.*"

Pontin, D. I., Bhattacharjee, A., & Galsgaard, K. 2007, *Physics of Plasmas*, 14, 052106

Pontin, D. I. Craig, I. J. D. 2005, *Physics of Plasmas*, 12, 072112

Pontin, D. I., Hornig, G., & Priest, E. R. 2004, *Geophysical and Astrophysical Fluid Dynamics*, 98, 407

Pontin, D. I., Hornig, G., & Priest, E. R. 2005, *Geophys. Astrophys. Fluid Dynanics*, 99, 77

Pontin, D. I., Wilmot-Smith, A. L., Hornig, G., & Galsgaard, K. 2011, *Astron. Astrophys.*, 525, A57+

Priest, E. R. Démoulin, P. 1995, *J. Geophys. Res.*, 100, 23443

Priest, E. R. Forbes, T. G. 2000, *Magnetic reconnection* (Cambridge, UK: Cambridge University Press)

Priest, E. R., Hornig, G., & Pontin, D. I. 2003, *J. Geophys. Res.*, 108, 1285

Priest, E. R. Pontin, D. I. 2009, *Physics of Plasmas*, 16, 122101

Rickard, G. J. Titov, V. S. 1996, *Astrophys. J.*, 472, 840

Schindler, K., Hesse, M., & Birn, J. 1988, *J. Geophys. Res.*, 93, 5547

Thornton, L. M. Parnell, C. E. 2010, *Solar Phys.*, 220

Titov, V. S. 2007, *Astrophys. J.*, 660, 863

Titov, V. S., Forbes, T. G., Priest, E. R., Mikić, Z., & Linker, J. A. 2009, "*Astrophys. J.*", 693, 1029

Titov, V. S., Galsgaard, K., & Neukirch, T. 2003, *Astrophys. J.*, 582, 1172

Wilmot-Smith, A. L., Hornig, G., & Pontin, D. I. 2009, *Astrophys. J.*, 696, 1339

Wilmot-Smith, A. L., Pontin, D. I., & Hornig, G. 2010, *Astron. Astrophys.*, 516, A5+

Astrophysical Dynamics: From Stars to Galaxies
Proceedings IAU Symposium No. 271, 2010
N. H. Brummell, A. S. Brun, M. Miesch & Y. Ponty, eds.

© International Astronomical Union 2011
doi:10.1017/S1743921311017662

Competing kinematic dynamo mechanisms in rotating convection with shear

Michael R. E. Proctor[1] and David W. Hughes[2]

[1]Department of Applied Mathematics and Theoretical Physics,
University of Cambridge, Cambridge CB3 0WA, U.K.
email: mrep@cam.ac.uk

[2]Department of Applied Mathematics, University of Leeds, Leeds LS2 9JT, U.K.
email: d.w.hughes@leeds.ac.uk

Abstract. Following earlier work by Hughes & Proctor (2009) on the role of velocity shear in convectively driven dynamos, we present preliminary results on the nature of dynamo action due to modified flows derived by filtration from the full convective flow. The results suggest that filtering the flow fields has surprisingly little effect on the dynamo growth rates.

Keywords. convection, instabilities, magnetic fields, MHD, Sun: activity, Sun: magnetic fields, Sun: rotation

1. Introduction

The principal problem in the dynamo theory for cosmical bodies such as the Sun is to explain the occurrence of magnetic fields on scales much larger than those of the underlying motions responsible for the dynamo. Such large-scale fields are often studied within the framework of *mean field electrodynamics* (Moffatt 1978, Krause & Rädler 1980), but there are significant problems in applying the theory quantitatively when (as is the case in the Sun) the magnetic Reynolds numbers are large even on the smallest scales of motion (see Cattaneo & Hughes 2009). In that circumstance, dynamo action is predominantly of small-scale type, with the magnetic fields having scales no larger than those of the flow. It is problematical to understand how such processes are compatible with the large-scale fields that are certainly observed in nature.

An appropriate geometry in which to investigate the difference between large-scale and small-scale dynamo processes is thermal convection in a rotating plane layer. This problem has been investigated in the regime of large magnetic Reynolds numbers by Cattaneo & Hughes (2006) and Hughes & Cattaneo (2008). They found that at moderate rotation rates the dynamos produced were always of small-scale type, with no evidence of any significant mean emf produced by the flow in the presence of a large-scale magnetic field.

However, the Sun and other bodies possess a vigorous large-scale shear flow, and it is possible to think of a number of ways in which adding a velocity shear to the convection problem may enhance the generation of the large-scale components of the field. For example, a coherent large-scale shear may change the nature of the correlations and, in so doing, enhance the α-effect; alternatively, even if the mean emfs remain very small, then a large shear may be able to compensate for a feeble α-effect (or a more complicated mean-field process) to make a viable two-scale dynamo. Finally, there is the very different possibility that the interactions between the large-scale shear and the (initially) small-scale convective flow lead to a large-scale velocity that can, of itself, act as a dynamo;

this would then be effectively a small-scale (i.e. one-scale) dynamo, but on the scale of the shear flow rather than that of the convection.

This situation was investigated by Hughes & Proctor (2009); hereinafter HP. They found (see Figure 1 below) that adding a horizontal shear to the convection problem of Cattaneo and Hughes did indeed enhance dynamo action, in particular leading to a viable dynamo even when the convecting flow without shear was not of itself a dynamo. They also demonstrated that adding the shear did nothing to improve the coherence of the emfs produced from a large-scale magnetic field by the flow. However, from inspection of the solutions, it was not possible to resolve the question as to which of the above alternative scenarios was the one responsible for dynamo action.

In this paper we try to answer this question by comparing the (kinematic) dynamo properties of the actual convective flows with those of related flows obtained by filtration in Fourier space to eliminate all small scales, or alternatively all large scales apart from the shear itself. Remarkably, we find from the preliminary results presented here that neither filtration process has very much effect on the dynamo growth rate, and so the results can be used to support both of the alternative hypotheses above. Further work is currently in progress to clarify the situation.

The plan of the paper is as follows. In the next section we set up the problem and explain the filtration process. The results are presented in §3 and the paper concludes with a discussion.

2. Problem description

2.1. *Governing equations*

The underlying problem is that described in HP. We consider a plane Boussinesq convective layer ($0 < x, y < \lambda d$, $0 < z < d$) with rotation about the vertical axis, as described in Cattaneo & Hughes (2006). This basic model is then extended by the inclusion of a horizontal flow of the (dimensional) form

$$U_0 = U_0 f(y/d)\hat{x}, \text{ where } f(y/d) = \cos\frac{2\pi y}{\lambda d}, \tag{2.1}$$

accomplished by replacing u with $u+U_0$ in the governing equations except for the viscous term (equivalent to forcing the flow via the momentum equation, but eliminating viscous transients). For the purposes of this paper we shall restrict attention to kinematic dynamo action, so that the back-reaction of the Lorentz forces on the convection is neglected, as is appropriate for very weak fields. Then the governing non-dimensional equations for the velocity u, temperature perturbation θ and magnetic field B are

$$(\partial_t - \sigma\nabla^2)u + u \cdot \nabla u + \sigma(S(f(y)\partial_x u + f'(y)u_y\hat{x} + Ta^{1/2}\hat{z} \times u) = -\nabla p + \sigma Ra\,\theta\hat{z}, \tag{2.2}$$

$$(\partial_t - \zeta\nabla^2)B + u \cdot \nabla B + Sf(y)\partial_x B = B \cdot \nabla u + Sf'(y)B_y\hat{x}, \tag{2.3}$$

$$(\partial_t - \nabla^2)\theta + u \cdot \nabla\theta + Sf(y)\partial_x\theta = u \cdot \hat{z}, \tag{2.4}$$

$$\nabla \cdot B = \nabla \cdot u = 0. \tag{2.5}$$

It should be noted that although a flow with a large-scale component (i.e. with the same spatial dependence as the 'target flow' (2.1)) does indeed occur, its amplitude may differ appreciably from U_0; the hydrodynamic state that ensues depends on interactions between the shear flow and convection and, possibly, on instabilities of the shear flow itself. Importantly, the scale of variation of this shear flow is much greater than all scales

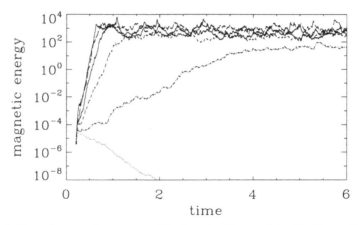

Figure 1. Evolution of magnetic energy with time for the model with the unfiltered flow and with the Lorentz force included, for various values of S (from HP). The non-dynamo case is $S = 1/3$. The case $S = 5/3$ considered here is third from bottom.

of the convection; this is essential if the results are to be explained within the mean field framework.

The solution depends on the following dimensionless parameters: the Rayleigh number $Ra = g\tilde{\alpha}\Delta T d^3/\kappa\nu$, proportional to the temperature difference across the layer, the Prandtl number $\sigma = \nu/\kappa$, the diffusivity ratio $\zeta = \eta/\kappa$, the Taylor number $Ta = 4\Omega^2 d^4/\nu^2$ and shear parameter $S = 2\pi U_0 \ell/d\lambda u_{\mathrm{rms}}$, where d is the layer depth, ΔT the applied temperature difference, g the gravitational acceleration, $\tilde{\alpha}$ the coefficient of thermal expansion, and ν, κ, η are respectively the kinematic viscosity, thermal conductivity and magnetic diffusivity; ℓ and u_{rms} are estimates, respectively, of the horizontal scale of the convection and of the typical velocity in the absence of shear. For the particular problem below we take $S = dU_0/300\kappa$. At $z = 0, 1$ we take $\theta = u_z = \partial_z u_{x,y} = B_z = \partial_z B_{x,y} = 0$, corresponding to fixed temperature, stress-free and perfectly conducting boundaries. We employ periodic boundary conditions in the horizontal directions.

We solve the system as an initial value problem for the convection, starting from the state of pure shear, with small random perturbations, and integrate in time until a statistically steady state is reached for the flow. We then add a weak seed magnetic field with no net flux, which will, on average, grow or decay exponentially.

In HP we investigated the effect of varying S in the case $\lambda = 5$, $Ra = 1.5 \times 10^5$, $Ta = 5 \times 10^5$, $\sigma = 1$, $\zeta = 0.2$. These parameters give vigorous convection, but there is no dynamo when $S = 0$. Figure 1 shows how the dynamo growth rate increases with increasing S. The figure also shows the equilibration of the dynamo due to Lorentz force back reaction: this is neglected in the present paper, so that according to (2.2)–(2.5), exponential growth continues indefinitely.

While it is certainly sufficient for the neglect of finite domain effects to take $\lambda = 5$ when $S = 0$ (see Hughes & Cattaneo 2008), in the presence of shear some flow structures become elongated, as shown in Figure 2 below. It remains a possibility that some of the results both of HP and the present work will be modified to some degree if computations are performed in wider periodic domains.

2.2. *Flow filtration*

Our aim in the present work is to learn more about the role of velocity shear in promoting dynamo action and the mechanism responsible for the enhancement of the growth rate. As discussed in the Introduction, the answer to such questions depends on what scales of the flow, apart from the shear, are significant in providing the interaction between the shear and the convection that powers the dynamo. Accordingly, we investigate the dynamo properties of flow fields obtained from the convective solution by a process of filtration in wavenumber space. Any Fourier mode obeying the boundary conditions has a part proportional to $\exp \pm i(\pi k_z z + 2\pi \lambda^{-1}(k_x x + k_y y))$. If we denote a cut-off wave number by k_{cut} then the filtration takes one of the following forms:
(a) short wavelength (SW) cutoff: set to zero the amplitudes of all modes for which $k = \max(|k_x|, |k_y|) > k_{\text{cut}}$;
(b) long wavelength (LW) cutoff: set to zero the amplitudes of all modes for which $k = \min(|k_x|, |k_y|) < k_{\text{cut}}$, *but retain the mode* $(0,1)$ *corresponding to the shear.*

To date we have performed no filtration on the vertical spectrum. This will be done in future work, though we do not consider it likely that truncating the vertical spectrum will have a significant effect.

Whatever the filtration adopted, the procedure is as follows:
(1) Solve the momentum and heat equations at full resolution;
(2) At each time step perform the filtering to produce a filtered velocity $\boldsymbol{u}_f$ together with the shear;
(3) Solve the induction equation (2.3) at full resolution with $\boldsymbol{u}$ replaced by $\boldsymbol{u}_f$.

3. Results

Our preliminary results are all for the case $S = 5/3$, which we believe to be typical among those for which there is a dynamo. We first show the effects of filtration on the flow. Figure 2 shows a contour plot of the vertical velocity distribution near the top of the layer, with an SW cutoff and $k_{\text{cut}} = 30$. The residual is shown in the right hand panel. There is little effect on the flow when k_{cut} is as large as this. Figure 3 shows the effect of

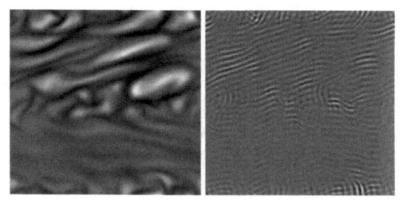

Figure 2. Left panel: Contour plot of vertical velocity near the top of the layer for a snapshot of an SW cutoff flow with $k_{\text{cut}} = 30$. Right panel: the residual difference between the full and filtered flows.

changing the cutoff. By the time the cutoff wave number has been reduced to $k_{\text{cut}} = 5$ the flow has been significantly altered. The (normalised) spectra of the magnetic fields

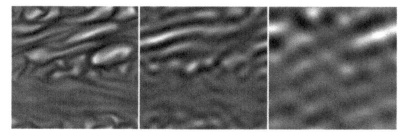

Figure 3. Effect of different SW cutoffs, for the same flow as in Figure 2. From left to right: $k_{\mathrm{cut}} = 30, 10, 5$. All plots are individually scaled.

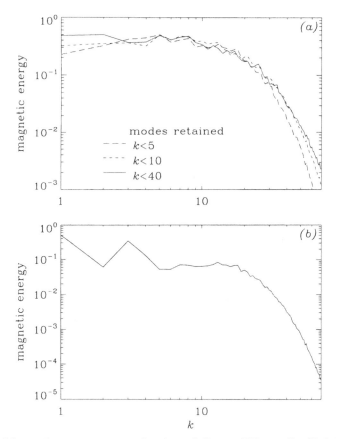

Figure 4. (*a*) Magnetic energy spectra for three different SW cutoffs. (*b*) Magnetic energy spectrum for an LW cutoff with $k_{\mathrm{cut}} = 10$.

generated by the filtered flows do not however seem to be much affected by the filtration. In Figure 4 are shown magnetic energy spectra in the interior 80% of the layer for several filtrations. For the SW cutoff (Figure 4(*a*)) there is scarcely any effect on the energy in the modes with $5 < k < 20$. Figure 4(*b*) shows a similar magnetic energy spectrum for a flow with an LW cutoff. Again there seems to be little effect of the filtration except at the very largest scales.

Because of the remarkable similarities in the spectra it might then be anticipated that the growth rates for the different filtrations would not be too dissimilar. In fact they are

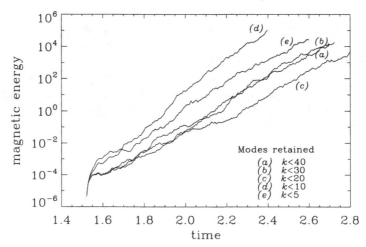

Figure 5. Time traces of magnetic energy for SW filtered flows with different values of k_{cut}.

hardly different at all. This is illustrated in Figures 5 and 6, which show the growth of magnetic energy for various filtrations. The first of these compares different SW cutoffs. It would seem that removing the high wave numbers makes very little difference to the growth of the instability — though the most severe filtration does seem to lead to a greater growth rate. (The only other significant effect is that for the highly filtered flows, the system takes a little longer to settle into its ultimate exponential growth phase.) These results might lead one to speculate that the dynamo is 'small-scale on the large scale'; so that only the largest scales of motion, those comparable with the field scales, are responsible for the instability.

One's confidence in the truth of this speculation does not however survive examination of Figure 6, which shows a comparison between the growth rates for the unfiltered flow and for an LW filtration with $k_{\mathrm{cut}} = 10$. Amazingly, this flow also gives almost the same growth rate as the SW filtrations shown in Figure 5. Looking at just this figure, one might conclude that the dynamo is in fact of mean field type, since there is now a significant separation of scales in the filtered flow between the shear and what is left of the convection! Clearly more work needs to be done to distinguish between the two explanations offered above, or indeed to synthesise them.

4. Discussion

This paper has presented some preliminary calculations on the effects of different scales of flow on the growth rate of the dynamo instability due to rotating convection with shear. The aim of these calculations has been to ascertain just how the dynamo is being driven. On the basis of the results of HP on the unfiltered flow, we may rule out the possibility that the shear improves the coherence of the cyclonic aspects of the convection and thus produces an enhanced α-effect. The remaining possibilities are that of a two-scale, $\alpha\Omega$ type dynamo with a strong 'Ω' and weak 'α', (or a two-scale dynamo of more complicated type) or, conversely, of a dynamo with no scale separation controlled by the flows with scales larger than the convection. Our results would appear to support either scenario!

Clearly more work has to be done to resolve the question. We certainly need to investigate larger values of the aspect ratio λ since there is some evidence that horizontal scales of convection are approaching the current box size when shear is significant. Then more

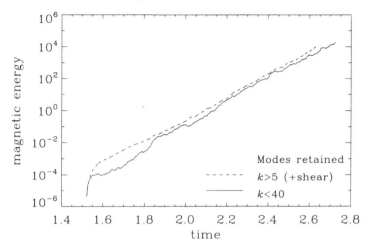

Figure 6. Increase of magnetic energy with time for (essentially) the original flow and for an LW filtration with $k_{\text{cut}} = 5$.

careful filtrations at different values of S are needed in order to pin down precisely which scales are really important, or whether in fact the shear controls everything and the observed growth rates are directly related to the shear rate, whatever the nature of the additional convective flow (this would of course reconcile the results shown in Figures 5 and 6). Investigation of the problem without rotation will lead to further insights as to whether the nature of the dynamo is two-scale (where rotation is likely to be important) or small-scale (where rotation is unimportant). It will also be of importance, particularly in view of the astrophysical implications, to repeat the calculations here and in HP for higher values of the magnetic Reynolds number, for which there is a small-scale dynamo in the absence of shear, in order to examine the role of shear in that case.

Acknowledgements

We are grateful for support from the UK STFC. The paper was prepared while the authors were invited participants in the ISIMA programme at the University of California, Santa Cruz. We are grateful to the organisers for making our attendance possible and providing an environment conducive to research.

References

Cattaneo, F. & Hughes, D. W. 2006, *J. Fluid Mech.*, 553, 401
Krause, F. & Rädler, K.-H. 1980, *Mean Field Magnetohydrodynamics and Dynamo Theory.* New York: Pergamon Press
Hughes, D. W. & Cattaneo, F. 2008, *J. Fluid Mech.*, 594, 445
Hughes, D. W. & Proctor, M. R. E. 2009, *Phys. Rev. Lett.*, 102:044501
Moffatt, H. K. 1978, *Magnetic Field Generation in Electrically Conducting Fluids.* Cambridge: University Press

Questions

Steven Tobias: When filtering the equations, which scales are important? Are these scales dependent on Rm; what is the value of Rm?

We find that paradoxically both large and small scale parts of the flow seem to be important in determining the growth rate. Rm depends on the shear rate, but for the underlying convection is of the order of 300.

Juri Toomre: What is the role of reintroducing structures by the shear since your x direction is periodic?

We have not investigated other boundary condition in x. Periodic boundaries are appropriate if the shear is taken to mimic zonal flows in the Sun, for example.

Geoff Vasil: How do you think that the critical shear would need to scale as a function of the Rossby number (for a fixed Ra/Ra_c)?

Our feeling is that the shear should be comparable with the rotation rate for a significant change in the growth rate.

Keith Moffatt: It seems to me very likely that this is an $\alpha\Omega$ dynamo. Do your results exclude this interpretation or not?

One of our filtrations gives support to the mean field hypothesis, while the other (with all small scales removed) is just as good a dynamo! So the jury is out.

Astrophysical Dynamics: From Stars to Galaxies
Proceedings IAU Symposium No. 271, 2010
N. H. Brummell, A. S. Brun, M. S. Miesch & Y. Ponty, eds.

© International Astronomical Union 2011
doi:10.1017/S1743921311017674

Solar and Stellar Dynamos

Nigel O. Weiss

Department of Applied Mathematics and Theoretical Physics,
University of Cambridge, Cambridge CB3 0WA, U.K.

Abstract. Records of the solar magnetic field extend back for millennia, and its surface properties have been observed for centuries, while helioseismology has recently revealed the Sun's internal rotation and the presence of a tachocline. Dynamo theory has developed to explain these observations, first with idealized models based on mean-field electrodynamics and, more recently, by direct numerical simulation, notably with the ASH code at Boulder. These results, which suggest that cyclic activity relies on the presence of the tachocline, and that its modulation is chaotic (rather than stochastic), will be critically reviewed. Similar theoretical approaches have been followed in order to explain the magnetic properties of other main-sequence stars, whose fields can be mapped by Zeeman-Doppler imaging. Of particular interest is the behaviour of fully convective, low-mass stars, which lack any tachocline but are nevertheless extremely active.

Keywords. magnetic fields, MHD, Sun: activity, Sun: magnetic fields, Sun: rotation, sunspots, stars: activity, stars: magnetic fields, stars: spots

1. Introduction

Half a century has passed since Parker (1955), Babcock (1961) and their predecessors set out the basic principles of the solar dynamo, and identified the key roles of differential rotation and cyclonic eddies. It is now accepted that turbulent convection in a star is able not only to amplify any pre-existing field but also to maintain both small-scale and large-scale fields indefinitely against ohmic decay. Mean field electrodynamics has been developed as a sophisticated theory, while space and ground-based observations have revealed the structure of solar and stellar magnetic fields in ever-increasing detail. Yet there is still no fully satisfactory model of the solar dynamo: mean-field dynamo theory has to rely on plausible parametrizations of the key effects, while computational models – despite massive efforts in which Juri Toomre has played a leading part– are still not able to reproduce the essential features of the solar cycle.

In this review I shall therefore consider general properties of solar and stellar dynamos and avoid detailed descriptions of specific models. First I shall summarize the main physical processes that are involved. Although it is generally accepted that strong toroidal fields are formed and stored near the base of the convection zone, where there is a steep radial gradient of angular velocity in the tachocline, there are two competing mean-field scenarios. Interface dynamos have all the action concentrated near the tachocline, but in flux-transport dynamos the poloidal fields are formed near the surface and transported down by a meridional circulation. I shall contrast what we can deduce from mean-field models with what we hope to learn from direct numerical simulations. The next section focuses on the modulation of cyclic activity in stars like the Sun, a subject that has become topical owing to the feeble start of the latest cycle, which raises the question of whether solar activity might be heading for a grand minimum like that in the seventeenth

century. Section 4 is devoted to dynamos in other stars, and the final section looks ahead
to the future.

2. Physics of the solar dynamo

The challenge to dynamo models is to explain the large-scale systematic features of
the solar cycle: the emergence of isolated, predominantly toroidal fields (antisymmetric
about the equator) in sunspots and active regions, the equatorward movement of the
activity zones, the reversal of polarities from one 11-year activity cycle to the next, and
the development of a weak polar field (see, for example, Solanki, Inhester & Schüssler
2006; Thomas & Weiss 2008).

2.1. *Relevant processes*

The generation of magnetic fields is ultimately governed by hydrodynamic processes in
stellar convection zones. The Sun's radiative interior is surrounded by a convection zone
that occupies the outer 30% by radius; in most of this region the angular velocity Ω is
constant on conical surfaces and there is a balance between Coriolis and baroclinic effects
that satisfies the thermal wind equation (Balbus *et al.* 2009). At the base of the convection
zone there is an abrupt transition in the tachocline (Hughes, Rosner & Weiss 2007) to
the almost uniformly rotating radiative zone, as shown in Figure 1. Superimposed on
this pattern are zonal shear flows ("torsional oscillations") that vary with the solar cycle,
appearing as a branch of enhanced rotational velocity that coincides with the activity
belt, together with a subsidiary poleward branch, as depicted in Figure 2 (Howe 2009).
As expected for a flow driven by the quadratic Lorentz force, the torsional oscillations
have an 11-year period, half that of the underlying magnetic cycle. (It is still worth

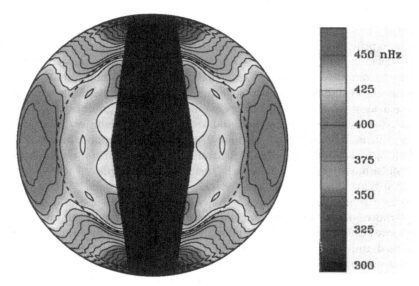

Figure 1. Solar differential rotation, as determined from the rotational splitting of acoustic
p-modes (Thompson *et al.* 2003; Christensen-Dalsgaard & Thompson 2007). The dashed line
denotes the base of the convection zone at a radius $r = 0.713R_\odot$. Near the surface the rotation
frequency $\Omega/2\pi$ is greatest at the equator and least at the poles; in the radiative zone Ω is
essentially uniform. The slender tachocline lies below the convection zone at the equator, with
an estimated thickness of only $0.03R_\odot$ but is slightly prolate and thicker at the poles. (Courtesy
of J. Christensen-Dalsgaard.)

emphasizing that any "oscillator" theory requires torsional oscillations with a 22-year period, which have never been seen. If the steady dipolar field in the radiative zone does protrude into the region where the dynamo operates, its only effect is to bias the cyclic field slightly towards one polarity – as appears to be the case (Boruta 1996).)

Given an initial dipolar field, such a pattern of differential rotation can readily stretch field lines to generate a toroidal field that is antisymmetric about the equator, and this process (the ω-effect of mean-field dynamo theory) is most effective at the tachocline. In a turbulent region, large-scale fields are expelled down the gradient of turbulent intensity, an effect that is enhanced at the base of the convection zone, owing to downward pumping of magnetic flux by the vigorous sinking plumes (Tobias *et al.* 2001; Weiss *et al.* 2004). Hence we expect the strong toroidal fields to be stored in the region immediately below the convection zone itself, with typical strengths of up to 10^4 G, corresponding to equipartition with the turbulent motion, although fields may be locally more intense. This is the most that differential rotation can create within the tachocline itself. Such fields will be liable to instabilities driven by magnetic buoyancy (Hughes 2007), which allow loops to enter the convection zone, where they can acted upon by cyclonic motions, as indicated in Figure 3(a). If the toroidal field were weak then different loops might be twisted through arbitrary angles, as sketched in Figure 3(b), so that their combined effects would cancel out – as demonstrated in the numerical calculations of Cattaneo & Hughes (2006, 2008). With a strong toroidal field, the loops are only slightly twisted and their contributions can combine constructively, as Parker originally suggested. It is this process that is parametrized by the α-effect in mean-field dynamo models. Field strengths will of course vary and may locally be much higher than the average. Some flux tubea, with fields an order of magnitude greater ($\sim 10^5$ G) can break out of the

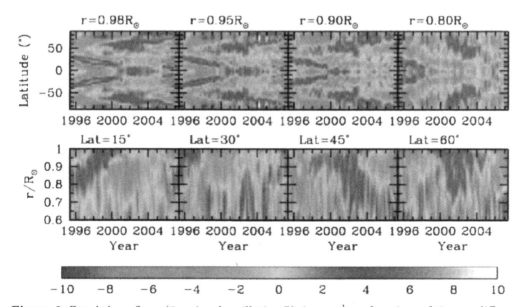

Figure 2. Zonal shear flows ("torsional oscillations"), in m s^{-1}, as functions of time at different latitudes and depths in the solar convection zone. The band of positive velocity tracks the sunspot zones at low latitudes but there is also a strong poleward branch. At low latitudes the shear flow first appears near the base of the convection zone. (From Antia, Basu & Chitre 2008, using GONG data.)

magnetic layer and, aided by magnetic buoyancy, rise unimpeded by Coriolis forces, to erupt as active regions or as sunspots at the surface.

This dynamo process relies, of course, on some form of diffusion that allows the individual loops to reconnect and so to generate a reversed poloidal field. However, the same diffusion leads to dissipation of magnetic energy and the critical question is whether the constructive effects of stretching and reconnection can overcome magnetic dissipation. The actual magnetic diffusivity within a star is very small (though much greater than the viscous diffusivity) and turbulent processes are intrinsically complicated (Tobias, Cattaneo & Boldyrev 2011). All models therefore rely on some parametrized description of turbulent diffusion, which may or may not be realistic. The least that one can require is that diffusion should be explicit, in the hope that its effects on reconnection and dissipation might be estimated and compared. This is essential, for there is a long history of over-optimistic model calculations which exhibited dynamo action that subsequently faded as the numerical procedure was refined.

Doppler measurements have revealed a poleward meridional flow in both hemispheres at the solar surface, with a peak speed of about $20 \mathrm{~m\,s}^{-1}$. Helioseismology has confirmed that this motion persists downwards for at least 15 Mm but below that depth its magnitude and direction become increasingly uncertain. If there is an equatorward flow of $1 \mathrm{~m\,s}^{-1}$ at the base of the convection zone, as has plausibly been suggested, it could significantly affect the latitudinal drift of activity with the solar cycle. In fact, the surface motion does itself vary with the activity cycle too (Basu & Antia 2010). Moreover, the latitudinal component may reverse more than once within the convection zone.

2.2. Location of the dynamo

Although it is widely accepted that the ω-effect is concentrated around the tachocline, where radial and latitudinal gradients of Ω have comparable effects, there are several competing locations for the α-effect (Rüdiger & Hollerbach 2004; Charbonneau 2005). In *interface* dynamos (Parker 1993), the strong toroidal fields become locally unstable to modes driven by magnetic buoyancy and, in the nonlinear regime, Coriolis effects eventually lead to the formation of a reversed poloidal field (Tobias & Weiss 2007; Jones, Thompson & Tobias 2010)i, as outlined above. In this picture, all the action is at the base of the convection zone and the flux loops that make their way to the surface and give

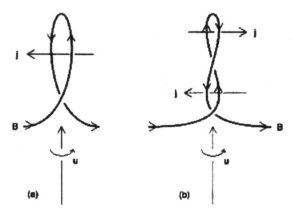

Figure 3. Sketches showing (a) twisting of a toroidal field by a cyclonic eddy to give a poloidal component, as envisaged by Parker, and (b) the consequences of repeated twisting if the toroidal field is too weak to limit the process. (From Jones, Thompson & Tobias 2010.)

rise to sunspots are just by-products, or epiphenomena (Cowling 1975), of the dynamo process.

An alternative (currently much favoured in the USA) is to assert that photospheric fields and meridional flows do indeed play an essential role in the process, as in *flux transport* dynamos. Those flux loops that emerge at low latitudes must have had 10^5 G fields in order to survive and they are observed to be slightly tilted with respect to lines of latitude, as expected from the cyclonic effect of the Coriolis force. Presumably the underlying flux tubes eventually detach themselves from the deep-seated toroidal field. As active regions decay, their fields are acted upon by the diffusive effect of supergranular convection and the erratic poleward meridional motion in such a way that reversed poloidal flux accumulates at high latitudes, whence it can be transported to the base of the convection zone by the meridional circulation.

The attraction of the flux transport model is that it relates the global dynamo to observable fields at the surface. This was a very plausible assumption when it was thought that the convection zone was shallow (Leighton 1969) but we now know that it is deep. I remain of the opinion that interface models are more reliable. It is hard to imagine how a weak poloidal field can be transported intact down to the tachocline without being tangled up or even destroyed by turbulent convection on the way; indeed, radial diffusion may be more effective than meridional transport. On the other hand, the weak polar fields, which reverse at sunspot maximum, do indeed appear to be a global feature, for they supply reliable predictions of the following activity maximum. Are they just accumulated rubble from the preceding cycle, or are they generated much deeper down?

2.3. *Modelling stellar dynamos*

Up to now, nearly all numerical models have been axisymmetric and relied on mean-field formalisms, with processes that are arbitrarily parametrized. Field growth has to be limited by some nonlinear effect and many variants have been tried. In the example illustrated in Figure 4, the field drives an azimuthal flow that corresponds to the observed zonal shear flows. By varying parameters and adding more effects both interface and flux

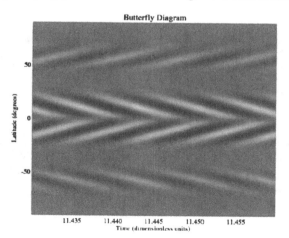

Figure 4. Butterfly diagram for a highly idealized mean-field dynamo, with both the α-effect and the ω-effect concentrated around the interface between the radiative and convective zones. Growth is limited by the nonlinear effect of the Lorentz force on differential rotation, giving rise to zonal shear flows such as are observed. Shown here is the toroidal field near the base of the convection zone, which is antisymmetric about the equator. The corresponding zonal shear flows have twice the frequency of the magnetic cycle. (From Bushby 2005.)

transport dynamo models can be adjusted to reproduce the principal observed features of the solar cycle, and many examples are available. It follows, however, that such models, though certainly instructive, are only illustrative and can have no detailed predictive power.

The obvious alternative is direct numerical simulation. The earliest attempt, by Gilman (1983) already revealed some of the difficulties: the behaviour of a nonlinear dynamo – whether fields have dipole or quadrupole symmetry, whether they are steady or cyclic, whether waves progress towards or away from the equator – is very sensitive to the parameter values that are assumed and small changes may have big effects. Although the geodynamo has been convincingly modelled (Glatzmaier and Roberts 1995; Christensen, Schmitt & Rempel 2009) there is as yet no fully convincing computational representation of the solar cycle. The greatest efforts have been put into the anelastic ASH code developed by Juri Toomre and his associates at Boulder (Miesch & Toomre 2009). The first requirement was to match the helioseismic measurements of differential rotation; that was solved by imposing a latitudinal entropy gradient at the tachocline (Miesch, Brun & Toomre 2006; Miesch *et al.* 2008). When a seed field was inserted, dynamo action was indeed found, yielding an antisymmetric toroidal field at the base of the convection zone, as shown in Figure 5 (Browning *et al.* 2006; Miesch & Toomre 2009). However, this large-scale field did not reverse and there was no indication of cyclic behaviour.

The ASH code introduces a laminar viscosity and resistivity, with a ratio (the magnetic Prandtl number) that is of order unity, as expected for turbulent diffusion. The actual coefficients are as small as accuracy allows, and get reduced as the numerical resolution is refined, in the hope that large-scale behaviour will eventually become independent of their precise numerical values. In a parallel calculation, with much lower resolution and a different numerical procedure, Ghizaru, Charbonneau & Smolarkiewicz (2010) have recently reported cyclic behaviour. The differential rotation in their model does not resemble that in the Sun. and the reversing toroidal fields are all at high latitudes. Moreover, both viscous and magnetic diffusion rely on the numerical scheme – a procedure that can lead to spurious results. There is an analogy here with studies of dynamo action

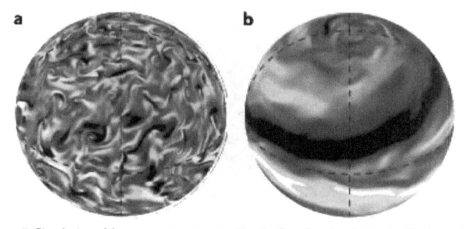

Figure 5. Simulation of dynamo action in a star like the Sun. Results obtained with the anelastic ASH code and a realistic representation of the actual differential rotation. The azimuthal field B_ϕ is shown, with positive and negative values denoted by orange-yellow and blue-black tones respectively. Snapshots (a) in the middle of the convection zone, with small-scale magnetic structures only, and (b) in the tachocline, with a dominant large-scale antisymmetric field. (From Miesch & Toomre 2009.)

driven by the magnetorotational instability in accretion discs, where numerical diffusion produces behaviour that depends on the resolution, and it is necessary to introduce laminar diffusivities in order to obtain meaningful results (Fromang & Papaloizou 2007; Fromang *et al.* 2007).

Small-scale convection in a plane layer, whether incompressible or stratified, stationary or rotating, is capable of maintaining magnetic fields if the magnetic Reynolds number is sufficiently large (Cattaneo 1999; Cattaneo & Hughes 2006; Vögler & Schüssler 2007; see Bushby in these Proceedings). In a star like the Sun, we should expect small-scale dynamo action at different depth-dependent scales, and fields produced on any of these scales, as well as any large-scale field, will be processed and amplified on all other scales. Hence it is not surprising that we now see magnetic flux emerging on all scales at the photosphere, as described by Title elsewhere in these Proceedings. In modelling large-scale dynamos it is implicitly assumed that the overall pattern is not critically dependent on details of the coexisting small-scale dynamos.

3. Modulation of cyclic activity

3.1. *Observations and predictions*

The level of activity varies from one activity cycle to the next, and the feeble start to the latest Cycle 24 has made such variability more topical; estimates of the amplitude of the new cycle keep falling as it progresses. Predictions of future behaviour depend critically on the length of the dataset that is used. Just as tomorrow's weather is likely to resemble today's, so the next cycle may resemble the last, or else the average of the last few cycles. A longer record shows irregular modulation on a timescale of a century, and then there is the Maunder Minimum of th seventeenth century, when sunspots almost completely disappeared (Weiss 2010).

To proceed further, we can use the proxy record supplied by the abundances of cosmogenic isotopes (^{12}C and ^{10}Be) that are created by galactic cosmic ray particles impinging on the Earth's atmosphere. Since these particles are deflected by magnetic fields in the heliosphere, the production rates of ^{12}C and ^{10}Be vary in antiphase with the solar cycle. The time-series derived from ^{10}Be abundances in polar ice cores provides a record of solar activity extending back for 9500 years that can be combined with direct measurements of neutron fluxes over the past few decades (Abreu *et al.* 2008). Figure 6 shows the last 2000 years of this record, smoothed to eliminate the basic solar cycle. It can be seen that solar activity has been unusually high for most of the last century, but that similar episodes have recurred aperiodically and some were more extreme. Taken over the entire record, values of Φ are normally distributed and we can define grand maxima and grand minima by setting levels that are only exceeded for 20% of the total time. Power spectra yield robust peaks corresponding to periods of 205 and 2300 years in both the ^{12}C and ^{10}Be records (Damon & Sonett 1991; Stuiver & Braziunas 1993; Beer 2000; Wagner *et al.* 2001).

Based on this record, we can adopt a statistical approach and estimate the life expectancy of the recent grand maximum, which has already lasted for more than 80 years, making it the third longest in the entire record. Fitting a gamma distribution to the lifetimes of the 66 grand maxima in the smoothed record, we find a further life expectancy of only 15 years. From the record there is then a 40% chance that activity will collapse into a grand minimum (Abreu *et al.* 2008; Weiss 2010). If so, we may expect a slight cooling effect on the climate, though not enough to cancel out the global warming caused by anthropogenic greenhouse gases.

3.2. *Origins of aperiodic modulation*

What causes this irregular modulation? Is it a purely stochastic effect, or is it an example of deterministic chaos? Economists have long regarded the solar cycle as an example of stochastic behaviour, and Barnes, Sargent & Tryon (1980) showed that an Autoregressive Moving Average (ARMA) model (constructed by combining a three-term recurrence relation with filtered Gaussian noise) provided a very plausible representation of solar activity (see also Brajša *et al.* 2009). If we seek a more causal description, we may suppose that the cyclic dynamo is disturbed by fluctuations in the underlying hydrodynamics of the convection zone, which we can regard as random. These might, for example, be variations in the meridional circulation, as suggested by Charbonneau & Dikpati (2000; Dikpati *et al.* 2010). The only snag is that these do have to be very substantial disturbances. Choudhuri & Karak (2010) recently used a mean-field dynamo model to simulate behaviour during the Maunder Minimum – but they did have to introduce an order-one perturbation to the system in order to achieve this. An interesting possibility is that there is hysteresis in the dynamo, as suggested by Schmitt, Schüssler & Ferriz-Mas (1996). They considered a dynamo that would only operate when the toroidal field B_T exceeds some critical value that allows instabilities driven by magnetic buoyancy to free flux tubes that are then acted on by Coriolis forces to form a poloidal field. If a stochastic perturbation causes B_T to drop below its critical value then the dynamo will switch off until another perturbation from some weaker dynamo process can switch it on again.

The alternative is that the modulation is deterministic. In the simplest case, the fluctuations may themselves be the product of an independent deterministic process – for example, that governing variations in the meridional circulation – and give rise to on-off

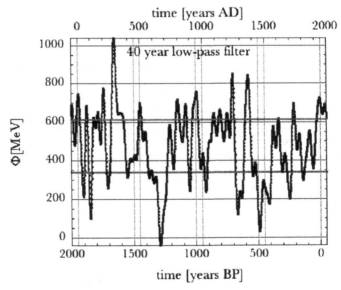

Figure 6. Variation of the solar modulation function Φ, a measure of solar activity derived from a composite record of ^{10}Be abundances in the Greenland GRIP ice core and a South Pole ice core and recent direct measurements, over the past 2000 years. The average value of Φ over the total period is 479 MeV and grand maxima and minima lie outside the horizontal lines. The recent grand maximum is now coming to an end, and both the Dalton minimum (around 1810) and the Maunder Minimum (1645–1710) are clearly visible, along with many similar episodes. Note the extremely high maximum around 300 AD, the longest in this record, which lasted for 95 years. (After Abreu *et al.* 2008.)

intermittency (Spiegel 1994). In the simplest models of deterministic switching, the amplitude of a nonlinear oscillation is controlled by the output of an independent chaotic oscillator (Platt, Spiegel & Tresser 1993). A more likely situation is that hydrodynamic processes are themselves influenced by the magnetic field, giving rise to in-out intermittency. The resulting bifurcation structure, illustrated in Figure 7, can be identified in low-order systems of nonlinear ordinary differential equations, with properties that are generic and therefore robust (Tobias, Weiss & Kirk 1995; Weiss 2010). Similar patterns appear in various mean-field dynamo models with differing nonlinearities. What appears essential for chaos is that the nonlinear coupling should introduce a time delay into the system, as first suggested by Yoshimura (1978). This is shown explicitly in the model of Jouve, Proctor & Lesur (2010), which is described elsewhere in these Proceedings.

A well-known feature of such systems is that a "ghost" of the torus survives after it is destroyed, giving rise to a peak in the power spectrum within the chaotic regime (Ott 1993; Tobias, Weiss & Kirk 1995). If modulation is stochastic, on the other hand, the spectrum should not contain such peaks; in particular, fluctuations in the meridional circulation are unlikely to generate periods longer than a few decades. The well-attested 205-year periodicity in the records of cosmogenic isotope abundances strongly suggests that the measured modulation of solar activity is indeed a product of chaotic modulation in a deterministic system.

4. Other stars

4.1. *Cool stars with deep convection zones*

Magnetic activity in lower main-sequence stars is strongly correlated with their normalized rotation rates, as measured by an inverse Rossby number (Donati & Landstreet 2009), as explained by Donati in these Proceedings. It is convenient to separate sun-like stars (spectral types F, G and K) from stars that are fully or almost fully convective (late M stars) and then to consider early-type stars separately.

There is a small population of slowly rotating stars like the Sun, with ages greater than 2–3 Gyr, that have been studied intensively over an interval of about 40 years and show similar cyclic activity (Baliunas *et al.* 1995; Baliunas, Sokoloff & Soon 1996). The cycle period decreases the more rapidly the star rotates; for a star of given mass, the cycle period $P \propto \Omega^{-(1+q)}$, where estimates of q vary from 0.25 to 1.0 (Noyes, Vaughan & Weiss 1984; Ossendrijver 1997; Saar & Brandenburg 1999; Saar 2002). These stars can

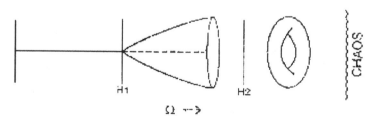

Figure 7. Bifurcation structure for transitions to chaos in large-scale magnetic activity. Although a star like the Sun actually spins down as it ages (Mestel 1999) it is appropriate to consider the effect of increasing its rotation velocity Ω from a value too low to allow the dynamo to operate. Then there are two successive Hopf bifurcations, the first leading to periodic cycles and the second to periodically modulated cycles; trajectories in phase space are attracted successively to a fixed point, to a limit cycle and to a two-torus. Then there are a series of bifurcations, involving frequency-locking and period-doubling, that lead to chaotically modulated cycles with a strange attractor in phase space.. (After Tobias *et al.* 1995.)

all be expected to possess tachoclines and to host dynamos like that in the Sun. It is to be expected that there are also other similar stars that are currently passing through grand minima and therefore relatively inactive.

Younger stars rotate much more rapidly and are correspondingly much more active, though activity appears to saturate for rotation periods less than a day. Here again, long-term measurements have revealed multiple cyclic behaviour but the corresponding periods do not follow the same pattern as the slow rotators. A key feature of these stars is the prevalence of huge spots, revealed by Zeeman-Doppler imaging, that cluster around the poles rather than at low latitudes as in the Sun (Thomas & Weiss 2008).

The Coriolis force becomes increasingly more important the more rapidly the star rotates. It is not to be expected that the pattern of rotation found in the Sun and shown in Figure 1 will apply also to these rapid rotators, nor do they need to possess a tachocline. Naively, we should expect rotation to be dominated by the Proudman-Taylor constraint, so that Ω is constant on cylindrical surfaces, at least outside the tangent cyclinder that encloses the radiative zone. This structure will profoundly affect the stellar dynamo (Bushby 2003). The Boulder group have computed the convection pattern in a star rotating at three times the solar rate and attempted to model the resulting dynamo (Brown *et al.* 2008, 2010). They find that the magnetic fields form persistent wreathlike structures, with predominantly azimuthal fields that are antisymmetric about the equator, as shown in Figure 8. Further calculations, with rotation at up to $10\Omega_\odot$ are discussed by Brown in these Proceedings.

4.2. *Fully convective stars*

The presence of a tachocline, or of a radiative core, is not an essential prerequisite for the existence of a dynamo. Indeed, Dobler, Stix & Brandenburg (2006) demonstrated dynamo action in a particular model of a fully convective star. In fact, stars much smaller than the Sun, of spectral class M, have very deep convection zones and, for masses less than 0.35 $M_\odot$, are fully convective. Yet these stars are known to be magnetically very active. Zeeman-Doppler imaging of 16 M-dwarfs reveals three different patterns of magnetic activity (Donati *et al.* 2008; Morin *et al.* 2008, 2010). Stars with masses in the range $0.8 > M/M_\odot > 0.5$ have significant radiative cores and exhibit weak, non-axisymmetric fields with strong azimuthal components and an average strength of a few hundred gauss. A star with mass less than 0.5 $M_\odot$ is effectively fully convective. Those stars in the range $0.5 > M/M_\odot > 0.2$ show very strong poloidal fields that are predominantly dipolar

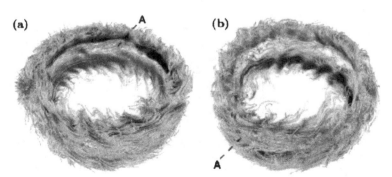

Figure 8. Magnetic wreaths in a model of the field in a rapidly rotating Sun. Colour denotes the strength and orientation of the predominant azimuthal field, and individual field lines are indicated. The two images show the two wreaths viewed from above and below the equator. The line A passes across the equator. (From Brown *et al.* 2010.)

and axisymmetric, with rms strengths of up to 2 kG (Reiners, Basri & Browning 2009). However, those M-dwarfs with $M/M_\odot < 0.2$ can display both strong poloidal and weak toroidal patterns, possibly even switching between them. Browning (2008) modelled dynamo action in a 0.3 $M_\odot$ M dwarf, rotating at the same rate as the Sun. He found that strong magnetic fields were generated, and that differential rotation was almost entirely quenched, leaving only a slight equatorial acceleration. The total magnetic energy was then comparable to the total kinetic energy in the uniformly rotating frame. When azimuthally averaged, the toroidal field gradually developed an antisymmetric structure, corresponding to a predominantly dipolar external field. Browning's more recent calculations, for stars rotating up to 10 times faster, are described in his contribution to these Proceedings.

4.3. *Upper main-sequence stars*

In stars more massive than the Sun the surface temperature is sufficiently high that hydrogen is already ionized and there is only a shallow convective layer caused by ionization of helium. However, the transition from energy generation by the *pp*-chain to the temperature-sensitive CNO-cycle leads to the appearance of a convective core. We naturally expect such a core, in a star rotating sufficiently rapidly, to act as a large-scale dynamo and to generate a magnetic field – though that field may not be visible at the surface of the star. Brun, Browning & Toomre (2005) attempted to model dynamo action in the core of an A-type star, with mass 2 $M_\odot$ and $1 \leqslant \Omega/\Omega_\odot \leqslant 4$. Similar dynamos must be responsible for the magnetic activity observed in some massive O and B type stars.

The best known magnetic stars are slow rotators of types A and B that exhibit peculiar anomalies in the abundances of certain elements, along with strong magnetic fields that vary periodically as the stars rotate. It is generally cosidered that these are fossil fields, predominantly dipolar but with an axis inclined to the rotation vector (Mestel 1999; Donati & Landstreet 2009). For such a field to be stable and to survive within a radiative zone it must have a mixed poloidal-toroidal structure (Tayler 1973; Wright 1973; Spruit 2002). Featherstone *et al.* (2009) have modelled the interaction between such a fossil field and a core dynamo in an Ap star, giving rise to complex magnetic topologies.

5. What next?

Dynamo theory, as applied to stars, is at present caught between two stages of development. The mean field approach has illuminated the main features of cyclic behaviour in the Sun – but it involves ad hoc assumptions, and there is no obvious way of establishing a unique model. Surely the time has now come to turn to direct numerical simulation to resolve these issues. Yet, despite prodigious efforts (led by Juri Toomre and his group at Boulder) there is still no convincing computational model of the solar cycle. Progress has been slow but steady, and it is reassuring that the latest calculations can now replicate the observed pattern of differential rotation. The next stage must be to reproduce the detailed behaviour of magnetic fields around the base of the convection zone and in the tachocline. That will require high radial resolution, accompanied by a corresponding reduction in the local values of viscous and magnetic diffusivities. For such a calculation to be credible it remains essential that diffusive processes should be treated explicitly, on the optimistic assumption that the large-scale dynamo (like linear fast dynamos) does not depend critically on fine details of the magnetic and kinetic energy cascades. It is extremely important therefore that this massive computational project should continue - -and there is surely a grand future ahead for Juri and his colleagues.

Acknowledgments

I have known Juri for many years, and enjoyed collaborating with him, so I am glad to have this opportunity of thanking him for his generosity and enthusiasm. I am grateful also for discussions and collaborations with Steve Tobias, Ed Spiegel, Mike Proctor, Chris Jones, David Hughes, Fausto Cattaneo, Paul Bushby and Jürg Beer.

References

Abreu, J. A., Beer, J., Steinhilber, S., Tobias, S. M., & Weiss, N. O. 2008, *Geophys. Res. Lett.*, 35, L20109

Antia, H. M., Basu, S., & Chitre, S. M. 2008, *ApJ*, 681, 680

Babcock, H. W. 1961, *ApJ*, 133, 572

Balbus, S. A., Bonart, J., Latter, H. & Weiss, N. O. 2009 *MNRAS*, 400, 176

Baliunas, S. L., Donahue, R. A., & Soon, W. H. *et al.* 1995, *ApJ*, 438, 269

Baliunas, S. L., Sokoloff, D., & Soon, W. H. 1996, *ApJ*, 457, L99

Barnes, J. A., Sargent, H. H., & Tryon, P. V. 1980, in *The Ancient Sun*, ed. R. O. Pepin, J. A. Eddy & R. B. Merrill, New York: Pergamon, p.159

Basu, S. & Antia, H. M. 2010 *ApJ*, 717, 488

Beer, J. 2000, *Space Sci. Rev.*, 94, 53

Boruta, N. 1996, *ApJ*, 458, 832

Brajša, R., Wöhl, H., & Hanslmeier, A. *et al.* 2009, *A&A*, 496, 855

Brown, B. P., Browning, M. K., Brun, A. S., Miesch, M. S., & Toomre, J. 2008, *ApJ*, 689. 1354

Brown, B. P., Browning, M. K., Brun, A. S., Miesch, M. S., & Toomre, J. 2010, *ApJ*, 711. 424

Browning, M. K. 2008, *ApJ*, 676, 1262

Browning, M. K., Miesch, M. S., Brun, A. S., & Toomre, J. 2006, *ApJ*, 648.L157

Brun, a. S., Browning, M. K., & Toomre, J. 2005, *ApJ*, 629, 461

Bushby, P. J. 2003, *MNRAS*, 342, L15

Bushby, P. J. 2005, *MNRAS*, 326, 218

Cattaneo, F. 1999, *ApJ*, 515, L39

Cattaneo, F. & Hughes, D. W. 2006, *J. Fluid Mech.*, 553, 401

Cattaneo, F. & Hughes, D. W. 2008, *J. Fluid Mech.*, 594, 495

Charbonneau, P. 2005, *Living Rev. Sol. Phys.*, 2, 2 (www.livingreviews.org/lrsp-2005-2)

Charbonneau, P. & Dikpati, M. 2000, *ApJ*, 543, 1027

Choudhuri, A. R. & Karak, B. B. 2009, *Res. A&A*, 9, 453

Christensen, U.R., Schmitt, D, & Rempel, M. 2009 *Space. Sci. Rev.*, 144, 105

Christensen-Dalsgaard, J. & Thompson, M. J. 2007, In *the Solar Tachocline*, ed. D. W. Hughes, R. Rosner & N. O. Weiss, Cambridge:Cambridge University Press, p. 53

Cowling, T. G. 1975. *Nature*, 255, 189

Damon, P. E. & Sonett, C. P. 1991, In *The Sun in Time*, ed. C. P. Sonett, M. S. Giampapa & M. S. Matthews, Tucson: University of Arizona Press, p. 360

Dikpati, M., Gilman, P. A., de Toma, G., & Ulrich, R. K. 2010, *Geophys. Res. Lett.*, 37, L14107

Dobler, W., Stix, M., & Brandenburg, A. 2006, *ApJ*, 638, 336

Donati, J.-F., Morin, J., & Petit, P. *et al.* 2008, *MNRAS*, 390, 545

Donati, J.-F. & Landstreet, J. D. 2009, *ARAA*, 47, 333

Featherstone, N.A., Browning, M.K., Brun, A. S. & Toomre, J. 2009. *ApJ*, 705, 1000

Fromang, S. & Papaloizou, J. 2007, *A&A*, 476, 1113

Fromang, S., Papaloizou, J., Lesur, G., & Heinemann, T. 2007, *A&A*, 476, 1123

Ghizaru, M., Charbonneau, P. & Smolarkiewicz P. K. 2010. *ApJ*, 715, L133

Gilman, P. A. 1983, *ApJSupp*, 53, 243

Glatzmaier, G. A. & Roberts, P. H. 1985, *Nature*, 377, 203

Hughes, D. W. 2007, In *The Solar Tachocline*, ed. D.W. Hughes, R. Rosner & N.O. Weiss, Cambridge: Cambridge University Press, p. 275

Hughes, D. W., Rosner, R., & Weiss, N. O., eds, 2007, *The Solar Tachocline*, Cambridge: Cambridge University Press

Howe , R. 2009, *Living Rev. Sol. Phys.*, 6, 1 (www.livingreviews.org/lrsp-2009-1)

Jones, C. A., Thompson, M. J., & Tobias, S. M. 2010, *Space. Sci. Rev.*, 152, 591

Jouve, L., Proctor, M. R. E., & Lesur, G. 2010, *A&A*, in press

Leighton, R. B. 1969, *ApJ*, 156, 1

Mestel, L. 1999, *Stellar Magnetism*, Oxford: Clarendon Press

Miesch, M. S., Brun, A. S., & Toomre, J. 2006, *ApJ*, 641, 618

Miesch, M. S., Brun, A. S., DeRosa, M. L., & Toomre, J. 2008, *ApJ*, 673, 557

Miesch, M. S. & Toomre, J. 2009, *AR Fluid Mech.*, 41, 317

Morin, J., Donati, J.-F., & Petit, P. *et al.* 2008, *MNRAS*, 390, 567

Morin, J., Donati, J.-F., & Petit, P. *et al.* 2010, *MNRAS*, in press

Noyes, R. W., Vaughan, A. H., & Weiss, N. O. 1984, *ApJ*, 287, 769

Ossendrijver, A. J. H. M. 1997, *A&A*, 323, 151

Ott, E. 1993, *Chaos in Dynamical Systems*, Cambridge: Cambridge University Press

Parker, E. N. 1955, *ApJ*, 122, 293

Parker, E. N. 1993, *ApJ*, 408, 707

Platt, N., Spiegel, E., & Tresser, C. 1993, *Geophys. Astrophys. Fluid Dyn.*, 73, 146

Reiners,A., Basri, G. & Browning, M. 2009, *ApJ*, 692, 538

Rüdiger, G. & Hollerbach, R. 2004, *The Magnetic Universe*, Weinheim: Wiley-VCH

Saar, S. H. 2002, in ASP Conf. Ser. 277, *Stellar Coronae in the Chandra and XMM-NEWTON Era*, ed. F. Favata and J. J. Drake, San Francisco: Astron. Soc. Pacific, p. 311

Saar, S. H. & Brandenburg, A. 1999, *ApJ*, 524, 295

Schmitt, D., Schüssler, M., & Ferriz-Mas, A. 1996, *A&A*, 311, L1

Solanki, S. K., Inhester, B., & Schüssler, M. 2006, *Rep. Prog. Phys.*, 69, 563

Spiegel, E. A. 1994, in *Lectures on Solar and Planetary Dynamos*, ed. M.R.E. Proctor & A.D. Gilbert, Cambridge: Cambridge University Press, p. 245

Spruit, H. C. 2002, *A&A*, 381, 923

Stuiver, M. & Braziunas, T. F. 1993, *Holocene*, 3, 289

Tayler, R. J. 1973, *MNRAS*, 161, 365

Thomas, J. H. & Weiss, N. O. 2008, *Sunspots and Starspots*, Cambridge: Cambridge University Press

Thompson, M. J., Christensen-Dalsgaard, J., Miesch, M. S., & Toomre, J. 2003, *ARAA*, 41, 599

Tobias, S. M., Brummell, N. H., Clune, T. L., & Toomre, J. 1998, *ApJ*, 502, L177

Tobias, S. M., Cattaneo, F., & Boldyrev, S. 2011, in *The Nature of Turbulence*, ed. P.A. Davidson, Y. Kaneda & K.R. Sreenivasan, Cambridge: Cambridge University Press

Tobias, S. M. & Weiss, N. O. 2007, In *The Solar Tachocline*, ed. D.W. Hughes, R. Rosner & N.O. Weiss, Cambridge: Cambridge University Press, p.319

Tobias, S. M., Weiss, N. O., & Kirk, V. 1995, *MNRAS*, 273, 1150

Vögler, A. & Schüssler, M. 2007, *A&A*, 465, L43

Wagner, G., Beer, J., Masarik, J., Kubik, P. W., Mende, W., Laj, C., Raisbeck, G. M., & Yiou, F. 2001, *Geophys. Res. Lett.*, 28, 303

Weiss, N. O. 2010, *A&G*, 51, 3.9

Weiss, N. O., Thomas, J. H., Brummell, N. H., & Tobias, S. M. 2004, *ApJ*, 600, 1073

Wright, G. A. E.. 1973, *MNRAS*, 161, 339

Yoshimura, H. 1978. *ApJ*, 226, 706

Discussion

BRANDENBURG: The number of spotless days during the current minimum is still not as low as it was in 1908. Are there other reasons suggesting that the current minimum might still be deeper than in 1908?

WEISS: Current predictions are that the next sunspot maximum will be similar to that in 1908. We cannot predict what will happen next. As a theoretician, I hope for a grand minimum, which could tell us something new about the solar dynamo – but I expect that our colleagues who observe the solar atmosphere would prefer a resumption of vigorous activity.

GOUGH: You commented that not all the stars have spots aligned with the rotation axis in such a way that one should expect alignment. Yet most of the Doppler images one sees of the stars have two spots whose axes lie almost in the equatorial plane, not even inclined to mid latitudes. Please could you comment on this?

WEISS: It depends which stars one looks at. Rapidly rotating late-type stars tend to have polar spots (unlike the slowly rotating Sun). The Ap stars are typically oblique rotators, with a magnetic axis strongly inclined to the rotation axis. Mestel has argued that dynamical processes in a rotating system act to promote this obliquity.

VASIL: Is there any historical evidence for a grand minimum starting right after a grand maximum ?

WEISS: Well, yes: there are a good many examples of a rapid descent from a grand maximum to a grand minimum (and vice versa) in the smoothed ^{10}Be record. The Maunder Minimum actually started quite abruptly (though it was not preceded by a grand maximum). Hevelius commented in 1668 that, while in 1643–1644 there were 100 to 200 spots visible in one hemisphere during a single year, "for a good many years recently, ten or more, I am certain that absolutely nothing of great significance (apart from some rather unimportant and small spots) has been observed either by us or by others".

Astrophysical Dynamics, From Stars to Galaxies
Proceedings IAU Symposium No. 271, 2010
N. Brummell, A. S. Brun, M. S. Miesch & Y. Ponty, eds.

© International Astronomical Union 2011
doi:10.1017/S1743921311017686

Magnetic Cycles and Meridional Circulation in Global Models of Solar Convection

Mark S. Miesch[1], Benjamin P. Brown[2], Matthew K. Browning[3], Allan Sacha Brun[4] and Juri Toomre[5]

[1]High Altitude Observatory, National Center for Atmospheric Research,
Boulder, CO, 80307-3000, USA
email: `miesch@ucar.edu`

[2]Dept. of Astronomy, Univ. of Wisconsin,
475 N. Charter St., Madison, WI 53706, USA

[3]Canadian Institute for Theoretical Astrophysics, Univ. of Toronto,
Toronto, ON M5S3H8, Canada

[4]DSM/IRFU/SAp, CEA-Saclay and UMR AIM, CEA-CNRS-Université Paris 7,
91191 Gif-sur-Yvette, France

[5]JILA and Dept. of Astrophysical & Planetary Sciences, Univ. of Colorado,
Boulder, CO 80309-0440, USA

Abstract. We review recent insights into the dynamics of the solar convection zone obtained from global numerical simulations, focusing on two recent developments in particular. The first is quasi-cyclic magnetic activity in a long-duration dynamo simulation. Although mean fields comprise only a few percent of the total magnetic energy they exhibit remarkable order, with multiple polarity reversals and systematic variability on time scales of 6-15 years. The second development concerns the maintenance of the meridional circulation. Recent high-resolution simulations have captured the subtle nonlinear dynamical balances with more fidelity than previous, more laminar models, yielding more coherent circulation patterns. These patterns are dominated by a single cell in each hemisphere, with poleward and equatorward flow in the upper and lower convection zone respectively. We briefly address the implications of and future of these modeling efforts.

Keywords. Sun: Interior, Convection, dynamo, MHD

1. Introduction

As the Solar and Heliospheric Observatory (SOHO) was undergoing its final stages of pre-launch preparations and as the telescopes of the Global Oscillations Network Group (GONG) were being deployed around the world, Juri Toomre looked ahead with characteristic vision and enthusiasm:

> The deductions that will be made in the near future from the helioseismic probing of the solar convection zone and the deeper interior are likely to provide a stimulus and to in turn be challenged by the major numerical turbulence simulations now proceeding apace with the developments in high performance computing (Toomre & Brummell 1995).

In the fifteen years since, the Michelson Doppler Imager (MDI) onboard SOHO and the GONG network have provided profound insights into the dynamics of the solar convection zone (Christensen-Dalsgaard 2002; Birch & Gizon 2005; Howe 2009). Meanwhile, high-resolution simulations of solar and stellar convection have become indispensable tools in interpreting and guiding helioseismic investigations (Miesch & Toomre 2009; Nordlund *et al.* 2009; Rempel *et al.* 2009). Juri has played a leading role in both endeavors.

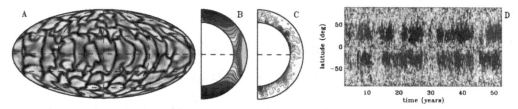

Figure 1. Dynamo action in a solar convection simulation. (A) Radial velocity field near the top of the convection zone ($r = 0.96R$) in a Molleweide projection (yellow denotes upflow, blue/black downflow). (B) Angular velocity Ω and (C) toroidal magnetic field $\langle B_\phi \rangle$, averaged over longitude and time (pink/blue denote fast/slow rotation, red/blue denote eastward/westward field). The time average for Ω spans the 50-year time interval shown while the time average for $\langle B_\phi \rangle$ focuses on a 6-day interval near $t = 25$ yr. (D) Longitudinally-averaged toroidal field $\langle B_\phi \rangle$ at the base of the convection zone versus latitude and time. Saturation levels for the color tables are (A) $\pm$ 70 m s^{-1}, (B) 375–470 nHz, and (C,D) $\pm$ 3kG. Peak values of $\langle B_\phi \rangle$ reach $\pm$ 10–14 kG.

In particular, it was Juri who led a team of young students and postdocs to develop what was to become known as the ASH (Anelastic Spherical Harmonic) code (Clune *et al.* 1999; Miesch *et al.* 2000). In last decade ASH has provided many novel insights into the dynamics of solar and stellar interiors, including the intricate structure of global-scale turbulent convection, the subtle nonlinear maintenance of differential rotation and meridional circulation, and the complex generation of mean and turbulent magnetic fields through hydromagnetic dynamo action (Miesch *et al.* 2000, 2006, 2008; Elliott *et al.* 2000; Brun & Toomre 2002; Brun *et al.* 2004, 2005; Browning *et al.* 2004, 2006; Browning 2008; Brown *et al.* 2008, 2010a,b; Brun & Palacios 2009; Featherstone *et al.* 2009).

In this paper we briefly review two recent development in ASH modeling, focusing on the current Sun in particular. These include the generation of coherent, organized, quasi-cyclic mean magnetic fields in an otherwise turbulent convection zone (§2) and the delicate maintenance of the solar meridional circulation (§3). Section 4 is a brief summary and look to the future. For ASH models of other stars and epochs, see the contributions by Brown and Browning (these proceedings).

2. Magnetic Self-Organization in a Convective Dynamo

Recent ASH simulations of convective dynamos in rapidly-rotating solar-like stars have revealed remarkable examples of magnetic self-organization, forming coherent toroidal bands of flux referred to as magnetic wreaths (Brown, these proceedings). There are typically two wreaths centered at latitudes of approximately $\pm 20 - 30$ degrees, with opposite polarity in the northern and southern hemispheres. They persist amid the intense turbulence of the convection zone, maintained by the convectively-driven rotational shear. Two simulations in particular highlight the possible temporal evolution exhibited by such dynamos. In Case D3, rotating at three times the solar rate (rotation period $P_{rot} = 9.3$ days), the wreathes persist indefinitely, buffeted by convective motions but otherwise stable for thousands of days (Brown *et al.* 2010a). In Case D5, rotating at five times the solar rate ($P_{rot} =5.6$ days), the wreathes undergo quasi-cyclic polarity reversals on a time scale of about 4 years (Brown *et al.* 2010b).

Such behavior is dramatically different from early ASH simulations of convective dynamos at the current solar rotation rate ($P_{rot} = 28$ days) by Brun *et al.* (2004). These were dominated by small-scale turbulent magnetic fields with complex, chaotic mean fields. The addition of a tachocline promoted the generation of stronger, more stable mean fields throughout the convection zone, with wreath-like equatorially antisymmetric toroidal bands in the stably-stratified region below the convection zone (Browning *et al.* 2006; Miesch *et al.* 2009). This tachocline simulation and a counterpart with different initial conditions Browning *et al.* 2007 were run for 10-30 years each but no polarity reversals were observed.

Motivated by these previous studies, we initiated a new simulation in order to investigate whether self-organization processes comparable to those exhibited by the rapid rotators D3 and D5 might occur also at the solar rotation rate. In order to enable long time integration, we chose a moderate resolution of N_r, N_θ, $N_\phi = 129, 256, 512$. This is the same vertical resolution but half the horizontal resolution of Case M3 from Brun *et al.* (2004). The lower horizontal resolution required higher viscous, thermal, and magnetic dissipation but by having the eddy dffusivity scale as $\eta(r) \propto \hat{\rho}(r)^{-1}$ rather than $\eta(r) \propto \hat{\rho}(r)^{-1/2}$ as in case M3, we were able to achieve comparable values of η at the base of the convection zone in the two cases; $\eta(r_b) = 2.03 \times 10^{11}$ cm^2 s^{-1} in Case M3 versus $\eta(r_b) = 2.88 \times 10^{11}$ cm^2 s^{-1} in the case reported presently, which we refer to as Case M4. Here $\hat{\rho}(r)$ is the background density profile and r_b is the base of the convection zone.

Other than the horizontal resolution and dissipation, the principle difference between M3 and M4 is the lower magnetic boundary condition. Case M3 matched to a potential field whereas Case M4 employs a perfect conductor. This has large implications for mean field generation; we find that perfectly conducting boundary conditions promote the generation of strong toroidal fields in wreath-building dynamos such as D3 and D5. Furthermore, we have also imposed a latitudinal entropy gradient at the lower boundary as described by Miesch *et al.* (2006) in order to promote a solar-like, conical angular velocity profile. This takes into account thermal coupling to the expected dynamical force balance in the tachocline without explicitly including the tachocline itself, which requires fine spatial and temporal resolution. Omission of the tachocline may well have important consequences as to the nature of the dynamo, but we focus here on the generation of large-scale fields by turbulent convection and rotational shear in the solar envelope as an essential step toward understanding the fundamental elements of the solar dynamo.

Results for case M4 are illustrated in Figure 1. The convective patterns (A) are similar to previous simulations of comparable resolution and the differential rotation (B) is solar-like, with nearly conical mid-latitude contours and a monotonic decrease in angular velocity of about 25% from equator to pole (475 nHz - 370 nHz). In contrast to case M3, this simulation generates coherent layers of toroidal magnetic flux near the base of the convection zone, with opposite polarity in each hemisphere (C). As the simulation evolves, the polarity of these flux layers reverses several times (D). The characteristic time scale for reversals appears to be about 14-15 years, although there is a failed reversal at $t \approx 30$ yr. The more rapid reversals early in the simulations may represent initial transients as the dynamo is becoming established. The magnetic diffusion time scale $\tau_\eta \sim r^2 \eta^{-1} \pi^{-2}$ ranges from 40 years at the inner boundary to 1.4 years at the outer boundary.

Given the generation of coherent mean fields, one might be tempted to refer to this as a large-scale dynamo (Brandenburg & Subramanian 2005). However, the magnetic energy spectrum does not peak at large scales, as one may expect from a turbulent α-effect or an inverse cascade of magnetic helicity. Rather, the magnetic energy peaks on scales smaller than the velocity field, with mean fields making up only 3% of the total magnetic energy. This is in contrast to the wreath-building rapid rotators D3 and D5 where approximately half (46-47%) of the magnetic energy is in the mean fields.

In cases D3 and D5 the wreathes form and persist in the midst of the turbulent convection zone whereas in case M4 the coherent toroidal fields are confined to the base of the convection zone. This can be attributed largely to the relative strength of the rotational shear. In cases D3 and D5, the bulk of the kinetic energy (relative to the rotating frame) is in the differential rotation (65% and 71% respectively). The stronger shear in turn is a consequence of the stronger rotational influence, which promotes angular momentum transport by means of the Coriolis-induced Reynolds stress. In case M4 the rate at which the wreathes are generated by rotational shear is comparable to or larger than the rate at which they are destroyed by convective mixing. Here only 35% of the kinetic energy is in the differential rotation, with 65% in the convection (the meridional circulation accounts for less than 1% of the KE in all three cases). The rotational shear is not strong enough to sustain the wreathes in the mid convection zone and horizontal flux is pumped downward by turbulent convective plumes. Toroidal flux accumulates and persists near the base of the convection zone where the vertical velocity drops to zero and where it is further amplified by rotational shear. The depth dependence of η also contributes to the localization of the wreathes near the base of the convection zone. As mentioned above, $\eta \propto \hat{\rho}^{-1}$ in case M4 whereas $\eta \propto \hat{\rho}^{-1/2}$ in cases D3 and D5.

It is notable that such toroidal flux layers develop even without a tachocline. In the penetrative simulations by Browning *et al.* (2006) radial shear in the tachocline does contribute to the formation of persistent toroidal flux layers but case M4 demonstrates that latitudinal shear in the lower convection zone is sufficient. This is consistent with recent mean-field dynamo models which suggest that latitudinal shear in the lower convection zone is more effective at generating strong, latitudinally-extended toroidal flux layers than the radial shear in the tachocline (e.g. Dikpati & Gilman (2006).

Latitudinal shear is sufficient to generate toroidal flux layers at the base of the convection zone but are they strong enough to spawn the buoyant flux structures responsible for photospheric active regions? The peak strength of the mean toroidal field in case M4 reaches about 10^4G. This is toward the lower end of estimated field strengths of 10^4-10^5G based on observations of bipolar active regions coupled with theoretical and numerical models of flux emergence (Fan 2004; Jouve & Brun 2009). However, local (pointwise) values of the longitudinal field B_ϕ typically reach 30-40kG in the magnetic layers and even stronger fields would be expected in higher-resolution simulations with less subgrid-scale diffusion. Convection may also promote the destabilization and rise of flux tubes, producing solar-like tilt angles of bipolar active regions even for relatively weak fields or order 15 kG (Weber *et al.* 2010). Case M4 does not exhibit buoyant flux structures, possibly due to insufficient resolution.

A prominent feature of the rapidly-rotating wreath-building dynamos D3 and D5 is an octupolar structure for the mean poloidal field (Brown 2010a,b). Poloidal separatrices lie at the poleward edge of the wreathes, although the magnetic topology can become more complex during reversals. By contrast, the mean poloidal field in case M4 is predominantly dipolar, often with weak, transient loops of opposite polarity at high latitudes.

We do not propose that this is a viable model of the solar dynamo. It does not exhibit equatorward-propagating activity bands or flux emergence comparable to active regions. Still, it is remarkable that the characteristic time scale for evolution of the mean fields is just over a decade, two orders of magnitude longer than the rotation period and the convection turnover time scale (both of order a month). Furthermore, quasi-cyclic polarity reversals occur on decadal time scales without flux emergence (required by the Babcock-Leighton mechanism), without a tachocline, and without significant flux transport by the meridional circulation. Cyclic variability on similar time scales was also found by Ghizaru *et al.* (2010) in convective dynamo simulations with a tachocline.

3. Maintenance of Meridional Circulation

With the recent surge in popularity of Flux-Transport solar dynamo models, the meridional circulation in the solar envelope has become a topic of great interest. According to the Flux-Transport paradigm, equatorward advection of torodial flux by the meridional circulation near the base of the solar convection zone is largely responsible for the observed butterfly diagram and thus regulates the time scale for the 22-year solar activity cycle (Dikpati & Gilman 2006; Rempel 2006; Jouve & Brun 2007; Charbonneau 2010). In the postulated advection-dominated regime under which many Flux-Transport models operate, the meridional circulation also dominates the transport of magnetic flux from the poloidal source region in the upper convection zone to the toroidal source region near the base of the convection zone.

Kinematic mean-field dynamo models cannot address how the solar meridional circulation is maintained; it is imposed as a model input. Non-kinematic mean-field models (e.g. Rempel 2005,2006) can give much insight into the underlying dynamics but they are necessarily based on theoretical models for the convective Reynolds stress, heat flux, and α-effect that need to be verified. 3D Convection simulations are essential for a deeper understanding of how the solar meridional circulation is established and what its structure and evolution may be.

The meridional circulation in the solar envelope is established in response to the convective angular momentum transport as follows

$$\langle \hat{\rho} \mathbf{v}_m \rangle_{\phi,t} \cdot \boldsymbol{\nabla} \mathcal{L} = \mathcal{F}, \tag{3.1}$$

where $\hat{\rho} = \langle \rho \rangle_{\phi,t}$ is the mean density stratification, $\mathbf{v}_m$ is the velocity in the meridional plane, and $\mathcal{L}$ is the specific angular momentum:

$$\mathcal{L} = \lambda^2 \Omega = r \sin \theta \left(r \sin \theta \Omega_0 + \langle v_\phi \rangle_{\phi,t} \right). \tag{3.2}$$

Angular brackets with subscripts $<>_{\phi,t}$ denote averages over longitude and time, Ω_0 is the rotation rate of the rotating coordinate system. Ω is the net rotation rate (including differential rotation), and $\lambda = r \sin \theta$ is the moment arm. The net torque $\mathcal{F}$ includes components arising from the convective Reynolds stress (RS), viscous diffusion (VD), and the Lorentz Force:

$$\mathcal{F} = -\boldsymbol{\nabla} \cdot \left(\mathbf{F}^{RS} + \mathbf{F}^{VD} + \mathbf{F}^{LF} \right). \tag{3.3}$$

where

$$\mathbf{F}^{RS} = \hat{\rho} \lambda \langle \mathbf{v}'_m v'_\phi \rangle_{\phi,t}, \quad \mathbf{F}^{VD} = -\hat{\rho} \lambda^2 \nu \boldsymbol{\nabla} \Omega \ , \text{ and } \quad \mathbf{F}^{LF} = -\frac{\lambda}{4\pi} \langle \mathbf{B}_m B_\phi \rangle_{\phi,t}. \tag{3.4}$$

Here ν is the kinematic viscosity, $\mathbf{B}_m$ is the magnetic field in the meridional plane, and primes denote variations about the mean, e.g. $v'_\phi = v_\phi - \langle v_\phi \rangle_{\phi,t}$.

Note that equation (3.1) is derived from the *zonal* component of the momentum equation yet largely determines the *meridional* flow. A convergence (divergence) of angular momentum flux, yielding a positive (negative) torque $\mathcal{F}$, induces a meridional flow across $\mathcal{L}$ isosurfaces directed toward (away from) the rotation axis. This is the concept of gyroscopic pumping; for further discussion and references see Miesch & Toomre (2009).

The principle component of $\mathcal{F}$ responsible for maintaining the solar differential rotation and thus the meridional circulation via gyroscopic pumping is that due to the convective Reynolds stress $\mathbf{F}^{RS}$. The viscous component $\mathbf{F}^{VD}$ is negligible in the Sun. However, in numerical simulations viscous diffusion can largely oppose the convective Reynolds stress. The meridional circulation then responds only to the residual torque, adversely influencing its structure and evolution. This also holds in the presence of the Lorentz

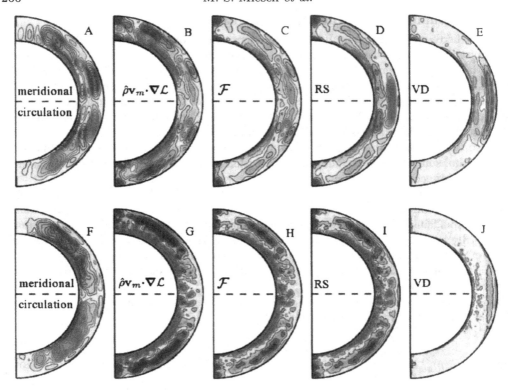

Figure 2. Maintenance of meridional circulation in two solar convection simulations. The top row (A–E) shows the dynamo simulation M4 discussed in §2 while the bottom row (F–J) shows a higher-resolution hydrodynamic simulation. Plotted are streamlines of the mass flux (A, F), the left (B,G) and right (C,H) hand sides of equation (3.1), and the contributions of the Reynolds stress (D,I) and the viscous diffusion (E,J) to the net torque $\mathcal{F}$, as expressed in eq. (3.3). All quantities are averaged over longitude and time (each over 300 days) and contour levels range from $\pm 2 \times 10^7$ g cm^{-1} s^{-2} in (B–E) and $\pm 5 \times 10^6$ g cm^{-1} s^{-2} in (G–J). Contour levels for the streamfunctions in (A) and (F) each range from $\pm 8 \times 10^{21}$ g s^{-1}, corresponding to typical meridional flow speeds of order 20 m s^{-1} and several m s^{-1} in the upper and lower convection zone respectively. Red and blue denote positive and negative values, which correspond to counter-clockwise and clockwise circulations in (A) and (F).

force. High resolution is required in order to minimize the artificial viscous diffusion and to therby achieve a more realistic zonal force balance.

The sensitivity of the meridional circulation profile to the zonal force balance is illustrated in Fig. 2 for two ASH convection simulations. The first (top row) is the moderate-resolution dynamo simulation M4 discussed in §2 (N_r, N_θ, N_ϕ = 129, 256, 512). The second (bottom row) is a higher-resolution non-magnetic case with a lower viscosity that we will refer to as Case H (N_r, N_θ, N_ϕ = 257, 512, 1024). The difference in ν is roughly a factor of four (mid convection zone values are 3.3×10^{12} cm^2 s^{-1} for case M4 versus 0.80×10^{12} cm^2 s^{-1} for case H, both varying with depth as $\hat{\rho}^{-1/2}$). Furthermore, the thermal diffusivity in case H is a factor of 2.7 less than in case M4 (mid convection zone values are 3.2×10^{12} cm^2 s^{-1} and 8.7×10^{12} cm^2 s^{-1}, again varying as $\hat{\rho}^{-1/2}$).

Comparison of panel pairs *B–C* and *G–H* indicate that the dynamical balance expressed by eq. (3.1) is approximately satisfied. In both cases, the net torque $\mathcal{F}$ is given by summing up the contributions from the Reynolds stress and the viscous diffusion such that panel C ≈ D + E and panel H = I + J. The Lorentz force also contributes to the net torque in panel C but its amplitude is small relative to the two terms shown (D,E).

There are two notable differences between the simulations M4 and H with regard to the maintenance of the meridional circulation. First, the nature of the Reynolds stress is different (D, I). Both simulations exhibit a convergence of $\mathbf{F}^{RS}$ at low latitudes that maintains the rotational shear but simulation H also exhbits a much more prominent inward angular momentum flux that is tending to accelerate the rotation in the lower convection zone relative to the upper convection zone. Furthermore, there is a radially outward angular momentum flux at low latitudes in case M4 that decelerates the lower convection zone. This is largely absent from case H. The second difference is the relative contribution of the viscous diffusion, which plays a smaller role in Case H due mainly to the smaller diffusivity ν.

The net result of these two differences is that Case H exhibits a substantial positive (negative) torque at mid-latitudes in the lower (upper) convection zone. The dynamical balance achieved in eq. (3.1) then induces a coherent single-celled meridional circulation pattern with poleward flow in the upper convection zone and equatorward flow in the lower convection zone (F). By contrast, the radially outward Reynolds stress near the equator in case M4 and the opposing influence of viscous diffusion produce a weaker, less coherent net torque $\mathcal{F}$. The circulation patterns are consequently more complex, with multiple cells in radius and latitude (A).

The circulation pattern in Case H (panel F) is predominantly single-celled but exhibits narrow counter-cells near the boundaries. These are likely artifacts of the boundary conditions. At the bottom boundary we impose a latitudinal entropy gradient as discussed by Miesch *et al.* (2006) in order to take into account thermal coupling with the tachocline. We also impose stress-free boundaries so the total angular momentum in the shell is conserved. Although both boundary conditions are justified, they are incompatible with thermal wind balance whereby baroclinic and Coriolis-induced torques offset one another. The imbalance induces a clockwise (counter-clockwise) circulation in the northern (southern) hemisphere near the boundary. Penetrative convection simulations generally exhibit an equatorward circulation throughout the overshoot region, induced by the turbulent alignment of downflow plumes with the rotation axis (Miesch *et al.* 2000).

A more realistic upper boundary condition would include coupling to solar surface convection which includes granulation, mesogranulation and supergranulation. In lieu of the large Rossby number (rotation period relative to the convective turnover time), one may expect such surface convection to efficiently mix angular momentum, exerting a retrograde zonal torque ($\mathcal{F} > 0$). The presence of such a retrograde torque is implied by the existence of the near-surface shear layer, a subsurface increase in rotation rate ($\partial\Omega/\partial r < 0$) detected in helioseismic inversions (Howe 2009). A retrograde torque in the solar surface layers would induce a poleward circulation according to eq. (3.1). Thus, the upper and lower counter-cells in the circulation profile of Fig. 2F are likely artificial.

4. Summary and Outlook

Inspired by helioseismology and fueled by continuing advances in high-performance computing, global convection simulations continue to shape our understanding of solar internal dynamics and the solar dynamo. Recent simulations have achieved quasi-cyclic magnetic activity on decadal time scales and meridional circulation profiles that are qualitatively consistent with the single-celled circulation profiles assumed in many kinematic mean-field dynamo models. A more thorough discussion of these simulations and their implications will appear in forthcoming papers.

Further progress is imminent. The next decade may see the first global convective dynamo simulations that spontaneously generate buoyant magnetic flux structures from rotational shear that is self-consistently maintained by the convection itself, providing unprecedented insights into the origins of solar magnetic activity. Achieving this milestone will require numerical algorithms capable of exploiting next-generation computing architectures with 10^5-10^6 processing cores. It is a daunting but glorious challenge; one that Juri will surely relish.

Acknowledgements

This research is supported by NASA through Heliophysics Theory Program grants NNG05G124G and NNX08AI57G, and NASA SR&T grant NNH09AK14I, with additional support for Brown through NSF Astronomy and Astrophysics postdoctoral fellowship AST 09-02004. Browning was supported by CITA and Brun was partly supported by the Programme National Soleil-Terre of CNRS/INSU (France), and by the STARS2 grant from the European Research Council. The simulations were carried out with NSF PACI support of PSC, SDSC, TACC and NCSA, and by NASA HEC support at the NASA Advanced Supercomputing Division (NAS) facility at NASA Ames Research Center.

References

Brandenburg, A. & Subramanian, K. 2005, *Phys. Rep.*, 417, 1

Brown, B. P., Browning, M. K., Brun, A. S., Miesch, M. S., & Toomre, J. 2008, *ApJ*, 689, 1354

Brown, B. P., Browning, M. K., Brun, A. S., Miesch, M. S., & Toomre, J. 2010, *ApJ*, 711, 424

Brown, B. P., Browning, M. K., Brun, A. S., Miesch, M. S., & Toomre, J. 2010, *ApJ*, submitted

Browning, M. K., Brun, A. S., & Toomre, J. 2004, *ApJ*, 601, 512

Browning, M. K., Miesch, M. S., Brun, A. S., & Toomre, J. 2006, *ApJ Let.*, 648, L157

Browning, M. K., Brun, A. S., Miesch, M. S., & Toomre, J. 2007, *Astron. Nachr.*, 328, 1100

Browning, M. K. 2008, *ApJ*, 676, 1262

Brun, A. S. & Toomre, J. 2002, *ApJ*, 570, 865

Brun, A. S., Miesch, M. S., & Toomre, J. 2004, *ApJ*, 614, 1073

Brun, A. S., Browning, M. K., & Toomre, J. 2005, *ApJ*, 629, 885

Brun, A. S. & Palacios, A. 2009, *ApJ*, 702, 1078

Charbonneau, P. 2010, *LRSP*, 7, http://www.livingreviews.org/lrsp-2010-3

Clune, T. L., Elliott, J. R., Miesch, M. S., Toomre, J., & Glatzmaier, G. A. 1999, *Parallel Computing*, 25, 361

Christensen-Dalsgaard, J. 2002, *Rev. Mod. Phys.*, 74, 1073

Dikpati, M. & Gilman, P. A., 2006, *ApJ*, 649, 498

Eliott, J. R., Miesch, M. S., & Toomre, J. 2000, *ApJ*, 533, 546

Fan, Y. 2001, *ApJ*, 546, 509

Fan, Y. 2004, *LRSP*, 1, http://www.livingreviews.org/lrsp-2004-1

Featherstone, N. A., Browning, M. K., Brun, A. S., & Toomre, J. 2009, *ApJ*, 705, 1000

Ghizaru, M., Charbonneau, P., & Smolarkiewicz, P.K. 2010, *ApJ Let.*, 715, L133

Gizon, L. & Birch, A. C. 2005, *LRSP*, 2, http://www.livingreviews.org/lrsp-2005-6

Howe, R. 2009, *LRSP*, 6, http://www.livingreviews.org/lrsp-2009-1

Jouve, L. & Brun, A. S. 2007, *A&A*, 474, 239

Jouve, L. & Brun, A. S. 2009, *ApJ*, 701, 1300

Miesch, M. S., Elliott, J. R., Toomre, J., Clune, T. L., Glatzmaier, G. A., & Gilman, P. A. 2000, *ApJ*, 532, 593

Miesch, M. S., Brun, A. S., & Toomre, J. 2006, *ApJ*, 641, 618

Miesch, M. S., Brun, A. S., DeRosa, M. L., & Toomre, J. 2008, *ApJ*, 673, 557

Miesch, M. S., Browning, M. K., Brun, A. S., Toomre, J., & Brown, B. P. 2009, in: M. Dikpati, T. Arentoft, I. González Hernández, C. Lindsey & F. Hill (eds.), *Proc. GONG 2008/SOHO*

XXI Meeting on Solar-Stellar Dynamos as Revealed by Helio- and Asteroseismology, ASP Conf. Ser., vol. 416, p. 443

Miesch, M. S. & Toomre, J. 2009, *Ann. Rev. Fluid Mech.*, 41, 317

Nordlund, A., Stein, R. F., & Asplund, M. 2009, *LRSP*, 6, http://www.livingreviews.org/lrsp-2005-2

Rempel, M. 2005, *Ap.J.*, 622, 1320

Rempel, M. 2006, *Ap.J.*, 647, 662

Rempel, M., Schüssler, M., Cameron, R. H., & Knölker 2009, *Science*, 325, 171.

Toomre, J. & Brummell, N. H. 1995, in: J. T. Hoeksema, V. Domingo, B. Fleck & B. Battrick (eds.), *Fourth SOHO Workshop: Helioseismology* (ESA: Noordwijk), p. 47

Weber, M, Fan, Y., & Miesch, M. S. 2010, in preparation

Discussion

ROGERS: I don't think counter-rotating cells at bottom boundary are entirely artificial because of hard boundaries, we see them in simulations with stable regions as well.

MIESCH: The sense of the meridional flow near the lower boundary arises from the stress-free mechanical boundary condition coupled to the imposed latitudinal entropy variation. Our 3D penetrative convection simulations generally have equatorward meridional flow in the overshoot region as a result of the turbulent alignment of downflow plumes and gyroscopic pumping associated with the convective angular momentum transport (Miesch *et al.* 2000). 2D simulations may exhibit different behavior. For further discussion, see the last two paragraphs of §3.

HUGHES: Do you think all the physics for magnetic buoyancy instability is incorporated in the anelastic approximation?

MIESCH: The essential assumption behind the anelastic approximation is that the Mach number is small. In order for this to break down in the deep solar interior within the context of flux emergence, it would imply MG fields and order-one thermal perturbations that can be ruled out by helioseismic structure inversions. The anelastic equation of state includes the influence of the pressure on density variations so it captures the physical mechanism underlying magnetic buoyancy. For a derivation of the Parker Instability in an anelastic system and nonlinear simulations of rising flux tubes see Fan (2001).

GOUGH: You commented that helioseismology indicates that the concentrated polar vortex hich you illustrated as a property of some your simulations poleward of about 85′ does not exist. Actually that is not so: helioseismology has nothing to say so close to the axis of rotation. But it does indicate that the angular velocity poleward of 70 is lower than a smooth extrapolation would suggest- slow rotation in the region where the large scale dipole like magnetic field emanates and from which the fast component of the solar wind comes. Do you find any indications of that in the simulations?

MIESCH: The short answer is no. Some of our simulations have a monotonic decrease in the angular velocity Ω that continues to the poles and some exhibit a slight increase in Ω toward the poles. In the former solutions, the slope is rather smooth, with no indication of an abrupt steepening above 85°.

Astrophysical dynamics – from stars to galaxies
Proceedings IAU Symposium No. 271, 2010
N. H. Brummell, A. S. Brun. M. S. Miesch & Y. Ponty, eds.

© International Astronomical Union 2011
doi:10.1017/S1743921311017698

MHD relaxation of fossil magnetic fields in stellar interiors

Stéphane Mathis[1,2], Vincent Duez[3] & Jonathan Braithwaite[3]

[1]Laboratoire AIM, CEA/DSM-CNRS-Université Paris Diderot, IRFU/SAp Centre de Saclay, F-91191 Gif-sur-Yvette, France
email: stephane.mathis@cea.fr

[2]Observatoire de Paris-LESIA 5, place Jules Janssen, F-92195 Meudon Cedex

[3]Argelander Institut für Astronomie, Universität Bonn, Auf dem Hügel 71, D-53111 Bonn, Germany
email: vduez@astro.uni-bonn.de; jonathan@astro.uni-bonn.de

Abstract. The understanding of fossil fields origin, topology, and stability is one of the corner stones of the stellar magnetism theory. On one hand, since they survive on secular time scales, they may modify the structure and the evolution of their host stars. On the other hand, they must have a complex stable structure since it has been demonstrated that the simplest purely poloidal or toroidal fields are unstable on dynamical time scales. In this context, the only stable stellar configurations found today are those resulting from numerical simulations by Braithwaite and collaborators who studied the evolution of an initial stochastic magnetic field, which relaxes with a selective decay of magnetic helicity and energy, on mixed stable configurations (poloidal and toroidal) that seem to be in equilibrium and then diffuse. In this talk, we report the semi-analytical investigation of such an equilibrium field in the axisymmetric case. We use variational methods, which describe selective decay of magnetic helicity and energy during MHD relaxation, and we identify a supplementary invariant due to the stable stratification of stellar radiation zones. This leads to states that generalize force-free Taylor's relaxation states studied in plasma laboratory experiments that become non force-free in the stellar case. Moreover, astrophysical applications are presented and the stability of obtained configurations is studied.

Keywords. magnetohydrodynamics (MHD), plasmas, stars: magnetic field

---

1. Introduction

Magnetic fields are now detected more and more often at the surface of main-sequence (and pre main-sequence) intermediate mass and massive stars, which have an external radiative envelope. Indeed, strong fields (300 G to 30 kG) are observed in some fraction of Herbig stars (Alecian *et al.* 2008), A stars (the Ap stars, see Aurière *et al.* 2007), as well as in B stars and in a handful of O stars (Grunhut *et al.* 2009). Furthermore, we cannot dismiss the possibility of a large-scale magnetic field being responsible for the quasi-uniform rotation of the bulk of the solar radiation zone, as revealed by p-modes helioseismology (Eff-Darwich *et al.* 2008). Finally, non convective compact objects display fields strength of $10^4 - 10^9$ G for white dwarfs and of $10^8 - 10^{15}$ G for neutron stars. Magnetic fields in stably stratified non convective stellar regions will thus be able to deeply modify our vision of stars evolution since their formation (Commerçon *et al.* 2010) to their late stages, for example for gravitational supernovae. Indeed, they will modify stellar internal dynamics, for example the transport of angular momentum and the resulting rotation history, and chemicals mixing (see Mathis & Zahn 2005).

The large-scale, ordered nature (often approximately dipolar) of such magnetic fields and the scaling of their strengths as a function of their host properties (according to the flux conservation scenario) favour a fossil hypothesis, whose origin has to be understood. One of the fundamental question is thus the understanding of the topology of these large-scale magnetic fields. To have survived since the star's formation or the PMS stage, a field must be stable on a dynamic (Alfvén) timescale. It was suggested by Prendergast (1956) that a stellar magnetic field in stable axisymmetric equilibrium must contain both poloidal (meridional) and toroidal (azimuthal) components, since both are unstable on their own (Tayler 1973; Wright 1973; Braithwaite 2006; Bonanno & Urpin 2008b). This was confirmed recently by numerical simulations by Braithwaite & Spruit (2004); Braithwaite & Nordlund (2006); Braithwaite (2008) who showed that initial stochastic helical fields evolve on an Alfvén timescale into stable configurations: axisymmetric and non-axisymmetric mixed poloidal-toroidal fields were found. This phenomenon well known in plasma physics is a MHD turbulent relaxation (*i.e.* self-organization process involving magnetic reconnections in resistive MHD). In this short paper, we present our physical understanding of such mechanism in stellar interiors focusing on the axisymmetric case. First, we show how to derive such magnetic configurations. Then, stability properties are studied. Astrophysical consequences and perspectives are finally discussed.

2. The relaxed non force-free configuration

In this work, we deal with axisymmetric, non force-free magnetic configurations (*i.e.* with a non-zero Lorentz force) in equilibrium inside stellar radiation zones, which result from an initial MHD relaxation of the field created by a PMS dynamo or the stellar formation. We first restrict ourselves to the non-rotating case, but results also apply to radiation regions in a state where rotation is uniform (Woltjer 1959), that could be the case if magnetic field is strong enough, and where meridional circulation can be neglected (*i.e.* if the star is near an equilibrium where the Lorentz torque vanishes, does not loose angular momentum, and have a stationary structure: see Mestel, Moss & Tayler 1988; Busse 1981; Zahn 1992; Decressin *et al.* 2009). The more general case including differential rotation (and induced meridional circulation) will be treated in a near future.

Several reasons inclined us to focus on non force-free relaxed equilibria instead of force-free ones, which are often studied in plasma laboratory experiments. First, Reisenegger (2009) reminds us that no configuration can be force-free everywhere. Although there do exist "force-free" configurations, they must be confined by some region or boundary layer with non-zero or singular Lorentz force. This induces discontinuities such as current sheets, which are unlikely to appear in nature except in a transient manner. Second, non force-free equilibria have been identified in plasma physics as the result of MHD relaxation (see for example Montgomery & Phillips 1988; Shaikh *et al.* 2008). Third, as shown by Duez & Mathis (2010), this family of equilibria is a generalization of Taylor states (force-free relaxed equilibria in plasma laboratory experiments; see Taylor 1974) in a stellar context, where the stable stratification of the medium plays a crucial role.

2.1. *The magnetic field in MHS equilibrium*

Let us describe the assumptions made in building the semi-analytical model of relaxed magnetohydrostatic (MHS) equilibrium described by Duez & Mathis (2010). The axisymmetric magnetic field $\boldsymbol{B}(r,\theta)$ is expressed as a function of a poloidal flux $\Psi(r,\theta)$, a toroidal potential $F(r,\theta)$, and the potential vector $\boldsymbol{A}(r,\theta)$ so that it is divergence-free by construction:

$$\boldsymbol{B} = \frac{1}{r\sin\theta}\left(\boldsymbol{\nabla}\Psi \times \hat{\boldsymbol{e}}_{\varphi} + F\,\hat{\boldsymbol{e}}_{\varphi}\right) = \boldsymbol{\nabla}\times\boldsymbol{A}, \qquad (2.1)$$

where in spherical coordinates the poloidal component (B_{P}) is in the meridional plane $(\hat{e}_r, \hat{e}_\theta)$ and the toroidal component (B_{T}) is along the azimuthal direction ($\hat{e}_\varphi$). The MHS equation expressing balance between the pressure gradient force, gravity and the Lorentz force is

$$0 = -\nabla P - \rho \nabla V + \frac{1}{\mu_0}(\nabla \times B) \times B, \qquad (2.2)$$

where V is the gravitational potential, which satisfies the Poisson equation: $\nabla^2 V = 4\pi G\rho$.

2.2. The non force-free relaxed equilibria family

2.2.1. MHD relaxation and variational method

Here, we focus on the minimum energy non force-free MHS equilibrium that a stably stratified radiation zone can reach. First, given the field strengths in real stars, the ratio of the Lorentz force to gravity is very low: stellar interiors are thus in a regime where $\beta = P/P_{\mathrm{Mag}} \gg 1$, $P_{\mathrm{Mag}} = B^2/(2\mu_0)$ being the magnetic pressure. Then, we identify the invariants governing the evolution of the reconnection phase, that leads to relaxed states in the non force-free case. The first one is the magnetic helicity

$$\mathcal{H} = \int_\mathcal{V} A \cdot B \, d\mathcal{V}, \qquad (2.3)$$

which is an ideal MHD invariant known to be roughly conserved at large scales during relaxation. The second one is the mass encompassed in poloidal magnetic surfaces

$$M_\Psi = \int_\mathcal{V} \Psi \rho \, d\mathcal{V}, \qquad (2.4)$$

conserved because of the stable stratification, which inhibits the radial movements and thus the transport of mass and flux in this direction. Note that this invariant can also be seen as a topological constraint. (see Moffatt 1985). Next, we assume a selective decay during relaxation (c.f. Biskamp 1997), in which the magnetic energy $E_{\mathrm{mag}} = \int_\mathcal{V} \frac{B^2}{2\mu_0} d\mathcal{V}$ (μ_0 being the vaccum magnetic permeability), and thus the total energy

$$E = E_{\mathrm{mag}} + \frac{1}{2} \int_\mathcal{V} \rho \, (V + 2\mathcal{U}) \, d\mathcal{V}, \qquad (2.5)$$

where $\mathcal{U}$ is the specific internal energy per unit mass, decays much faster than $\mathcal{H}$ and M_Ψ, so that they can be considered constant on an energetic decay e-folding time. This is due to the stable stratification and to the different orders of spatial derivatives involved in the variation of E_{mag} and $\mathcal{H}$:

$$\frac{dE_{\mathrm{mag}}}{dt} = -\int_\mathcal{V} \eta j^2 \, d\mathcal{V} \quad \text{and} \quad \frac{d\mathcal{H}}{dt} = -\int_\mathcal{V} \eta j \cdot B \, d\mathcal{V}, \qquad (2.6)$$

where η is the magnetic diffusivity and $\mu_0 j = \nabla \times B$, j being the current. The reached equilibrium is thus the one of minimum energy for given magnetic helicity and mass encompassed in magnetic flux tubes. This can be determined applying a variational method where we minimize E with respect to $\mathcal{H}$ and M_Ψ

$$\delta E + a_\mathcal{H} \mathcal{H} + a_{M_\Psi} M_\Psi = 0, \qquad (2.7)$$

where $a_\mathcal{H}$ and a_{M_Ψ} are Lagrangian multipliers. This allows to derive the elliptic linear partial differential equation governing Ψ (Woltjer 1959; Montgomery & Phillips 1989; Duez & Mathis 2010):

$$\Delta^* \Psi + \frac{\lambda_1^2}{R^2} \Psi = -\mu_0 \, \bar{\rho} \, r^2 \sin^2 \theta \, \beta_0. \qquad (2.8)$$

Here, $\bar{\rho}$ is the density in the non-magnetic case, $\Delta^*\Psi \equiv \partial_{rr}\Psi + \sin\theta\,\partial_\theta\left(\partial_\theta\Psi/\sin\theta\right)/r^2$ the Grad-Shafranov operator in spherical coordinates, λ_1 the eigenvalue to be determined, R a characteristic radius, and β_0 is constrained by the field's intensity. We have identified $a_{\mathcal{H}} = -\frac{1}{\mu_0}\frac{\lambda_1}{R}$ and $a_{M_\Psi} = -\beta_0$. Note that if M_Ψ is not taken into account, we recover force-free minimum magnetic energy equilibria for a given helicity derived by Taylor (1974).

This equation is similar to the Grad-Shafranov equation used to find MHS equilibria in magnetically confined plasmas (Grad & Rubin 1958; Shafranov 1966), the source term being here related to the stellar structure through $\bar{\rho}$ (for a discussion of the general form of this equation in astrophysics, see Heinemann & Olbert 1978). Furthermore, this equilibrium is in a barotropic state (in the *hydrodynamic* meaning of the term, *i.e.* isobar and iso-density surfaces coincide) where the field is explicitly coupled with stellar structure through $\nabla \times (\boldsymbol{F_\mathcal{L}}/\bar{\rho}) = \boldsymbol{0}$, where $\boldsymbol{F_\mathcal{L}}$ is the Lorentz force. This is a generalization of Prendergast's equilibrium taking into account compressibility, first studied in polytropic cases by Woltjer (1960).

2.2.2. *Solution*

The boundary conditions have now to be discussed. In Duez, Mathis & Turck-Chièze (2010) and Duez & Mathis (2010), we considered the general case of a field confined between two radii, owing to the possible presence of both a convective core and a convective envelope and to ensure the conservation of magnetic helicity. We here choose to cancel both radial and latitudinal fields at the surface, to avoid any current sheets, conserving once again magnetic helicity; the possible effects of the convective core on the large-scale surrounding field are neglected. Using Green's function method we finally obtain the purely dipolar, general solutions indexed by i:

$$
\begin{aligned}
\Psi_i\left(r,\theta\right) \;=\; & -\mu_0\beta_0\frac{\lambda_1^i}{R}r\left\{ j_1\left(\lambda_1^i\frac{r}{R}\right)\int_r^R\left[y_1\left(\lambda_1^i\frac{\xi}{R}\right)\bar{\rho}\xi^3\right]\mathrm{d}\xi\right.\\
& \left. +y_1\left(\lambda_1^i\frac{r}{R}\right)\int_0^r\left[j_1\left(\lambda_1^i\frac{\xi}{R}\right)\bar{\rho}\xi^3\right]\mathrm{d}\xi\right\}\sin^2\theta,
\end{aligned}
\tag{2.9}
$$

R being the upper boundary confining the magnetic field; λ_1^i are the set of eigenvalues indexed by i allowing to verify the boundary conditions. The functions j_l and y_l are respectively the spherical Bessel functions of the first and the second kind.

As shown in Duez & Mathis (2010), the first radial mode is the lowest energy state for given $\mathcal{H}$ and M_Ψ; we thus focus here only on this mode $i = 1$. The toroidal magnetic field is then given using $F(\Psi) = \lambda_1^1\Psi/R$. Furthermore, this state is ruled by the following helicity-energy relation

$$
\mathcal{H} = \frac{2\mu_0 R}{\lambda_1^1}\left(E_{\mathrm{mag}} - \frac{1}{2}\beta_0 M_\Psi\right),
\tag{2.10}
$$

which generalizes the one known in plasma physics for Taylor states to the stellar non force-free case.

In the case of a stably stratified $n = 3$ polytrope (a good approximation to an upper main-sequence star radiative envelope) where we set $R = 0.85\,R_*$, we have $\lambda_1^1 \simeq 32.95$ (represented in FIG. 1), while for a constant density profile, we have $\lambda_1^1 \simeq 5.76$.

2.2.3. *Comparison with numerical simulations*

Let us now compare our analytical configuration to those obtained using numerical simulations (see Braithwaite & Spruit 2004; Braithwaite & Nordlund 2006; Braithwaite

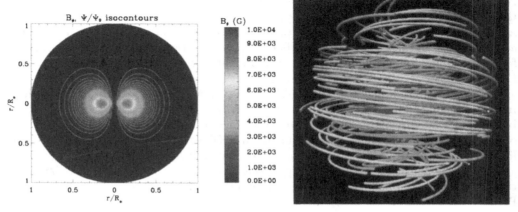

Figure 1. Left: toroidal magnetic field strength in colorscale (arbitrary field's strength) and normalized isocontours of the poloidal flux function (Ψ) in meridional cut for the lowest energy equilibrium configuration ($\lambda_1^1 \simeq 33$); the neutral line is located at $r \simeq 0.23\,R_*$. Right: magnetic field lines representing this mixed field configuration in 3-D looking from the side (the colorscale is a function of the density). Taken from Duez, Braithwaite & Mathis (2010).

2008) in more details. Braithwaite and collaborators performed numerical magnetohydro-dynamical simulations of the relaxation of an initially random magnetic field in a stably stratified star. Then, this initial magnetic field is always found to relax on the Alfvén time scale into a stable magneto-hydrostatic equilibrium mixed configuration consisting of twisted flux tube(s). Two families are then identified: in the first, the equilibria configurations are roughly axisymmetric with one flux tube forming a circle around the equator, such as in our configuration; in the second family, the relaxed fields are non-axisymmetric consisting of one or more flux tubes forming a complex structure with their axis lying at roughly constant depth under the surface of the star. Whether an axisymmetric or non-axisymmetric equilibrium forms depends on the initial condition chosen for the radial profile of the initial stochastic field strength $||\mathbf{B}|| \propto \bar{\rho}^p$: a centrally concentrated one evolves into an axisymmetric equilibrium as in our configuration while a more spread-out field with a stronger connection to the atmosphere relaxes into a non-axisymmetric one. Braithwaite (2008) indicates that, if using an ideal-gas star modeled initially with a polytrope of index $n = 3$, the threshold is $p \approx 1/2$.

Moreover, as shown in Fig. 7 in Braithwaite (2008), the selective decay of the magnetic helicity ($\mathcal{H}$) and of the magnetic energy (E_{mag}) assumed in §2.1.1. occurs during the initial relaxation with a stronger decrease in E_{mag} than that of $\mathcal{H}$. Furthermore, the transport of flux and mass in the radial direction is inhibited because of the stable stratification and the mass encompassed in poloidal magnetic surfaces is conserved (*i.e.* M_Ψ). The obtained configuration is of course non force-free.

Finally, note that our analytical configuration for which $E_{\mathrm{mag;P}}/E_{\mathrm{mag}} \approx 5.23 \times 10^{-2}$ (where $E_{\mathrm{mag;P}} = \int_\mathcal{V} B_\mathrm{P}^2 / (2\mu_0)\,\mathrm{d}\mathcal{V}$) verifies the stability criterion derived by Braithwaite (2009) for axisymmetric configurations: $\mathcal{A}\,E_{\mathrm{mag}}/E_{\mathrm{grav}} < E_{\mathrm{mag;P}}/E_{\mathrm{mag}} \leqslant 0.8$, where E_{grav} is the gravitational energy in the star, and $\mathcal{A}$ a dimensionless factor whose value is ~ 10 in a main-sequence star and $\sim 10^3$ in a neutron star, while we expect $E_{\mathrm{mag}}/E_{\mathrm{grav}} < 10^{-6}$ in a realistic star (see for example Duez, Mathis & Turck-Chièze 2010). Our analytical solution is thus similar to the axisymmetric non force-free relaxed solution family obtained by Braithwaite & Spruit (2004) and Braithwaite & Nordlund (2006).

These configurations can thus be relevant to model initial equilibrium conditions for evolutionary calculations involving large-scale fossil fields in stellar radiation zones (see for example Mathis & Zahn 2005; Brun & Zahn 2006; Garaud & Guervilly 2009).

3. Stability: numerical method

3.1. *The numerical model*

Analytical methods have been powerful to prove linear instabilities of magnetic configurations but unable to study nonlinear phases and to demonstrate stability. For this reason, we now turn on numerical simulations. The setup of the numerical model is similar to that in Braithwaite & Nordlund (2006), where a fuller account can be found. We use the STAGGER code (Nordlund & Galsgaard 1995), a high-order finite-difference Cartesian MHD code containing a "hyper-diffusion" scheme. The resolution is 192^3. We model the star as a self-gravitating ball of ideal gas ($\gamma = 5/3$) with radial density and pressure profiles initially obeying the polytropic (thus barotropic) relation $P \propto \rho^{1+(1/n)}$, with index $n = 3$, which models an upper main-sequence star radiative envelope. It seems unlikely that a different EOS will make even much quantitative difference to the results; the important point is the stable stratification. We use this model to compare the dynamical evolution of the mixed (poloidal-toroidal) configuration to that of its purely poloidal and toroidal components on their own, both of which are unstable as mentioned above. We should therefore see these instabilities, growing on an Alfvén timescale. To test the stability of the configurations, we add a random "white noise" perturbation to the density field. The perturbation in density (1% in amplitude) contains length scales ranging from R_* to $0.08R_*$, the latter being double the Nyquist wavelength. This is roughly equivalent to azimuthal wavenumbers up to $m = 38$ at a radius of $R_*/2$.

3.2. *Results*

3.2.1. *Purely poloidal component*

The simulation is run for around ten Alfvén crossing times τ_A, over which time the instability grows, becomes nonlinear and results in the destruction of most of the original magnetic energy. The magnetic field amplitude is plotted at the left of FIG. 2, split into components according to azimuthal wavenumber m; obviously at $t = 0$ all the energy is in the axisymmetric $m = 0$ part. Note the clear transition at $t \approx 2\,\tau_A$ from the linear phase to the nonlinear, reconnective phase.

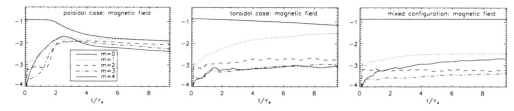

Figure 2. Time evolution of the (log) amplitudes in azimuthal modes $m = 0$ to 4 averaged over the stellar volume of the magnetic field in the simulations with the purely poloidal field (left), purely toroidal field (middle) and the mixed field (right). Initially, all the magnetic energy is in the $m = 0$ mode since the initial conditions are axisymmetric. Taken from Duez, Braithwaite & Mathis (2010).

3.2.2. *Purely toroidal component*

The middle plate of FIG. 2 shows the evolution of the toroidal field; clearly, the $m = 1$ mode of the Tayler instability is dominant. After its linear growth, the instability manifests itself in the nonlinear regime (cf. Brun 2007; Elstner *et al.* 2008) mainly in the movement of spherical shells relative to one another. We expect eventual complete destruction of the field on a longer timescale.

3.2.3. *Mixed configuration*

The mixed poloidal-toroidal configuration exhibits completely different behaviour. The magnetic amplitudes are plotted on the right of FIG. 2, where we see an absence of growing modes. The kinetic energy present results simply from the initial perturbation and the oscillations and waves it sets up. No significant change in the configuration during the evolution is seen. To better examine the potentially unstable regions, we use Tayler's stability criteria (Tayler 1973) for purely toroidal fields and estimate the stabilisation from the poloidal component, following Braithwaite (2009). In FIG. 3, we plot this criteria for modes $m = 0$ and $m = 1$. The $m = 0$ mode is unstable almost everywhere and the $m = 1$ mode is unstable in a large cone around the poles; however the poloidal field stabilises these modes in most of the meridional plane except near the equatorial plane where it merely stabilises all wavelengths small enough to fit into the available space. Moreover, we can examine closely the behaviour of the field in the vicinity of the magnetic axis, where it can be approximated as the addition of an axial and a toroidal field (cylindrical geometry). Bonanno & Urpin (2008a) outlined that in this case magnetic configurations can be subject to non-axisymmetric resonant instability. They determined the dependency of the Tayler instability maximum growth rate as a function of the azimuthal wave-number m and of the ratio ε of the axial field to the toroidal one. In our case, close to the center the flux function exhibits a behaviour in $\Psi \propto r^2$, so the azimuthal field is proportional to $s = r \sin \theta$ corresponding to the Bonanno *et al.*'s parameter $\alpha = 1$. As underlined by the authors, in that case the maximum growth rate changes remarkably slowly with m for all modes with $m \geqslant 2$ and the instability is weakly non-anisotropic. If we take as a value for s_1 the radius of the neutral line or the one where the azimuthal field is strongest, we obtain respectively $\varepsilon = 0.64$ or $\varepsilon = 0.79$. According to their study (see Bonanno & Urpin 2008a, FIG. 7), we fulfill the stability criterion for the modes $m = 0, 1$ and 2. Our results are therefore in agreement with their linear analysis.

4. Conclusion and perspectives

Using semi-analytic methods, we derived (with an appropriate choice of boundary conditions) then tested an axisymmetric non force-free relaxed magnetostatic equilibrium, which could exist in any non-convective stellar region: the radiative core of solar-type stars, the external envelope of massive stars, and compact objects. Using numerical simulations, we demonstrate the ability of the set-up to recover well-known instabilities in purely poloidal and toroidal cases, then find stability of the mixed configuration under all imaginable perturbations. We show the agreement of the result with linear analysis, highlighting the stabilizing influence of the poloidal field on the toroidal one, especially in the region close to the symmetry axis where purely toroidal fields usually develop kink-type instabilities in priority. This is the first time that the stability of an analytically-derived stellar magnetic configuration has been confirmed numerically (Duez, Braithwaite & Mathis 2010).

This result has strong astrophysical implications: the configuration, as described in Duez & Mathis (2010), provides a good initial condition to magneto-rotational transport

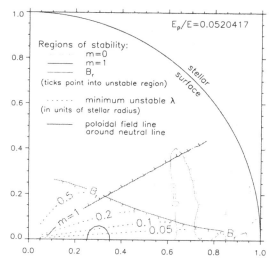

Figure 3. Half of the meridional plane, showing the regions stable against the $m = 0$ and 1 Tayler modes in the absence of the poloidal component, and their stabilisation by the radial component B_r. Taken from Duez, Braithwaite & Mathis (2010).

to be included in next generation stellar evolution codes and to magnetohydrodynamic simulations – where up to now the initial field would have been chosen arbitrarily; furthermore it will help to appreciate the internal magnetic structure of neutron stars, and various astrophysical processes involving magnetars.

Finally, we have to take into account differential rotation and induced meridional circulation in MHD relaxation theory in a near future.

Acknowledgments: S. Mathis thanks the organizers for this very nice conference. He also thanks A. Pouquet, K. Moffatt, J. Toomre, R. Arlt and A.-S. Brun for fruitful discussions. This work was supported in part by PNPS (CNRS/INSU).

References

Alecian, E., *et al.* 2008, *A&A*, 481, L99
Aurière, M., *et al.* 2007, *A&A*, 475, 1053
Biskamp, D. 1997, *Nonlinear Magnetohydrodynamics (Cambridge, UK: Cambridge University Press)*
Bonanno, A. & Urpin, V. 2008a, *A&A*, 488, 1
—. 2008b, *A&A*, 477, 35
Braithwaite, J. 2006, *A&A*, 449, 451
—. 2008, *MNRAS*, 386, 1947
—. 2009, *MNRAS*, 397, 763
Braithwaite, J. & Spruit, H. C. 2004, *Nature*, 431, 819
Braithwaite, J. & Nordlund, Å. 2006, *A&A*, 450, 1077
Brun, A. S. 2007, *AN*, 328, 1137
Brun, A. S. & Zahn, J.-P. 2006, *A&A*, 457, 665
Busse, F. H. 1981, *Geophysical and Astrophysical Fluid Dynamics*, 17, 215
Commerçon, B., Hennebelle, P., Audit, E., Chabrier, G., & Teyssier, R. 2010, *A&A*, 510, L3
Decressin, T., Mathis, S., Palacios, A., Siess, L., Talon, S., Charbonnel, C., & Zahn, J.-P. 2009, *A&A*, 495, 271
Duez, V. & Mathis, S. 2010, *A&A*, 517, A58

Duez, V., Braithwaite, J., & Mathis, S. 2010, *ApJ*, accepted

Duez, V., Mathis, S., & Turck-Chièze, S. 2010, *MNRAS*, 402, 271

Eff-Darwich, A., Korzennik, S. G., Jiménez-Reyes, S. J., & García, R. A. 2008, *ApJ*, 679, 1636

Elstner, D., Bonanno, A., & Rüdiger, G. 2008, *A&A*, 329, 717

Grad, H. & Rubin, H. 1958, *Proceedings of the Second United Nations International Conference on the Peaceful Uses of Atomic Energy*, Vol. 31, IAEA, Geneva, 190–197

Garaud, P. & Guervilly, C. 2009, *ApJ*, 695, 799

Grunhut, J. H., et al. 2009, *MNRAS*, 400, L94

Heinemann, M. & Olbert, S. 1978, *Journal of Geophysical Research*, 83, 2457

Mathis, S. & Zahn, J.-P. 2005, *A&A*, 440, 653

Mestel, L., Moss, D., & Tayler, R. J. 1988, *MNRAS*, 231, 873

Moffatt, H. K. 1985, *J. Fluid Mechanics*, 159, 359

Montgomery, D. & Phillips, L. 1988, *Phys. Rev. A*, 38, 2953

Montgomery, D. & Phillips, L. 1989, *Physica D*, 37, 215

Nordlund, Å. & Galsgaard, K. 1995, A 3D MHD code for Parallel Computers, Tech. rep., http://www.astro.ku.dk/~aake/papers/95.ps.gz

Prendergast, K. H. 1956, *ApJ*, 123, 498

Reisenegger, A. 2009, *A&A*, 499, 557

Shafranov, V. D. 1966, *Reviews of Plasma Physics*, 2, 103

Shaikh, D., Dasgupta, B., Hu, Q., & Zank, G. P. 2008, *J. Plasma Physics*, 75, 273

Spruit, H. C. 2002, *A&A*, 381, 923

Tayler, R. J. 1973, *MNRAS*, 161, 365

Taylor, J. B. 1974, *Phys. Rev. Lett.*, 33, 1139

Woltjer, L. 1959, *ApJ*, 130, 405

—. 1960, *ApJ*, 131, 227

Wright, G. A. E. 1973, *MNRAS*, 162, 339

Zahn, J.-P. 1992, *A&A*, 265, 115

Discussion

C. FOREST: What are the boundary conditions on relaxation? Is there a mechanism for helicity injection or extraction at tachocline?

S. MATHIS: In this work, we consider initially confined relaxed magnetic configurations as a first step. These then open due to ohmic dissipation. The case of open boundary conditions should be treated in a near future.

J. TOOMRE: In your work, you don't take into account the differential rotation and the induced meridional circulation. What could be the modification of relaxed configurations if those are treated?

S. MATHIS: In this work, we first study non-rotating purely magnetic equilibria as Braithwaite and collaborators. The obtained states are inchanged by the presence of a uniform rotation. However, if there is a differential rotation, a meridional circulation will be induced that will modify the obtained relaxed configurations (see Mestel, Moss & Tayler 1988).

Astrophysical Dynamics: From Stars to Galaxies
Proceedings IAU Symposium No. 271, 2010
N. H. Brummell, A. S. Brun, M. S. Miesch, & Y. Ponty, eds.

© International Astronomical Union 2011
doi:10.1017/S1743921311017704

From convective to stellar dynamos

Axel Brandenburg[1,2], Petri J. Käpylä[1,3], Maarit J. Korpi[3]

[1]NORDITA, Roslagstullsbacken 23, SE-10691 Stockholm, Sweden
[2]Department of Astronomy, Stockholm University, SE-10691 Stockholm, Sweden
[3]Department of Physics, PO Box 64, FI-00014 University of Helsinki, Finland

Abstract. Convectively driven dynamos with rotation generating magnetic fields on scales large compared with the scale of the turbulent eddies are being reviewed. It is argued that such fields can be understood as the result of an α effect. Simulations in Cartesian domains show that such large-scale magnetic fields saturate on a time scale compatible with the resistive one, suggesting that the magnitude of the α effect is here still constrained by approximate magnetic helicity conservation. It is argued that, in the absence of shear and/or any other known large-scale dynamo effects, these simulations prove the existence of turbulent α^2-type dynamos. Finally, recent results are discussed in the context of solar and stellar dynamos.

Keywords. MHD – turbulence – Sun: magnetic fields

1. Introduction

Stars with outer convection zones are known to display magnetic activity, often in a cyclic fashion like in the Sun. Such activity can generally be explained by a turbulent dynamo influenced by rotation and stratification to produce the anisotropies required for the generation of large-scale magnetic fields. The basic theory is reasonably well understood (Moffatt 1978; Parker 1979; Krause & Rädler 1980), but there continues to be substantial controversy until the present day. A major stumbling block has been the understanding of what is known as catastrophic quenching (Vainshtein & Cattaneo 1992; Cattaneo & Hughes 1996) and resistively limited saturation (Brandenburg 2001), as well as the very existence of the α effect in convection even without nonlinearity (Cattaneo & Hughes 2006; Hughes & Cattaneo 2008).

The first two issues have been reviewed in detail by Brandenburg & Subramanian (2005). The purpose of the present paper is to review recent progress concerning convective dynamos. However, in view of applications to solar and stellar dynamos, it is important to realize that we are still lacking simulations that reproduce the salient features of the solar dynamo. We should therefore keep our eyes open for new phenomena that may emerge as simulations become more realistic.

2. Excitation conditions of small-scale and large-scale dynamos

Small-scale and large-scale dynamos are quite different in nature. The difference becomes most evident in the nonlinearly saturated regime, provided one allows for what we call scale separation, which means that the size of the domain is large compared with the scale of the largest (energy-carrying) eddies of the turbulence. In Figure 1 we show spectra highlighting the remarkable difference between the two types of dynamos. Conversely, if there is insufficient scale separation, a large-scale dynamo becomes impossible and both types of simulations would look very similar, as has been demonstrated by Haugen *et al.* (2004); see their Fig. 23.

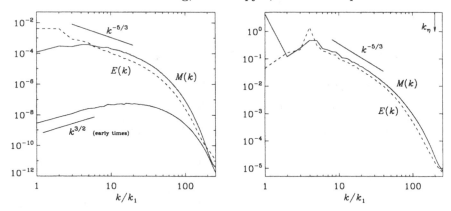

Figure 1. Kinetic and magnetic energy spectra in a turbulence simulation without net helicity (left) and with net helicity (right) for a magnetic Prandtl number of unity and a mesh size is 512^3 meshpoints. Notice the pronounced peak of $M(k)$ at $k = k_1$ in the case with helicity. Adapted from Brandenburg & Subramanian (2005) and Brandenburg (2009), respectively.

There can be different types of large-scale dynamos. The most frequently studied ones are the α^2 and $\alpha\Omega$ type dynamos. These are dynamos that can produce large-scale magnetic fields owing to the presence of kinetic helicity in the turbulence, giving rise to an α effect. The presence of shear can further modify the dynamo, making it usually easier to excite and favoring oscillatory over non-oscillatory solutions.

Shear can be a typical result of rotation of a gaseous body in the presence of anisotropic turbulence (Rüdiger 1980, 1989). Shear alone is often found to give rise to large-scale fields – even if the turbulence is non-helical (Brandenburg 2005; Yousef *et al.* 2008a,b; Brandenburg *et al.* 2008). The nature of such dynamo action is still a matter of debate and ranges from *incoherent* $\alpha\Omega$ dynamos (Vishniac & Brandenburg 1997; Proctor 2007) to shear–current dynamos (Rogachevskii & Kleeorin 2003, 2004).

Returning to the α^2 and $\alpha\Omega$ dynamos, it is important to realize that their excitation conditions are generally quite different. The onset of small-scale dynamo action depends generally on the value of the magnetic Reynolds number,

$$\mathrm{Re}_M = u_{\mathrm{rms}}/\eta k_{\mathrm{f}}, \tag{2.1}$$

where u_{rms} is the typical rms velocity of the turbulence, η is the microscopic magnetic diffusivity, and k_{f} is the forcing or integral wavenumber, i.e. the wavenumber of the energy-carrying motions. This is roughly where the peak of the energy spectrum is located. The critical value, $R_{\mathrm{m,crit}}$, above which dynamo action commences, depends on the value of the magnetic Prandtl number, $\mathrm{Pr}_M = \nu/\eta$, where ν is the microscopic kinematic viscosity and is about 35 for $\mathrm{Pr}_M = 1$ (Novikov *et al.* 1983; Subramanian 1999; Haugen *et al.* 2004), but increases to values around and above 400 for Pr_M somewhere between 0.2 and 0.1 (Schekochihin *et al.* 2005). There is now also evidence that $R_{\mathrm{m,crit}}$ may actually show a peak at $\mathrm{Pr}_M = 0.1$ and might then drop to slightly lower values for $\mathrm{Pr}_M = 0.05$ and below (Iskakov 2007). This unusual behavior is connected with a change of the "roughness" of the velocity field (Boldyrev & Cattaneo 2004) and the occurrence of a bottleneck effect in the velocity spectrum (Falkovich 1994; Dobler *et al.* 2003), which means that the velocity has maximum roughness for $\mathrm{Pr}_M \approx 0.1$ when the resistive scale coincides with the position of the bottleneck.

The situation is quite different with large-scale dynamos that operate completely independently of the value of Pr_M, as long as Re_M is large enough. Already in Brandenburg

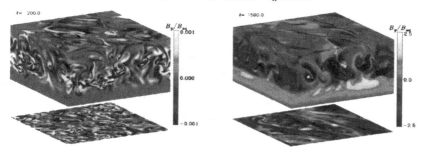

Figure 2. Snapshots of B_y in the early phase (*left*) and saturated phase (*right*) of a convective dynamo with shear. Adapted from Käpylä *et al.* (2008); see also http://www.helsinki.fi/~kapyla/movies.html

(2001) the critical value of Re_M was found to be around unity regardless of whether $\mathrm{Pr}_M = 1$ or 0.1. This finding was then extended by Mininni (2007) and Brandenburg (2009), who demonstrated dynamo action down to $\mathrm{Pr}_M = 0.005$ and 0.001, respectively, or up to $\mathrm{Pr}_M = 1000$ (Brandenburg 2011). The conclusion is that large-scale dynamo action depends solely on the *dynamo number*, which is given by $D = C_\alpha$ for α^2 dynamos and by $D = C_\alpha C_S$ for $\alpha\Omega$ (or α–shear) dynamos. Here,

$$C_\alpha = \alpha_0/\eta_T k_1 \quad \text{and} \quad C_S = S/\eta_T k_1^2, \tag{2.2}$$

where $\eta_T = \eta_t + \eta$ is the sum of turbulent and microscopic magnetic diffusivities, α_0 is a typical value of the α effect, and S is the shear rate (i.e. a typical value of the velocity gradient). For C_α and C_S, we use standard estimates:

$$\alpha_0 \approx -\frac{\tau}{3}\overline{\boldsymbol{\omega}\cdot\boldsymbol{u}} \approx -\frac{\epsilon_f}{3u_{\mathrm{rms}}k_f}k_f u_{\mathrm{rms}}^2 = -\tfrac{1}{3}\epsilon_f u_{\mathrm{rms}}, \tag{2.3}$$

where $\epsilon_f = \overline{\boldsymbol{\omega}\cdot\boldsymbol{u}}/k_f u_{\mathrm{rms}}^2$ is a measure of the relative kinetic helicity, $\tau = (u_{\mathrm{rms}}k_f)^{-1}$ is the turnover time, and

$$\eta_t \approx \frac{\tau}{3}\overline{u^2} \approx u_{\mathrm{rms}}/3k_f. \tag{2.4}$$

With this we find

$$C_\alpha = -\frac{\tfrac{1}{3}\epsilon_f u_{\mathrm{rms}}/k_1}{u_{\mathrm{rms}}/3k_f + \eta} = -\iota\epsilon_f \frac{k_f}{k_1} \tag{2.5}$$

where

$$\iota = 1/\left(1 + 3\mathrm{Re}_M^{-1}\right) \tag{2.6}$$

is a correction factor that is close to unity for $\mathrm{Re}_M \gg 1$. Furthermore, we have

$$C_S = \frac{S/k_1^2}{u_{\mathrm{rms}}/3k_f + \eta} = \frac{3\iota S}{u_{\mathrm{rms}}k_f}\left(\frac{k_f}{k_1}\right)^2 = 3\iota\mathrm{Sh}\left(\frac{k_f}{k_1}\right)^2, \tag{2.7}$$

where we have defined the shear parameter $\mathrm{Sh} = S/u_{\mathrm{rms}}k_f$. Note that, especially in the presence of shear, the possibility of dynamo action is strongly connected with the scale separation ratio. Indeed,

$$D = -3\iota\epsilon_f\mathrm{Sh}\left(\frac{k_f}{k_1}\right)^3 \tag{2.8}$$

depends cubicly on the scale separation ratio. This explains why $\alpha\Omega$ dynamos are often much easier to obtain than α^2 dynamos.

3. Large-scale dynamos in Cartesian domains

3.1. *Dynamos of $\alpha\Omega$ type*

Given the alarming signs of earlier investigations by Cattaneo & Hughes (2006) and Hughes & Cattaneo (2008), it was quite unclear whether the α effect even exists in simulations with convection. At the time, several possible reasons were put forward, including the absence of stratification; see, for example, Brandenburg (2009). In the wake of this initial frustration, it was quite surprising when large-scale dynamo action was found in rotating convection in the presence of shear (Käpylä *et al.* 2008); see Figure 2. Similar results were obtained by Hughes & Proctor (2009). This controversy was still ongoing at the conference "Turbulence and Dynamos" in Stockholm in March 2008† where Hughes‡ argued that no convective large-scale dynamos exist, while Käpylä¶ showed results from low Reynolds number convection with shear where large-scale fields were indeed obtained. In an attempt to clarify the still conflicting results regarding the actual value of α in rotating convection, Käpylä *et al.* (2010a) pointed out that for a nonuniform mean field, the mean current density cannot be neglected. In this case, the turbulent magnetic diffusivity contributes and explains the small net electromotive force measured by Hughes & Proctor (2009) by imposing a uniform field.

3.2. *Dynamos of α^2 type*

The simulations mentioned above do not provide conclusive evidence for the existence of an α effect in rotating convection, because it is in principle possible that the dynamo could be the result of an incoherent $\alpha\Omega$ dynamo or a shear–current dynamo. In the absence of shear, however, there is no viable alternative to an α^2 dynamo. It is therefore important to consider the conceptually simpler case without imposed shear, as was also emphasized by Hughes *et al.* (2011), who noted that this was not done by Käpylä *et al.* (2010a), who just considered the case of a sinusoidal shear profile. For this reason, we discuss in the following the papers of Käpylä *et al.* (2009a) and, in particular, Käpylä *et al.* (2009b), where large-scale dynamo action was studied in non-shearing convection at sufficiently large Coriolis numbers.

Before trying to simulate an α^2 dynamo for rotating convection, it is instructive to obtain guidance from numerically obtained measurements of the α and turbulent diffusivity tensors. This can be done using the test-field method (Schrinner *et al.* 2005, 2007), which has been applied to turbulence in a number of recent papers (Brandenburg 2005; Brandenburg *et al.* 2008). The result is shown in Figure 5. Using this method, Käpylä *et al.* (2009a) noted that the magnitude of the relevant components of the α tensor vary only weakly with Coriolis number, $\mathrm{Co} = 2\Omega/u_{\mathrm{rms}}k_{\mathrm{f}}$, where Ω is angular velocity, while η_{t} diminishes with increasing values of Co. This was a clear indication that dynamos of α^2 type might become possible once Co is large enough. We emphasize this point, because it is one of the several examples where mean-field theory has proven its predictive power.

Consequently, in a subsequent investigation, Käpylä *et al.* (2009b) carried out simulations for large enough values of Co and did indeed find dynamo action of large-scale type when $\mathrm{Co} \gtrsim 10$. The large-scale field became even more pronounced as the aspect ratio was increased. In Figure 3 we present horizontal spectra of magnetic and kinetic energies. What is important here is the fact that, even though the magnetic energy is less (by factor 5) than the kinetic energy at what we estimate to be k_{f} (about $5k_1$), the magnetic energy strongly *exceeds* the kinetic energy at the scale of the domain. This seems to

---

† http://agenda.albanova.se/conferenceDisplay.py?confId=325
‡ http://videos.nordita.org/conference/Turbulence2008/hires/March17/Part1.WMV
¶ http://videos.nordita.org/conference/Turbulence2008/hires/March19/Part5.WMV

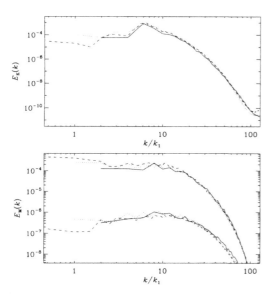

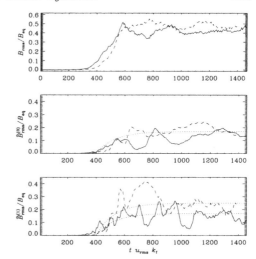

Figure 4. Top panel: rms-values of the total magnetic field as functions of time for vertical field (solid lines) and perfect conductor boundary conditions (dashed lines). The two lower panels show the sums of the rms-values of the Fourier amplitudes of B_x and B_y for $k/k_1 = 0$ (middle panel) and $k/k_1 = 1$ (bottom panel). The dotted lines in the two lower panels show a saturation predictor according to the model of Brandenburg (2001). Adapted from Käpylä et al. (2009b).

Figure 3. Two-dimensional power spectra of velocity (upper panel) and magnetic field (lower panel) as functions of system size. In the lower panel the upper curves show the spectra from the saturated state whereas the lower curves show the spectra from the kinematic state multiplied by 10^7. Adapted from Käpylä et al. (2009b).

exclude alternative explanations whereby the magnetic field at smaller wavenumbers might just be a trivial result of diffusion in wavenumber space. Instead, we argue that this is strong evidence for the physical reality of an α^2 dynamo driven by rotating convection.

In agreement with virtually all earlier work on large-scale dynamos of α^2 type the saturation time of the dynamo is comparable with the resistive time. Indeed, Brandenburg (2001) found that in the absence of strong magnetic helicity fluxes, the saturation of an α^2 dynamo follows a switch-on behavior where, after saturation, the mean field is given by

$$\frac{\overline{B}^2}{B_{\mathrm{eq}}^2} \approx \frac{k_{\mathrm{f}}}{k_{\mathrm{m}}}\left[1 - e^{-2\eta k_{\mathrm{m}}^2\,(t-t_{\mathrm{s}})}\right]. \tag{3.1}$$

This is also seen in the present case; see Figure 4, where we overplot the prediction from Equation (3.1).

It is likely that diffusive magnetic helicity fluxes are present in the convection simulations discussed above (Brandenburg et al. 2009). Those fluxes could in principle give rise to faster saturation times than what is seen in Figure 3.1. This question has been addressed quantitatively by Mitra et al. (2010a) and Hubbard & Brandenburg (2010), who note that at the magnetic Reynolds numbers accessible so far, diffusive helicity fluxes are still quite weak compared with the resistive processes. Based on their scalings for different values of Re_M, they estimate that resistive saturation effects would only begin to be alleviated for Re_M well in excess of values around 1000 or even 10^4; see also Fig. 10 of Candelaresi et al. (2011). Convection simulations with closed magnetic boundaries do

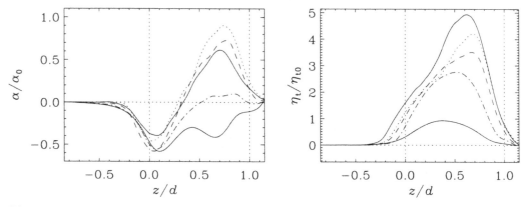

Figure 5. Normalized profiles of α (left panel) and η_t (right panel) from kinematic test field simulations. The vertical dotted lines at $z/d = 0$ and $z/d = 1$ indicate the base and top of the convectively unstable layer, respectively. Adapted from Käpylä *et al.* (2009b).

seem to suffer from catastrophic quenching at least up to $\mathrm{Re}_M \approx 240$ (Käpylä *et al.* 2010b). Reaching much higher values to study this issue further in the near future is not possible. While this is certainly somewhat discouraging news, it does highlight the importance of studying detailed scaling properties of large-scale dynamos rather than producing solitary examples of dynamos at large resolution, hoping that they represent the Sun.

4. Bi-helical magnetic field

An important property of α effect dynamos is the fact that they produce bi-helical magnetic fields. This means that one expects to see magnetic helicity fluxes with opposite signs at small and large scales. While this is now well established theoretically (Blackman & Brandenburg 2003; Yousef & Brandenburg 2003), there is still no widespread observational evidence for this. Large-scale magnetic helicity can be estimated using synoptic maps of the azimuthally averaged radial magnetic field of the Sun; see Fig. 1 of Brandenburg *et al.* (2003). Small-scale magnetic helicity fluxes have been inferred from magnetic field measurements in active regions and coronal mass ejections (Berger & Ruzmaikin 2000).

There is also some evidence from measurements of magnetic helicity in the solar wind. Using the assumption of homogeneity, Matthaeus *et al.* (1982) were able to determine magnetic helicity spectra for the solar wind. Preliminary analysis of more recent solar wind data from the Ulysses spacecraft does indeed suggest that the field in the solar wind is also bi-helical. Further details on this are presented in a dedicated publication (Brandenburg *et al.* 2011).

5. Concluding remarks

We still do not know exactly how the solar dynamo works. If it is of $\alpha\Omega$ type, given that the α effect is positive in the northern hemisphere, and using the fact that radial shear is positive in the bulk of the convection zone, one would expect poleward migration of the dynamo wave. This is also what three-dimensional simulations in spherical shells have shown repeatedly over several decades starting with the early work of Gilman (1983), and now again in the spherical wedge simulations of Käpylä *et al.* (2010c). For rapid rotation, however, polarity flips of toroidal magnetic field can also occur more abruptly, as has been demonstrated by Brown *et al.* (2010) and Ghizaru *et al.* (2010), which is beginning to be reminiscent of polarity reversals in the geodynamo (Glatzmaier & Roberts 1995), but is different from what we know about the solar dynamo. In this connection it is worth recalling yet another recent surprise: oscillatory solutions with equatorward migration are also possible in the absence of any differential rotation provided the dynamo is somehow bounded between highly conducting media at high latitudes (Mitra *et al.* 2010b). It is obviously unclear whether this has any bearing on the solar dynamo, but it reminds us of the possibility of surprises.

Other proposals for how the solar dynamo might work include the flux transport dynamo (Durney 1995; Choudhuri *et al.* 1995; Dikpati & Charbonneau 1999), and the possibility of a dynamo shaped by the negative radial angular velocity gradient in the near-surface shear layer (Brandenburg 2005). Neither of these two scenarios have been seen in three-dimensional turbulence simulations. The former suffers from the difficulty of obtaining a sufficiently coherent meridional circulation that does not break up into smaller circulation patterns, while the latter may suffer from the difficulty of resolving the small-scale turbulence in the near-surface shear layer. A possible step forward might therefore be a combined effort utilizing a range of different simulations in spherical and Cartesian geometries on the one hand, and improved mean-field theory on the other. Clearly, in order to improve mean-field theory it is essential to seek guidance from direct simulations, as has already been done in recent years with increasing success.

Acknowledgements

We acknowledge the allocation of computing resources provided by the Swedish National Allocations Committee at the Center for Parallel Computers at the Royal Institute of Technology in Stockholm and the National Supercomputer Centers in Linköping as well as the Norwegian National Allocations Committee at the Bergen Center for Computational Science and the computing facilities hosted by CSC - IT Center for Science Ltd. in Espoo, Finland, who are administered by the Finnish Ministry of Education. This work was supported in part by the European Research Council under the AstroDyn Research Project 227952, the Swedish Research Council grant 621-2007-4064, and the Finnish Academy grants 121431, 136189, and 112020.

References

Berger, M. A. & Ruzmaikin, A. 2000; J. Geophys. Res., 105, 10481
Blackman, E. G. & Brandenburg, A. 2003, ApJ, 584, L99
Boldyrev, S. & Cattaneo, F. 2004, Phys. Rev. Lett., 92, 144501
Brandenburg, A. 2001, ApJ, 550, 824
Brandenburg, A. 2005, ApJ, 625, 539
Brandenburg, A. 2005, Astron. Nachr., 326, 787
Brandenburg, A. 2009, ApJ, 697, 1206
Brandenburg, A. 2009, Space Sci. Rev., 144, 87
Brandenburg, A. 2011, Astron. Nachr., 332, 51
Brandenburg, A. & Subramanian, K. 2005, Phys. Rep., 417, 1
Brandenburg, A., Blackman, E. G., & Sarson, G. R. 2003, Adv. Space Sci., 32, 1835
Brandenburg, A., Rädler, K.-H., Rheinhardt, M., & Käpylä, P. J. 2008, ApJ, 676, 740
Brandenburg, A., Candelaresi, S., & Chatterjee, P. 2009, MNRAS, 398, 1414
Brandenburg, A., Subramanian, K., Balogh, A., & Goldstein, M. L. 2011, arXiv:1101.1709
Brown, B. P., Browning, M. K., Brun, A. S., Miesch, M. S., & Toomre, J. 2010, ApJ, 711, 424
Candelaresi, S., Hubbard, A., Brandenburg, A., & Mitra, D. 2011, Phys. Plasmas, 18, 012903

Cattaneo, F. & Hughes, D. W. 1996, Phys. Rev. E, 54, R4532

Cattaneo, F. & Hughes, D. W. 2006, J. Fluid Mech., 553, 401

Choudhuri, A. R., Schüssler, M. & Dikpati, M. 1995, A&A, 303, L29

Dikpati, M. & Charbonneau, P. 1999, ApJ, 518, 508

Dobler, W., Haugen, N. E. L., Yousef, T. A. & Brandenburg, A. 2003, Phys. Rev. E, 68, 026304

Durney, B. R. 1995, Solar Phys., 160, 213

Falkovich, G. 1994, Phys. Fluids, 6, 1411

Ghizaru, M., Charbonneau, P. & Smolarkiewicz, P. K. 2010, ApJ, 715, L133

Gilman, P. A. 1983, ApJS, 53, 243

Glatzmaier, G. A. & Roberts, P. H. 1995, Nature, 377, 203

Haugen, N. E. L., Brandenburg, A. & Dobler, W. 2004, Phys. Rev. E, 70, 016308

Hubbard, A. & Brandenburg, A. 2010, Geophys. Astrophys. Fluid Dyn., 104, 577

Hughes, D. W. & Cattaneo, F. 2008, J. Fluid Mech., 594, 445

Hughes, D. W. & Proctor, M. R. E. 2009, Phys. Rev. Lett., 102, 044501

Hughes, D. W., Proctor, M. R. E. & Cattaneo, F. 2011, arXiv:1103.0754

Iskakov, A. B., Schekochihin, A. A., Cowley, S. C., McWilliams, J. C. & Proctor, M. R. E. 2007, Phys. Rev. Lett., 98, 208501

Käpylä, P. J., Korpi, M. J., & Brandenburg, A. 2008, A&A, 491, 353

Käpylä, P. J., Korpi, M. J., & Brandenburg, A. 2009a, A&A, 500, 633

Käpylä, P. J., Korpi, M. J., & Brandenburg, A. 2009, ApJ, 697, 1153

Käpylä, P. J., Korpi, M. J., & Brandenburg, A. 2010a, MNRAS, 402, 1458

Käpylä, P. J., Korpi, M. J., & Brandenburg, A. 2010b, A&A, 518, A22

Käpylä, P. J., Korpi, M. J., & Brandenburg, A., Mitra, D., & Tavakol, R. 2010b, Astron. Nachr., 331, 73

Krause, F. & Rädler, K.-H. 1980, Mean-field magnetohydrodynamics and dynamo theory (Pergamon Press, Oxford)

Matthaeus, W. H., Goldstein, M. L., & Smith, C. 1982, Phys. Rev. Lett., 48, 1256

Mininni, P. D. 2007, Phys. Rev. E, 76, 026316

Mitra, D., Candelaresi, S., Chatterjee, P., Tavakol, R., & Brandenburg, A. 2010a, Astron. Nachr., 331, 130

Mitra, D., Tavakol, R., Käpylä, P. J., & Brandenburg, A. 2010b, ApJ, 719, L1

Moffatt, H.K. 1978, Magnetic field generation in electrically conducting fluids (Cambridge University Press, Cambridge)

Novikov, V. G., Ruzmaikin, A. A., & Sokoloff, D. D. 1983, Sov. Phys. JETP, 58, 527

Parker, E. N. 1979, Cosmical magnetic fields (Oxford University Press, New York)

Proctor, M. R. E. 2007, MNRAS, 382, L39

Rogachevskii, I. & Kleeorin, N. 2003, Phys. Rev. E, 68, 036301

Rogachevskii, I. & Kleeorin, N. 2004, Phys. Rev. E, 70, 046310

Rüdiger, G. 1980, Geophys. Astrophys. Fluid Dyn., 16, 239

Rüdiger, G. 1989, Differential rotation and stellar convection: Sun and solar-type stars (Gordon & Breach, New York)

Schekochihin, A. A., Haugen, N. E. L., Brandenburg, A., Cowley, S. C., Maron, J. L., & McWilliams, J. C. 2005, ApJ, 625, L115

Schrinner, M., Rädler, K.-H., Schmitt, D., Rheinhardt, M. & Christensen, U. 2005, Astron. Nachr., 326, 245

Schrinner, M., Rädler, K.-H., Schmitt, D., Rheinhardt, M., & Christensen, U. R. 2007, Geophys. Astrophys. Fluid Dyn., 101, 81

Subramanian, K. 1999, Phys. Rev. Lett., 83, 2957

Subramanian, K. & Brandenburg, A. 2006, ApJ, 648, L71

Vainshtein, S. I. & Cattaneo, F. 1992, ApJ, 393, 165

Vishniac, E. T. & Brandenburg, A. 1997, ApJ, 475, 263

Warnecke, J. & Brandenburg, A. 2010, A&A, 523, A19

Yousef, T. A. & Brandenburg, A. 2003, A&A, 407, 7

Yousef, T. A., Heinemann, T., Schekochihin, A. A., Kleeorin, N., Rogachevskii, I., Iskakov, A. B., Cowley, S. C. & McWilliams, J. C. 2008a, Phys. Rev. Lett., 100, 184501

Yousef, T. A., Heinemann, T., Rincon, F., Schekochihin, A. A., Kleeorin, N., Rogachevskii, I., Cowley, S. C., & McWilliams, J. C. 2008b, Astron. Nachr., 329, 737

Discussion

M. PROCTOR: What do you do about the gauge when calculating magnetic helicity, and does it make any difference to the answers obtained?

A. BRANDENBURG: Yes, the magnetic helicity density is in general gauge-dependent. However, if there is sufficient scale separation between mean and fluctuating fields, the magnetic helicity density computed from the fluctuating field can be shown to be gauge-invariant (Subramanian & Brandenburg 2006). Hubbard & Brandenburg (2010) have recently confirmed this in a simulation where the magnetic helicity from the mean field was strongly gauge dependent, but that from the fluctuating field was not.

T. ROGERS: To calculate magnetic helicity you assumed that the solar wind was isotropic, which observations show it is not. How would this affect the results you present?

A. BRANDENBURG: Since we have to adopt the Taylor hypothesis, only the two magnetic field components perpendicular to the radial direction enter the calculation. The field in the plane perpendicular to the radial direction is still fairly isotropic, so I guess our results are still meaningful. To clarify the significance of our results further, it might be useful to compute magnetic helicity from simulations of anisotropic MHD turbulence with one preferred direction.

D. HUGHES: In your simulations that show the generation of large-scale fields on a long time, what is the timescale for the generation of the fields? If it is ohmic then it is not surprising.

A. BRANDENBURG: The initial exponential growth occurs always on a dynamical time scale, but full saturation is only obtained on a resistive time scale. We know that this problem can only be alleviated by magnetic helicity fluxes, which are quite weak in our Cartesian simulations. Nevertheless, our simulations prove the point that the α effect works in rotating convection, which was until now quite unclear.

C. FOREST: How do you model the boundary conditions? Open or Closed?

A. BRANDENBURG: At the bottom we adopt a perfect conductor boundary condition and at the top we assume that the horizontal field vanishes. This pseudo-vacuum condition is numerically more robust than a proper vacuum condition. However, it would be more realistic to couple the convection simulation to a force-free model, as has been done for forced turbulence simulations in the paper by Warnecke & Brandenburg (2010), which is also presented here as a poster.

Astrophysical Dynamics: From Stars to Galaxies
Proceedings IAU Symposium No. 271, 2010
N.H. Brummell, A.S. Brun, M.Miesch & Y. Ponty, eds.

© International Astronomical Union 2011
doi:10.1017/S1743921311017716

Can short time delays influence the variability of the solar cycle?

Laurène Jouve, Michael R. E. Proctor and Geoffroy Lesur

DAMTP, Centre for Mathematical Sciences, Wilberforce Road, CB3 0WA CAMBRIDGE, UK

Abstract. We present the effects of introducing results of 3D MHD simulations of buoyant magnetic fields in the solar convection zone in 2D mean-field Babcock-Leighton models. In particular, we take into account the time delay introduced by the rise time of the toroidal structures from the base of the convection zone to the solar surface. We find that the delays produce large temporal modulation of the cycle amplitude even when strong and thus rapidly rising flux tubes are considered. The study of a reduced model reveals that aperiodic modulations of the solar cycle appear after a sequence of period doubling bifurcations typical of non-linear systems. We also discuss the memory of such systems and the conclusions which may be drawn concerning the actual solar cycle variability.

Keywords. Magnetic fields - Sun: activity - Sun: interior - Methods: numerical

1. Introduction: time delay dynamics and solar dynamo models

In the framework of mean-field dynamo theory, several possibilities have been studied to explain the variability of the solar cycle. They mainly fell into two categories: stochastic forcing or dynamical nonlinearities. Indeed, the solar convection zone is highly turbulent and it would be surprising if the dynamo processes acting inside this turbulent plasma were nicely regular. The influence of stochastic fluctuations in the mean-field dynamo coefficients has been studied in various models (e.g. Hoyng 1988, Weiss & Tobias 2000, Charbonneau & Dikpati 2000). Moreover, the dynamical feedback of the strong dynamo-generated magnetic fields is likely to be significant enough to produce non-linear effects on the activity cycle. A number of models have introduced these non-linear effects (Tobias 1997, Moss & Brooke 2000, Bushby 2006, Rempel 2006) and have resulted in the production of grand minima-like periods or other strong modulations of the cyclic activity. However, time delays have hardly been considered, even if they were known to increase the order of the governing equations and thus likely to produce interesting long-term evolutions.

Yoshimura (1978) was the first to introduce time delays both on the toroidal and poloidal potential equations. His values were quite arbitrary but he made the point that those time delays alone could act to produce long-term modulation of the periodic solutions. In Babcock-Leighton flux-transport models, a meridional flow is introduced and flux emergence from the base of the convection zone to the surface is implicitly present. In this particular model, the introduction of time delays and their consequences have also been studied (e.g. Wilmot-Smith *et al.* 2006) but the delays were mainly due to the advection time by meridional flow. Indeed, the time-scale of the buoyant rise of flux tubes was considered to be so small compared to the cycle period (and to the meridional flow turnover time) that this particular step was assumed to be instantaneous. However, we address here the question of the influence of magnetic field dependent delays on the cycle produced by Babcock-Leighton models and especially on its potential modulation.

2. What about the rise time of buoyant magnetic fields?

2.1. *The modified source term for poloidal field*

In Babcock-Leighton dynamo models, the poloidal field owes its origin to the tilt of magnetic loops emerging at the solar surface. Since these emerging loops are thought to rise from the base of the convection zone through magnetic buoyancy, we see that we can directly relate the way we model the Babcock-Leighton (BL) source term and the results of 3D calculations of Jouve & Brun (2009).

In the standard model, the source term is confined in a thin layer at the surface and is made to be antisymmetric with respect to the equator, due to the sign of the Coriolis force which changes from one hemisphere to the other. These features are retained in our modified version.

However, the standard term is proportional to the toroidal field at the base of the convection zone at the same time, implying an instantaneous rise of the flux tubes from the base to the surface where they create tilted active regions. The 3D calculations showed that the rise velocity and thus the rise time depend on the field strength at the base of the CZ, we thus introduce in our modified version of the source term, a magnetic field-dependent time delay in the toroidal field at the base of the convection zone. We thus take into account the time delay between the formation of strong toroidal structures at the base of the convection zone and the surface regeneration of poloidal field.

The modified expression of the source term is thus

$$
S(r, \theta, B_\phi) = \frac{1}{2}\left[1 + \mathrm{erf}\left(\frac{r - r_2}{d}\right)\right]\left[1 - \mathrm{erf}\left(\frac{r - 1}{d}\right)\right] \times \left[1 + \left(\frac{B_\phi(r_c, \theta, t - \tau_B)}{B_0}\right)^2\right]^{-1}
$$
$$
\times \cos\theta \sin\theta \; B_\phi(r_c, \theta, t - \tau_B) \tag{2.1}
$$

where $r_c = 0.7$, $r_2 = 0.95$, $d = 0.01$, $B_0 = 10^4$, with the time delay τ_B proportional to the inverse of the magnetic energy at the base of the convection zone

$$
\tau_B(\theta, t) = \tau_0/B_\phi(r_c, \theta, t)^2 \tag{2.2}
$$

In the 3D simulations, the approximate rise time for a 6×10^5 G field is about 4 times that of a 3×10^5 G field, agreeing with the formulation 2.2. This takes into account the more significant effects of convective downdrafts and Ohmic diffusion in the "weak B" case. Note that the time delay is then dependent both on space and time.

Finally, a threshold has been introduced on the source term for both very strong (above 10^4 G) and very weak fields (below 100 G). We thus take into account the fact that very strong toroidal structures are not influenced by the Coriolis force enough to gain a significant tilt as they reach the surface and we prevent unrealistically large delays to appear in our model.

2.2. *Long-term modulation of the cycle amplitude*

The main result of our calculations (as shown in Jouve, Proctor & Lesur 2010) is that increasing the time delays in our model tends to produce a long-term modulation of the magnetic activity, even in the limit of small delays compared to the full cycle. Figure 1 illustrates the behaviour of the toroidal magnetic energy at the base of the convection zone and at the latitude of $25°$ when the delay is increased from 0 to 14 days on 5×10^4 G fields. The major effect of the delays is to modulate the amplitude of the cycle to such a point that the weakest cycle on the particular period calculated here reaches in terms

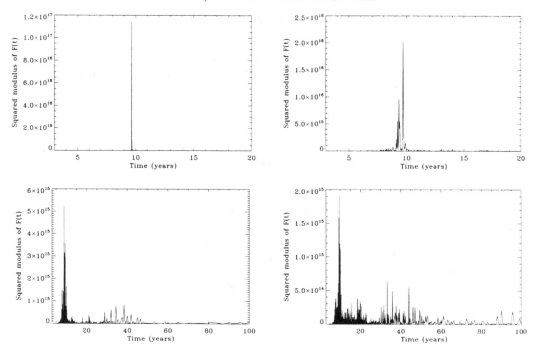

Figure 1. Powerspectra for the following values of the delay: no delay, 14 days delay on 5 kG-fields, on 10 kG-fields and on 50 kG-fields. The peak at T≈10 years is clearly visible in every panels, the only difference is the appearance of extra frequencies for increasing delays. Note the change of scale for the abscissa in the two lower panels.

of peak energy only 30% of the peak energy of the strongest cycle. If we consider the powerspectra of the various cases (shown on Fig. 1), we can study if any particular secondary frequencies are excited and how this depends on the delay. The first striking point which can be noted from analysing this figure is that the fundamental cycle period is retained in all the models. The most significant peak is indeed always located at about 10 years. More interestingly, some additional periods seem to appear, particularly at higher values than the fundamental component. This shows our small time delays produce modulations of the cycle on time scales much longer than the cycle period. In particular, the last panel of Fig. 1 seems to show some secondary periods between 30 and 50 years, as well as a peak around 20 years, possibly resulting from a previous period doubling which led to this chaotic behaviour.

2.3. A reduced model showing the route to chaos

In Jouve, Proctor & Lesur (2010), we develop a reduced model to make a tentative explanation of this observed modulated activity and the chaotic solution we obtain in the limit of strong delays. We summarize here the formulation of the model and the associated results.

The reduced model we choose to study does not contain any explicit delays, it rather introduces a new variable Q_t which will represent the delayed toroidal field and will be responsible for the regeneration of the poloidal component. For simplification, the equations are written in one dimension and we use a Fourier expansion in this direction, with wavenumber $k = 1$. We thus get the following set of complex ODEs:

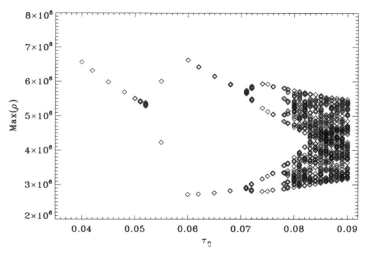

Figure 2. Bifurcation diagram obtained with a reduced model of the delayed equations. The maximum value of the modulus of B_t is plotted here against the intensity of the delay.

$$\frac{dA_t}{dt} + ivA_t = \frac{SQ_t}{1 + \lambda|Q_t|^2} - \eta A_t \tag{2.3}$$

$$\frac{dB_t}{dt} + ivB_t = i\Omega A_t - \eta B_t \tag{2.4}$$

$$\frac{dQ_t}{dt} = \frac{1}{\tau}(B_t - Q_t) \tag{2.5}$$

with $\tau = \tau_0/(1 + |B_t|^2)$.

Thanks to its phase invariance, this system can be further reduced from 6 to 5 degrees of freedom and the evolution of our solution can now be accurately followed when the delay is increased.

On Fig. 2, the peak amplitude of the modulus of B_t is followed for successive discrete values of the delay. This procedure results in a bifurcation diagram showing a succession of period doublings leading to a chaotic behaviour when the delay on the strongest fields represents about 10% of the cycle period. Up till now, the nonlinearities at the origin of the sequence of bifurcations in most models were always related to the dynamical feedback of the Lorentz force on the flow. The striking result here is that the modulation of the cycle amplitude can also arise in models where the only nonlinearities are the quenching term in the Babcock-Leighton source and the magnetic field-dependent time delays. It has here to be noted though that the time delays also appear in the quenching term, possibly amplifying the nonlinear effects in the model.

2.4. *Memory of the system*

In this subsection, we go back to the full spherical 2D model and study the so-called memory of the system between various cycles. This analysis has been shown to be efficient at distinguishing between advection-dominated and diffusion-dominated models as discussed in Yeates, Nandy & Mackay (2008). In their work, the peak surface radial flux for cycle n is compared to the peak toroidal flux for cycles n, $n + 1$, $n + 2$ and $n + 3$.

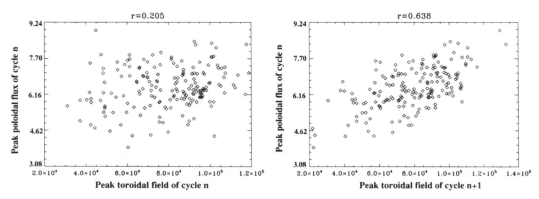

Figure 3. Correlation analysis between successive cycles. Only the relations between cycles n/n (left panel) and n/n+1 (right panel) are shown here. The poloidal flux unit is 10^{24} Mx.

They find that in an advection-dominated model where the BL source term is stochastically perturbed, the fixed time delay imposed by the advection by the meridional flow is so dominant that the memory of the system persists for several cycles.

On the contrary, in the diffusion-dominated regime, the transport of magnetic fields by diffusion may act to short-circuit the conveyor belt related to the meridional flow profile. The time-scale of the meridional flow is thus less dominant and the only visible correlation is between cycles n and $n + 1$. We note here that the peak toroidal field precedes the peak surface radial field for the same cycle, which has the same sign. The poloidal field then produces the toroidal field for cycle $n + 1$ with the opposite sign. It is thus natural to find a strong correlation between cycles n and $n + 1$ since it relies on the regeneration of toroidal fields by the differential rotation, a very robust (and unperturbed) effect.

In our model, it might also be interesting to look at the correlation between successive cycles. Indeed, we are still in the advection-dominated regime but we have now introduced a new time-scale by imposing our delays.

On Fig. 3, we plot the peak poloidal flux at the surface at high latitudes (between $60°$ and $90°$) against the unsigned peak toroidal field at the base of the convection zone at low latitudes (between 0 and $40°$). The linear Pearson correlation coefficient has been computed for the relation between the poloidal component of cycle n and the toroidal field of cycle n, $n + 1$, $n + 2$ and $n + 3$. Figure 3 represents the relations between n and n and n and $n + 1$. The only significant correlation we find is between cycles n and $n + 1$ but is much less significant between cycle n and the later ones. Indeed, the correlation coefficient is about 0.64 for cycles n and $n + 1$ and drops to values around 0.1 and 0.2 for the other cycles. The strong relation between cycles n and $n + 1$ is again explained by the robust Ω-effect shearing the poloidal flux of cycle n into the toroidal field of opposite sign of cycle $n + 1$. This mechanism has not been modified by the introduction of time-delays and the correlation is thus naturally retained. On the contrary, the chaotic behaviour of our magnetic field prevents the memory of the system to persist for several cycles. After the toroidal field of cycle $n + 1$ has been generated, the flux emergence producing the new poloidal field (which is then responsible for the toroidal field of cycle $n + 2$) is perturbed by the time delays and the correlation between cycle n and $n + 2$, which was very significant in the advection-dominated regime, is mostly lost. We thus conclude that advection-dominated models may well lose their memory in the same way as the diffusion dominated ones, provided that additional time scales coexist with the typical time scale of the meridional flow (which is typically 2 to 3 cycles).

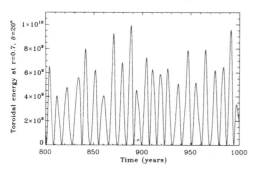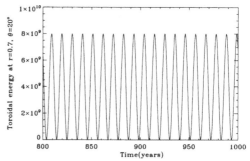

Figure 4. Toroidal energy at a particular point in space for a model with magnetic field-dependent delays and for a calculation where the delay has been fixed to about 1 year.

3. Variations on the 2D model: stabilising effects?

In this section, we try to point out the main factors responsible for the modulation of the cycle. The reduced model helped us to gain some insight on the mechanisms leading to a chaotic behaviour, we now turn to investigate the similarities with the 2D model in terms of stabilising or destabilising effects.

3.1. *Fixed vs varying delays*

The striking results of a long-term modulation due to small time-delays compared to the cycle period can be partly explained by the dependence of the delays on the magnetic field strength. Indeed, in the reduced model, it is clear that small but varying delays produce long-term modulation much sooner than fixed delays. The same feature is recovered in the 2D model. Figure 4 shows the time evolution of the toroidal magnetic field at a particular point in space in a case where the delay varies according to Eq. 2.2 and in a case where the delay is fixed to about 1 year for any field strength. It is clear that a fixed delay about one order of magnitude smaller than the cycle period has a very weak effect on the long-term activity, as expected. On the contrary, when the delay is made to vary with the field strength, reaching values between 0 and 5 years, the effect on the long-term modulation is much stronger, as shown on the first panel of Fig. 4. This particular case corresponds to the last panel of Fig. 1 which illustrates the existence of various dominant frequencies in the time series of the same toroidal field.

If the fixed time delay is increased further to values comparable to the cycle period for example, we recover that some long-term modulation is also produced and that a series of bifurcations may be identified. We thus confirm the results of Yoshimura (1978) that significant time delays modify the order of the equations and give solutions which are far from being harmonic in time. However, the purpose of our work was to show that even small delays can also produce similar long-term modulation, provided that those delays vary with the variable (the toroidal magnetic field) itself.

3.2. *The quenching factor*

The analytical study of the reduced model and particularly of the stability of the harmonic solution (described in detail in Jouve, Proctor & Lesur 2010) shows that the strength of the quenching term also plays a significant role in the model. In particular, the threshold for the destabilisation of the harmonic solution is strongly modified when B_0 (or equivalently λ, see Eq. 2.1 and 2.3) is varied. As a result, for the threshold delay at which the harmonic solution with $B_0 = 10^4$ loses stability, the solution with $B_0 = 10^3$

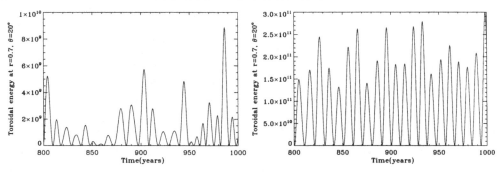

Figure 5. Influence of the quenching strength on the long-term modulation of the cyclic activity. The left panel corresponds to a value of $B_0 = 8 \times 10^3$ and the right panel to $B_0 = 5 \times 10^4$.

will still be stable and the one with $B_0 = 10^5$ will already have undergone the period doubling bifurcations leading to chaos.

Similar results are found with the 2D model when B_0 is varied, at least qualitatively. Figure 5 shows the time evolution of the toroidal magnetic energy in cases where $B_0 = 8 \times 10^3$ (left panel) and $B_0 = 5 \times 10^4$ (right panel). Indeed, when the non-linearity is high (corresponding to a value of $B_0 = 8 \times 10^3$), the modulation of the cycle amplitude is very strong, the energy of the strongest peaks reaching about 45 times the value for the weakest cycles. Another striking point is that not only is the amplitude modulated but the cycle period also seems to vary from one cycle to another. In particular, the very weak cycles between $t = 950$ and $t = 970$ exhibit periods of only 60% of the cycle peaking at $t = 880$ for instance. On the contrary, when B_0 is higher, the quenching term is less efficient and the solution seems much less modulated, the ratio between the highest peaks and the weakest in this particular time interval being of order 2. These results can be explained by the presence of the delays in the non-linear quenching term. Indeed, not only is the poloidal regenerated by the delayed toroidal field, it is also saturated by this same field. This particular feedback is thus probably responsible for the differences in the modulation of the activity even when the delay is kept identical, and thus small compared to the period.

3.3. Adding a tachocline α-effect

We may now introduce a variation on the model itself by adding an α-effect in the tachocline. Indeed, there is no clear reason why a regeneration term for poloidal fields should only be confined at the solar surface, the more "classical" α-effect due to helical motions in the convection zone might also be responsible for the stretching of toroidal field and its conversion into a poloidal component. Some studies (Mason, Hughes & Tobias 2002, Mann & Proctor 2009) have shown that even a weak α-effect at the base of the convection zone might be dominant over the BL source term. For example, Mann & Proctor (2009) have shown that a tachocline effect of 3 orders of magnitude less than a surface source term still controlled the behaviour of the magnetic field in the case where a non-local BL effect is considered. It is thus natural to think that adding a tachocline α-effect, however weak, could modify the effect of the time delays on the long-term modulation. Indeed, the time delays are only present to take into account the spatial segregation between the source terms of both components of the magnetic field, there is thus no reason to introduce a delay for a tachocline α-effect, which acts locally on the toroidal field. Moreover, this process being easily dominant on the field evolution, we might see it as a stabilising effect.

We have thus introduced an additional α-effect concentrated in a thin layer between 0.68 and 0.78 in the most perturbed model we had (14 days-delay on 5×10^4 G fields). Our preliminary calculations, with a tachocline α-effect about 2 orders of magnitude smaller than the BL source term (and thus supposed to be dominant according to Mann & Proctor results) do not seem to show any stabilisation of the cycle modulation. The evolution of the toroidal magnetic field still seems to be chaotic, with some bursts of activity coexisting with periods of grand minima. This might be understood by the fact that if the time delays in the BL term have a slight effect on the modulation of the toroidal field amplitude, it may be in fact amplified by the presence of the tachocline α-effect which will act on the modulated toroidal field to produce the new poloidal flux. As a consequence, the small time delays and the weak BL source term may still be significant enough to produce some modulation which will in turn be amplified by the additional instantaneous (or undelayed) α-effect. Moreover, Mann & Proctor (2009) showed that adding a fast meridional flow had the tendency to reduce the efficiency of the tachocline α-effect, suggesting that we may still be dominated by the BL source term in our particular case. This still needs some more investigation, especially if we want to consider a distributed α-effect instead, which may also be subject to delays.

4. Conclusions

Delayed differential equations appear to have some interesting properties on the dynamical system point of view. In the particular case of the mean-field dynamo equations, time delays can be related to the advection time by the meridional flow and the rise time of toroidal structures from the base of the convection zone to the solar surface. Both can be responsible for long-term modulation of the cyclic magnetic activity. The effect of time delays due to the buoyant rise of flux tubes is of particular interest since those delays are thought to be small compared to the cycle period (about a few percents typically). However, the fact that those delays are dependent on the magnetic field makes the governing equations belong to the class of state-dependent delayed partial differential equations. Their properties and their differences with fixed delayed equations are still not completely understood and in our particular case, we showed that they indeed produce solutions with various temporal evolutions.

A considerable step forward would obviously be to develop a self-consistent global model with buoyant toroidal structures built up and making their way from the base of the convection zone where they become unstable to the photosphere where they create active regions. Unfortunately, this has not been achieved yet due to numerous physical and numerical difficulties. In the mean time, we have shown here that the combination of 3D MHD models simulating a particular step of the dynamo cycle, 2D mean-field calculations and reduced models help us to progress on the influence of isolated physical processes on the magnetic cycle. In particular, we have shown that they can even help reproducing a feature of the solar cycle (its variability) which was not present in the standard model before. We will continue to develop these ideas in future work.

References

Bushby, P. J., 2006, *Mon. Not. R. Astron. Soc.*, 371, 772-780
Charbonneau, P. & Dikpati, M. 2000, *Astrophysical Journal*, 543, 1027
Charbonneau, P., St-Jean, C. & Zacharias, P., 2005, *Astrophysical Journal*, 619, 613
Hoyng, P. 1988, *Astrophysical Journal*, 332, 857
Jouve, L. & Brun, A. S. 2009, *Astrophysical Journal*, 701, 1300

Jouve, L., Proctor, M. R. E. & Lesur, G. 2010, *Astronomy & Astrophysics*, 519, A68
Mann, P. D. & Proctor, M. R. E. 2009, *Mon. Not. R. Astron. Soc.*, 399, 99
Mason, J., Hughes, D. & Tobias, S. 2002, *Astrophysical Journal*, 580, L89
Moss, D. & Brooke, J., 2000, *Mon. Not. R. Astron. Soc.*, 315, 521-533
Ossendrijver, A. J. H. & Hoyng, P. 1996, *Astronomy & Astrophysics*, 313, 959
Proctor, M. R. E. 1977, *Journal of Fluid Mechanics*, 80, 769
Rempel, M., 2006, *Astrophysical Journal*, 647, 662
Tobias, S. M. 1997, *Astronomy & Astrophysics*, 322, 1007
Weiss, N. O. & Tobias, S. M. 2000, *Space Science Reviews*, 94, 99
Wilmot-Smith, A. L., Nandy, D., Hornig, G. & Martens, P. C. H. 2006, *Astrophysical Journal*, 652, 696
Yeates, A. R., Nandy, D. & Mackay, D. H. 2008, *Astrophysical Journal*, 673, 544
Yoshimura, H. 1978, *Astrophysical Journal*, 226, 706

Discussion

T. Rogers: The time delay implied by the mean field models to get aperiodic modulation was around 10% of the cycle period, but the rise time implied by the simulations was a few days. This is a big difference, how do you reconcile the two?

L. Jouve: In the 3D model, the field strengths considered are quite high (more than 2×10^5 G in order to rise radially and not parallel to the rotation axis). We thus get rise times which are very short (about a few days). However, the time delay depends on the inverse of $|B|^2$ and thus if a 10^5 G field is delayed by 5 days, a 10^4 G field will be delayed by about 16 months, which corresponds to the values chosen for the 2D model. It is more difficult to relate thoses values to the reduced model which mostly serves illustrative purposes. Nevertheless, a **fixed** delay of 10% of the cycle is still inefficient to produce long-term modulation, so can still be considered small.

G. Vasil: Does the nonlinear time delay needs to be respect to the buoyancy terms or could it just as easily be attached to some other process in the system?

L. Jouve: The non-linear term in the time delay (proportional to $1/B^2$) was chosen to represent the magnetic buoyancy effects. Other formulations (with various non-linear functions of the magnetic field) were not tested but the intensity of this term is here shown to have a significant effect on the threshold for destabilization of the periodic solution. We could imagine representing the meridional flow by a time-delay and since concentrations of strong fields are known to modify this flow, this time delay could also be non-linearly dependent on B. This nonlinear delay would thus be linked to the transport by meridional flows instead of buoyancy. This has not yet been tested but could be interesting to consider.

S. Tobias: Do you see signs of a secondary frequency and if so, how does it depend on the delay?

L. Jouve: The spectra show that it is essentially smaller frequencies (or larger periods) which are excited when the delay is increased. In our most perturbed case, particular periods of about 20 years and 40 years seem to appear, possibly indicating previous period doublings which led to the observed chaotic behaviour.

Astrophysical Dynamics: From Stars to Galaxies
Proceedings IAU Symposium No. 271, 2010
N. H. Brummell, A. S. Brun, M. S. Miesch & Y. Ponty, eds.
© International Astronomical Union 2011
doi:10.1017/S1743921311017728

Nonlinear Dynamos

David Galloway

School of Mathematics and Statistics, University of Sydney, NSW 2006, Australia
email: dave@maths.usyd.edu.au

Abstract. This paper discusses nonlinear dynamos where the nonlinearity arises directly via the Lorentz force in the Navier-Stokes equation, and leads to a situation where the Lorentz force and the velocity and the magnetic field are in direct competition over substantial regions of the flow domain. Filamentary and non-filamentary dynamos are contrasted, and the concept of Alfvénic dynamos with almost equal magnetic and kinetic energies is reviewed via examples. So far these remain in the category of toy models; the paper concludes with a discussion of whether similar dynamos are likely to exist in astrophysical objects, and whether they can model the solar cycle.

Keywords. magnetic fields, MHD, Sun: magnetic fields, stars: magnetic fields

1. Introduction

Nonlinear dynamo calculations can be divided into three main classes, each with a well-defined purpose and a different set of advantages and shortcomings. The most astrophysically applicable theories typically use some form of mean-field electrodynamics; these include the so-called α–ω models (Krause & Rädler 1981), Parker's dynamo wave model (Parker 1955), the solar Babcock-Leighton dynamo and its flux-transport variants (see eg Dikpati & Charbonneau 1999), and Braginskii's nearly axisymmetric model for the Earth (Braginskii 1964). These theories all started life as kinematic models that took a prescribed velocity field and tested whether it gave growing or decaying solutions for the magnetic field; the resulting mathematical problem was linear and various parametrised terms arose which were in essence turbulent closures. The dynamical feedback of the the ensuing Lorentz force was neglected, so the basic theoretical outputs were an eigenvalue and its associated eigenfunction. The sign of the real part of the eigenvalue says whether there is a dynamo or not, the complex part if present determines the period, and the eigenfunction gives the configuration of the generated magnetic field. There is no information about the final field strength of such a dynamo, as the field grows exponentially without bound. Dynamics and nonlinearity are typically added via ad hoc parametrisations, a procedure which has been the subject of vigorous debate (see eg Vainshtein & Cattaneo 1992). The advantages of this approach are versatility and direct applicability in a wide variety of astrophysical objects. The disadvantage is that the underlying physics is questionable and relies on assumptions which are not satisfied in practice. The situation is reminiscent of the use of mixing-length theory for stellar convection; one knows that the theory is at best only qualitatively right, but somehow this does not really matter and stellar evolution theory proceeds very satisfactorily notwithstanding.

The second class of calculation is direct numerical simulation. The patron saint of these Proceedings has been a major contributor in this area, encouraging successive waves of students and postdocs by leading them in the solution of a sequence of important and realistic problems, first in convection, then in magnetoconvection, and most recently in dynamo theory.

The third approach consists of the solution of model problems which though not meant to apply literally to a specific astrophysical object, aspire nonetheless to isolate and elucidate the fundamental processes which may be at work. The important thing is that these models include the crucial physics, yet are simple enough to understand and interpret. This is the classic applied mathematician's approach, as epitomised in Lord Rayleigh's *Theory of Sound*. Nowadays the tongue-in-cheek terminology is that these are "toy models". The hope is that the processes studied can, if they look promising, be incorporated into direct numerical simulation of a specific object. This happens all too rarely, but it is a real treat when it works.

Magnetoconvection calculations constitute a side-class: they are not fundamentally different from the dynamo case, except that typically a mean field is imposed across the computational domain. The latter is normally guaranteed to be a conserved quantity, and the convection acts to amplify the magnetic energy by a large factor. Such a process is not regarded as a dynamo, because if the mean field were turned off, all field would eventually drain away by ohmic dissipation. To be a real dynamo, the field must be self-excited, and over the whole domain its average must be zero, because $\nabla \cdot \mathbf{B} = 0$. There can be patches of "mean field" in the sense of mean field theory, but they must average to zero over the whole domain. This point is often not properly understood. Any magnetoconvective calculation aimed at exhibiting genuine dynamo action must use initial conditions with no overall mean component.

This isolation of models into what is almost a philosophical classification was beautifully summarised by Spiegel (1977), who described it in terms of a political analogy; the three categories above are left-wing, centrist, and right-wing respectively. In this paper I will be concerned mainly with the right-wing end of the spectrum, but this is my own preference and is not meant to imply the other approaches are less worthwhile—with difficult problems attack on all fronts is vital!

In the next section, I will discuss filamentary dynamos, citing the example of ABC dynamos and showing the problems that arise when considering the generation of astrophysically significant fields. In subsequent sections I will then ask the question whether all dynamo fields have to be filamentary, and give a counterexample, the Archontis dynamo, where this is not the case. At the end I will try to be more astrophysically responsible and speculate about the forms nonlinear dynamos might take in real objects.

2. Filamentary dynamos

We start by writing down the induction and momentum equations for an electrically conducting fluid with both viscous and magnetic diffusivities, assuming a simple ohmic diffusion. For simplicity we take the incompressible case and use the standard notation for the various quantities appearing. The equations to be solved are then the induction equation

$$\frac{\partial \mathbf{B}}{\partial t} = \nabla \times (\mathbf{u} \times \mathbf{B}) + \eta \nabla^2 \mathbf{B}$$

and the momentum equation

$$\frac{\partial \mathbf{u}}{\partial t} - \mathbf{u} \times (\nabla \times \mathbf{u}) = -\nabla \left(p/\rho + \frac{1}{2} u^2 \right) + \mathbf{j} \times \mathbf{B} + \mathbf{F} + \nu \nabla^2 \mathbf{u},$$

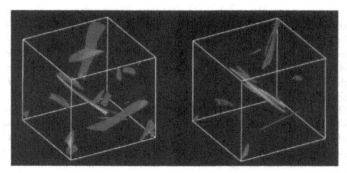

Figure 1. The ABC dynamo with $(1,1,1)$ forcing, for $\nu = 1/5, \eta = 1/400$. Shown are isosurfaces where the magnetic energy attains 20% of its maximum value at that instant, for two separate times approximately 40 turnover units apart.

with $\nabla \cdot \mathbf{u} = \nabla \cdot \mathbf{B} = 0$, and $\mu_0 \mathbf{j} = \nabla \times \mathbf{B}$. The driving force $\mathbf{F}$ is necessary to sustain $\mathbf{u}$ and $\mathbf{B}$ against viscous and ohmic dissipation. In astrophysics it would arise from some natural physical mechanism such as buoyancy, but in model calculations it is often externally imposed with a specific form. The assumption of incompressibility makes for considerable mathematical and computational simplification, and those anelastic or fully compressible computations which have been done show remarkably few differences compared to the incompressible case, at least as far as dynamo theory is concerned.

The ABC dynamo is a prototype example where the resulting magnetic field is typically filamentary. Here everything is 2π-periodic and a force

$$\mathbf{F} = \nu(A \sin z + C \cos y, B \sin x + A \cos z, C \sin y + B \cos x)$$

is supplied. In the absence of magnetic fields the ABC flow (with $\mathbf{u}$ equal to the above formula without the factor ν) is a solution to the Navier-Stokes equation with this forcing field. In the case $A : B : C = 1 : 1 : 1$, this solution is hydrodynamically unstable when the kinetic Reynolds number R_e exceeds 13.09 (Galloway & Frisch, 1987).

Figures 1 and 2 show the nature of the magnetic field generated for the case $\nu = 1/5, \eta = 1/400$ (this calculation was computed some time ago by Olga Podvigina; solutions were earlier given by Galanti, Sulem & Pouquet 1992). The viscosity is high enough that the non-magnetic case has the (1,1,1) ABC flow as a stable solution; the magnetic diffusivity is low enough to yield a dynamo when a divergence-free small seed field with zero mean is added. The solution evolves quite rapidly to a time-dependent but statistically steady state with a total magnetic energy around twice the total kinetic energy. The isosurfaces of magnetic energy show the field is highly filamentary.

Astrophysical dynamos operate in the regime where both the magnetic and kinetic Reynolds numbers are very large, so it is natural to ask if these solutions where the total magnetic and kinetic energies are comparable can persist to high kinetic Reynolds number. The observation that the fields are filamentary suggests we attempt to establish scaling laws: two possibilities are that the filaments have thickness $R_m^{-1/2} L$ and length L, where L is the characteristic size of the domain, or that the filaments form magnetic plugs with order L thickness and length but order $R_m^{-1/2} L$ edges in which the dissipation takes place. This gives formulae

$$\frac{\text{Total Magnetic Energy}}{\text{Total Kinetic Energy}} \simeq \frac{1}{R_e} \quad \text{(filaments of thickness } R_m^{-1/2})$$

$$\frac{\text{Total Magnetic Energy}}{\text{Total Kinetic Energy}} \simeq \frac{1}{R_e}\left(\frac{\nu}{\eta}\right)^{1/2} \quad \text{(thickness order } L \text{ with } R_m^{-1/2} \text{ edges) .}$$

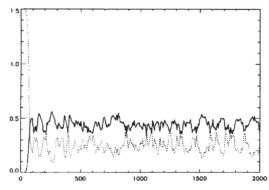

Figure 2. The same case as in Figure 1, but now showing the evolution of total magnetic energy (solid) and total kinetic energy (dashed), for a time long compared with the turnover and both diffusion timescales.

Here R_e and R_m are the kinetic and magnetic Reynolds numbers. The above formulae are derived in Galloway (2003), but related results for other flow types were known well before then (Vainshtein & Cattaneo 1992, Brummell, Cattaneo & Tobias 2001). The bad news about all this is that at high R_e the relative fraction of magnetic energy is very small, in contrast with the substantial fractions which are observed in many astrophysical objects. This has lead since 1990 to much questioning of how dynamos, mean field or otherwise, can be viable in astrophysics. The convective, intermittent, numerical photospheric dynamo of Cattaneo (1999) is a good illustration of this. This debate has continued until the present day, and in my opinion has yet to be satisfactorily resolved.

3. Non-filamentary dynamos

For a long time, numerical dynamos based on the full nonlinear equations rather than mean-field parametrisations seemed to suggest that all such dynamos were filamentary. The first example where this was not the case appeared in the PhD thesis of Archontis (2000), who studied a problem similar to the ABC-forced dynamo, but with the cosines omitted. The corresponding velocity field was known to give a very effective and probably fast kinematic dynamo (Galloway & Proctor 1992). However, this velocity field is *not* a solution to the Navier-Stokes equation with the forcing term $\mathbf{F} = \nu(\sin z, \sin x, \sin y)$, because the nonlinear term is now unbalanced. With no magnetic field, the flow initially adopts the right form if started from rest, but then quickly evolves into something quite different as soon as the nonlinear term starts to bite.

Once a divergence-free random seed magnetic field is added, something quite remarkable happens. Figure 3 shows the time evolution of the total magnetic and kinetic energies over a long time. The "something quite different" referred to above is apparently an effective dynamo and leads to a time-dependent state which nonetheless looks quasi-steady. Presumably this dynamo is filamentary and resembles the ABC dynamo discussed in the last section. However, after several diffusion times spent in this state, a transition occurs to another state where the energies are almost equal to one another. Inspection of this state shows that $\mathbf{u}$ and $\mathbf{B}$ are in fact *locally* almost equal to one another throughout nearly all of the domain, when expressed in Alfvénic units. The technicalities of the limiting behaviour as both diffusivities tend to zero are described in Cameron & Galloway (2006a). The fields $\mathbf{u}$ and $\mathbf{B}$ tend to exactly the same thing, and in the limit both are close, *but not equal*, to $0.5(\sin z, \sin x, \sin y)$. The discrepancy is of the order of a few percent in a selection of Fourier modes, but why these modes are present and need to be there has so far defied understanding—they are clearly the result of some unknown solvability condition. Also amazing is that the solution is steady and stable—we have tested this

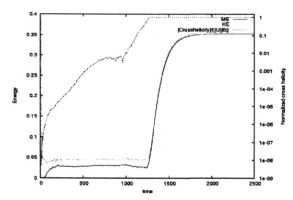

Figure 3. The evolution of total magnetic energy (red) and total kinetic energy (green), for the Archontis dynamo described in the text, over a time long compared with the turnover and both diffusion timescales. Also shown is the evolution of the normalised cross-helicity, better referred to as the magnetic alignment.

down to diffusivities of 1/2000, and Gilbert, Ponty & Zheligovsky (2010) report that the stability also persists when they are 1/10000. This latter paper represents the current state of the art in understanding the Archontis dynamo; there is also an extensive description in Archontis, Dorch & Nordlund (2007). Note that these last authors use a fully compressible code (which seems to make little difference), and they employ a feedback mechanism which modulates the forcing term in an (unnecessary) attempt to keep the energies at a fixed level. In fact, the feedback appears to destabilise things because it acts on the Alfvénic timescale and can cause resonances. (We have reproduced this by using their feedback term in our incompressible code.)

To understand what causes the transition from the early quasi-steady behaviour to the evolved $\mathbf{u} \simeq \mathbf{B}$ state, it is useful to plot the integral of $\mathbf{u}.\mathbf{B}$ divided by the square roots of the integrals of u^2 and B^2 over the periodicity cube. This quantity is sometimes called the normalised cross-helicity, though the terminology is unfortunate and and it is more informative to regard it as a measure of the degree to which $\mathbf{u}$ and $\mathbf{B}$ are aligned—it is the average of $\cos\theta$ where θ is the angle between $\mathbf{u}$ and $\mathbf{B}$. When the alignment is 1, the two fields are perfectly lined up, and then the nonlinear $\mathbf{u} \times (\nabla \times \mathbf{u})$ and $\mathbf{J} \times \mathbf{B}$ terms can cancel one another out in the momentum equation. The topmost curve in Figure 3 shows the evolution of the alignment; all through the dormant apparently quasi-steady phase it is steadily growing, and once its value is appreciable it triggers a transition to a completely aligned state. By adding $\mathbf{B}$ dotted with the momentum equation to $\mathbf{u}$ dotted with the induction equation and integrating over the box, one finds that the time rate of change of the alignment has a source term equal to the integral of $\nu\mathbf{F} \cdot \mathbf{B}$ together with viscous and magnetic diffusive terms. Given the form of $\mathbf{F}$, this shows that it is the amplitude of the $(\sin z, \sin x, \sin y)$ mode in the magnetic field which causes the alignment either to grow towards $+1$ if it is positive, or to decrease towards -1 if it is negative. (Note there is a symmetry between solutions with $\mathbf{u} = \pm\mathbf{B}$). It also explains why the alignment is a slow process taking place on a diffusive timescale—the initial dynamo may be fast but the final dynamo is slow! However, alignment is a well-known process in MHD turbulence, and in that context it may not always be slow—see Matthaeus *et al.* (2008).

The fact that both fields are close to $0.5(\sin z, \sin x, \sin y)$ means that the solution has a very regular, non-turbulent form This is shown in Figure 4, which plots the stable and unstable manifolds connecting some of the stagnation points.

Can further solutions with $\mathbf{u} \simeq \pm\mathbf{B}$ be constructed? In fact, if one is not fussy about how unrealistic a force is to be employed, there is a simple recipe by which such dynamos can be constructed to order. Take any neutrally stable solution $\mathbf{B_0}$ to the kinematic problem with prescribed velocity $\mathbf{u_0}$ and magnetic diffusivity η_0. We can now generate an equilibrium solution to the full dynamo problem (including the momentum equation) for $\eta = \epsilon\eta_0$. This is $\mathbf{u} = \epsilon\mathbf{u_0} + \mathbf{B_0}$, with $\mathbf{B} = \mathbf{B_0}$, and $\mathbf{F} =$ whatever it has to be in order that the momentum equation

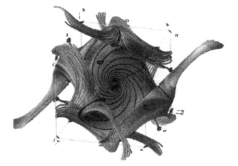

Figure 4. A plot showing the beautiful structure of the velocity and magnetic fields for the final steady state of the Archontis dynamo; both fields are almost indistinguishable from one another. Plotted are trajectories of the solution which form heteroclinic orbits connecting some of the stagnation points.

is satisfied. (The actual formula is given in Cameron & Galloway (2006a), along with a slightly more general scaling transformation.) As $\epsilon \to 0$ we end up with an Alfvénic solution.

Variations on this theme were followed up in Cameron & Galloway (2006b), particularly for the Gibson (1968) three-sphere rotor dynamo. We also looked there at a family of forcings

$$\mathbf{F} = \nu(\sin z + \epsilon \cos y, \sin x + \epsilon \cos z, \sin y + \epsilon \cos x)$$

where $\epsilon = 0$ gives the Archontis dynamo and $\epsilon = 1$ gives the (1,1,1) ABC dynamo. This enables the study of the transition from the non-filamentary to the filamentary cases, as well as a few other issues not dealt with here.

4. Discussion

These non-filamentary dynamos are fascinating objects, but the obvious question is whether they are isolated freaks or whether such behaviour (with $\mathbf{u} \simeq \mathbf{B}$) can be expected to occur as a typical feature of an astrophysical dynamo, at least over some fraction of the flow domain. When the forcing and both diffusivities are formally absent, any steady $\mathbf{u}, \mathbf{B}$ pair with $\mathbf{u} = \pm\mathbf{B}$ (in Alfvénic units) satisfies the induction and momentum equations; these are often referred to as Alfvénic solutions. This is because the nonlinearities in the momentum equation cancel one another out exactly. Friedlander & Vishik (1995) proved the remarkable result that *all* these solutions are neutrally stable. This is in complete contrast to the non-magnetic case, where one finds that typical solutions to the Euler equation are unstable. When small diffusivities are added, one can conjecture that some fraction of these Alfvénic solutions end up being stable attractors. The challenge is to ascertain which ones emerge in practice—we have seen how hard it is to determine the solvability conditions that enable the Archontis dynamo to home in on one particular solution which is closely related, but not identical, to the forcing function.

The natural next step is to examine the numerical calculations done to date and see whether there is any evidence of alignment processes at work. The best evidence I have seen occurs in the ASH-code simulations of faster-rotating Suns due to Brown, Browning, Brun, Miesch and Toomre, which are found elsewhere in these proceedings (see also Brown *et al.* 2010; the solutions are referred to as wreath-building dynamos, and are clearly strongly influenced by differential rotation). There are many hints in other calculations, but nothing as convincing as what is seen in the "toys". One important point about the latter is that they have boundary conditions which are compatible with an aligned outcome; the Archontis dynamo for instance is periodic. Any real astrophysical object will have boundary conditions which are *not* compatible, and it is not clear to what extent boundary effects can wreck any alignment mechanism. In the Sun, the non-compatible boundary conditions are enough to ensure that the magnetic field seen at the photosphere is highly intermittent. The question is whether this persists throughout the Sun or whether somewhere, deeper down, there may be a region where alignment mechanisms are significant for field generation. The tachocline is the obvious place to consider, and Robert Cameron and myself have been trying to construct an aligned dynamo which works there (see the discussion item by Proctor).

This has been a personal view of what I think are some interesting issues in nonlinear dynamo theory. I have not discussed several topics to which other authors would attach great importance, such as scale separation or the significance of various kinds of helicity—they understand these much better than I do. I would like to end by thanking Robert Cameron for his conviction that non-filamentary dynamos are possible, and Juri Toomre for his example in formulating and fostering the solution of problems which address fundamental issues and propel the subject forward rather than spread confusion.

References

Archontis, V. 2000, PhD Thesis, University of Copenhagen, Linear, Non-Linear and Turbulent dynamos

Archontis, V., Dorch, B. & Nordlund, A. 2007 *A&A* 472, 715

Braginskii, V.I. 1964 *Sov. Phys. JETP* 20,726

Brown, B. P., Browning, M. K., Brun, A. S., Miesch, M. S. & Toomre, J 2010 *ApJ* 711, 424

Brummell, N., Cattaneo, F. & Tobias, S. M. 2001, *Fluid Dynam. Res.*, 28,237

Cameron, R. & Galloway, D. 2006 *Mon. Not. R. Astron. Soc*, 365, 735

Cameron, R. & Galloway, D. 2006 *Mon. Not. R. Astron. Soc*, 367, 1163

Cattaneo, F. 1999, *ApJ*, L39, 515

Dikpati, M. & Charbonneau, P. (1999) *ApJ* 518, 508

Friedlander, S. J. & Vishik, M. M (1995) *Chaos* 5,416

Galanti, B. & Sulem, P. L., Pouquet,A 1992, *Geophys. Astrophys. Fluid Dyn.* 66, 183

Galloway, D. J. 2003, in: A. Ferriz-Mas & M.Nunez (eds.), *Advances in Nonlinear Dynamos* (Taylor & Francis), pp. 37

Galloway, D. J. & Proctor, M. R. E. 1992, *Nature* 356, 691

Gibson, R. D 1968 *Q.J. Mech. Appl. Math.* 21, 257

Gilbert, A.D., Ponty, Y. & Zheligovsky, V. 2010 *Geophys. Astrophys. Fluid Dyn.*, published electronically as DOI: 10.1080/03091929.2010.513332, paper version to appear

Krause, F. & Rädler, K. H. 1980 *Mean-Field Magnetohydrodynamics and Dynamo Theory* (Berlin, Akademie-Verlag)

Matthaeus, W. H., Pouquet, A., Mininni, P. D., Dmitruk, P. & Breech, B. 2008 *Phys. Rev. Lett.* 100, 085003

Parker, E. N. 1955 *ApJ* 122,293

Spiegel, E. A. 1977, in: E. A. Spiegel & J. P. Zahn (eds.), *Problems of Stellar Convection* (Springer-Verlag), 1

Vainshtein, S. & Cattaneo, F. 1992, *ApJ*, 393, 199

Discussion

BRANDENBURG: When you allow for scale separation (bigger box than the size of one cell), you do not get flux alignment that can lead to a Yoshizawa effect; see Sur & Brandenburg 2009 (MNRAS 299, 273)

GALLOWAY: I have deliberately chosen to stay off the subject of scale separation in this talk but I realise it is another interesting aspect and I refer readers to the paper you cite above.

POUQUET: Why do you say that the growth of $V - B$ alignment is a slow (diffusion time) process?

GALLOWAY: One can write down an equation for the evolution of the integral of **u.B** over the fundamental periodic cube. After converting various terms to surface integrals which vanish because of the periodicity, one is left on the right hand side just with terms which are proportional to the diffusivities. Thus the total must evolves slowly. This does not preclude the idea that their could be cancelling patches with opposite sign which evolve on the fast timescale (see the next question).

PROCTOR: Do you find Archontis type dynamos with "shocks" across which **u.B** changes sign?

GALLOWAY: We have not seen such a phenomenon in any of the relatively small number of calculations conducted so far. However, Robert Cameron and myself have a proposal for a tachocline dynamo which needs what you refer to in order that the magnetic field changes sign every 11 years while the differential rotation does not. So we very much hope that what you suggest is possible! (This work is being submitted to MNRAS.)

Astrophysical Dynamics: From Stars to Galaxies
Proceedings IAU Symposium No. 271, 2010
N.H. Brummell, A.S. Brun, M.S. Miesch & Y. Ponty, eds.

© International Astronomical Union 2011
doi:10.1017/S174392131101773X

Lack of universality in MHD turbulence, and the possible emergence of a new paradigm?

Annick Pouquet[1], Marc-Etienne Brachet[2], Ed Lee[3], Pablo Mininni[1,4], Duane Rosenberg[1], and Vadim Uritsky[5]

[1] Geophysical Turbulence Program, National Center for Atmospheric Research,
PO Box 3000, Boulder CO 80304 USA
email: `pouquet@ucar.edu` *and* `duaner@ucar.edu`

[2] École Normale Supérieure, 24 rue Lhomond, 75005 Paris, France
email: `brachet@physique.ens.fr`

[3] Centrum voor Plasma-Astrofysica, Departement Wiskunde, Katholieke Universiteit Leuven,
Celestijnenlaan 200B, B-3001 Leuven, Belgium
email: `edlee@ucar.edu`

[4] Departamento de Física, Facultad de Ciencias Exactas y Naturales,
Universidad de Buenos Aires, Ciudad Universitaria, 1428 Buenos Aires, Argentina
email: `mininni@df.uba.ar`

[5] Physics and Astronomy Department, University of Calgary, Calgary, AB T2N1N4 Canada
email: `uritsky@ucalgary.ca`

Abstract. We review some of the recent results obtained in MHD turbulence, as encountered in many astrophysical objects. We focus attention on the lack of universality in such flows, including in the simplest case (no externally imposed magnetic field, no forcing, unit magnetic Prandtl number). Several parameters can foster such a breakdown of classical Kolmogorov scaling, such as the presence of velocity-magnetic field correlations, or of magnetic helicity and the role of the interplay between nonlinear eddies and Alfvén waves. A link with avalanche processes is also discussed. These findings have led to the conjecture of the emergence of a new paradigm for MHD turbulence, as a possibly unsettled competition between several dynamical phenomena.

Keywords. turbulence, magnetic fields, (magnetohydrodynamics:) MHD, waves, Sun: flares, (Sun:) solar wind, stars: activity, ISM: magnetic fields

1. Introduction

Observational data, old and new, concerning structures and statistical properties of magnetic fields, in particular in the solar photosphere and the solar wind, unambiguously speak to their ubiquity and dynamical importance. The large scales, which contain the energy, can be analyzed in the framework of the MHD limit but small scales need an approach incorporating plasma effects. One open question is: How independent are the small-scale kinetic effects and the large-scale fluid behavior? This is important in many applications of astrophysical interest, in particular in reconnection processes whereby energy is being lost to the fields (velocity and induction), resulting in particle acceleration, in heating (e.g., of the solar corona) and in dissipation of energy at a finite rate in the limit of large Reynolds number (see Biskamp & Welter 1989 and Politano *et al.* 1989 for the two-dimensional (2D) case, and Mininni & Pouquet 2009 in 3D). One of the major problems in the reconnection theory is understanding the physical mechanism of fast magnetic reconnection which seems to operate in weakly collisional space plasmas. The classical models of Sweet-Parker and Petschek approach this question in the realm of resistive MHD, but the plasmas mentioned above in fact obey a generalized Ohm's law in which the Hall current for fast collisionless reconnection plays a crucial role (see e.g. Wang *et al.* 2000 and references therein). However, numerous studies have confirmed recently that there is really no need for kinetic effects, e.g. a Hall current for reconnection (of, say, a Harris sheet, a well-known example of magnetic field reversal using a simple *tanh* profile) to occur provided the Reynolds number is high enough. The origin of current sheets is to be found in the nonlinear coupling of many interacting scales as happens in any turbulent flow, be it shear layers in fluids or vorticity and current sheets in MHD (Matthaeus & Lamkin 1986, Servidio *et al.* 2010). Hence, there is a renewed interest in the large scale properties of conducting fluids in the MHD framework.

The interactions between widely separated scales, or nonlocal interactions (the lack of locality referring to the nonlinear coupling of modes in Fourier space), is an issue that can be addressed in the context of MHD, without reference to plasma effects at small scale. It has been thoroughly reviewed recently (Mininni 2011), and it will not be dealt with here beyond stating the important fact that, as expected on the basis of numerous studies in the mid sixties, MHD is more nonlocal than fluids; for example, reconnection (which gives rise to change in topology) involves a global structure modification because of alterations to the small scales (for Hall MHD non-locality, see Mininni *et al.* 2007). Furthermore, we shall not discuss either the generation of magnetic fields through dynamo action, as it is the topic of several contributions to this symposium by other authors.

We thus now review some of the results that have been obtained in recent years for MHD turbulence. Two points stand out: (i) There is now a wealth of evidence, theoretical, observational and numerical, from several groups, that universality breaks down in MHD (where universality could be defined as having one answer to one question such as the energy distribution among Fourier modes in MHD); this takes place depending on how fast the interactions of eddies and waves happen, including in the simplest case (incompressible, no Hall term, no ambipolar drift, no forcing, no uniform magnetic field); and (ii) The nature of the structures that develop in MHD flows, such as rolled-up current and vorticity sheets, as have been observed in the Solar Wind, where rotational discontinuities, flux transfer events and plasmoids are all common. We will also discuss what can be done about the limitations in power of present-day computers. Specifically, we will comment on some of the models that may allow us to explore, for example, the regimes of realistic magnetic Prandtl numbers, i.e. very small in the solar convection zone and photosphere or very large in the interstellar medium, both unreachable to this day using direct numerical simulations (DNS) of the primitive MHD equations.

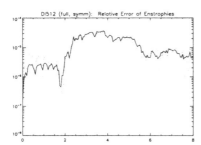
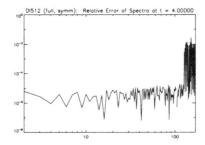

Figure 1. *Left:* Temporal evolution of the normalized difference of the total energy (solid line) and of its kinetic E_v and magnetic E_m components in a full DNS and in the symmetric code for the TG flow at the same Reynolds number. *Right:* Spectrum of that difference shortly after the peak of dissipation ($t = 5$). In both cases, the discrepancies are $\sim 10^{-5}$ except at the truncation.

2. Three energy spectra in MHD

There has been a dispute in the literature for a long time as to whether the total energy spectrum $E_T(k)$ in MHD follows a Kolmogorov (1941) law $\sim k^{-5/3}$ (hereafter K41) as for incompressible fluids, including in the anisotropic case in the presence of a strong imposed magnetic field $\mathbf{B}_0$ as proposed in Sridhar & Goldreich (1994) (see also Montgomery & Matthaeus 1995, Ng & Bhattacharjee 1996), or whether the interplay between Alfvén waves and nonlinear eddies slows down the nonlinear transfer to small scales, leading to a $-3/2$ spectrum as proposed by Iroshnikov (1965) and Kraichnan (1967) (hereafter IK), again with an anisotropic version which happens to coincide with the theory developed in the case of weak turbulence (hereafter WT), leading (phenomenologically) to $E_2(k_\perp, k_\parallel) \sim k_\perp^{-2} k_\parallel^{-1/2}$, perpendicular and parallel referring to the direction of $\mathbf{B}_0$, and with in the isotropic case $k_\perp \sim k_\parallel \sim k$. The differences in spectral indices are small and therefore hard to measure in the Solar Wind or in DNS (note that for energy spectra to converge both at small and large scales, a turbulent spectrum must have a spectral index $1 < \alpha < 3$, with $E(k) \sim k^{-\alpha}$). This includes the two-dimensional MHD case which was thoroughly investigated in the mid eighties by several teams and which corresponds, to lowest order, to reduced (anisotropic) MHD whereby the flow is reduced to the 2D plane in the presence of a strong $\mathbf{B}_0$. But these investigations were still somewhat inconclusive, with either IK (Biskamp & Welter 1989, and Politano *et al.* 1989) or K41 scaling (see Mininni *et al.* 2005, in the context of a regularization model of MHD in 2D).

It is first Dmitruk *et al.* (2003) that showed that different spectra may occur in the case of reduced MHD; Müller & Grappin (2005) then showed this breaking of universality in MHD turbulence, with either a K41 or IK scaling depending on the strength of the imposed field, conclusions that were confirmed by Mason *et al.* (2008). We recently did a study devoid of two of the main constraints of these previous works: we include no forcing, and $\mathbf{B}_0 \equiv 0$; however, in order to obtain a sizable resolution in the inertial range in a parametric study, we impose the symmetries of a specific velocity field, the Taylor-Green vortex (or TG flow) at all times using a specialized code; such a flow has been studied extensively in the past (Brachet *et al.* 1983) and is akin to an experimental configuration using two counter-rotating cylinders. We couple the TG velocity in a series of three sets of numerical experiments, to three different magnetic fields, also fulfilling the same symmetries, with at $t = 0$, equal kinetic and magnetic energy, identically zero magnetic helicity $H_m = <\mathbf{A} \cdot \mathbf{B}>$ (with $\mathbf{B} = \nabla \times \mathbf{A}$) and negligible velocity-magnetic field correlation ($H_c = <\mathbf{v} \cdot \mathbf{B}>$ is less than 4% in normalized value for the three runs). Hence, from a statistical point of view, all three flows are identical since they have the same quadratic invariants, differing only in detailed phase factors. However, they develop markedly different spectra, one following K41, one IK and one WT (Lee *et al.* 2010).

The imposed symmetries do not seem to alter the dynamical evolution of the flow (see Fig. 1) and the results show as well that, in the case of the IK spectrum, indeed the relative energy $E_v(k) - E_m(k)$ follows a k^{-2} law (see Fig. 2), as a simple argument can show using as a small parameter the ratio of the Alfvén to the eddy turnover timescale

$$R_\tau(\ell) = \tau_A(\ell)/\tau_{NL}(\ell). \tag{2.1}$$

Fig. 2 (right) gives the anisotropic second order structure function (corresponding to the energy spectrum, with $S_2 \sim \ell^s$, $E(k) \sim k^{-\alpha}$ and $\alpha = s + 1$). The flow is well resolved until $\ell \sim 0.04$ (with $S_2 \sim l^2$), and there is a clear inertial range in the perpendicular structure function (solid line) until $\ell \sim 0.7$ corresponding for this flow to a weak turbulence spectrum; the scaling of the parallel data is not as good but could possibly follow $E(k_\parallel) \sim k_\parallel^{-3/2}$. Note that perpendicular and parallel refer to the local uniform magnetic

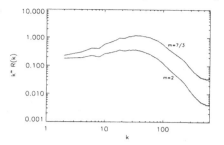 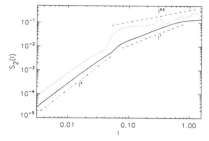

Figure 2. *Left:* Fourier spectrum of $E_v - E_m$ at peak of dissipation. *Right:* Anisotropic second-order structure function S_2 as a function of distance ℓ; solid and dash lines correspond to perpendicular and parallel to the mean local magnetic field; at small scales, the flow is well resolved ($S_2 \sim \ell^2$), and at large scale, $S_2(\ell_\perp) \sim \ell_\perp$, corresponding to weak MHD turbulence.

field $\mathbf{B}_{0,local}$ in the absence of imposed uniform field as in this run; $\mathbf{B}_{0,local}$ is computed, at a cost, as an average of the fluctuating field in a sphere of radius the integral length scale and centered on the spatial location at which the structure functions are evaluated.

As a last point, we might mention that these new highly resolved numerical simulations point out to a lack of observance of what is called critical balance (Goldreich & Sridhar 1995). Indeed, the ratio of time scales in the above equation depends on the scale ℓ since τ_A and τ_{NL} have different scaling. It was assumed in Sridhar & Goldreich (1994) that $R_\tau(\ell) \sim \ell^0 = C$ with $C = 1$ (note that in Galtier *et al.* (2005), $C \neq 1$ leaving room for scaling laws that differ from K41). However, the numerical data shows that the constancy of $R(\ell)$ with scale is not fulfilled and in fact there is weak evidence for it following what one can expect when evaluating the eddy turn-over time in the usual manner based on the observed scaling for the energy spectrum (see Lee *et al.* 2010). Note that this points out to the non-uniformity of the approximation behind weak turbulence theory for which $R(\ell)$ is assumed small at all scales.

The lack of universality in MHD seems to be in agreement with several recent observations in the Solar Wind. It also corroborates the findings of different scaling laws for structure functions of varying order in solar active regions, as a function of their intensity (see e.g. Yurchyshyn *et al.* 2005), with more intermittency the stronger the flare; a similar tendency has been found in the turbulent component of the ultraviolet auroral emission representing multiscale magnetic disturbances in the Earth's magnetosphere (Uritsky *et al.* 2001), and in DNS (Müller & Biskamp 2003). However, many questions remain open. For example, in the case of the TG study, would the results given above persist when the symmetries of the flow are not maintained? Would such results persist in time (they may differ in fact)? As the Reynolds number is increased, even when keeping the magnetic Prandtl number equal to unity? And are they the stationary solutions for such flows in the presence of forcing? These questions will require many investigations, but the fact remains that there may indeed be different energy spectra in MHD, at least given the imposed constraints of the flows studied in the several papers mentioned above.

3. The role of helicity

The two other quadratic invariants of the MHD equations in 3D in the absence of dissipation are H_c and H_m defined above. What role do they play? Again, old and new results shed some light on the dynamics of MHD when taking into account the helical part of the velocity and magnetic field two-point correlation functions.

3.1. *Non-zero cross-correlations between the velocity* **v** *and the magnetic field* **b**

The lack of universality in MHD turbulence is a well-known phenomenon when one varies the amount of correlation between the velocity and magnetic field: at high correlation, the nonlinear interactions are weakened and the spectral indices of the Elsässer variables $\mathbf{z}^{\pm} = \mathbf{v} \pm \mathbf{b}$, with spectra $E^{\pm} \sim k^{-m^{\pm}}$, obey $m^{+} + m^{-} = 3$. Note that it can be shown analytically in the context of weak MHD turbulence that $m^{+} + m^{-} = 4$ in that case (in other words, these results should be seen in the context of IK and/or WT). This has been observed numerically for example in two dimensions using both DNS and two-point closure models of turbulence (see e.g. Pouquet (1993, 1996) for reviews).

Recently, Boldyrev (2006) (see also Boldyrev 2005) gave a very interesting and novel interpretation of such a scaling in MHD turbulence, finding a way to incorporate in a simple manner the effect of H_c. In the context of the exact laws that can be written in MHD at the level of third-order correlators (Politano & Pouquet 1998), one could consider that the angle $\theta(\ell)$ between **v** and **b** plays a role in such scaling, an hypothesis which in the simplest case (of equal scaling for all variables) leads to the IK spectrum and to a dependence $\theta(\ell) \sim \ell^{1/4}$ (omitting anisotropy for simplicity here), as observed in several numerical simulations (Mason *et al.* 2006).

3.2. *Revisiting the dynamics of magnetic helicity and its role*

One can predict, on the basis of its conservation, that magnetic helicity will cascade à la Kolmogorov, that it will be to large scales because of ideal dynamics (Frisch *et al.* 1975), and that it will follow $H_m(k) \sim k^{-2}$ (thus $E_m(k) \sim k^{-1}$ in the maximal helical case, leading to the necessity of logarithmic corrections and to the importance of nonlocal effects, as discussed in Pouquet *et al.* 1976). Note that a -2 spectrum obtains irrespective of whether one takes a K41 or IK approach. The data for such an inverse cascade is ambiguous today; older studies at lower resolution or using two-point closures of turbulence are in agreement with such predictions but recent numerical results under a variety of conditions and at high resolution (up to 1536^3 grid points), and thus high Reynolds numbers, are in striking disagreement (Müller 2008, Malapaka 2009, Mininni & Pouquet 2010), with in some cases a direct cascade of H_m. The magnetic helicity cascade can be viewed as a competition between the Alfvén effect leading to a tendency toward equipartition in the small scales (a phenomenon that occurs faster the smaller the scale and faster the stronger the magnetic field in the large scales), and what one can call helicity effects dealing with an inverse transfer of magnetic helicity and thus magnetic energy to large scales (Pouquet *et al.* 1976). So, what is happening?

One might ask whether an invariant can present a dual (direct *and* inverse) cascade with dual constant flux, of one sign at large scales and the other sign at small scales, even though one direction would probably be dominant. It is the case for energy in rotating turbulence (Mininni & Pouquet 2010), with both an inverse (quasi-2D) and a direct cascade, the latter because rotation is felt less at smaller scales and isotropy recovers beyond what is called the Ozmidov scale at which the inertial wave time and the eddy turn-over time equilibrate. Magnetic helicity presents a dual cascade as well. It certainly has dual transfer, as does H_c, to small and large scales as shown in low resolution decay computations (Pouquet & Patterson 1978); more importantly, highly resolved MHD flows using hyperdiffusivities and small-scale forcing clearly show a dual cascade of H_m with dual constant flux (Malapaka 2009) and with a steeper spectrum that seems to be governed by a partial Alfvénization of the flow, the inertial indices of spectra varying in such a way that $H_v/kE_v \sim k^2 H_m/E_m$ is fulfilled, compatible with a dynamo regime as well. Observe that $kH_m/E_m \sim 1/k$, as for Navier-Stokes (for which $H_v \sim E_v \sim k^{-5/3}$), implying a slow recovery of mirror symmetry at small scale.

Moreover, a study of the statistical equilibria of incompressible three-dimensional MHD turbulence (Stribling & Matthaeus 1990) indicates that the cascades of magnetic helicity and cross helicity, H_m and H_c, are different. Whereas H_m may undergo a complete condensation (by which is meant that it will condense on the largest scale available to the system and possibly rebound to smaller scales), the case for H_c is not so clear: a priori, having the same physical dimension as energy, it is expected to cascade to the small scales but it may also exhibit for some values of the parameters (here, the "temperatures" associated with the three global invariants), a quasi-condensation at large scales.

How do such duality and differences in cascades and fluxes affect the dynamics in a way that we can model in simple ways? This question may be particularly relevant in the framework of recent observations in the solar wind where an inverse cascade of cross-correlation is diagnosed for highly aligned **v**–**b** fields (Smith *et al.* 2009), whereas it is direct at lower levels of correlation. These topics clearly need further investigations.

4. Modeling of MHD flows

There is a necessity to study fundamental phenomena in astrophysical environments but in a range of parameters too wide for DNS at high resolution. Space limitations do not permit us to deal at length with this important issue (see e.g. a recent review of the use of specific models in Pouquet *et al.* 2010, and for a general review of Large Eddy Simulations, Meneveau & Katz 2000). Hybrid models are one possibility, combining kinetic and continuum physics. But it cannot be stressed enough that we are close to hitting a brick wall until new technological venues are found: miniaturisation of components is at roughly ten times the atomic scale, and the power concentration in chips is approaching that of a rocket engine. So it is best to think in terms of a three-pronged approach, combining (i) DNS – giving the most accurate description of the phenomena at hand for the primitive equations, (ii) quasi-DNS – where at a given grid size, Reynolds numbers higher, say an order of magnitude at most, than in the DNS on the same grid, can be achieved by "proper" filtering (e.g., Lagrangian method of regularization of the primitive equations, preserving the Hamiltonian structure of the equations albeit in a different norm), and (iii) sub-grid scale modeling where one introduces a wealth of turbulent transport coefficients stemming from theoretical or phenomenological studies, the alpha effect of dynamo generation possibly being the best known example in MHD in astrophysics. Other models can and should be developed and studied, and all such approaches should be combined and contrasted in a quasi-steady state, with hopefully a positive slope toward enlightenment. As the number of parameters increases with the complexity of the physics/chemistry, these developments will have to occur, the small (or large) magnetic Prandtl number regime being one such successful example. The imposing of symmetries of the TG flow is another approach, as are the studies of ideal dynamics as a precursor to dissipative phenomena (see e.g. Krstulovic *et al.* 2009 and references therein). Another important development to be seen in the future will be the use of more sophisticated numerical methods, not only in their scaling properties to a large number of processors but also in their adaptivity (dynamic or static, e.g. with embedded grids), keeping in mind that a high level of accuracy may prove important in the assessment of extreme phenomena such as reconnection events requiring the evaluation of steep gradients.

5. Evidence for self-organized criticality and avalanches in MHD

Small scale structures are dominated by current and vorticity sheets which, at high Reynolds number, undergo several types of instability at any given time (see Fig. 3).

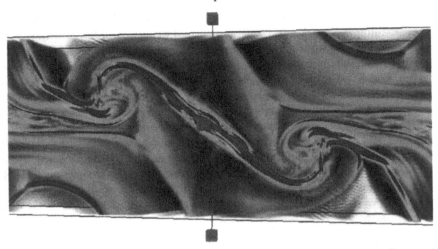

Figure 3. Zoom on current density at t = 4, TG flow with an IK spectrum, 2048^3 equivalent grid resolution. Strong current sheets appear at different stages of their evolution: some are straight, some are curved, some are undergoing complex roll-up and some display secondary islands (at the center) that could be associated with avalanche-like cascades of reconnecting events.

These dissipative events may have statistics that follow self-organized criticality (SOC), as demonstrated on an avalanche model in Lu & Hamilton (2001), and as analyzed in 3D-DNS data in Uritsky *et al.* (2010) (note that no such avalanche behavior was discernible in the inertial range structures). The coexistence of SOC and turbulent regimes has also been recently documented for the solar corona (Uritsky *et al.* 2007, 2009).

We can speculate on one of the possible sources of burstiness in the dissipation and its relation with avalanches. Starting from the pioneering work of Onsager (1949), Lee (1952) and Kraichnan (1975), one can think of the dynamical evolution of a turbulent flow being due to nonlinear interactions with weak forcing and weak dissipation balancing each other. Solutions of the ideal truncated equations obtain at late times with, in the simplest instance, equipartition between all the modes, and with zero energy flux. At intermediate times and intermediate scales, one observes turbulent dynamics with non-zero flux (Cichowlas *et al.* 2005), the "dissipation" of large-scale energy being associated with a turbulent eddy viscosity due to the thermalized modes at small scale (see Krstulovic *et al.* (2009, 2010) for similar results for helical flows, and for 2D MHD). This dynamics has also been observed in viscous cases, e.g., in Navier-Stokes fluids and in MHD at high resolution, where the resulting flow can be decomposed, using wavelets, into a set of coherent structures with a spectrum close to Kolmogorov, and a large number of modes at small scales and in thermal equilibrium (Okamoto *et al.* 2007 and references therein).

Related to these results, it has been known for some time, and in different instantiations of turbulent flows, that the energy flux of a given sign on average, has in fact huge fluctuations of both signs and of amplitudes much larger than the mean (see for example Bandi *et al.* 2009 and references therein, and Graham *et al.* 2010 for studies of regions with zero flux in models of turbulence). These large fluctuations in the flux can be attributed to the balance between forcing and dissipation mentioned above, and to the two components (one thermalized and random, one turbulent and coherent) identified in turbulent flows at small scales. The interplay between the two components can result in a bursty flux transfer of energy to the small scales, as observed in particular when looking at dissipation and reconnection events (Politano *et al.* 1989). These bursts are the needed excursions that lead the system away from thermal equilibrium; thus, they may give rise to a state of criticality in order to dissipate the energy accumulated over various lapses of time through the injection mechanism. Some of these events will trigger in turn other events, by pushing around structures which then can make contact with other structures that may in turn destabilize, leading to secondary island formation and ejection of plasmoids. It will be of interest to study SOC behavior more fully in a turbulent flow, including in the presence of a Hall current, sometimes interpreted as a dissipative range although it is nonlinear and leads to a self-similar energy distribution. Will SOC behavior be identified in the Hall regime, or will it be relegated, as in MHD, to the exponentially decreasing part of the Fourier spectrum?

Note that, in our analysis of structures in several MHD configurations, it is the flow with strong **v**–**b** correlations that exhibits SOC signatures: its critical exponents lie in the range of some well-known SOC universality classes. In general, it is possible that the lack of SOC universality in the inertial range may be caused by distinct universality classes describing SOC avalanches in the sub-inertial range. If we were to accept that SOC physics does control growth and decay of dissipative structures at these scales, then the (presumably) inverse transfer associated with avalanches at these scales would counter-stream with the direct MHD cascade. The two (small scale inverse and large scale direct) will meet roughly at the dissipation scale which may be more or less "penetrable" for the energy flux in Fourier space depending on the SOC universality class which itself may depend on the nature of the turbulence (e.g., 2D or 3D, fluid or MHD, ...).

6. Concluding remarks

In view of the lack of universality of energy and helicity spectra for MHD turbulence, in the presence or not of cross correlation or magnetic helicity, one may wonder what's left? Exact laws, in the simplest case under the hypotheses of isotropy, incompressibility, homogeneity, stationarity and high Reynolds number, may be one answer (see Politano & Pouquet 1998 for MHD, Politano *et al.* 2003 for magnetic helicity, and Galtier 2010 for Hall MHD). In fact it has been shown, in the context of a model insuring regularization of the primitive equations (see §4), that the different scaling expected because of the presence of new terms in the modified "4/5th" (Kolmogorov) exact law for this model indeed lead to a succession of energy spectral ranges that are compatible with the theory (Graham *et al.* 2009). So the exact laws do contain information about possibly different scaling in the specific case of MHD as well; they also show the strong link between the conservation of total energy E_T and cross-correlation H_c (the two are coupled), and the fact that these two equations for E_T and H_c may introduce a priori two time-scales (not taking into account magnetic helicity itself): this opens the door for more complex dynamics than in the incompressible fluid case. As conjectured by several authors in different ways, it is the interplay between the invariants (with given rates of transfer) and their associated time-scales (the Alfvén time, the eddy turn-over time and their combinations) that may well govern the dynamics. Moreover, anisotropy in MHD is bound to play a role, opening even more the door to a variety of dynamical equilibria.

MHD is not the only case where breaking of universality has been observed. For example, for neutral flows in the presence of rotation, one obtains a dual cascade to the small scales of kinetic energy (with spectral index e) and of (kinetic) helicity $H_v = < \mathbf{v} \cdot \omega >$ (with $\omega = \nabla \times \mathbf{v}$ the voriticity), with spectral index h, and with $e + h = 4$, at least at sufficiently high rotation at a given Reynolds number (see Mininni & Pouquet 2010, Baerenzung *et al.* 2008, and Baerenzung *et al.* 2010). This can be understood in terms of an argument stemming from the dual considerations of (i) a cascade to small scales dominated by the helicity (the energy being cascaded to large scales because of the quasi bi-dimensionalization of the flow), and (ii) the transfer to small scales being mediated by inertial waves, like in MHD where Alfvén waves play a similar role.

In order to study further the cascades of kinetic and magnetic energy and helicity as well as cross-correlation, and the dynamical equilibria that can be reached between them, one needs large scale separation and large Reynolds number R_V. Models of MHD turbulence (see §4) may help unravel further these competing dynamics. Do we obtain $H_m \sim k^{-4}$ in the limit of high R_V? What role is played by the scales larger than the forcing (or the scale of the initial conditions)? Are there dual cascades in turbulent flows besides the one mentioned previously? (Note that it is already known that the magnetic energy is transferred both to large scale (following the magnetic helicity) and to small scales, through stretching by vorticity gradients; however, strictly speaking, E_m is not an invariant but in the case when $E_m \gg E_v$, the total energy cascade is dominated by the magnetic energy.) Furthermore, what happens when the inverse cascade of magnetic helicity reaches the size of the vessel? Is there a rebound of H_m and what effect would it have on the dynamics at smaller scales? Answers to these questions are not known. Note also that they are linked to the 2D Navier-Stokes case, underlying in fact the current debate around the interpretation of atmospheric data with a dual (k^{-3} at large scale and $k^{-5/3}$ at small scale, see, e.g., Gkioulekas & Tung 2007 and references therein).

Acknowledgements

We dedicate this paper to Juri Toomre who has been an inspiration to several generations of researchers and who has conducted with his team magnificent computations of solar and stellar convection, leading the way to a deeper understanding of astrophysical flows. The computations reported here were performed, for the most part, at NCAR which is sponsored by the National Science Foundation.

References

Biskamp, D. & Welter, H. 1989, *Phys. Plasmas*, B1, 1964

Politano, H., Pouquet , A. & Sulem, P. L. 1989, *Phys. Fluids B*, 1, 2330

Mininni, P. D. & Pouquet, A. 2009, *Phys. Rev. E*, 80, 025401(R)

Wang, X. G., Bhattacharjee, A. & Ma, Z. W. 2000, *J. Geophys. Res. – Space Physics*, 105, 27633

Matthaeus, W. H. & Lamkin, S. L. 1986, *Phys. Fluids*, 29, 2513

Servidio, S., Matthaeus, W. H., Shay, M. A., Dmitruk, P., Cassak, P. A. & Wan, M. 2010, *Phys. Plasmas*, 17, 032315

Mininni, P. D. 2011, January issue, *Ann. Rev. Fluid Mech.*, 43

Mininni, P. D., Alexakis, A. & Pouquet, A. 2007, *J. Plasma Phys.*,73, 377

Sridhar, S. & Goldreich, P. 1994, *Astrophys. J.*, 432, 612

Montgomery, D. & Matthaeus, W. H. 1995, *Astrophys. J.*, 447, 706

Ng, C. S. & Bhattacharjee, A. 1996, *Astrophys. J.*, 465, 845

Iroshnikov, P. S. 1963, *Sov. Astron.*, 7, 566

Kraichnan, R. H. 1965, *Phys. Fluids,* 8, 1385

Mininni, P., Montgomery, D. & Pouquet, A. 2005, *Phys. Fluids*, 17, 035112

Dmitruk, P., Matthaeus, W. H. & Gòmez, D. 2003, *Phys. Plasmas*, 10, 3584

Müller, W. C. & Grappin, R. 2005, *Phys. Rev. Lett.*, 95, 114502

Mason, J., Cattaneo, F. & Boldyrev, S. 2008, *Phys. Rev. E*, 77, 036403

Brachet, M. E., Meiron, D. I., Orszag, S. A., Nickel, B. G., Morf, R. H. & Frisch, U. 1983, *J. Fluid Mech.*, 130, 411

Lee, E., Brachet, M. E., Pouquet, A., Mininni, P. & Rosenberg, D. 2010, *Phys. Rev.E*, 81, 016318

Goldreich, P. & Shridar, S. 1995, *Astrophys. J.*, 438, 763

Galtier, S., Pouquet, A. & Mangeney, A. 2005, *Phys. Plasmas*, 12, 092310

Yurchyshyn, V., Yashiro, S., Abramenko,V., Wang, H. & Gopalswamy, N. 2005, *Astrophys. J.*, 619, 599

Uritsky, V. M., Pudovkin, M I. & Steen, A. 2001, *J. Atmosph. Solar-Terrestrial Phys.*, 63, 1415

Müller, W. C. & Biskamp, D. 2003, *Phys. Rev. E*, 67, 066302

Pouquet, A. 1993, *Magnetohydrodynamic Turbulence*, Les Houches Session **XLVII**, 139; Zahn, J.P. & Zinn–Justin, J., Elsevier, eds.

Pouquet, A. 1996, *Turbulence, Statistics and Structures: an Introduction*, Springer–Verlag, Lecture Notes in Physics "Plasma Astrophysics" **468**, 163, Chiuderi, C. & Einaudi, G., eds.

Boldyrev, S. 2006 *Phys. Rev. Lett.*, 96, 115002

Boldyrev, S. 2005, *Astrophys. J. Letters*, 626, L37

Politano, H. & Pouquet, A. 1998, *Geophys. Res. Lett.*, 25, 273

Mason, J., Cattaneo, F. & Boldyrev, S. 2006, *Phys. Rev. Lett.*, 97, 255002

Frisch, U., Pouquet, A., Léorat, J. & Mazure, A. 1975, *J. Fluid Mech.*, 68, 769

Pouquet, A., Frisch, U. & Léorat, J. 1976 *J. Fluid Mech.*, 77, 321

Müller, W. C. 2008, *Private communication*

Malapaka, S. K. 2009, PhD Thesis, "A study of magnetic helicity in decaying and forced 3D-MHD turbulence," University of Bayreuth

Mininni, P. D. & Pouquet, A. 2010, *Phys. Fluids*, 22, 035106

Pouquet, A. & Patterson, G. S. 1978, *J. Fluid Mech.*, 85, 305

Stribling, T. & Matthaeus, W. H. 1990, *Phys. Fluids*, B2, 1979

Smith, C. W., Stawarz, J. E., Vasquez, B. J., Forman, M. A. & MacBride, B. T. 2009, *Phys. Rev. Lett.*, 103, 201101

Pouquet, A., Baerenzung, J., Pietarila Graham, J., Mininni, P. D., Politano, H. & Ponty, Y. 2010, Notes on Numerical Fluid Mechanics, and Multidisciplinary Design, Springer Verlag, M. Deville, J.P. Sagaut & T. Hiep eds., (see also arXiv:0904.4860)

Meneveau, C. & Katz, J. 2000, *Ann. Rev. Fluid Mech.*, 32, 1

Krstulovic, G., Mininni, P. D., Brachet, M. E. & Pouquet, A. 2009, *Phys. Rev. E*, 79, 056304

Lu, E. T. & Hamilton, R. J. 1991, *Astrophys. J. Letters*, 380, L89

Uritsky, V., Pouquet, A., Rosenberg, D. & Mininni, P. D. 2010, "Structures in magnetohydrodynamics turbulence: detection and scaling," submitted to *Phys. Rev. E*, see arxiv:1007.0433

Uritsky, V. M., Paczuski, M., Davila, J. M. & Jones, S. I. 2007, *Phys. Rev. Lett.*, 99, 025001

Uritsky, V. M., Davila, J. M. & Jones, S. I. 2009, Phys. Rev. Lett., 103, 039502

Onsager, L. 1949, Supplemento al vol. VI, Serie IX del Nuovo-Cimento, 2, 279

Lee, T. D. 1952, Quart. Appl. Math, 10, 69

Kraichnan, R. H. 1975, J. Fluid Mech., 67, 155

Cichowlas, M., Bonaiti, P., Debbasc, F. & Brachet, M. E. 2005, Phys. Rev. Lett., 95, 264502

Krstulovic, G., Brachet, M. E. & Pouquet, A. 2010, "Ideal two-dimensional dynamics of the Euler and MHD equations," in preparation

Okamoto, N., Yoshimatsu, K., Schneider, K., Farge, M. & Kaneda, Y. 2007, Phys. Fluids, 19, 115109

Bandi, M. M., Chumakov, S. G. & Connaughton, C. 2009, "On the probability distribution of power fluctuations in turbulence," submitted, see arXiv:0901.0743

Pietarila Graham, J., Holm, D., Mininni, P. D. & Pouquet, A. 2010, "The effect of subfilter-scale physics on regularization models," to appear, Springer -Verlag, Conference on Quality and Reliability in Large Eddy Simulations ; see also arxiv.org/abs/1003.0335v1

Politano, H., Gomez, T. & Pouquet, A. 2003, *Phys. Rev. E*, 68, 026315

Galtier, S. 2010, *Phys. Rev. E*, 77, 015302(R)

Pietarila Graham, J., Mininni, P. D. & Pouquet, A. 2009, *Phys. Rev. E*, 80, 016313

Baerenzung, J., Politano, H.,. Ponty, Y. & Pouquet, A. 2008, *Phys. Rev. E*, 78, 026310

Baerenzung, J., Mininni, P. D., Pouquet, A. & Rosenberg, D. 2010, "Spectral Modeling of Turbulent Flows and the Role of Helicity in the presence of rotation," see also arXiv:0912.3414

Gkioulekas, E. & Tung, K. K. 2007, *Discrete Contin. Dyn. Syst. Ser. B*, 7, 293

Discussion

S. Tobias: This lack of universality in MHD turbulence is very interesting, but also very worrying in astrophysical/geophysical flows that often arises due to the presence of coherent structures but you find it in homogeneous isotropic turbulence. Is there any "phase information" you can pick out in your models to distinguish the initial conditions that behave so differently?

A. Pouquet: Well, there are subtleties. The MHD turbulence we generally study is isotropic insofar as we do not impose a mean field, but it is locally anisotropic at small scale. And the velocity and magnetic fields we use to force are large-scale ordered fields, for the velocity corresponding to a flow between two counter-rotating cylinders (the von Kàrman flow) and which can be viewed as modeling large-scale instabilities as can be found indeed in astrophysical flows. But to answer you about phase, no, I cannot pick up any information to explain why the magnetic energy grows in one case rather substantially (by a factor close to 6) and does not grow or decay as much (vis-à-vis its kinetic counterpart) in the other cases, except to say that one of the flows is "conducting" insofar as the current in the so-called impermeable domain (from which the whole flow is reconstructed when implementing the symmetries) is orthogonal to the walls, whereas in the other cases it is insulating, with the current in the walls of the impermeable domain. However, I might add that in that first case of strong domination of magnetic energy, at $t = 0$, we have $\mathbf{b} = \omega$. Also note that, when observed, the excess of magnetic energy is clearly taking place in the gravest mode accessible to the system, and that when E_m/E_v decays in the first flow, the energy spectrum shows sighs of being less steep. In other words, these classes of universality may well be not an intrinsic property of a given flow (initial condition) but may rather result from the internal complex dynamics of turbulent flows. Of course, this last result is obtained for late times at which the total energy and thus the Reynolds numbers have decreased measurably. In order to pursue such investigations, models such as those mentioned in §4, and the so-called shell models of turbulence as well (see e.g., Plunian & Stepanov 2006) may prove quite useful.

A.S. Brun: Are you seeing evidence of the current sheet vortex in the high order structure functions?

A. Pouquet: Small-scale structures, such as current and vorticity sheets and rolls are the basis of the intermittency of turbulence, and as such are probably responsible for the variation with order of the anomalous exponents of structure functions, as shown e.g. in an analysis using wavelets (Okamoto *et al.* 2007). As I discussed in my talk, in solar active regions, there is observational evidence that the functional form of these exponents varies with the intensity of the flare (Yurchyshyn *et al.* 2005), and numerically as well it varies with the intensity of the imposed magnetic field (Müller & Biskamp 2003). How different are the small-scale structures for an IK, K41 or WT spectrum? This is of course a good but difficult question, that perhaps can be addressed today using high-resolution direct numerical simulations (DNS). But what are we asking for precisely? Probably one could start by examining the structures associated with the extreme values of the velocity and magnetic field gradients.

A. Brandenburg: If τ_A/τ_{NL} is a function of k, you should expect kinetic and magnetic energy spectra to become non parallel. Can this be true also asymptotically?

Figure 4. Snapshots of a zoom on the current density (left) [as in Fig. 3 but with a different perspective], and of the enstrophy density ω^2 (right) at the same time ($t = 4$) as in Fig. 3 and for the same Taylor-Green (TG) flow with an Iroshnikov-Kraichnan spectrum (see Lee *et al.* 2010); 2048^3 equivalent grid resolution with a code implementing the symmetries of the TG flow generalized to MHD (see §2). Note the overall spatial correlation between both sets of structures, but with more complexity in the vorticity field.

A. POUQUET: Yes indeed, there are measurable differences between the velocity and the magnetic filed inertial indices, as measured both in the solar wind (Podesta *et al.* 2007) and in highly resolved DNS (Mininni & Pouquet 2010). Now, I suppose that you mean asymptotically in Reynolds number R_V (let us assume for the sake of simplicity that the magnetic Prandtl number is equal to unity, which is the case for most of the numerical simulations I am describing here). But we have another parameter, which is time. It takes a time $\sim 1/K_0$ to reach equipartition at wavenumber K_0, but as I increase R_V by keeping the dissipation scale fixed, I have more and more wavenumbers available at large scale for which equipartition is not yet reached, leading to the possibility that the velocity and magnetic field energy spectra differ even at high Reynolds numbers. That does lead again to the question, though, of whether the breaking-down of universality we have observed in decaying flows occurs as well in the statistically steady case in the absence of an imposed uniform field.

H.K. MOFFATT: The roll-up of current sheets suggests that these must coincide with vortex sheets which are still subject to Kelvin-Helmholtz instability (partially stabilized by the magnetic field). Can you explain the tendency for the formation of these current-vortex sheets?

A. POUQUET: Absolutely, vortex sheets roll-up as well, but the current and vorticity structures may not coincide exactly in space (see Fig. 4). Indeed, it is known that in two dimensions, at a neutral X point, the current is a dipole whereas the vorticity has a more complex structure (it is a quadrupole). In fact, when looking at the normalized correlation between **v** and **b**, one observes that the large scales can be highly correlated (and/or anti-correlated, even if the total correlation integrated over the domain is weak) but that at the boundaries between large energy-containing regions or eddies, the structure of the correlation coefficient is quite complex (see, e.g, Meneguzzi *et al.* 1996). What these rolls are precisely has not been studied much. There is a possibility that they fit the description of so-called Alfvén vortices (see Kadomtsev & Pogutse 1974), as observed in the solar wind by Alexandrova *et al.* (2006); but to my knowledge a Kelvin-Helmholtz instability of the development of these rolls (probably simpler when using the Elsässer fields since one can expect a more symmetric form in these variables) has not been performed yet. When doing so, one may want to take into consideration that we observe that the magnetic field lies within the current sheets which roll around its main direction. We have also observed at least in some cases that a clear rotational discontinuity develops (Lee *et al.* 2008), with two nearby sheets coming into close contact and with the magnetic field in each sheet in different directions (note that rotational discontinuities have been documented in the solar wind, see e.g. Whang 2004).

References

Plunian, F. & Stepanov, R., 2006 *J. Turbulence*, 7, 39

Podesta, J. J., Roberts, D. A. & Goldstein, M. L. 2007, *Astrophys. J.*, 664, 543

Meneguzzi, M., Politano, H., Pouquet, A. & Zolver, M. 1996, *J. Comp. Phys.*, 123, 32

Kadomtsev, B. B. & Pogutse, O. P. 1974, *Z. Eksp. Teor. Fiz.*, 65, 575

Alexandrova, O., Mangeney, A., Maksimovic, M., Cornilleau-Wehrlin, N., Bosqued, J.-M. & André, M. 2006, *J. Geophys. Res.*, 111, A12208

Lee, E., Brachet, M., Pouquet, A., Mininni, P. & Rosenberg, D. 2008, *Phys. Rev.E*, 78, 066401

Whang, Y. C. 2004, *Nonlinear Processes Geophys.*, 11, 259

Astrophysical Dynamics: from Stars to Galaxies
Proceedings IAU Symposium No. 271, 2010
N.H. Brummell, A.S. Brun, M.S. Miesch & Y.Ponty, eds.

© International Astronomical Union 2011
doi:10.1017/S1743921311017741

Overshooting above a convection zone

Kwing L. Chan[1], Tao Cai[1], and Harinder P. Singh[2]

[1] Department of Mathematics, Hong Kong University of Science & Technology,
Clear Water Bay, Hong Kong, China
email: maklchan@ust.hk

[2] Department of Physics & Astrophysics, University of Delhi,
Delhi 110007, India
email: hsingh@physics.du.ac.in

Abstract. As compressible convection has inherent up/down asymmetry, overshooting above and below a convection zone behave differently. In downward overshooting, the narrow downflow columns dynamically play an important role. It is customary, and reasonable, to use the downward flux of kinetic energy as a proxy for overshooting. In the upward situation, the flux of kinetic energy can take on different signs near the upper boundary of the convection zone, and its magnitude is generally small. It cannot make a good proxy for overshooting. This paper discusses the results of a set of numerical experiments that investigate the problem of overshooting above a convection zone. Particle tracing and color advection are used to follow the mixing process. The overshoot region above a convection zone is found to contain multiple counter cell layers.

Keywords. stars, convection, overshooting

---

1. Introduction

Convective overshooting is an important process in stars. Its main effect is to mix materials across a convection zone and a neighboring stable zone. Due to the up-down (with respect to the direction of gravity) asymmetry associated with stratification, overshooting above and overshooting below have different behaviors. The attention of the current paper is on overshooting above a convection zone.

Overshooting is a topic that Juri (Toomre) has worked on for a long time and made seminal contributions. Before I came to this meeting, I tried to count his papers on this subject. Table 1 gives a list which is only part of the full set. In this list, the 1986, 1994, and 2002 papers discuss configurations closest to the one of the present paper. These three articles, however, only discuss overshooting below convection zones (let us call it 'under-shooting' here). Most other studies (e.g. Singh *et al.* 1998, Pal *et al.* 2007) also focus on under-shooting, and the penetration of the downward flux of kinetic energy into the stable zone is ubiquitous used as the proxy of mixing. In this paper, we will show that this flux is not a good indicator of overshooting above a convection zone (or simply 'overshooting').

2. Model

We perform three-dimensional simulation of turbulent compressible motions of an ideal gas (ratio of specific heat = 5/3) in a rectangular box containing a convection zone in the lower part (6.3 pressure scale heights, 60% of the domain depth) and a stable zone in the upper part (4 pressure scale heights, 40% depth). The two layers are initially polytropic. The polytropic index of the convection zone is 1.5 (the adiabatic gradient is 0.4), while that of the stable zone is 5.67 (corresponding to a radiative gradient of 0.15).

Table 1. Juri's overshooting-related papers.

Year	Authors	Title	Journal
1976	Toomre, Zahn, Latour, & Spiegel	Stellar convection theoy. II. Single-mode stydy of the second convection zone in an A-type star	ApJ, 207, 545
1981	Latour, Toomre, & Zahn	Stellar convection theoy. III. Dynamical coupling of the two convection zones in A-type stars by penetrative motions	ApJ, 248, 1081
1983	Latour, Toomre, & Zahn	Nonlinear anelastic modal theory for solar convection	Solar Phys., 82, 387
1984	Massaguer, Latour, Toomre, & Zahn	Penetrative cellular convection in a stratified atmoshere	A&A, 140, 1
1986	Hurlburt, Toomre, & Massaguer	Nonlinear compressible convection penetating into stable layers producing internal gravity waves	ApJ, 311, 563
1994	Hurlburt, Toomre, Massaguer,& Zahn	Penetration below a convective zone	ApJ, 421, 245
1998	Tobias, Brummell, Ckune, & Toomre	Pumping of magnetic fields by turbulent penetrative convection	ApJ, 502, L177
2000	Miesch, Elliott, Toomre, Clune, Glatzmaier, & Gilman	Three-dimensional spherical simulations of solar convection: Differential toation and pattern evolution achieved with laminar and turbulent states	ApJ, 532, 593
2001	Tobias, Brummell, Ckune, & Toomre	Transport and storage of magnetic field by overshooting turbulent compressible convection	ApJ, 549, 1183
2002	Brummell, Clune & Toomre	Penetration and overshooting in turbulent compressible convection	ApJ, 570, 825
2004	Browning, Brun, & Toomre	Simulation of core convection in rotating A-type stars: Differential rotation and overshooting	ApJ, 601, 512
2007	Browning, Brun, Miesch, & Toomre	Dynamo action in simulations of penetrative solar convection with an imposed tachocline	AN, 328, 1100

The domain is periodic in the two horizontal directions. The top and bottom boundaries are stress-free and impenetrable. The temperature at the top and the energy flux at the bottom are uniform and fixed.

The numerical method, the code, and the setup are similar to those used in Singh *et al.* (1994). The main differences are the larger aspect ratio and the enhanced resolution. Table 2 lists these parameters for the computed cases. Cases A - C have sequentially doubled input energy fluxes (F). The aspect ratio of Cases D and E is two times that of Cases A and B. Case F has doubled horizontal resolution of Case B. The variables are in units which make the initial temperature, density, pressure, and total depth all equal to 1.

Table 2. Computed cases.

Case	Flux	Grids	Aspect Ratio
A	0.053	$162 \times 162 \times 122$	3
B	0.105	$162 \times 162 \times 122$	3
C	0.210	$162 \times 162 \times 122$	3
D	0.053	$322 \times 322 \times 122$	6
E	0.105	$322 \times 322 \times 122$	6
F	0.105	$322 \times 322 \times 122$	3

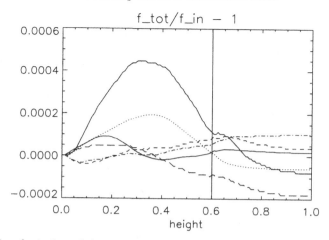

Figure 1. Relative deviation of the mean total energy flux from the input flux $(F_{tot} - F)/F$. The dotted, dashed, dot-dashed, triple-dot-dashed, long dashed, and solid curves represent Cases A-F respectively.

3. Results

Thermal relaxation of a fluid layer containing a deep stable zone generally takes very long time. It typically requires ten of millions of steps to complete a calculation. To ensure accuracy of the statistical information, a high degree of thermal and dynamical relaxation is needed. The calculations presented here all satisfy the condition that the relative variations of the mean total energy flux F_{tot} (horizontally and temporally averaged) are less than 0.1% (see Figure 1). After relaxation, flow statistics are computed, and cross-zone mixing is studied with a particle tracing scheme and a color advection scheme.

The particle scheme uses third-order interpolation to obtain velocities of intergrid particles and a fourth-order Runge-Kutta method to perform time marching. The initial particle density n is proportional to the mean density profile (a function of height z) inside the convection zone, and is set to 0 in the stable layer. The horizontal locations of the particles are randomly distributed. The total number of particles is about 180,000. Figure 2 shows two side views of the particle distributions in the box (Case A). It

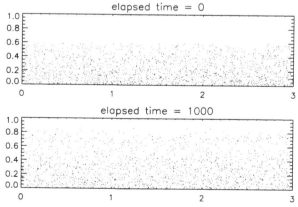

Figure 2. Side views of particle distributions (Case A) at two instances. Only 1% of the total number of particles used in the computation are shown. The upper panel illustrates the initial distribution. The lower panel illustrates the distribution at time = 1000.

illustrates the vertical profiles of the particles at the initial instance ($t = 0$, upper panel) and at a much later time ($t = 1000$, about 110 turnover times later, lower panel). At the beginning, all particles were below the height $z_I = 0.6$ (location of the interface between the convection and stable zones); at the later instance, particles spreaded above $z = 0.8$.

The color scheme solves a continuity equation for the color density c: $\partial c/\partial t + \vec{\nabla} \cdot \vec{v} c = 0$. where $\vec{v}$ is the fluid velocity. c is a passive scalar that does not affect the dynamics. The initial distribution of c is everywhere proportional to the density ρ inside the convection zone, and is 0 outside. The vertical profiles of the horizontally averaged specific particle number density ($= n/\rho$) and the specific color ($= c/\rho$) are shown in Figure 3 by the solid and dashed curves respectively. The upper panel shows the initial distributions (almost coincide), and the lower panel shows the distributions at time 1000. While $\overline{c/\rho}$ decreases almost monotonically (not exactly) outside the convection zone, $\overline{n/\rho}$ shows some humps (the over-bar denotes horizontal averaging). The humps may be caused by the over-representation of low density fluid (larger volume and upward moving) in the initial particle distribution. The bias is maximal near the bottom of the convection zone; the dashed curve in the lower panel shows a depletion on the left. The effect of particle transport on $\overline{n/\rho}$ is amplified in the overshoot region as mass density is smaller there. At any rate, the important features to note are the extends of advection and the appearance of some dips/steps in the curves. The extends of overshooting and the locations of the dips in the two wiggling curves agree very well. What causes these dips? Before answering this question, we need to look at the behaviors of some more quantities.

Particle tracing is an expensive computation, and the interpretation of results is often complicated by the mixed presence of true advection and numerical truncation errors. It is customary that some easily computed quantities are used as proxies of mixing. In the case of undershooting, the flux of kinetic energy F_k is most frequently used. In the case of overshooting, however, this flux cannot be used. The problem is illustrated in Figure 4 in which the relative values of the kinetic energy flux (F_k/F, dashed curve), the difference between the radiative/diffusive flux and the total flux ($(F_r - F)/F$, dot-dashed curve), and the enthalpy flux (F_e/F, solid curve) in the stable zone of Case A are shown. F_k/F appears flat as its magnitude in this zone is everywhere less than 5×10^{-5}. Furthermore, F_r and F_e are not of much use as their extends are confined below $z = 0.7$ (remember that the particle and color spreads are beyond $z = 0.8$). The triple-dot-dashed curve

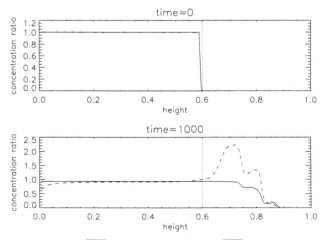

Figure 3. Vertical profiles of $\overline{n/\rho}$ (dashed curves) and $\overline{n/\rho}$ (solid curves) at $t = 0$ (upper panel) and $t = 1000$ (lower panel).

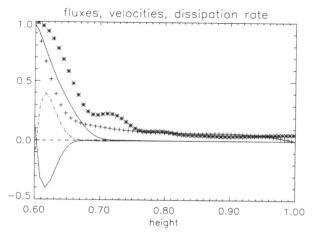

Figure 4. Normalized fluxes, dissipation rate, and rms velocities. The dashed, dot-dashed, solid, and triple-dot-dashed curves represent F_k/F, $(F_r - F)/F$, F_e/F, and $\mathcal{E}/\mathcal{E}_I$, respectively. The pluses and asterisks represent $\mathrm{rms}(v_z)/\mathrm{rms}(v_z)_I$ and $\mathrm{rms}(v_h)/\mathrm{rms}(v_h)_I$.

shows the normalized kinetic energy dissipation rate $\mathcal{E}/\mathcal{E}_I$. The subscript I denotes value at the interface between the convection and stable zones ($z_I = 0.6$). The extend of this curve shows that dissipation of the overshooting motions reaches beyond $z = 0.7$.

In Figure 4, the normalized root-mean-square (rms) velocities $\mathrm{rms}(v_z)/\mathrm{rms}(v_z)_I$ and $\mathrm{rms}(v_h)/\mathrm{rms}(v_h)_I$ are shown by the pluses and asterisks, respectively. v_z and v_h are the vertical and horizontal velocities. The rms values are computed through averaging horizontally and temporally. These quantities cannot be used as proxies as they do not decay over long distances (this is connected to the generation of gravity waves). The curve describing $\mathrm{rms}(v_h)$ even indicates some growing trend at large distances ($\mathrm{rms}(v_z)$ is forced to 0 by the upper boundary condition). An eye-catching behavior of this curve is the presence of wiggles. This is suggestive of a connection with the wiggles in the particle/color advection curves. There are at least two ways to interpret these wiggles: One is the occurrence of layers of counter cells; another is the nodes of standing gravity waves.

Figure 5 plots the autocorrelation coefficients of v_z (solid curve) and the temperature deviation $T'(= T - \overline{T})$ (dashed curve) for Case A. The autocorrelation is made between two horizontal planes. One is fixed at the interface of the convection and stable zones, and the other is at an arbitrary height. Near the zone interface, both autocorrelations are positive, but the autocorrelation of v_z stays positive for a longer distance. The positive correlation means that an upward moving fluid parcel can maintain its upward motion (statistically). The originally positive temperature correlation quickly turns negative as an upward moving fluid parcel cools faster than the environment in the stable zone (and the density deviation becomes positive). Here, buoyancy breaking plays a significant role in decelerating the fluid parcel. As the Prandtl number in the stable zone is generally small, the memory of temperature deviation does not last as long as that of velocity. The temperature correlation damps out quicker than the velocity correlation. The pluses show the autocorrelation coefficient of v_z for Case D which is two times wider than Case A in each horizontal direction. Apparently the width does not seem to affect the results much.

The zeros of the velocity correlation can be interpreted as the boundaries of counter cell layers or as nodes of gravity waves. It is important to note that the locations of

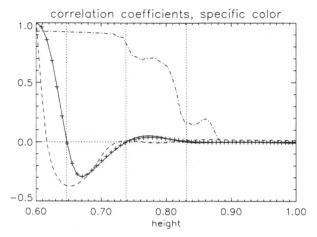

Figure 5. Autocorrelation coefficients and distribution of specific color for Case A. The auto-correlation coefficients of the vertical velocity and the temperature deviation are represented by the solid and dashed curves, respectively. The autocorrelation coefficient of vertical velocity of Case D is shown as pluses. The specific color distribution at $t = 1000$ is shown by the dot-dashed curve.

the zeros of the autocorrelation of vertical velocity correspond to the drops/dips of the color/particle distributions. To assist comparison, the color distribution at $t = 1000$ is shown again in Figure 5, by the dot-dashed curve. The vertical dotted lines identify the locations of the correlation zeros ($z = 0.647, 0.738, 0.831$; hereafter referred as z_1, z_2, z_3, respectively). The first zero is within the region where turbulent dissipation is still significant. Turbulent motion quickly carries the color through the location of the first zero. The reversal of the sign of velocity correlaiton has little effect on the spread of color. The second zero, on the other hand, is associated with a barrier of the mixing process. It takes a while for the color to cross it, but once through the mixing can propagate faster until it hits another barrier near the location of the third zero of velocity correlation.

Independent of whether the zeros correspond to counter cell boundaries or wave nodes, it is clear that color can cross them only through some small-scale dissipative process. If the numerical results represent a realistic situation, there could be some mild 'turbulence' generated (or remnant turbulence) at the zero locations to facilitate the color crossing. On the other hand, both the color scheme and the particle advection scheme contain numerical errors. The barrier crossing could just be a result of the artificial effects. Our numerical tests indicate that the numerical diffusion of color is sufficiently small to make the color profile trustworthy for $t \leqslant 1000$ (Case A). But the situation may not remain so ideal when t becomes greater than a few thousand. In the present study, we pick the second zero of the velocity correlation as the indicator for the extend of overshooting. There are several reasons for this choice. First, the magnitude of the auto correlation damps out at large distances and the number of detectable zeros are no more than a few. The maximal distance of these zeros from the interface is no more than two or three times the distance of the second zero. Second, the first zero is not a significant barrier. Remnant turbulence remains substantial at heights below the second zero (see later discussion). The second zero forms the first true barrier to mixing. There is a third reason, but it is to be discussed later.

In the stable zone, the velocity response to convective hammering can survive for quite some distance from the zone interface (as shown by the profiles of rms(v_h) and rms(v_z) in Figure 3). What are the natures of these motions at different heights? We analyze

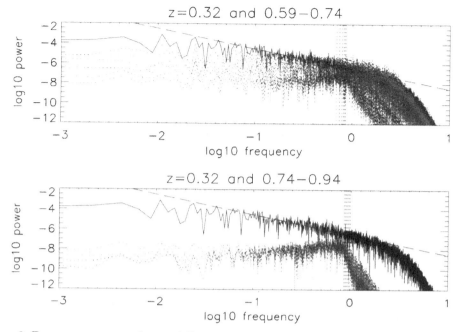

Figure 6. Frequency spectra of v_z at different height levels. To assist comparison, the spectrum at the mid-layer of the convection zone is shown by the solid curve. In the upper panel, the dotted curves show spectra in the height range $z_1 \leqslant z \leqslant z_2$. In the lower panel, the dotted curves show spectra in the height range $z_2 \leqslant z \leqslant 0.94$.

this problem by computing the frequency spectra of the motions at different levels. On each level a long time series of v_z is recorded and Fourier transformed to obtain the frequency spectrum. Figure 6 shows the changes of spectra over different characteristic ranges of heights (Case A). In both the upper and lower panels, the uppermost spectrum (solid curve) is at a level close to the middle of the convection zone. The dashed straight line has a slope of -2 which is the slope of the frequency spectrum corresponding to a Kolmogorov energy spectrum. Apparently, a short range of such exists (Chan & Sofia, 1996). The lower curves (dotted) in the upper panel show spectra at seven levels within the range of heights 0.59 to 0.74. The dotted curves in the lower panel show spectra at seven levels within the range of heights 0.74 to 0.94. The vertical lines show values of the Brunt-Vaisala frequency (f_{BV}) at the different levels.

Recall that the second zero of the autocorrelation of v_z is at $z_2 = 0.74$. The upper and lower panels of Figure 6 depict, respectively, the change of spectra below and above this second zero. Below the z_2 level, as z increases, there is a fast drop in spectral power, and the high frequency extends of the spectra above f_{BV} shrinks quickly. Above the z_2 level, most of the spectral powers (over 90%) are already stored in frequencies below f_{BV}, and the change of spectra is much slower. This supports the interpretaion that most of the motions above z_2 are associated with gravity waves. Movies of the temperature field show that the waves propagate in slant directions. Despite of the impenetrable boundary imposed at the top of the domain, there is no evidence of significant standing waves. We thus favor the interpretation that the zeros of v_z correlation are associated with horizontal boundaries of counter cells. Temperature and particle movies indicate that the counter cells have large aspect ratios.

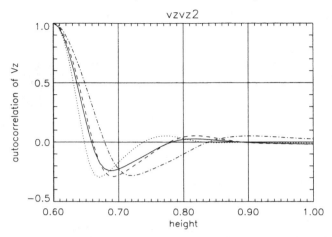

Figure 7. Autocorrelation coefficients of v_z for Cases A (dotted curve), B (dashed curve), C (dot-dashed curve), and F (solid curve).

Since we have computed a few cases with different input energy fluxes, we can try to look for a scaling relationship between the flux and the locations of the correlation zeros. Figure 7 shows the autocorrelation curves of v_z for Cases A (dotted), B (dashed), C (dot-dashed), and F (solid). As the flux increases form A to C, the strength of turbulence in the convection zone increases. As a result, the distances of the zeros from z_I also increase. Case F is a high-resolution version of Case B. The locations of its correlation zeros agree well with those of Case B, but there are small drops in the amplitude of correlation. The differences have little effect in the layer below z_2, but the impact at heights above z_2 may not be so easily discarded. In numerical simulations, higher resolution usually means lower viscosity. If the lower velocity correlation is associted with the lower viscosity, the upper layer counter cells could vanish in the very low viscosity limit. This is the third reason we use z_2 as the proxy for estimating the extend of overshooting, not the farther zeros. It is expected that the full extend of mixing should be within a factor of 2 of $z_2 - z_I$.

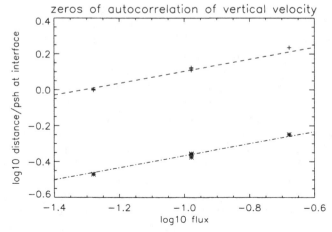

Figure 8. $log_{10}(z_1 - z_I)$ (asterisks) and $log_{10}(z_2 - z_I)$ (pluses) versus $log_{10}(F)$ for all cases.

The logarithmic values of $z_1 - z_I$ and $z_2 - z_I$ are plotted versus the logarithmic values of the input fluxes for all the cases in Figure 8. Dashed straight lines with slope $1/3$ pass through the points of all cases rather well (the upper and lower lines are for z_2 and z_1 respectively). Based on this fitting, the dependence of the distance $l = z_2 - z_I$ on F can be written as

$$l/h_* \approx 11 \times (F/F_*)^{1/3}$$

where h_* is the presure scale height at the interface of the convection and stable zones, $F_* = p_*(p_*/\rho_*)^{1/2}$, and p_*, ρ_* are the pressure and density at the interface.

4. Summary

The process of overshooting above a convection zone is studied by numerical models. Particle tracing and color advection schemes are used to study the effect of overshooting on mixing, a matter of primary interest in application. For overshooting above a convection zone, the flux profiles, including the flux of kinetic energy profile, are inadequate for estimating the extend of mixing. Through detailed analysis of the statistical quantities, frequency spectra of the velocity field at different heights, and visual examination of the particle and temperature movies, we conclude that the overshooting region contains some layers of thin counter cells. The boundaries between cell layers appear as barriers of mixing. These cell layer boundaries can be located by the zeros of the autocorrelation of the vertical velocity (with one layer fixed at the interface between the convection and stable zones). We argue that the distance of the second correlation zero to the zone interface (l) makes a good proxy to estimate the extend of overshooting/mixing. At the very least, l can be viewed as a lower bound of overshooting. A scaling relationship between this distance and the total flux can be found.

Acknowledgement

KLC thanks the Hong Kong Research Grants Council for support. We thank Prof. Kun Xu for suggesting the color method and Prof. Jimmy Fung for helpful discussions.

References

Chan, K. L. & Sofia, S. 1996, *ApJ*, 466, 372
Pal, P. S., Singh, H. P., Chan, K. L. & Srivastava, N. P. 2007, *Ap&SS*, 307, 399
Singh, H. P., Roxborgh, I. W. & Chan, K. L. 1994, *A&A*, 281, L73
Singh, H. P., Roxborgh, I. W. & Chan, K. L. 1998, *A&A*, 340, 178

Astrophysical Dynamics: From Stars to Galaxies
Proceedings IAU Symposium No. 271, 2010
N.H. Brummell, A.S. Brun, M.S. Miesch & Y.Ponty, eds.

© International Astronomical Union 2011
doi:10.1017/S1743921311017753

Turbulent dynamos

B. Dubrulle[1]

[1]DRECAM/SPEC/CEA Saclay, and CNRS (URA2464), F-91190 Gif sur Yvette Cedex, France
email: berengere.dubrulle@cea.fr

Abstract. We discuss the problem of turbulent dynamo and illustrate it through numerical simulations and few results from the VKS experiment.

Keywords. MHD, Turbulence

1. Introduction

A great amount of the work of Juri concerns the solar dynamo, that converts turbulent convective motions into magnetic energy. It is described quite generally by the coupled set of equations:

$$\partial_t \mathbf{v} + (\mathbf{v} \cdot \nabla)\mathbf{v} = -\frac{1}{\rho_0}\nabla P + \mathbf{j} \times \mathbf{B} + \nu\Delta\mathbf{v} + \mathbf{f},$$
$$\partial_t \mathbf{B} = \nabla \times (\mathbf{v} \times (\mathbf{B}) + \eta\Delta\mathbf{B}, \tag{1.1}$$

where $\mathbf{v}$ is the fluid velocity, P is the pressure, $\mathbf{B}$ is the Alfven velocity (or equivalently $\sqrt{\rho\mu_0}\,\mathbf{B}$ is the magnetic field), $j = \nabla\times\mathbf{B}$ is the magnetic current, ν and η are the viscosity and the magnetic diffusivity, and ρ_0 is the (constant) fluid density. Four interesting dimensionless parameter are necessary to understand them: i) the Reynolds number $Re = LV/\nu$, where L and V are typical length and velocity; ii) the magnetic Reynolds number $Rm = LV/\eta$ (or equivalently the Prandtl number $Pm = Rm/Re$); iii) the Rossby number $Ro = U/L\Omega$, where Ω is the rotation rate (or equivalently the rotation number $\theta = 1/Ro$); iv) the interaction parameter $N = RmB^2/V^2$. It characterizes the ratio of the Lorentz force over the velocity advective term, and is a measure of the "non-linearity" of the MHD system: when $N \ll 1$, the magnetic field does not react back to the velocity field, and the two equations decouple.

In stars, both Rm and Re are large. In numerical simulations, one can reach values of Rm up to 100 (Pm varying from 1 to 0.01). In present days laboratory experiments with liquid metals, Rm cannot exceed 100, with $Pm = 2 \times 10^{-5}$. They are however quite interesting to understand the dynamo process in stars, as we discuss now.

1.1. The VKS experiments

The VKS experiment is based on stirring a von Karman flow of liquid sodium through impellers counter-rotating at frequencies F_1 and F_2. When monitoring the two frequencies, one can vary the rotation number $\theta = (F_1 - F_2)/(F_1 + F_2)$ between -1 and 1- corresponding to regime of turbulence predominant over rotation. We have shown on Figure 1 the operating regime of the VKS experiments, to be compared with other natural objects. One sees that VKS operates in rotation number equivalent to that of sun-like-stars, but that another operating set up should be used to reach the parameters regime of the earth. This is interesting because most cosmic dynamos are turbulent, resulting in serious difficulties of modelling as we now discuss.

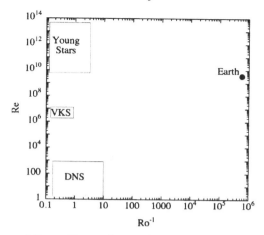

Figure 1. Typical range of Reynolds number and Rossby number for young stars, the earth, VKS2 experiment and DNS simulations.

2. Laminar vs turbulent dynamo

We first focus on the "linear" (also called "kinematic") case, where $N \ll 1$, so that v can be considered as independent of B. We decompose the velocity field into a mean (time averaged) $\overline{U}$ and a fluctuating part u', with $\delta =< u'^2 > / < \overline{U}^2 >$, the $<>$ denoting spatial average. Since the velocity field is independent of B, we can study the evolution of B from the induction equations, that reads:

$$\partial_t \mathbf{B} = \nabla \times (\overline{\mathbf{U}} \times \mathbf{B}) + \nabla \times (\mathbf{u}' \times \mathbf{B}) + \eta \Delta \mathbf{B}. \tag{2.1}$$

This is a linear equation. Since $\overline{U}$ is by construction time-independent, it admits exponentially growing or decaying solutions in the absence of the second term of the r.h.s., like in any classical instability problem. The natural non-dimensional parameter to quantify the importance of the fluctuating term is $\epsilon = \delta - 1$. Therefore, when $\epsilon \ll 1$, we have a "laminar" instability, with exponential growth or decay. The frontier in between the two behavior is the dynamo threshold, that will be close to the instability threshold computed only the mean flow.

For ϵ of order unity, the fluctuating term becomes important, and the equation now includes a time-dependent, stochastic like behavior. The instability is now akin to an instability in presence of a multiplicative noise, and requires special tools to be detailed later.

From the behavior of the parameter $\delta - 1$ detailed in previous Section, we see that the dynamo is probably laminar for Taylor-Green flow at $Re < 20$, or for VKS with an annulus and rotation, while it is probably turbulent for a TG flow with $Re > 50$ and for VKS without an annulus.

2.1. *Laminar dynamo*

Laminar dynamo are countless. Some, like Ponomarenko or Robert's dynamo, can even be studied analytically. Here we focus on the TG and VKS laminar dynamos. In the case of the TG flow, the laminar dynamo is characterized by two windows of instability (Ponty *et al.* (2007); Laval *et al.* (2006)) (Fig. 2): the dynamo takes place for $Rm_{c1} < Rm < Rm_{c2}$ and for $Rm > Rm_{c3}$. The three critical magnetic Reynolds number have been computed for mean flows measured at different Reynolds number $6 < Re < 100$ and were found to be roughly independent of Re. With the forcing adopted in Laval *et al.* (2006), one finds: $Rm_{c1} \sim 6$, $Rm_{c2} \sim 13$ and $Rm_{c3} \sim 25$.

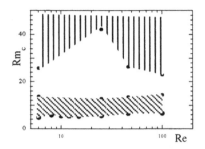

Figure 2. Windows of kinematic dynamo action with a time-averaged TG flow, as a function of the Reynolds number Re. The dashed area corresponds to region of the parameter space where kinematic dynamo is observed, corresponding to positive values of the Lyapunov exponent.

Because it was at the heart of VKS optimization, the laminar dynamo has been studied with different codes, and different boundary conditions or propeller shape and size. The lower threshold were obtained for TM73 propellers. In addition, it was found that the addition of a layer of sodium at rest produces a significant reduction of the dynamo threshold from $Rm_c \sim 180$ to $Rm_c \sim 40$ (Avalos-Zuniga *et al.* (2003); Ravelet *et al.* (2005)), and that the moving sodium behinds the propeller had a tendency to increase the dynamo threshold (Stefani *et al.* (2006); Laguerre *et al.* (20006)). This is summarized in Fig. 3.

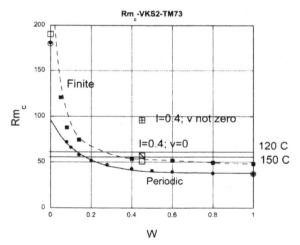

Figure 3. Critical value of the magnetic Reynolds number as a function of the percentage of sodium at rest W from kinematic simulation with time-averaged von Karman flow with inox TM73 propellers rotating in the contra direction. Note that the kinematic simulation with iron propellers have not yet been done. Filled circle (Ravelet *et al.* (2005)) and open circle (courtesy C. Nore) : with periodic axial boundary conditions; Filled square : with finite axial boundary conditions (Laguerre *et al.* (20006)); open square (resp. square with cross): when taking into account the thin layer of fluid at rest (resp. stirred) behind the impellers (Stefani *et al.* (2006)). The two solid line delimit the largest Rm that can be reached in the VKS2 experiment at the lowest (resp. largest) operating temperature 120 C (resp. 150 C).

Specifically, the various threshold found with the kinetic simulation based on the time-averaged velocity field with a layer of resting sodium of size $w = 0.4$ are:

- $Rm_c = 43 \pm 1$ for periodic axial boundary conditions in a homogeneous conducting domain (Ravelet *et al.* (2005));
- $Rm_c = 49 \pm 2$ for finite axial boundary conditions;
- $Rm_c = 57$ (resp. 95) when taking into account the thin layer of fluid at rest (resp. stirred) behind the impellers (Stefani *et al.* (2006));
- $Rm_c = 46$ without the fluid behind the impellers for more realistic conditions: finite axial boundary condition, 5 mm copper shell separating the flow and the static conducting layer, copper container.
- $Rm_c = 55$ (resp. $Rm_c = 150$) for these conditions with the fluid behind the impellers at rest (resp. stirred). These results are given by Laguerre *et al.* (20006).

In the experiments, dynamo has been observed with iron propellers, with a threshold $Rm_c \sim 32$ in contra rotation. With inox propellers, no dynamo has been observed in contrarotation. However, induction measurements with an external applied field B_a can be used to estimate a dynamo threshold via the response B_i as $B_a/B_i \sim \Lambda \sim Rm - Rm_c$. Linear fit to the induction curve B_a/B_i then gives (see Fig. 4):

- $Rm_c = 127$ for TM73 inox propellers with no resting sodium and no annulus (VKS2b campaign);
- $Rm_c = 67$ for TM73 inox propellers with $w = 0.4$ of resting sodium and no annulus (VKS2a campaign);
- $Rm_c = 53$) for TM73 inox propellers with $w = 0.4$ of resting sodium and an annulus (VKS2f campaign);
- $Rm_c = 32$ for TM73 iron propellers with $w = 0.4$ of resting sodium and an annulus (VKS2g campaign).

The decrease of Rm_c seen between 2b and 2a suggests that indeed the resting sodium is favorable to dynamo action. The difference between 2a and 2f threshold suggests that the turbulence (described by the parameter δ) has an impact on the dynamo threshold. This is the subject of the next section.

2.2. *Turbulent dynamo*

We consider now a situation where fluctuation are non-negligible. A first natural approach is to identify a small parameter ϵ in the problem, and try and solve the full problem by perturbation theory. Specifically, one consider first the time-averaged of Eq. (2.1):

$$\partial_t \overline{\mathbf{B}} = \nabla \times (\overline{\mathbf{U}} \times \overline{\mathbf{B}}) + \nabla \times (\overline{\mathbf{u}' \times \mathbf{b}'}) + \eta \Delta \overline{\mathbf{B}}. \tag{2.2}$$

The main idea is to find the shape of $\overline{\mathbf{u}' \times \mathbf{b}'}$ as a function of $\overline{B}$ through the perturbation expansion.

An historically successful approach is to consider an ideal case where there is a scale separation between the typical scale l of (u', b') and the typical scale L of $\overline{B}$. The natural expansion parameter is therefore $\epsilon = l/L \ll 1$, or equivalently $\nabla \overline{B}$. Without any computations, one can then infer that

$$\epsilon_{ijk}\overline{u'_j \times b'_i} = \alpha_{ij}\overline{B_j} + \beta_{ijk}\nabla_j\overline{B_k} + O(\epsilon^2), \tag{2.3}$$

where α_{ij} and β_{ijk} are two tensors that depend on the velocity field and that can be computed through classical perturbation procedure applied to Eq. (2.1) (see e.g. Dubrulle & Frisch (1991)). When plugged back into (2.2), this expansion gives:

$$\partial_t \overline{\mathbf{B}} = \nabla \times (\overline{\mathbf{U}} \times \overline{\mathbf{B}}) + \nabla_j \alpha_{ijk}\overline{\mathbf{B_k}} + \nabla_j \nabla_k (\beta_{ijkl} + \eta\delta_{jk}\delta_{il})\overline{\mathbf{B_l}}, \tag{2.4}$$

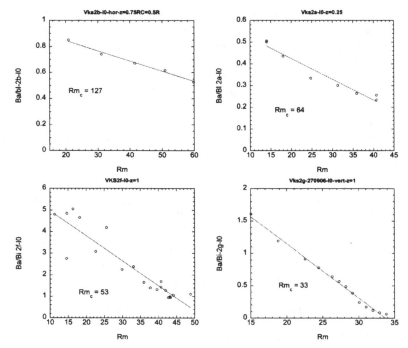

Figure 4. Estimate of dynamo threshold from induction measurements in VKS2 through the quantity B_a/B_i in four different configuration : Upper left : 2b : no resting sodium, TM73 inox propellers, no annulus; Upper right : 2a : 40% of resting sodium, TM73 inox propellers, no annulus; Lower left : 2f : 40 % of resting sodium, TM73 inox propellers, annulus; Lower right : 2g : 40 % of resting sodium, TM73 iron propellers, annulus. The lines are linear fit, providing the value of Rm_c indicated on each plot.

where $\alpha_{ijk} = \epsilon_{ijm}\alpha_{mk}$ and $\beta_{ijkl} = \epsilon_{ikm}\beta_{mjl}$. α is the famous alpha coefficient, while β is a turbulent diffusivity tensor, that need not be definite positive. In the absence of mean flow, this equation usually leads to a large scale instability via the alpha effect.

3. Stochastic Theory

Non-perturbation computations can be performed both analytically and numerically by replacing the true velocity fluctuations by some well chosen noise. Of course, real turbulence is characterized by temporal and spatial correlation that cannot be captured by such a simple noise. One can however hope that first order effects can be captured by our simple model. The reader can judge by himself from the final comparison. In any case, the advantage of these stochastic computations is twofold: first, they allow for non-perturbation analytical and numerical computations; second, their numerical cost is equivalent to the cost of a kinematic simulation. Simulation of 64^3 can then prove sufficient to explore a range of fluctuations equivalent to $Re = 10^7$, (i.e. that would require 10^{15} grid points).

We therefore now consider that u' is a white noise, characterized by:

$$\langle u'_i(x,t)u'_j(x',t')\rangle = 2G_{ij}(x,x')\delta(t-t'), \tag{3.1}$$

where the brackets denote ensemble average, over the realizations of the noise. Equation (2.1) then takes the shape of a stochastic partial differential equation for B, with multiplicative noise. The problems associated with this type of noise can be understood by looking at a simple unidimensional model:

$$\dot{x} = \mu x + \xi x, \tag{3.2}$$

where $\xi(t)\xi(t') = 2D\delta(t - t')$. In the absence of noise, x is exponentially increasing (unstable) as soon as $\mu > 0$. In the presence of noise, we can take different moments of the equation and get the following hierarchy:

$$\begin{aligned} <\dot{x}> &= (D + \mu) \\ <\dot{x}^2> &= 2(2D + \mu) \end{aligned} \tag{3.3}$$

so that the $< x >$ (resp. $< x^2 >$) is unstable for $\mu > -D$ (resp. $\mu > -2D$). Therefore, its seems that the instability threshold depends on the moment we consider! One can in fact prove that this pathology is due to the absence of non-linear terms, and that in fact the correct threshold that would prevail with non-linear term is captured by considering the Lyapunov:

$$\Lambda = \partial_t < \ln x > . \tag{3.4}$$

Due to the convexity of the log, $\Lambda \leqslant \partial_t \ln < x >$, so that Λ is always smaller than the growth rate. The system is unstable as soon as $\Lambda > 0$.

Analytical computation of the stochastic model have been done by Leprovost & Dubrulle (2005). In order to make the computations tractable, two approximation were made: i) a saturating term was added to the induction equation as $-cB^2 B_i$ because of symmetry consideration. In some sense, this modification is akin to an amplitude equation, and the cubic shape for the non-linear term could be viewed as the only one allowed by the symmetries. Such a procedure is motivated by the observation that the precise form of the nonlinear term does not affect the threshold value. ii) The diffusivity was ignored.

Using standard techniques (Boldyrev (2001)), one can then derive the evolution equation for $P(\mathbf{B}, x, t)$, the probability of having the field $\mathbf{B}$ at point x and time t:

$$\begin{aligned} \partial_t P &= -\bar{U}_k \partial_k P - (\partial_k \bar{U}_i)\partial_{B_i}[B_k P] + \partial_k[\beta_{kl}\partial_l P] \\ &\quad + c\partial_{B_i}[B^2 B_i P] + 2\partial_{B_i}[B_k\alpha_{lik}\partial_l P] \\ &\quad + \mu_{ijkl}\partial_{B_i}[B_j\partial_{B_k}(B_l P)] , \end{aligned} \tag{3.5}$$

with the following turbulent tensors:

$$\beta_{kl} = \langle u'_k u'_l \rangle, \quad \alpha_{ijk} = \langle u'_i \partial_k u'_j \rangle \quad \text{and} \quad \mu_{ijkl} = \langle \partial_j u'_i \partial_l u'_k \rangle . \tag{3.6}$$

Due to incompressibility, the following relations hold: $\alpha^{kii} = \mu^{iikl} = \mu^{ijkk} = 0$.

The physical meaning of these tensors can be found by analogy with the "Mean-Field Dynamo theory" (Krause& Rädler (1980); Moffatt (1978)). Indeed, consider the equation for the evolution of the mean field, obtained by multiplication of equation (3.5) by B_i and integration:

$$\begin{aligned} \partial_t \langle B_i \rangle &= -\bar{U}_k \partial_k \langle B^i \rangle + (\partial_k \bar{U}_i)\langle B_k \rangle - 2\alpha_{kil}\partial_k \langle B_l \rangle \\ &\quad + \beta_{kl}\partial_k \partial_l \langle B_i \rangle - c\langle B^2 B_i \rangle. \end{aligned} \tag{3.7}$$

This equation resembled the classical Mean Field Equation of dynamo theory, with generalized (anisotropic) "α" and "β". The first effect leads to a large scale instability for the mean-field, while the second one is akin to a turbulent diffusivity. Note that the tensor μ does not appear at this level.

The Lyapunov exponent can be computed in a similar way from (3.5) by changing variable $B_i = Be_i$, then multiplying the resulting equation by $B^{d-1} \ln B$ and integrating with respect to B. This yields:

$$\Lambda \equiv \partial_t \langle \ln B \rangle = \langle \partial_k \bar{U}_i e_i e_k \rangle_\phi + \langle \mu_{ijkl} (\Delta_{ik} e_j e_l + \Delta_{kj} e_i e_l) \rangle_\phi, \tag{3.8}$$

where we used $\Delta_{ij} = \partial_{n_i}(n_j) = \delta_{ij} - e_i e_j$ an "angular Dirac tensor", and the symbol $\langle \bullet \rangle_\phi$ denotes verages over the angular variables.

The condition for dynamo action in this model is $\Lambda > 0$. In the limit of zero noise, the term proportional to μ is negligible and one recovers the classical criterion for instability in a laminar dynamo in the infinite Prandtl number limit. Indeed, in such a case, the magnetic field will mainly grow in the direction given by the largest eigenvalue λ_{max} of $S_{ij} = \partial_j \bar{U}_i$, so that

$$\Lambda \approx \langle \partial_k \bar{U}_i e_i e_k \rangle_\phi = \lambda_{max}. \tag{3.9}$$

There will be dynamo only if $\lambda_{max} > 0$.

Consider now a situation where you increase the noise level. Two different influences on the sign of Λ then result: one through the factor proportional by μ. According to the sign of this factor, it can therefore favor or hinder the dynamo. In isotropic homogeneous turbulence, μ is positive, so that it is in general favorable to dynamo action. Moreover, being proportional to derivatives of the noise, this term gets larger as the typical scale of the noise is small.

However, there exists another less obvious -and adverse- influence of the noise: the disorientation effect. Indeed, noise changes the distribution of magnetic field orientation. In the absence of noise, the latter tends to be oriented towards the most unstable direction. However, noise constantly drives the magnetic field away from this favorable direction, sometimes even driving it towards a stable direction, where the magnetic field exponentially decreases. The net result is a decrease of the effective growth rate of the magnetic field. A phenomenological way to quantify this effect is through the parameters $\delta - 1$ and δ_2. Indeed, the largest these coefficient are, the further away the instantaneous velocity field is from the averaged field, and the largest the disorientation effect can be. This effect is more important when the noise it at largest scale, since in that case the disorientation effect is more pronounced-one can get farther from the mean flow.

From this discussion, one expects large scale noise to be adverse to dynamo action-through the disorientation mechanism, while small scale noise should be favorable to dynamo action-through the μ effect.

The previous analytical computation were tractable only in the limit $\eta \to 0$ ($Rm \to \infty$. To investigate the more realistic case of finite diffusivity, one may resort to numerical simulations. This has been done by Laval et al. (2006) and Dubrulle et al. (2007) for the case of the Taylor-Green flow, without inclusion of the non-linear term in the induction equation. Two kinds of noise were tested: shortly correlated noise, like in the analytical case, and Markovian noise, with finite correlation time τ_c that can be varied from 0 to several eddy-turnover time. Two typical noise scale were also tested, one at the largest available scale of the system $k = 1$, and one of the order of the magnetic diffusive scale $k = 16$.

The results are summarized in Fig. 5 for time-correlated noises at large and small scale. In the case of the large scale noise, one sees that the two dynamo windows are lifted up by the noise, resulting in an increase of the dynamo threshold. It can also be shown that this effect becomes more pronounced as the correlation time of the noise increase until the mean eddy turn over time is reached. Above this, the effect does not change anymore. From the previous discussion, we can attribute this increase of the threshold

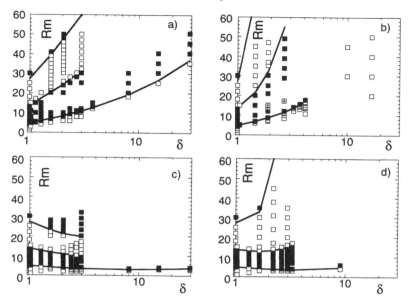

Figure 5. Parameter space for noise at $Re = 6$ for different noise parameters : a) $\tau_c = 0$, $k_I = 1$; b) $\tau_c = 0.03$, $k_I = 1$; c) $\tau_c = 0$, $k_I = 16$; d) $\tau_c = 0.03$, $k_I = 16$. Open square : no-dynamo case; Square with cross : undecided state; Filled square : dynamo case. The full lines are zero-Lyapunov lines.

to the disorientation effect. In contrast, when the noise it at small scale, the dynamo threshold is -slightly- decreased with respect to the laminar case. This is probably a benefit of the μ effect.

The influence of the noise onto the first dynamo threshold can be summarized by plotting the critical magnetic Reynolds numbers as a function of the noise intensity (Fig. 6-a). Large scale (resp. small-scale) noise tends to increase (resp. decrease) the dynamo threshold. For small noise intensities, the correction $Rm_c^{turb} - Rm_c^{MF}$ is linear in $\delta - 1$, in agreement with the perturbation theory (Pétrélis (2002)). Furthermore, one sees that for small scale noise, the decrease in the dynamo threshold is almost independent of the noise correlation time τ_c, while for the large scale noise, the increase is proportional to τ_c at small τ_c. At $\tau_c > 1$, all curves $Rm_c(\delta)$ collapse onto the same curve. We have further investigated this behavior to understand its origin. Increasing δ first increases of the flow "turbulent viscosity" $\overline{v_{rms}l_{int}}$ with respect to its mean flow value $V_{rms}L_{int}$. This effect can be corrected by considering $Rm_c^* = Rm_c V_{rms} L_{int}/\overline{v_{rms}l_{int}}$. Second, an increase of δ produces an increase of the fluctuations of kinetic energy, quantified by δ_2. This last effect is more pronounced at $k_I = 1$ than at $k_I = 16$. It is amplified through increasing noise correlation time. We thus re-analyzed our data by plotting Rm_c^* as a function of δ_2 (Fig. 6-b). All results tend to collapse onto a single curve, independently of the noise injection scale and correlation time. This curve tends to a constant equal to Rm_c^{MF} at low δ_2. This means that the magnetic diffusivity needed to achieved dynamo action in the mean flow is not affected by spatial velocity fluctuations. This is achieved for small scale noise, or large scale noise with small correlation time scale. In contrast, the curve diverges for δ_2 of the order of 0.2, meaning that time-fluctuations of the kinetic energy superseding 20 percent of the total energy annihilate the dynamo.

An obvious stationary solution of (3.5) is a Dirac function, representing a solution with vanishing magnetic field. Another stationary solution can be found for B such that

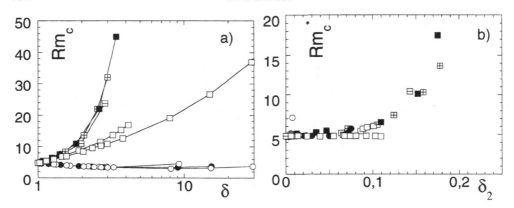

Figure 6. Evolution of the dynamo threshold for KS simulations with $\bar{u}(Re = 6)$. a) Rm_c as a function of δ and b) Rm_c^* as a function of δ_2 for different noise parameters : $k = 1$: square $\tau_c = 0$; boxdot : $\tau_c = 0.1$ sec; boxminus : $\tau_c = 1$ sec; boxplus : $\tau_c = 8$ sec; black square $\tau_c = 50$ sec; $k = 16$: circle : $\tau_c = 0$; odot : $\tau_c = 0.1$ sec; bullet : $\tau_c = 50$ sec.

$B_i = Be_i$ by setting $\partial_t P = 0$ in (3.5), with solution:

$$P(B) = \frac{1}{Z}B^{\Lambda/a-1}\exp\left[-\frac{c}{2a}B^2\right],\qquad(3.10)$$

where Z is a normalization constant and $a = \langle\mu_{ijkl}e_ie_je_ke_l\rangle_\phi$. This solution can represent a meaningful probability density function-and therefore a dynamo case- only if it can be normalized. Condition of integrability at infinity of (3.10) requires a be positive. This illustrates the importance of the non-linear term which is essential to ensure vanishing of the probability density at infinity. Condition of integrability near zero requires $\Lambda > 0$ be positive.

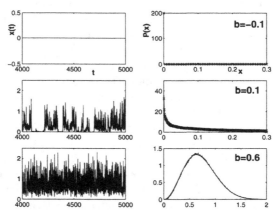

Figure 7. Result of the surrogate 1D model $\partial_t x = [b + \xi(t)]x - \gamma x^3$: On the left side we show time series for $a = 0.2$, $\gamma = 1$ and 3 different values of the parameter b. On the right side, the corresponding PDF and the theoretical curve corresponding to equation (3.10), with $\Lambda = b$.

This is the dynamo condition identified before, that is obtained using the mean field as control parameter. However, the shape of the PDF traces an interesting new paradigm for the turbulent dynamo (Fig. 7). Indeed, in the range $0 < \Lambda < a$, the PDF is maximum at zero, meaning that the most probable value for the magnetic field is zero. This is the signature of an "intermittent" dynamo, with periods of large magnetic field followed by quiescent periods, in a way reminiscent to "on-off" intermittency. Above this second threshold, $\Lambda > a$, the PDF exhibits a non-zero value for its most probable value, meaning a more classical "turbulent stationary dynamo", with fluctuations of the magnetic field around a finite value. Note that the transition from one regime to another can be mediated by the value of $\delta - 1$: as this parameter is increased, the disorientation effect becomes more and more important, and Λ decreases. This remark is corroborated by recent stochastic computations of Aumaitre *et al.* (2005), who showed that the intermittent behavior could be switched off by changing the value of the noise spectrum at zero frequency, i.e. by removing large scale noise. Note also that this new paradigm cannot be tested in the previous TG computations, since they did not include any non-linearities. A more serious question is also: Is this new paradigm an artifact of our synthetic turbulence, or is it something that one can actually see? To check this, one needs to resort to numerical simulations or experiments.

4. Taylor-Green Numerical simulations

Direct numerical simulation of Eq. (1.1) for Taylor-Green forcing have been made in Ponty *et al.* (2005), Laval *et al.* (2006), Ponty *et al.* (2007) and Dubrulle *et al.* (2007). The dynamo threshold Rm_c has been computed for different values of Re. The result is shown in Fig. 8. One sees that the dynamo threshold increases with Re until a value of the order $Re \sim 100$ where it seems to saturate. Dubrulle *et al.* (2007) also performed computations at low Re but larger and larger Rm to try and detect a possible signature of the second laminar window of instability. At $Re = 6$, they detected a transition from an intermittent dynamo at $Rm = 25$, to a dynamo with a mean field at $Rm = 100$ (see Fig. 9). Moreover, one can see a remarkable correlation between the dynamo windows predicted by the stochastic numerical simulation and the direct numerical simulation. This is an indication that maybe the stochastic model does capture the main features of the turbulent dynamo transition.

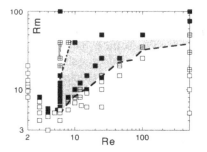

Figure 8. Comparison between DNS and KS simulations with $Re = 6$ with $k_I = 1$, $\tau_c \geqslant 0.3$. Squares refer to DNS-MHD and LES-MHD simulations, and shaded areas to windows of dynamo action for kinematic-stochastic simulations at $Re = 6$ with $k_I = 1$, $\tau_c \geqslant 0.3$. Note the tiny dynamo window near $Re = 6$, $Rm = 40$. Open square : no-dynamo case; Square with cross : intermittent dynamo; Filled square : dynamo case; square with line : undecided state; $- Rm_c^{turb}$; $-- Rm_c^{MF}$; $-\cdot-\cdot$ end of the first dynamo window; $\cdots$ beginning of the second dynamo window.

5. Experiments

Various configurations have been tested in VKS with the TM73 propellers and the layer of sodium at rest: with and without annulus, and with inox or iron propellers. In the inox

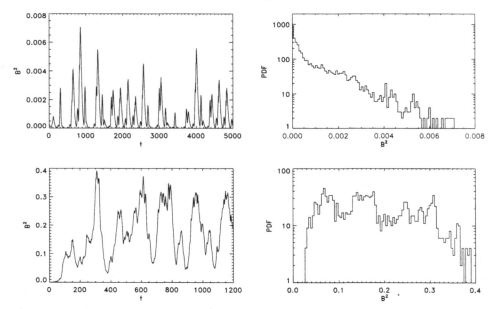

Figure 9. Example of dynamos in TG flow. Left : Magnetic energy as a function of time. Right : PDF of the magnetic energy. Upper panel : Intermittent dynamo $Re = 6$, $Rm = 40$. Lower panel : Turbulent dynamo $Re = 25$, $Rm = 50$.

case, no dynamo has been observed (Ravelet *et al.* (2008)). However, a critical magnetic Reynolds number could be estimated from induction measurements for configurations with positive and negative rotation. In the induction regime, the disorientation effect could be directly measured by following strong local magnetic field perturbations (Volk *et al.* (2006)).

In the iron case, with an annulus in the midplane, different types of dynamo have been identified, in the rotating and non-rotating case (Berhanu *et al.* 2007; Monchaux *et al.* (2007); Monchaux *et al.* (2008)). Among them, intermittent dynamo have been observed around $\theta = 0.2$ (see Fig. 10).

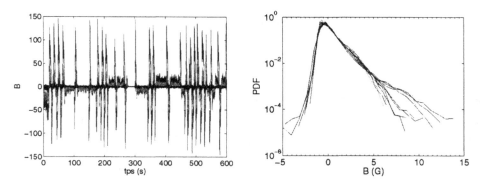

Figure 10. Example of intermittent dynamo observed in the VKS2 with TM73 iron propellers at $\theta = 0.17$, $Rm = 32$. Left : Components of the magnetic field as a function of time (red : B_z, green B_θ; blue : B_r. Right : Corresponding PDF of the magnetic field components.

Note that the intermittent dynamo observed near $\theta = 0.2$ is characterized by the largest value of δ. In the previous Section we argued that it was probably a good condition to observe, if any, the intermittent dynamo.

Regarding threshold for dynamo instability (transition towards stationary dynamo), it has been accurately measured so far in 3 cases: at $\theta = 0$, with impellers rotating in the $(+)$ or $(-)$ direction with respect to the pales curvature; at $\theta = -1$ with impellers rotating in the $(+)$ direction. Using the values of δ measured in water, we can check whether the trend observed in TG numerical simulation (higher threshold for larger values of δ or δ_2, see Fig. 6) is also valid here. With the presently available data, the trend is indeed respected (Fig. 11, but more data is needed to confirm this point.

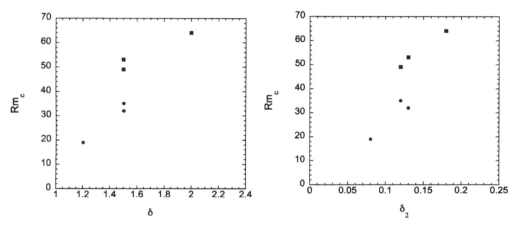

Figure 11. Critical magnetic Reynolds number Rm_c as a function of δ (left) and δ_2 (right) in the VKS2 experiment for inox (square) and iron propellers (circle). The δ_2 have been estimated from the water model experiment. The Rm_c are measured in the case of iron propellers, and estimated using induction measurements for the inox propellers.

Acknowledgements

The numerical data in the paper are obtained through a collaboration with J-Ph. Laval, Y. Ponty, F. Daviaud, P. Blaineau and J-F. Pinton. The data from the von Karman water experiments have been obtained in Saclay in the group of F. Daviaud by A. Chiffaudel, F. Daviaud, L. Marié, F. Ravelet, R. Monchaux, P. Diribarne. The data from von Karman sodium experiment have been obtained in Cadarache by the VKS collaboration (CEA Saclay : S. Aumaitre, A. Chiffaudel, F. Daviaud, B. Dubrulle, C. Gasquet, L. Marie, R. Monchaux, V. Padilla and F. Ravelet; ENS Paris : M. Berhanu, S. Fauve, N. Mordant and F. Pétrélis; ENS Lyon : M. Bourgoin, Ph. Odier, J-F. Pinton, N. Plihon, R. Volk). The data interpretation is personal and does not engage the responsibility of any of these people.

References

Aumaitre, S., Pétrélis, F. & Mallick, K., 2005, *Phys. Rev. Lett.*, 95,064101
Avalos-Zuniga, R., Plunian, F. & Gailitis, A., 2003, *PRE*, 68, 066307
Berhanu, M., Monchaux, R., Fauve, S., Mordant, N., Pétrélis, F., Chiffaudel, A., Daviaud, F., Dubrulle, B., Marié, F. Ravelet, L., Bourgoin, M., Odier, Ph., Pinton, J.-F., & Volk, R. 2007, *Eur. Phys. Rev. Lett.*, 77, 59001
Boldyrev, S. 2001, *Astrophys. J.*, 562, 1081
Dubrulle, B. & Frisch, U. 1991, *Phys. Rev. A*, 43, 5355

Dubrulle, B., Blaineau, P., Mafra Lopez, O., Daviaud, F., Laval, J-P., & Dolganov, R., 2007, *New J. Phys.*, 9, 308

Krause, F. & Rädler, K.-H., *Mean field MHD and dynamo theory*, 1980, Pergamon press.

Laguerre, R., Nore, C., Leorat, J., & Guermond, J-L., 2006, *C. R. Mécanique*, 334, 593

Laval, J.-P., Blaineau, P., Leprovost, N., Dubrulle, B., Daviaud, F., 2006, *Phys. Rev. Let.*, 96, 204503

Leprovost, N. & Dubrulle, B., 2005, *Eur. Phys. J. B*, 44, 395

Moffatt, H. K., 1978 *Magnetic field generation in electrically conducting fluids* (Cambridge University Press).

Monchaux, R., Berhanu, M., Bourgoin, M., Odier, Ph., Moulin, M., Pinton, J.-F., Volk, R., Fauve, S., Mordant, N., Pétrélis, F., Chiffaudel, A., Daviaud, F., Dubrulle, B., Gasquet, C., Marié, L. & Ravelet, F. 2007, *Phys. Rev. Lett.*, 98, 044502

Monchaux, Romain, Berhanu, Michael, Aumaitre, Sebastien, Chiffaudel, Arnaud, Daviaud, Francois, Dubrulle, Berengere, Florent, Stephan Ravelet, Nicolas, Fauve, Francois Mordant, Mickael Petrelis, Philippe, Bourgoin, Jean-Francois Pinton, Odier, Plihon,Nicolas, & Volk, Romain, 2009, *Phys. Fluids*, 21, 035108

Pétrélis, F., 2002, *PhD Thesis Paris VI*

Ponty, Y., Mininni, P. D., Pinton, J.-F., Politano, H., & Pouquet, A. 2005, *Phys. Rev. Lett.*, 94, 164512

Ponty, Y., Mininni, P. D., Pinton, J.-F., Politano, H. & Pouquet, A. 2007, *New J. Phys.*, 9, 296

Ravelet, F., Chiffaudel, A., Daviaud, F. & Leorat, J. 2005, *Phys. Fluids*, 17, 17104

Ravelet, F., Berhanu, M., Monchaux, R., Aumaitre, S., Chiffaudel, A., Daviaud, F., Dubrulle, B., Bourgoin, M., Odier, Ph. , Plihon, N., Pinton, J.-F., Volk, R., Fauve, S., Mordant, N., & Petrelis, F., 2008,*Phys. Rev. Lett.*, 101, 074502

Stefani, F., Xu, M., Gerbeth, G., Ravelet, F., Chiffaudel, A., Daviaud, F., & Leorat, J., 2006, *Eur. J. Mech. B*, 25, 894

Volk, R., Ravelet, F., Monchaux, R., Berhanu, M., Chiffaudel, A., Daviaud, F., Fauve, S., Mordant, N., Odier, Ph., Pétrélis, F. & Pinton, J.-F., 2006, *Phys. Rev. Lett.*, 97, 074501

Discussion

E. NTORMOUSI: How far are your experiments from similarity to solar dynamo?

B. DUBRULLE: Our device is cylindrical and not spherical (Smile). The energy source is through impellers rather than convection. But apart from that, we have regimes with periodical or aperiodical dynamos, exactly like in stars. So we think we may gain a lot of insight on the solar dynamo from studying our device.

C. FOREST: What is the status of the soft iron propellors in getting or not dynamo action in VKS2?

B. DUBRULLE: For the present time, we only get dynamo when at least one iron propellor is rotating. However, we have observed situations where dynamo depends on the sense of rotation of the iron impellers. This change mainly influences the sodium fluid properties. So this demonstrates that VKS is more than a pure "disk-dynamo" and that the fluid effect is very important.

Astrophysical Dynamics: From Stars to Galaxies
Proceedings IAU Symposium No. 271, 2010
N. H. Brummell, A. S. Brun, M. S. Miesch & Y. Ponty, eds.

© International Astronomical Union 2011
doi:10.1017/S1743921311017765

Juri Toomre
and the art of modeling convection zones

Jean-Paul Zahn

LUTH, Observatoire de Paris, CNRS UMR 8102, Université Paris Diderot
5 place Jules Janssen, 92195 Meudon, France
email: Jean-Paul.Zahn@obspm.fr

Abstract. Thermal convection plays a very important role in the structure and evolution of stars, as it is one of the main physical processes that transport heat from their interior where it is released, to the surface where it is radiated into space. Much progress has been achieved in modeling that process during the past 60 years, and I shall recall here how Juri Toomre has greatly contributed to it.

Keywords. convection; turbulence; stars: interiors.

1. Early work on stellar convection

It all began with the modeling of fully convective stars which, to a good approximation, are adiabatically stratified and are therefore polytropes of degree 3/2. If all stars were such polytropes, stellar astrophysics would be a very simple matter, captured by a differential equation of second order that we owe to Lane and Emden. But thanks to Eddington (1926), the pendulum swung to the other side, although his opinion was a bit extreme. I cannot resist quoting him: "We shall not enter further into the historic problem of convective equilibrium since modern researches show that the hypothesis is untenable. In stellar conditions the main process of transport of heat is by radiation and other modes of transfer may be neglected."

As often, the truth was lying in between. Unsöld (1930) pointed out that, due to the ionization of hydrogen, the solar atmosphere would be unstable just below the surface, according to the Schwarzschild criterion; and indeed, solar granulation was clearly the signature of convective motions, as it had been already suggested by Emden. Soon solar models were built with a convective envelope and a stable radiative interior (cf. Cowling 1936). To treat that convection, and to describe the transition from radiative to convective transport of heat, Ludwig Biermann (1938) borrowed from Prandtl the concept of mixing-length - the mean free path of convective eddies, much as that of molecules in the kinetic theory of gas. This concept was later applied by Erika Böhm-Vitense to produce the first realistic model of the Sun (Vitense 1953); she chose to scale the mixing-length ℓ with the pressure scale-height H_P, and she showed that the depth of the solar convection zone depends sensitively on the value of the mixing-length parameter $\alpha = \ell/H_P$.

Since that parameter cannot be derived from first principles, it must be calibrated by adjusting evolutionary models to the Sun, after they have been run to solar age. But it is still a matter of debate how this calibration should be extended to stars of different mass and other evolutionary states.

The shortcomings of the mixing-length treatment are well known; beyond the uncertainties afflicting the mixing-length parameter, there is the fact that the convective flux is a local function of the superadiabatic gradient, which does to not allow for penetration or overshoot. In spite of its deficiencies, MLT is still widely used, because it is so simple to

Figure 1. Woods Hole Summer School 1969. Standing (left to right): Buschi (posted), Stern, Prendergast, Gough, Keller, Kraichnan, Spiegel (being supported by) Veronis, Toomre (pillared), Harrison, Backus, Malkus. Seated: Zahn, Defouw, Gans, McKee, Trasco, Perdang, Barker, Auré, Thayer, Mészáros.

implement in a stellar evolution code. Valuable attempts have been made to improve the prescription by introducing higher order correlations, notably by Da-Run Xiong (1985) and Vittorio Canuto (1992), but these involve several additional parameters which - like α - cannot be derived from first principles. For most purposes that go beyond standard stellar evolution, a more realistic description of convection is required, and Juri Toomre has dedicated most of his scientific carreer to accomplish this task.

2. A step forward : the modal treatment

I met Juri in July 1969, on Cape Cod, where we were participating in a summer school with Douglas Gough, at the Woods Hole Oceanographic Institution. This school was founded in 1959 by Willem Malkus, George Veronis, Ed Spiegel and a few others, "with the aim of introducing a then relatively new topic in mathematical physics, geophysical fluid dynamics, to graduate students in physical sciences", as one can read on its website. The school is attended by a dozen or so graduate students, and as many senior researchers. It lasts ten weeks: two weeks of lectures, and thereafter the students work on a project under the guidance of a senior participant. Once in a while, the subject of the school is on astrophysics, as it was the case in 1969. Figure 1 shows the participants of that session; Ed and Juri were absent when the picture was taken - admire how they have been included nevertheless!

It was there in Woods Hole that I learned from the work undertaken by Ed with Douglas and Juri, to give a more physical representation of thermal convection, rooted

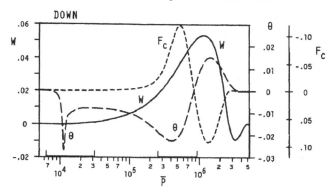

Figure 2. The modal treatment applied to convection in an A-type star. Variation with depth (or mean pressure $\overline{P}$) of the amplitude functions for the vertical momentum density W and the relative variation Θ of temperature; also displayed is the convective heat flux F_C. The horizontal plane is tiled with hexagonal cells in which the flow is predominantly downwards ($W > 0$) at cell center. Such convection of large horizontal scale is buoyancy driven in the second unstable zone and penetrates with vigor into the stable region below, resulting in countercells at greater depth. The motions also extend upward through the interzone and establish a link between the two unstable zones: the entire domain is thoroughly mixed. (From Latour *et al.* 1981, courtesy *ApJ*.)

in the basic Navier Stokes equations. At that time, it was out of question to treat the problem in full three dimensions; to capture nevertheless some of the inherent three-dimensionality, their idea was to project the temperature field on a set of horizontal functions $f_k(x, y)$, such that

$$T(x, y, z, t) = \overline{T}(z, t) + \sum_k f_k(x, y)\, T_k(z, t), \qquad (2.1)$$

and likewise for the vertical velocity. The horizontal functions are chosen to be orthogonal and periodic in space, thus obeying the harmonic equation

$$\nabla^2 f_k(x, y) = -a_k^2\, f_k(x, y), \qquad (2.2)$$

with a_k being the modulus of the wavenumber. Their products, which intervene in the non-linear terms of the governing equations, are characterized by the coupling constants $C^{klm} = \frac{1}{2} < f_k\, f_l\, f_m >$. Le problem reduces then to a set of coupled differential equations in t and z, whose solutions are the amplitude functions $T_k(z, t)$ and the mean temperature profile $\overline{T}(z, t)$.

Ed, Douglas and Juri had applied this method, the so-called *modal approach*, to laboratory convection, in the Boussinesq approximation which treats the fluid as if it were almost incompressible, and they had successfully resolved the thin boundary layers that control the heat flux (Gough *et al.* 1975; Toomre *et al.* 1977) - a numerical challenge. Now Douglas was leaving for Cambridge, and it was my turn to join the team, which we later nick-named "The Convective Collective". We were hosted by the Goddard Institute for Space Studies, in New York City. Sponsored by NASA, this Institute was equipped with the most powerful computer of that time, an IBM 360/95. Only two of these had been built, both for NASA, and IBM was proud to announce that they were able to perform "over 330 millions of 14 digit multiplications in one minute". This means that these super-computers were 100 times slower than a current laptop, or 10^8 slower than today's petaflop computers. Ever since these New York years, Juri has managed to surf on the wave of the most powerful computers of their time.†

We were granted a generous access to this computer by Robert Jastrow, the director of GISS. Our goal was to apply the modal approach to the convection zone of a star, and for the first attempt we chose an A-type star, which possesses two superposed unstable regions, due respectively to the ionization of hydrogen and helium. This problem looked easier to handle than that of the solar convection zone, because the density contrast is less extreme in such a star, and convection carries there a modest fraction of the heat flux. To filter out the acoustic waves, whose inclusion would have required much higher computer resources, we adopted the anelastic approximation, which had just been discussed by Douglas Gough (Gough 1969). Since we could not afford more than one or two horizontal wavenumbers in our modal expansions, the solutions were developing somewhat artificial internal boundary layers, in an attempt to build up an inertial cascade. We spent many nights watching the solutions on the screen of the IBM 2250 monitor, interacting with the light-pen to adjust the spatial resolution to the thinness of these boundary layers. Quite independently from the initial conditions, our solutions would settle in a stationary flow pattern, a far cry certainly from the turbulent regime that occurs in a real star.

Nevertheless, we had faith in our main result, namely that the convective motions overshoot the unstable regions by a substantial amount, sufficient to link the two unstable zones of an A-type star (Fig. 2) and thus to establish a single mixed subsurface region. This was later confirmed by more realistic simulations; it has observable consequences on the surface composition of such stars, which had not been explained until then. Our first results were published in two papers of the same issue of ApJ (Latour *et al.* 1976; Toomre *et al.* 1976), completed by a third paper in 1981 (Latour *et al.* 1981). In the meantime my student Jean Latour had joined our team, and even after our departure from New York, we would come back regularly during several years to pursue our computations at GISS. (Juri's parents, who were living on Long Island, kindly offered us to stay at their home on Long Island.)

In the summer of 1976, I organized with Ed Spiegel an IAU colloquium on "Problems of Stellar Convection", at the Nice observatory. It was attended by most people working in the field, among them the great pioneers Ludwig Biermann, Erika Böhm-Vitense and Paul Ledoux. Some promising young scientists were present too, such as Åke Nordlund who presented us a clever "thumbnail movie" of two-dimensional convection. There we saw also the first 3-dimensional simulation of compressible convection over multiple scale-heights, which had just been obtained by Eric Graham on the GISS computer (Graham 1977). And Fritz Busse showed us how rotation may interact with convection.

Clearly the time had come for more serious simulations than our crude modal solutions, and the computers had grown ready for the task.

3. A serious improvement: convection in two dimensions

In the fall of 1971 Juri had moved to the Joint Institute for Laboratory Astrophysics in Boulder to take a professorship at the University of Colorado. Rapidly he built up around him a team of students and postdocs dedicated to computational astrophysics. He first tackled the two-dimensional modeling of stratified convection, where the horizontal scale of the convective flows was no longer imposed as in the modal approach, but resulted

---

† This was also the computer on which Juri ran the simulations of the famous paper with his brother Alar on "Galactic Bridges and Tails" (Toomre & Toomre 1972).

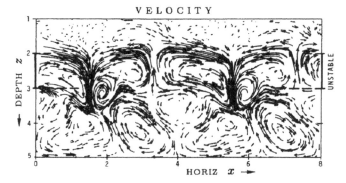

Figure 3. Two-dimensional simulation of stratified convection; the temperature contrast measures 5 between top and bottom of the computational domain. The unstable layer is located between two stable regions; its depth is taken as unit length. The downdrafts are much stronger than the upwellings, and they occupy less horizontal space. Note the deep penetration below into the stable region. (From Hurlburt *et al.* 1986, courtesy *ApJ*)

of the non-linear interactions of the problem. Moreover, the solutions are then no longer stationary, but display strong time dependence.

A snapshot of such a simulation is shown in Fig. 3, where an unstable layer is embedded between two stable regions (Hurlburt, Toomre & Massaguer 1986). The striking property of the convective flows is that they do not remain confined in the unstable region, but that they dig deep into the stable region below, emitting gravity waves. This penetration is due mainly to strong downdrafts, which occupy a rather small fraction of the area, as required by the conservation of mass. The depth of penetration depends on the strength of the stable stratification, but also quite sensitively on the Peclet number, which measures the ratio between the advective and diffusive transports. This behavior was later confirmed by three-dimensional simulations, where the down-flows take the form of plumes (Brummell, Clune & Toomre 2002). However, the exact penetration depth is still in debate, because the Péclet number which characterizes the base of the solar convection zone, $\approx 10^6$, remains out of reach of the numerical simulations, even with present-day supercomputers.

Juri was much aware of the need to validate the numerical simulations, because of their inherent limitations, through observations or laboratory experiments. Quite naturally he became extremely interested in the five minutes oscillations that had been observed on the Sun (Deubner 1975); it was clear that they could be used to probe the solar interior. His first paper on the subject was published in 1981 with Frank Hill and Larry November (Hill *et al.* 1981); it was followed by many others until now, in collaboration with most scientists working in the field. At the same time Juri engaged in active lobbying for helioseismology, both space-borne and ground based. All these efforts culminated in 1990 in a six months session he organized with Douglas Gough at the Institute for Theoretical Physics in Santa Barbara. All those who were working in the subject gathered there to spark off a new era in solar physics - it is worth browsing through the proceedings (Gough & Toomre 1990). It is there that I learned that the transition from differential rotation in the convection zone to quasi uniform rotation in the radiative interior occurs in a surprisingly thin layer, which we later called the tachocline, when I proposed with Ed Spiegel its first, purely hydrodynamic model (Spiegel & Zahn 1992). It is there also that Juri incited me to have a closer look on how the convective motions penetrate into a stable region. I guess that most participants benefited likewise from this stimulating meeting. And Juri and Douglas reached their goal: they obtained that the space mission SoHO (Solar & Heliospheric Observatory), a joint project between ESA and NASA, be

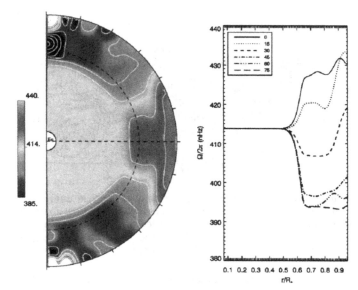

Figure 4. The solar interior rotation according to the ASH code (taking temporal and azimuthal averages). The computational domain encompasses $r/R_\odot = 0.07$ to 0.97, the numerical resolution is $N_r \times N_\theta \times N_\varphi = 770 \times 256 \times 512$. On the right panel radial cuts at selected latitudes (after averaging the north and south hemispheres) reveal the strong shear in the tachocline naturally realized in the model. (From Brun *et al.* 2011, courtesy *ApJ*)

equipped with the instrument MDI (Michelson Doppler Imager), which was especially designed for high resolution helioseismology, while the network GONG was deployed with six sites covering the whole Earth, to ensure continuous monitoring of the Sun. These programs were extremely successful, and they gave a tremendous impetus to solar research.

For the same purpose of validating computer simulations, Juri engaged with John Hart and a few other colleagues in a microgravity experiment put on board of Spacelab 3 which was launched in May 1985 (Hart, Toomre *et al.* 1986). Thermal convection was triggered in a rotating, differentially heated hemispherical shell, with a radial buoyancy force created by an alternating electrostatic field. Excellent agreement was found between the convective patterns observed in the experiment and those obtained by Gary Glatzmaier with a three-dimensional numerical code.

4. Convection on a global scale, in three dimensions

It was Glatzmaier's code, written while he was preparing his PhD under the supervision of Peter Gilman, which represented the first serious attempt to model a whole convection zone in spherical geometry. Juri saw that this was the way to proceed, and with Mark Miesch and Tom Clune they redesigned Gary's code "to take full advantage of the higher resolution possible on currently available parallel super-computing platforms", as they announced at the AAS meeting in June 1996. Since then the code kept being adapted to the latest breed of super-computers, a task involving a growing number of collaborators that Juri was accreting. In this code, which is described in Clune *et al.* (1999), the acoustic waves are again filtered out by the anelastic approximation, and the horizontal surfaces are mapped by spherical harmonics while the vertical profiles are projected on Chebyshev polynomials. Hence the acronym ASH, for Anelastic Spherical Harmonics.

Figure 5. The Convective Collective reunited in 2004 at the Isaac Newton Institute in Cambridge, for a meeting on the solar tachocline (picture by A.S. Brun).

ASH succeeded where earlier calculations had failed: at last the rotation profile of the convection zone was compatible with that determined through helioseismology: the equator rotated faster than the poles and the differential rotation depended little on depth. Moreover, the cause of the equatorial acceleration was now identified: it is due to the transport of angular momentum from pole to equator through the Reynolds stresses (Brun & Toomre 2002). The code has been applied since to various other problems, such as the young (thus fast rotating) suns and core convection in massive stars. The original hydrodynamic version has been extended to include the magnetic field (Brun, Miesch & Toomre 2004), a significant improvement which enables it to treat the dynamo problem and to predict the evolution of a fossil magnetic field.

However one weakness of most of these simulations is that they require boundary conditions to be applied on the computational domain, which for numerical reasons cannot extend to the surface and is often confined to the convectively unstable region. It was found that the results are rather sensitive to a pole-equator variation of the temperature imposed at the lower boundary (Miesch, Brun & Toomre 2006). This called clearly for including the tachocline in the computational domain, and the latest simulations reach even much deeper into the radiative core. Figure 4 displays such a state of the art calculation, where the tachocline is achieved without any artefact (Brun, Miesch & Toomre 2011).

For completeness, I should mention that Juri participated also in many simulations performed in a cartesian box (cf. Cattaneo *et al.* 1991; Brummell, Clune & Toomre 2002), which can be applied to specific problems or specific regions of a convection zone: subsurface layers, the lower boundary where penetration occurs. This was the path chosen by most other teams, with some putting their emphasis on a realistic treatment of radiative transfer, in order to predict the solar granulation pattern or the profile of spectral lines (cf. Stein & Nordlund 1989; Asplund *et al.* 2000). But to me Juri remains above all the master of global simulation.

The tachocline is still arousing much interest. It certainly plays a key role in the solar dynamo, by shearing the poloidal field, pumped down from above, into toroidal field (the omega-mechanism), as it has been demonstrated through 3-D simulations (Browning *et al.* 2006). But it is still not clear why that layer is so thin. To confront the various explanations that have been elaborated, a meeting was organized in 2004 by David Hughes, Bob Rosner and Nigel Weiss, at the Isaac Newton Institute in Cambridge. For the first time in thirty five years, the four members of the Convective Collective were reunited. In the meanwhile, they had drifted their way. Ed became fascinated in dynamical systems and in more or less strange attractors. Douglas became the leading figure in theorizing helioseismology - and now asteroseismology. I turned to rotational mixing in stellar radiation zones. Only Juri stayed faithful to his early engagement: to provide the best representation of stellar convection with the most powerful computers. And his achievements are truly impressive, as we will again be reminded during this symposium, a tribute that his students and friends dedicate to him for his 70th birthday.

References

Asplund, M., Nordlund, A. A., Trampedach, R., Allende Preto, C., & Stein, R. F. 2000, *A&A*, 359, 729

Biermann, L. 1938 *AN*, 264, 361

Browning, M., Miesch, M. S., Brun, A. S., & Toomre, J. 2006 *ApJ*, 648, 157

Brummell, N. H., Clune, T. L., & Toomre, J. 2002 *ApJ*, 570, 825

Brun, A. S., Miesch, M. S., & Toomre, J. 2004 *ApJ*, 614, 1073

Brun, A. S., Miesch, M. S., & Toomre, J. 2011 *ApJ* (submitted)

Brun, A. S. & Toomre, J. 2002 *ApJ*, 570, 865

Canuto, V. M. 1992 *ApJ*, 392, 218

Cattaneo, F., Brummell, N. H., Toomre, J., Malagoli, A., & Hurlburt, N. E. 1991, *ApJ*, 370, 282

Clune, T. L., Elliott, J. R., Glatzmaier, G. A., Miesch, M. S., & Toomre, J. 1999 *Parallel Comput.*, 25, 361

Cowling, T. G. 1936 *AN*, 258, 133

Deubner, F. L. 1965, *A&A*, 44, 371

Gough, D. O. 1969 *J. Atmospheric Sciences*, 26, 448

Gough, D. O., Spiegel, E.A., & Toomre, J. 1975 *J. Fluid Mech.*, 68, 695

Gough, D. O. & Toomre, J. (ed.) 1991 *Helioseismology - probing the interior of a star; LNP* vol. 388

Graham, E. 1977, *Problems of Stellar Convection*, IAU Coll. 38 (ed. E.A. Spiegel & J.-P. Zahn), p. 151

Hart, J. E., Toomre, J., Deane, A. E., Hurlburt, N. E., Glatzmaier, G. A., Fichtl, G. H., Leslie, F., Fowlis, W. W., & Gilman, P. A. 1986 *Science*, 243, 61

Hill, F., Toomre, J., & November, L. J. 1981 *BAAS*, 13, 860

Hurlburt, N. E., Toomre, J., & Massaguer, J.-M. 1986 *ApJ*, 311, 563

Latour, J., Spiegel, E. A., Toomre, J., & Zahn, J.-P. 1976 *ApJ*, 207, 233

Latour, J., Toomre, J., & Zahn, J.-P. 1981 *ApJ*, 248, 1081

Miesch, M. S., Brun, A. S., & Toomre, J. 2006 *ApJ*, 641, 618

Miesch, M. S., Clune, T. C., Toomre, J., & Glatzmaier, G. A. 1996, *BAAS*, 28, 936

Spiegel, E. A. & Zahn, J.-P. 1992, *A&A*, 265, 106

Stein, R. F. & Nordlund, Å. 1989 *ApJ*, 342, L95

Toomre, J., Gough, D.O., & Spiegel, E.A., 1977 *J. Fluid Mech.*, 125, 99

Toomre, J., Zahn, J.-P., Latour, J., & Spiegel, E. A., 1976 *ApJ*, 207, 545

Unsöld, A. 1930 *ZfA*, 1, 138

Vitense, E. 1953 *ZfA*, 32, 135

Xiong, D. R. 1985 *A&A*, 150, 133

Astrophysical Dynamics: From Galaxies to Stars
Proceedings IAU Symposium No. 271, 2010
N.H. Brummell, A.S. Brun, M.S. Miesch & Y. Ponty, eds.

© International Astronomical Union 2011
doi:10.1017/S1743921311017777

Onward from solar convection to dynamos in cores of massive stars

Juri Toomre

JILA & Dept of Astrophysical and Planetary Sciences, University of Colorado, Boulder, CO
80309-0440, USA
email: jtoomre@jila.colorado.edu

Abstract. We reflect upon a few of the research challenges in stellar convection and dynamo theory that are likely to be addressed in the next five or so years. These deal firstly with the Sun and continuing study of the two boundary layers at the top and bottom of its convection zone, namely the tachocline and the near-surface shear layer, both of which are likely to have significant roles in how the solar dynamo may be operating. Another direction concerns studying core convection and dynamo action within the central regions of more massive A, B and O-type stars, for the magnetism may have a key role in controlling the winds from these stars, thus influencing their ultimate fate. Such studies of the interior dynamics of massive stars are becoming tractable with recent advances in codes and supercomputers, and should also be pursued with some vigor.

Keywords. Sun: tachocline, Sun: near-surface shear layer, Sun: dynamo action, stars: massive star convection

---

1. Coupling of stellar convection, rotation, magnetism and winds

The last decade has seen a major resurgence of interest in many university departments concerning the structure and evolution of a broad range of stars. Contributing to this are the abundant successes in detecting extra-solar planets, and thus the need to understand in some detail the dynamical properties of the central stars in such systems as one ponders the possibilities of life elsewhere. Similarly, significant observational advances in spectropolarimetry and in Doppler imaging are permitting deductions about differential rotation and magnetic structures at the surfaces of a variety of stars, thus calling out for detailed theoretical interpretations and explanations. Further, asteroseismology is showing its ability to infer interior properties of stars, and when carried out on very large samples as permitted with Kepler observations, is raising many fascinating puzzles that require major new theoretical efforts to analyze the evolving interior dynamics of stars. Central to all of these topics is the need to understand the coupling of convection, rotation, shear and magnetism in stellar settings, which raises formidable theoretical challenges since the flows and structures are highly turbulent and involve a vast range of physical scales.

Convection plays many roles in the lives of stars. Moderate and low mass stars on the main sequence (when they burn hydrogen in their cores) possess convective envelopes and radiative interiors, whereas more massive stars have convective cores and radiative envelopes. Such turbulent convection in the outer envelopes of the less massive F, G, K and M stars, when influenced by the rotation of the star, can yield magnetic dynamo action that is likely responsible for much of the magnetic activity that is observed in their atmospheres. The effects of dynamo action in the cores of more massive O, B and A stars is less certain, for their extensive radiative envelopes may be effective in hiding contemporary magnetic field building as it proceeds. Thus the magnetic spots that are observed to rotate into view on some of these stars may well be mainly a signature of surviving primordial fields. However, if sufficiently strong and thus buoyant magnetic structures can be built by core dynamos, then such fields have the possibility of rising to the surface to be detected. Whether the contemporary core fields get out or not, they are likely to have an important role in the final stages of stellar evolution. Late in their lives as the massive stars expel their envelopes to expose their very hot cores and thus leave behind white dwarfs, or even after supernova explosions that leave behind neutron stars, that dynamo action may help to explain or contribute to the very strong magnetic fields that are observed in the remnants.

As another issue, the presence of strong magnetic fields close to the surface of all stars may have pivotal effects upon the strength and character of stellar winds that both carry away mass and angular momentum as the stars evolve. This may well influence whether some of the massive stars have lost enough mass to end their lives relatively quietly as white dwarfs, or whether they have a more fiery end as a core collapse supernova that leaves behind a neutron star or even a black hole. The moral from all this is that stellar convection, rotation, magnetism and winds are all closely linked in the life stories of most stars. The past decade or two has seen substantial progress in beginning to deal with these interlinked and highly nonlinear dynamical processes, as this conference is amply revealing. These decades have also provided splendid high-resolution imagery from Hubble Space Telescope that emphasize the richness of structure in the winds that must be attributable to both rotation and magnetism. It may be appropriate to reflect upon a few of these developments that should be pursued with vigor in the near future.

2. Ascending role of computational astrophysics

Computational astrophysics has become the third arm of research that complements both basic theory and observations, attaining this position through the ability to conduct ever more realistic 3–D simulations due to rapid advances in both supercomputers and in codes and algorithms that can exploit such machinery. These advances have enabled the simulations of convection and dynamo action in the outer envelope of the Sun and in their more rapidly rotating younger relatives, as discussed in graphic detail in our meeting. These are having a pivotal role in coming to understand the elements involved in achieving the differential rotation of such stars and of the vigorous magnetic dynamo that can result, even yielding magnetic fields that can take the form of striking wreaths that persist or ones that reverse their sense cyclically or episodically. The challenge here is to get the modeling of global-scale convection within spherical shells, such as carried out with the anelastic spherical harmonic (ASH) code, to regimes with ever smaller diffusivities and thus more turbulent conditions. Further, one needs more realistic treatments of the two boundary layers that are present in the Sun, namely the tachocline of rotational shear as revealed by helioseismology at the base of the convection zone and the near-surface shear layer just below the surface.

3. Solar tachocline

Dealing with the tachocline requires resolving penetrative convection entering a region of very stable stratification in which the downward-directed plumes can excite a rich medley of gravity waves, with all of this somehow establishing the pronounced rotational shear that is observed. Keeping that shear from gradually spreading into the deeper radiative interior that is deduced to be in solid-body rotation may well require magnetic fields, possibly involving very modest fields of primordial origin that survive in the deep interior, along with a magnetic boundary layer within the tachocline itself to isolate the deep from the fields and shear built within the convection zone itself. Amidst all such needs to keep that boundary layer seemingly narrow, this tachocline is likely to be the key factor in the reversing cycles of erupting magnetism observed as sunspots. How the tachocline is able to order and shear toroidal magnetic fields into structures that become buoyantly unstable in a self-consistent manner is the dominant challenge to dynamo theory. This has a reasonable hope of being sorted out through global 3-D simulations in roughly the next five years or so. It requires focused attention to attain realistic solar stratifications at high spatial resolution and much lower diffusivities in order to get into regimes that have a likelihood of building sufficiently strong and ordered toroidal fields. A recent new ingredient is that the lower reaches of the convection zone may be able generate leaky wreaths of fairly strong toroidal fields that already have a preferred different sense in the two hemispheres, as we have seen for younger Suns rotating more rapidly (Brown *et al.* 2010, 2011). This behavior may well carry over to the Sun itself as diffusive processes are lessened, with downward magnetic pumping serving to bring fields into the tachocline region to be further amplified and the more random elements gradually eliminated (Browning *et al.* 2006). It will be fascinating to discover just what sets the roughly 11-year periods for reversals, whether it be the meridional circulation times favored by flux-transport and some mean-field models, or possibly intrinsic competing generation processes in the wreaths or in the tachocline itself. We are beginning to see 3-D global solar dynamo simulations that are now possessing cycles (Ghizaru *et al.* 2010; Miesch *et al.* 2011), and these serve to encourage much more work to understand their sensitivities.

4. Solar near-surface shear layer

Seeking to understand the origin and role of the Sun's near-surface shear layer raises challenges that are quite different from that of the boundary layer at the base of the convection zone. Here we have intense competition and likely collisions between the fast small scales of descending granulation and supergranulation, driven by rapid cooling at the solar surface, with the ascending and sweeping large scales of giant cells of convection. Those giant cells sense rotation strongly and are key to achieving the differential rotation seen in the bulk of the convection zone (Miesch et al. 2008). Yet such current 3-D global simulations of solar convection with ASH do not take into account the descending plume patterns from the surface, nor do the localized-domain surface convection simulations (Stein *et al.* 2009) account for ascending or shearing motions from the deep. Thus neither achieves a near-surface shear layer, though helioseismology reveals its rotational signature in roughly the outer 5% in radius quite clearly. The forbidding barrier to either local or global models is the vast range of spatial and time scales that are inherent in solar convection in its upper reaches. This may be overcome by turning to intermediate domain simulations in depth that focus on spherical shell segments, being informed from above by the statistics of the descending plume networks of supergranulation, and from below by the ascending and shearing flows of giant cells. The required simulation tools are now becoming available, as with the compressible spherical segment (CSS) code (Augustson *et al.* 2011b), and hold out the promise that properties of the near-surface shear layer and its multi-scale flow interactions may soon be capable of starting to be resolved. Such an approach will also be essential to devise new upper boundary conditions for the global simulations, as with ASH, that capture the more gradual flow decelerations and deflections that are likely to result as the rising giant cells meet the intense small scales from above. The inclusion of near-surface dynamics may have subtle influences on

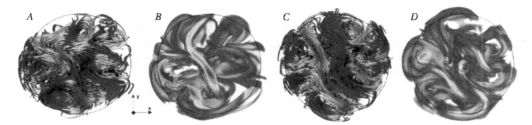

Figure 1. Evolving flow streamlines (A, C) and magnetic energy density (B, D) in the super equipartition A-type star convective core dynamo. Shown is the equatorial plane in two time instants separated by 50 days. Violet tones denote positive motions in the y-direction, and yellow tones negative motions. Regions of strong magnetic energy are shown in yellow/green tones. Large core-crossing flows carry broad swaths of magnetic field with them.

the character of differential rotation and meridional circulations achieved in the global modeling, and thus some account of the upper boundary layer must be pursued. Further, it is the near-surface shear layer that can now be probed in uninterrupted detail to study subsurface flows and their interaction with magnetism using local-domain helioseismology, as now feasible using Doppler and magnetic imaging data from the Solar Dynamics Observatory and its HMI and AIA instruments.

5. Marching toward convection and dynamos in massive stars

Stars more massive than about 1.5 solar masses (the A, B and O-type stars) possess radiative envelopes surrounding a convective core. Long suspected of harboring core dynamos, a subset of these massive stars also exhibit striking surface magnetism. These peculiar A- and B-type stars have strong patches of surface magnetism thought to arise where large-scale subsurface magnetic fields extend through the stellar surface. Such subsurface fields likely evolved from magnetic field initially threading the collapsing molecular cloud that formed the star. Our initial investigations into the nature of core dynamos of these massive stars focused on A-type stars lacking such an external field. In its absence, these core dynamos can generate magnetic fields with energies in near equipartition with the convective motions (Brun, Browning & Toomre 2005). Recent work with ASH began to examine the coupling of such a convective core dynamo to a primordial magnetic field threading the radiative envelope. Dynamo action achieved in the presence of such a field is much stronger than that found in its absence, yielding magnetic energies roughly ten-fold that with no primordial field (Featherstone *et al.* 2009). This is a remarkable finding, for here one has achieved a super-equipartition dynamo state that was not thought to be realizable, and may well be characteristic of many core dynamos. The super-equipartition dynamo is characterized by the presence of four to six meandering convective rolls whose motions frequently cross the rotation axis, linking distant regions of the core as they do so (Figure 1). The effects of stellar rotation, here at $4 \, \Omega_\odot$, are a crucial element in building these structures. Strong axial circulations in the cores of these rolls lead to plunging events that couple the northern and southern hemispheres. These motions serve to advect magnetic field from regions of strong shearing and generation, mixing it throughout the bulk of the core. This work suggests that new initiatives should be undertaken in the next few years to explore the limiting strength of such toroidal fields, here already at about 500 kG, as less diffusive approaches are employed in codes such as ASH. Even stronger fields may become magnetically buoyant and have the

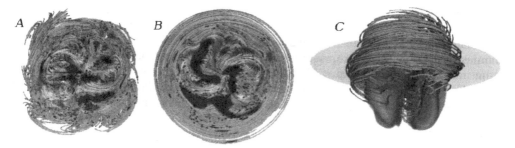

Figure 2. Convection and dynamo action in the core of a B star of mass 10 M$_\odot$ rotating at 8 Ω$_\odot$. Shown in (A) are a snapshot of velocity streamlines within the core, cut along the equator viewing the southern hemisphere. Rapid flows in yellow tones cross the core, and several columnar flows are evident. In contrast, (B) shows the same of the northern hemisphere, which exhibits strong differential rotation. In (C) are magnetic field streamlines in green and pink (for polarity) in perspective side view (equator shaded), emphasizing the dominance of magnetism in the northern hemisphere in this solution.

potential to erupt from the convective core toward the surface, thus providing a possible linkage between contemporary dynamo action in the deep with fields seen at the surface.

While O and B-type stars make up only a small fraction of the total stars in a typical galaxy, they are very luminous and have short lives. Indeed, a 10 solar mass B star has a luminosity about 7200 L$_\odot$, and lives on the main sequence (fusion burning hydrogen in its core) for only about 15 million years (compared to 10 billion for the Sun). Many O and B stars exhibit strong stellar winds, and many end up in supernova explosions, enriching the interstellar medium with heavy elements. They thus influence star formation and galactic structure throughout their brief lives, thereby having critical roles in many of the processes that sustain and evolve their host galaxies. Understanding how the convective dynamics occurring within the cores of massive stars alters their lifetimes is of interest to many astrophysical communities. Since the exact conditions necessary for self-sustained magnetism in O and B stars is not known, we need detailed study of the coupling of the radiative zone and the convective core in O and B stars. If the core dynamos are able to get magnetic fields into the atmospheres of these stars, again by field possibly erupting from the core, then their winds may be influenced by the presence of such magnetic fields, both in terms of mass loss and stellar spindown. As our preliminary simulations of B star core convection are revealing (Augustson *et al.* 2011a), sustained dynamo action leading to strong magnetic fields within the core are indeed realized (Figure 2), but for the case shown here there are pronounced asymmetries in the field strengths achieved in the northern and southern hemispheres. Other rotation rates lead to more symmetric states. These studies are showing that its is now feasible to begin exploring the interior dynamics of massive stars in some detail through 3-D simulations. This subject is ripe for investigation given the codes and computers that are becoming available, but above all motivated by the realization that the massive stars are the dominant recyclers in any galaxy. The near future for such research looks most promising, as one has seen with both A and B stars, and thus the study of massive star dynamics may become one of the areas of vigorous study.

6. Positive optimism that may be deserved

All the work implicitly discussed briefly here has been carried out as joyful collaborations with many friends dealing with helioseismology and with solar and stellar convection and dynamo theory, a good number of whom are at this meeting. I thank them for their enthusiastic participation and wisdom in such work, and hope that the research investments made so far will continue to bear splendid fruit as the topic of astrophysical fluid dynamics (AFD) continues to mature and expand into areas that seemed overly challenging not so long ago. Coming to really understand the operation of the solar global dynamo in a reasonably self-consistent manner is getting tantalizing closer to hand, and we hope that this does not prove to be a shimmering mirage, given the surprises that highly turbulent systems are able to provide. Yet the Sun seems to know how to operate its magnetic cycles in a largely regular fashion, assuming that we can overlook the occasional hiccups that may be extended minima. It is but for us to try to sort out the elements involved in such dynamo action, with the ever enhanced computational resources providing us with experimental tools that may be up to the task. The presence of two boundary layers in the solar convection zone certainly warns us of subtle effects that need to be considered with some care, but the next five years holds out good promise that we may achieve substantial steps forward. The suggested investments to study interior dynamics within the more massive stars on the main sequence are in earlier stages, yet they benefit greatly from the tools and experiences that have come about from trying to understand the functioning of our nearest modest star. The O and B stars with their great luminosities and short lives, and often with strong winds, are now becoming quite tractable to detailed simulation. This may well become one of the renaissance research areas in AFD, complementing in many ways the intense efforts to understand how these stars late in their lives may come to explosive conclusions, but not always. The core convection and its dynamos may well provide keys to the masses that survive late in such stellar lives.

This research was supported recently by NASA Heliophysics Theory grant NNX08AI57G and by NSF Teragrid supercomputing resource grant TG-MCA93S005 and NASA project GID 26133.

References

Augustson, K. C., Brun, A. S., & Toomre, J. 2011a, these proceedings
Augustson, K., Rast, M., Trampedach, R., & Toomre, J. 2011b, *J. Phys. Conf. Ser.* 271, 012070
Brown, B. P., Browning, M. K., Brun, A. S., Miesch, M. S., & Toomre, J. 2010, *ApJ*, 711, 424
Brown, B. P., Miesch, M. S., Browning, M. K., Brun, A. S., & Toomre, J. 2011, *ApJ*, 731, 69
Browning, M. K., Miesch, M. S., Brun, A. S., & Toomre, J. 2006, *ApJ*, 648, L157
Brun, A. S., Browning, M. K., & Toomre, J. 2005, *ApJ*, 629, 461
Featherstone, N. A., Browning, M. K., Brun, A. S., & Toomre, J. 2009, *ApJ*, 705, 1000
Ghizaru, M., Charbonneau, P., & Smolarkiewicz, P. K. 2010, *ApJ*, 715, L133
Miesch, M. S., Brown, B. P., Browning, M. K., Brun, A. S., & Toomre, J. 2011, these proceedings
Miesch, M. S., Brun, A. S., DeRosa, M. I., & Toomre, J. 2008, *ApJ*, 673, 557
Stein, R. F., Georgobiani, D., Schafenberger, W., Nordlund, A. A., & Benson, D. 2009, *AIP Conf. Ser.* 1094, 764

POSTERS

Astrophysical Dynamics: From Stars to Galaxies
Proceedings IAU Symposium No. 271, 2010
N. H. Brummell, A.S.Brun, M.S. Miesch & Y. Ponty, eds.

© International Astronomical Union 2011
doi:10.1017/S1743921311017789

Reduced MHD and Astrophysical Fluid Dynamics

Wayne Arter

EURATOM/CCFE Fusion Association, Culham Science Centre,
Abingdon, Oxon. OX14 3DB, UK
email: `wayne.arter@ccfe.ac.uk`

Abstract. Recent work has shown a relationship between between the equations of Reduced Magnetohydrodynamics (RMHD), used to model magnetic fusion laboratory experiments, and incompressible magnetoconvection (IMC), employed in the simulation of astrophysical fluid dynamics (AFD), which means that the two systems are mathematically equivalent in certain geometries. Limitations on the modelling of RMHD, which were found over twenty years ago, are reviewed for an AFD audience, together with hitherto unpublished material on the role of finite-time singularities in the discrete equations used to model fluid dynamical systems. Possible implications for turbulence modelling are mentioned.

Keywords. methods: numerical, convection, MHD, turbulence

1. Introduction

The author (Arter 2010) has recently shown that there is a close relationship between the equations of high-β Reduced Magnetohydrodynamics (RMHD), specified in e.g. Hazeltine & Meiss (1992, § 7) and incompressible magnetoconvection (IMC) described in e.g. Proctor (2005). Indeed, the two systems are mathematically equivalent in certain geometries. This paper describes implications of the equivalence for the simulation of AFD applications (of which IMC is one example), from what is known about RMHD, which is used to model magnetic fusion laboratory experiments.

The equivalence between the two systems arises in models with two space dimensions. In RMHD, the presence of a strong magnetic field $\mathbf{B_0}$, leads to an ordering in which fluid flow is incompressible in planes normal to $\mathbf{B_0}$. If further, at leading order, field lines are curved in the manner typical of laboratory magnetic confinement configurations, an effective gravity appears in the model proportional to the local field line curvature. In IMC, both incompressibility and the corresponding buoyancy force term due to gravity arise from the Boussinesq approximation, which is well described in Tritton (1998, § 14A). There are important physical differences, in that in RMHD, the energy source is primarily the background field $\mathbf{B_0}$, and the 'gravity' force may be stabilising or destabilising, whereas in IMC, thermal buoyancy is the main driver of instability.

One aspect pioneered in the modelling of laboratory plasmas was the study of the limitations on numerical simulation when dissipation is small, or equivalently at large Reynolds number. Such a regime is dominated by nonlinear effects, so the exact nature of the primary instability is less important.

2. Nonlinear instability in RMHD

The paper by Eastwood & Arter (1986) pointed out that the numerical divergence observed in simulations of RMHD could be explained by nonlinear numerical instability.

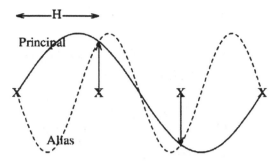

Figure 1. Quadratic nonlinearities cause wavelength $\lambda_0 = 3H$ to generate a component with $\lambda_1 = 3H/2$. When sampled on the mesh (crosses), these both contribute (shown by arrows) to mesh-resolved wavelength $3H$.

The numerical algorithms used in these codes do not conserve energy and further numerical work by Eastwood & Arter (1987a) confirmed that field energies actually grew explosively as $(t_0 - t)^{-2}$ prior to code termination at time $t = t_0$. Increasing the dissipation in the models removed the divergence and led to physically distinct, oscillatory behaviour. A physical explanation for the process is illustrated in Figure 1.

Although this 'alias-feedback' instability is an important issue for old RMHD codes, the above argument is neither rigorous nor quantitative. Since instability occurs even with vanishingly small timesteps, semi-discrete models, continuous in time, for RMHD were studied by Eastwood & Arter (1986). The 2-D limit of constant pressure was considered, so the results also apply directly to an IMC model. Computer algebra was used to produce semi-discrete models with a spectral discretisation in one co-ordinate and a f.d. (finite difference) representation in the other. The number of degrees of freedom was very limited, typically to three mesh-points and five wavenumbers.

Results were obtained for the resulting RMHD/IMC discretisations which can be understood in terms of the following heuristic model, from Arter & Eastwood (1987). Suppose v and b represent the velocity and magnetic field respectively, then all the nonlinearities are represented in

$$\dot{b} = -b + vb,$$
$$\dot{v} = -p_m v + v^2 + b^2 \qquad (2.1)$$

where the Prandtl number $p_m = \nu/\eta$ is the ratio of viscous diffusivity ν to magnetic diffusivity η. It will be seen that the origin in (b, v) space is attracting for sufficiently small initial conditions. This decay is the physically expected behaviour in the absence of any forcing terms, but as may easily be shown for larger initial (b, v), there is blow up at finite time $t = t_0$, asymptotically:

$$v \sim (t_0 - t)^{-1}, \quad b\sqrt{\ln b} \sim (t_0 - t)^{-1} \qquad (2.2)$$

Analysis gives the stability criterion

$$Rm_2^2 + \frac{1}{p_m} Sm_2^2 < \mathcal{O}(1) \qquad (2.3)$$

where the dimensionless parameter known as the mesh Reynolds number $Rm_s = H\|u\|_s/\nu$, is calculated in terms of the s-norm of the initial u and similarly the mesh Lundquist number $Sm_s = H\|u_{\text{Alfven}}\|_s/\eta$.

Since the obvious objection to all the above original analysis is that it is based on models which are low order and/or heuristic, further investigations were pursued.

3. Nonlinear instability in Burgers' Equation

The approach adopted was to solve a simpler problem than RMHD, but to do so rigorously and consider large Nth order systems, corresponding to discretisations on N grid-points. The simplest nonlinear model for fluid dynamics is Burgers' equation for advection and diffusion of the velocity field $u(x, t)$:

$$u_t + u u_x = u_{xx} \tag{3.1}$$

It is possible to solve Eq. (3.1) analytically by means of the Cole-Hopf transformation, which shows that all spatially periodic solutions must decay to $u = \text{const.}$, but also gives the weak, steady solution $u = \tan x$.

Burgers' equation may be written in semi-discrete form as

$$u_j = N_j(u_0, ..., u_{N-1}) + L_j(u_0, ..., u_{N-1}), \quad j = 0, .., N-1, \tag{3.2}$$

where specifically

$$N_j = -u_j(u_{j+1} - u_{j-1}), \quad L_j = u_{j+1} - 2u_j + u_{j-1} \tag{3.3}$$

The relevant dimensionless parameter is again Rm_s, where $||u||_s$, is evaluated using the initial velocity values u_j on the spatial grid.

Investigations by Eastwood & Arter (1987b) and Davidson (1995) showed that the above choice of N_j and L_j led to generic behaviour, unless the nonlinearity N_j was such as to conserve energy in the absence of dissipation, i.e. if it arose either from a specially constructed f.d. method, or from a spectral method. As anticipated in Section 1, the exact form of the boundary conditions was also found to be unimportant, and all results produced are for periodic boundaries $u_0 = u_N$ unless stated otherwise.

It was found that, as in the low order models, both decaying solutions and solutions diverging as $(t_0 - t)^{-1}$ were possible depending on initial conditions. In addition, the boundary in phase-space between the two types was found to be associated with the presence of fixed points of the Nth order system Eq. (3.2).

One of Davidson's major contributions was to give analytic expressions for these fixed points in N dimensions. For the case of periodic boundaries with a zero mean flow:

$$u_j = \tan\left(\frac{q\pi}{N}\right) \tan\left(\frac{q\pi}{N}\bar{j}\right) \tag{3.4}$$

$$\bar{j} = j + \frac{1}{2} \ (N \text{ even}), \text{ else } \bar{j} = j$$

for integer q. This ties up with the previously obtained numerical fixed point solution as shown in Figure 2. Eq. (3.4) also generalises to the case when there is a mean flow, suppose this is of amplitude $\tan\left(\frac{\pi}{N}\right) \tan\left(\frac{\epsilon}{2}\right)$ for some real ϵ, then Davidson gives

$$u_j = \tan\left(\frac{\pi}{N}\right) \tan\left(\frac{\pi}{N}\bar{j} + \frac{\epsilon\pi}{2N}\right) \tag{3.5}$$

The following solutions by Davidson reflect the fact that qualitatively similar phase space behaviour is found for the case of zero gradient boundary conditions, viz.

$$u_j = -\tanh\left(\frac{A}{N}\right) \tanh\left(\frac{A}{2N}j + \phi\right) \tag{3.6}$$

$$u_j = -\tanh\left(\frac{A}{N}\right) \coth\left(\frac{A}{N}\bar{j} + \phi\right)$$

Of note is the fact that whilst they lie close to the edge of the stability boundary, the fixed points do not determine it precisely. There are solutions with as much as 12%

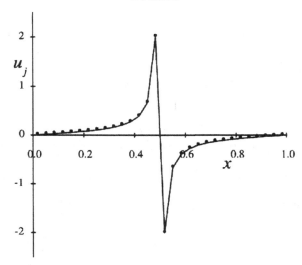

Figure 2. u_j plotted against $x = (j + \frac{1}{2})/N$ for the fixed point of the discrete Burgers' equation with $N = 30$. The solid line joins points of the analytic expression Eq. (3.4), $q = 1$.

smaller 2-norm or energy (although no smaller) than that possessed by the points Eq. (3.4) which do diverge. In addition, since the fixed points have at least one large jump, it is important to specify which norm is used to measure the initial velocity field. For example, as $N \to \infty$, the unstable fixed point approaches zero in the energy norm as $Rm_2 \to 2\pi/\sqrt{N}$, whereas it remains finite in the maximum norm $Rm_\infty \to 2$.

4. Nonlinear instability in Fluid Dynamics

Extrapolating results from the simple, compressible one-dimensional system represented by Burgers' equation to higher dimensional, incompressible flows is questionable. Hence, Davidson (1995) analysed discretisations of the viscous Euler equations for two-dimensional fluid flow. The main thrust of the work related to the 2-D f.d. scheme expressible as

$$\frac{d\omega_{j,k}}{dt} + (\omega_{j+1,k} - \omega_{j-1,k})(\psi_{j,k+1} - \psi_{j,k-1}) \tag{4.1}$$
$$- (\psi_{j+1,k} - \psi_{j-1,k})(\omega_{j,k+1} - \omega_{j,k-1})$$
$$= \omega_{j+1,k} + \omega_{j-1,k} + \omega_{j,k+1} + \omega_{j,k-1} - 4\omega_{j,k}$$

and

$$\psi_{j+1,k} + \psi_{j-1,k} + \psi_{j,k+1} + \psi_{j,k-1} - 4\psi_{j,k} = -\omega_{j,k} \tag{4.2}$$

where ω is the vorticity and ψ the velocity stream function.

Problems were studied in the cases where either zero or periodic boundary conditions were imposed in each co-ordinate direction, for mesh sizes up to 18×18, i.e. 18 points in each direction. It was also established that similar results applied for other f.d., pseudospectral (p.s.), mixed f.d./p.s and mixed f.d./spectral schemes for meshes up 15×15. The form of the phase space was typically found to resemble that for Burgers' equation, with an attracting region enclosing the origin. One result for arbitrary N was established, namely that solutions were proven to converge (i.e. decay) if the initial discrete velocity

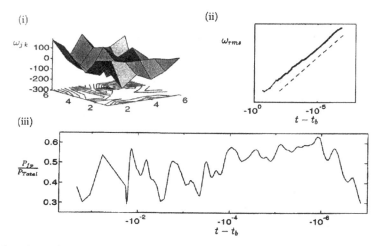

Figure 3. 6×6 mesh investigation, showing (i) initial conditions, (ii) blow-up in vorticity measure on log-log plot, and (iii) the power in the fixed point direction relative to the total power as the calculation diverges.

components satisfied

$$|u_{j,k}| \leqslant 1 \text{ and } |v_{j,k}| \leqslant 1 \qquad (4.3)$$

corresponding to a restriction on Rm_∞.

Although a general formula for fixed points could not be found, it was possible to deduce exact expressions from some of the numerically calculated points, e.g. in the case of a 6×6 mesh:

$$\omega_{j,k}(fp) = \begin{pmatrix} 30 & 0 & 0 & 0 & 0 & 0 \\ 30 & 0 & 0 & 0 & 0 & 0 \\ 0 & -30 & -30 & 0 & 0 & 0 \\ 0 & 0 & 0 & 30 & 0 & 0 \\ 0 & 0 & 0 & 30 & 0 & 0 \\ 0 & 0 & 0 & 0 & -30 & -30 \end{pmatrix} \qquad (4.4)$$

A measure of the complexity of the work is the fact that the 6×6 system of equations, which is the smallest model exhibiting behaviour typical of larger systems, is of 34th order. Figure 3 shows an indicative numerical calculation in this case. with a solution diverging as $(t_0 - t)^{-1}$. The principal difference between the vorticity equation and Burgers' equation seems to be that, whereas the singular solutions to Burgers' are stable, those for the vorticity equation admit unstable directions (Davidson, 1995). This is supported by the fact that in Figure 3(iii) not all the power finishes in the direction of the fixed point. As in the case of Burgers' equation, the most dangerous fixed points have large jumps. Davidson has estimated that the fixed point satisfies $Rm_2 \approx 59.7/\sqrt{N}$ as $N \to \infty$ whereas $Rm_\infty \to 41.7$. Criteria of the same form with similar magnitude estimates were also found for the other schemes mentioned above. In respect of Rm_∞ criteria, the stability of the vorticity equation is better than Burgers', but it remains uncertain whether this also holds true for RMHD. These stability results should be compared with the criterion for the accuracy of fluid calculations derived by Kreiss, see the review by Arter (1995). Kreiss's criterion may be put into the form $Rm_\infty \leqslant \pi^2$ (Davidson, 1995).

5. Discussion

Alias-feedback instabilities are not believed to be a problem for most present-day codes, which are normally designed so that the total energy cannot increase due to nonlinear interactions. In any event, satisfying accuracy criteria which limit the mesh Reynolds' and Lundquist numbers to order unity, should prevent the nonlinear numerical instability.

Nonetheless, the work described herein, particularly that due to J. Davidson, is still of value in that it indicates how difficult it is likely to be to establish by purely numerical means, the presence of finite time singular behaviour in fluid dynamics, see e.g. Hou (2009), Moffatt (2009) for background. It has also been shown that results obtained for more general flows are qualitatively similar to ones obtained for Burgers' equation. Now, applied to Burgers' equation, the energy conserving schemes like spectral methods are *incapable* of representing the steady singular solutions. As such a solution does not exist, it cannot be attracting and so a finite time singularity in the limit of vanishing timestep cannot be found by these methods.

Therefore the suggested way forward is to seek additional singular solutions in 2-D and 3-D of schemes for fluid dynamics which are not semi-conservative. These discrete solutions might be indicative, like the discrete *tan* and *coth* solutions found for Burgers' equation, of the functional form of weak solutions to Navier-Stokes'. It may then be possible to understand if or how such weak solutions can be produced, and whether they are unstable like the 2-D singular solutions found by Davidson (1995).

Acknowledgements

This work was funded by the United Kingdom Engineering and Physical Sciences Research Council under grant EP/G003955 and the European Communities under the contract of Association between EURATOM and CCFE. The views and opinions expressed herein do not necessarily reflect those of the European Commission.

References

Arter, W., *Plasma Physics and Controlled Fusion*, To be submitted, 2010.

Arter, W., *Reports on Progress in Physics*, 58:1–59, 1995.

Arter, W. & Eastwood, J. W., *Transport Theory and Statistical Physics*, 16:433–446, 1987.

Davidson, J., Dynamics of semi-discretised fluid flow, 1995. Dissertation submitted to the University of Cambridge.

Eastwood, J. W. & Arter, W., *Physical Review Letters*, 57:2528–2531, 1986.

Eastwood, J. W. & Arter, W., *Physics of Fluids*, 30:2774–2783, 1987a.

Eastwood, J. W. & Arter, W., *I.M.A. Journal of Numerical Analysis*, 7:205–222, 1987b.

Hazeltine, R. D. & Meiss, J. D., *Plasma Confinement*. Addison-Wesley, Redwood City, 1992.

Hou, T. Y., *Acta Numerica*, 18:277–346, 2009.

Moffatt, H. K., In R.L. Ricca, editor, *Lectures on Topological Fluid Mechanics*, pages 157–166. Springer Verlag, 2009.

Proctor, M. R. E., In A.M. Soward, C.A. Jones, D.W. Hughes, and N.O. Weiss, editors, *Fluid dynamics and dynamos in astrophysics and geophysics*, pages 235–278. CRC, 2005.

Tritton, D. J., *Physical Fluid Dynamics, 2nd Edition*. Clarendon Press, Oxford, 1988.

Astrophysical Dynamics: From Galaxies to Stars
Proceedings IAU Symposium No. 271, 2010
N.H. Brummell, A.S. Brun, M.S. Miesch & Y.Ponty

© International Astronomical Union 2011
doi:10.1017/S1743921311017790

Convection and dynamo action in B stars

Kyle C. Augustson[1], Allan S. Brun[2], and Juri Toomre[1]

[1] JILA & APS, University of Colorado, 440 UCB, Boulder, CO, 80309-0440, USA

[2] UMR AIM CEA-CNRS-Univ. P7 CEA Saclay, 91191 Gif-sur-Yvette Cedex, France

Abstract. Main-sequence massive stars possess convective cores that likely harbor strong dynamo action. To assess the role of core convection in building magnetic fields within these stars, we employ the 3-D anelastic spherical harmonic (ASH) code to model turbulent dynamics within a 10 $M_\odot$ main-sequence (MS) B-type star rotating at 4 $\Omega_\odot$. We find that strong (900 kG) magnetic fields arise within the turbulence of the core and penetrate into the stably stratified radiative zone. These fields exhibit complex, time-dependent behavior including reversals in magnetic polarity and shifts between which hemisphere dominates the total magnetic energy.

Keywords. stars: early type, rotation, magnetic fields

Surface magnetic fields have been found on many MS massive stars (e.g. Donati & Landstreet 2009). To some degree both the fossil fields and dynamo generated fields in these stars must coexist, although how they interact to produce and maintain these surface fields is unclear. Recent work has shed some light on the interaction between a super-equipartition core dynamo and fossil magnetic fields in A-type stars (Brun et al. 2005, Featherstone et al. 2009). We extend this work to a much more luminous 10 $M_\odot$ star with a rotation period of seven days (4 $\Omega_\odot$) which is typical for active MS B stars.

Using the 3-D ASH code, we study convection and dynamo action realized in the core and part of the surrounding radiative envelope of this 7200 $L_\odot$ B star. ASH is a mature modeling tool which solves the anelastic MHD equations of motion in a rotating spherical shell using a pseudo-spectral method (e.g. Brun et al. 2004). The mean structure in this ZAMS star is obtained from a stellar evolution code. We capture the full spherical geometry with a radial domain that occupies 0.6 R_* (covering 7 pressure scale heights), with the inner 0.2 R_* being convectively unstable. The innermost 0.02 R_* is excluded to avoid the coordinate singularity at the origin in the ASH code. The upper and lower radial boundary conditions are stress-free and impenetrable for the velocity field and perfect conductor (lower) and potential field (upper) for the magnetic field.

The intricate and time-varying flows established in this simulation are largely aligned with the rotation axis. These columnar convection cells break the spherical symmetry due to equator-crossing meridional circulations and a north-south asymmetric differential rotation. A central columnar flow (occupying the inner 0.1 R_* at the equator) stretches north-south across the entire core, rotates retrograde to the reference frame, and gently flares out to about 25° in latitude at the core boundary (Fig. 1a). Along the rotation axis within this column are strong vortical flows. Outside the central column, there are typically five columnar convection cells that rotate prograde to the reference frame. These cells transport angular momentum between the central column and the overshooting region, where there is a weak prograde equatorial flow. These flows maintain a mean rotation rate that increases monotonically from the center of the star to become nearly uniform within the radiative envelope, with an overall radial differential rotation of 25%.

A strong dynamo operates within the core, generating magnetic fields with peak strengths reaching 900 kG (200 kG rms). These fields form equatorward tilted strands

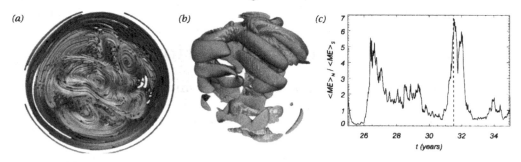

Figure 1. (a) Velocity streamlines within the core, cut along the equator. Fast core-crossing flows and several columnar flows are visible. Orange tones indicate fast flow speeds ($|v| > 300\,\mathrm{m\,s^{-1}}$, peak $1000\,\mathrm{m\,s^{-1}}$), slower speeds in blue tones. (b) An isocontour rendering of magnetic energy showing the dominance of the northern hemisphere and equatorward tilted magnetic structures (rotation axis vertical). (c) North to south ratio of hemispherical averages of ME shown for a decade of time evolution; (a) and (b) are rendered at 31.5 years (dashed line).

that encircle the core (Fig. 1b). The fluctuating component of the magnetic field comprises 76% of the total magnetic energy (ME) in the core, while 21% remains in the mean toroidal field and 3% in the mean poloidal field. On average the total ME is 55% of the convective kinetic energy, but there are intervals where it approaches 86% indicating that the ME is nearing equipartition.

The time evolution of magnetic field is complex and multi-periodic. When averaged over several decades, the ME of the northern hemisphere is 1.7 times greater than that of the southern hemisphere. There are intervals, however, when the southern hemisphere comes to dominate the magnetic energy (Fig. 1c), but only by a factor of at most 2.5. The northern hemisphere, on the other hand, dominates the ME for periods of up to two years by a factor as great as 6.7. These magnetic field configurations have quadrupolar and dipolar components that are nearly equal and opposite, which have been shown to exist when there is weak equatorial symmetry breaking (Gallet & Pétrélis 2009).

The greatest extent of convective overshooting into the stable radiative envelope occurs at mid-latitudes. The sustained overshooting pushes magnetic field and lower entropy fluid into the stable layer, making the core prolate and stochastically exciting gravity waves. The strongest magnetic fields (900 kG) and fastest flows (1 km s^{-1}) typically occur along the edge of the central column and are maximum where this column transects the core boundary. As this field is advected into the overshooting region it is combed into a large-scale toroidal field ($\sim$30 kG) by the flows in the stable region. Therefore, in this region, the velocity and magnetic fields are nearly aligned creating a force-free state.

To better understand the hemispherical dynamo state achieved within this B-star model, simulations at varying rotation rates and lower diffusivities must be run. Minimal diffusion is especially important if we are to capture the buoyant magnetic structures that likely arise from the strongest fields in these models.

This work was supported by NASA grants NNX08AI57G and NNX10AM74H and by NSF Teragrid supercomputing resource grant TG-MCA93S005.

References

Brun, A. S., Miesch, M. S., & Toomre, J. 2004, *ApJ*, 614, 1073
Brun, A. S., Browning, M. K., & Toomre, J. 2005, *ApJ*, 629, 461
Donati, J.-F. & Landstreet, J. D. 2009, *ARAA*, 47, 333
Featherstone, N. A., Browning, M. K., Brun, A. S., & Toomre, J. 2009, *ApJ*, 705, 1000
Gallet, B. & Pétrélis, F. 2009, *Phys. Rev. E*, 80, 3

Astrophysical Dynamics: From Stars to Galaxies
Proceedings IAU Symposium No. 271, 2010
N. Brummell, A. S. Brun, M. S. Miesch & Y. Ponty, eds.

© International Astronomical Union 2011
doi:10.1017/S1743921311017807

Internal wave breaking and the fate of planets around solar-type stars

Adrian J. Barker and Gordon I. Ogilvie

Department of Applied Mathematics & Theoretical Physics,
University of Cambridge, Cambridge, CB3 0WA, UK
email: ajb268@cam.ac.uk

Abstract. Internal gravity waves are excited at the interface of convection and radiation zones of a solar-type star, by the tidal forcing of a short-period planet. The fate of these waves as they approach the centre of the star depends on their amplitude. We discuss the results of numerical simulations of these waves approaching the centre of a star, and the resulting evolution of the spin of the central regions of the star and the orbit of the planet. If the waves break, we find efficient tidal dissipation, which is not present if the waves perfectly reflect from the centre. This highlights an important amplitude dependence of the (stellar) tidal quality factor Q', which has implications for the survival of planets on short-period orbits around solar-type stars, with radiative cores.

Keywords. hydrodynamics, instabilities, waves, stars: planetary systems, stars: rotation, binaries: close.

1. Introduction

Tidal interactions are important in determining the fate of short-period extrasolar planets and the spins of their host stars. The extent of the spin-orbit evolution that results from tides depends on the dissipative properties of the body. These are usually parametrized by a dimensionless quality factor Q', which is an inverse measure of the dissipation, and which in principle depends on tidal frequency, the internal structure of the body, and the amplitude of the tidal forcing. The mechanisms of tidal dissipation that contribute to Q' in fluid bodies are not well understood. We can generally decompose the response to tidal forcing into an equilibrium tide, which is a quasi-hydrostatic bulge, and a dynamical tide, which is a residual wave-like response. Dynamical tides in radiation zones of solar-type stars take the form of internal (inertia-) gravity waves (IGWs), which have frequencies below the buoyancy frequency N. These have previously been proposed to contribute to Q' for early-type stars (e.g. Zahn, 1975). We consider a nonlinear mechanism of tidal dissipation in solar-type stars, extending an idea by Goodman & Dickson (1998). A short-period planet excites IGWs at the base of the convection zone, where $N \sim 1/P$, where P is its orbital period. These waves propagate downwards into the radiation zone, until they reach the centre of the star, where they are geometrically focused and can become nonlinear. If their amplitudes are sufficient, convective overturning occurs, and the wave breaks. This has consequences for the tidal torque, and the stellar Q'. We study this mechanism, primarily using numerical simulations.

2. Numerical results and their implications

We solve the Boussinesq-type system of equations derived in Barker & Ogilvie (2010), which are valid in the central few per cent of the radius of a solar-type star, where $g \propto r$ and $N = Cr$. The parameter C measures the strength of the stable stratification at the centre, and takes a value $C_\odot \approx 8.0 \times 10^{-11} \mathrm{m}^{-1}\mathrm{s}^{-1}$, for the current Sun. A Cartesian spectral code is used to solve these equations, with our model being an initially

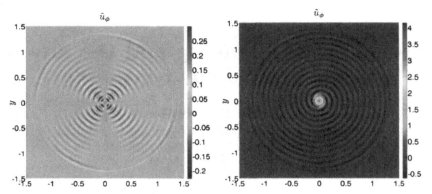

Figure 1. Plot of the normalised azimuthal velocity near the centre of a solar-type star in two simulations with small (left) and large (right) amplitude forcing.

non-rotating, 2D cylindrically symmetric star†. We artificially excite $m = 2$ waves in the outer parts of the computational domain, which are the dominant response for a planet on a circular, equatorial orbit. Fig. 1 shows the azimuthal velocity, normalised to the farfield radial phase speed of gravity waves in the simulation for a calculation with small-amplitude (left) and large-amplitude (right) forcing.

If waves are excited with small-amplitude forcing, such that they do not overturn the entropy stratification near the centre, then the waves reflect perfectly from the centre of the star, and global modes can form in the radiation zone. In this case efficient tidal dissipation only occurs at discrete resonances (Terquem *et al.*, 1998).

If the criterion

$$\left(\frac{C}{C_\odot}\right)^{\frac{5}{2}} \left(\frac{m_p}{M_J}\right) \left(\frac{M_\odot}{m_\star}\right) \left(\frac{P}{1\,\text{day}}\right)^{\frac{1}{6}} \gtrsim 3.3, \tag{2.1}$$

is satisfied, where $m_{\star,p}$ are the stellar and planetary masses, waves are excited with large enough amplitudes so that isentropes are overturned by fluid motions in the wave, within the innermost wavelength. This leads to wave breaking and the deposition of angular momentum, which spins up the mean flow to the orbital angular velocity. This results in the formation of a critical layer, at which ingoing wave angular momentum is efficiently absorbed, and the star is spun up from the inside out. This results in efficient dissipation, leading to (for waves launched where N^2 increases linearly from the interface),

$$Q'_\star \approx 1.5 \times 10^5 \left(\frac{P}{1\,\text{day}}\right)^{\frac{8}{3}}, \tag{2.2}$$

which results in a rapid and accelerating planetary inspiral, on a timescale on the order of Myr, for a one-day Jupiter-mass planet around the current Sun.

This is a potentially important nonlinear mechanism of tidal dissipation, which only operates in (solar-type) G or K stars, with radiative cores. This process requires massive planets or older/more centrally condensed stars. It is not in conflict with current observations of extrasolar planets, and may explain the absence of massive close-in planets around G-stars. This would not operate in F-type stars, with convective cores, and so its absence may partly explain the survival of massive close-in planets around F-stars, e.g. WASP-18.

References

Barker, A. J. & Ogilvie, G. I. 2010, *MNRAS*, 404, 1849

Goodman, J. & Dickson, E. S. 1998, *ApJ*, 507, 938

Terquem, C., Papaloizou, J. C. B., Nelson, R. P., & Lin, D. N. C. 1998, *ApJ*, 502, 788

Zahn, J.-P. 1975, *A&A*, 41, 329

---

† 3D simulations with a spherically symmetric background have since been performed, which confirm these results.

Astrophysical Dynamics: From Stars to Galaxies
Proceedings IAU Symposium No. 271, 2010
N.H.Brummell, A.S.Brun, M.S. Miesch & Y.Ponty, eds.
© International Astronomical Union 2011
doi:10.1017/S1743921311017819

Hunting down giant cells in deep stellar convective zones

Nicolas Bessolaz and Allan Sacha Brun

CEA Saclay, DSM/IRFU/SAp, 91191 Gif-sur-Yvette Cedex, France
email: Nicolas.Bessolaz@cea.fr

Abstract. 3D high resolution simulations for the convective zone of a 4Myr old 0.7 $M_\odot$ pre-main sequence star in gravitational contraction are carried out with different radial density contrast using the pseudo spectral ASH code (Brun *et al.* 2004). We extract giant cells signal from the complex surface convective patterns by using a wavelet analysis. We then characterize them by estimating their lifetime and rotation rate according to the density contrast.

Keywords. convection - stars : pre-main sequence, low-mass, rotation - turbulence

1. Introduction

Understanding the dynamical properties of deep convective zones is of fundamental interest to explain the evolution and redistribution of heat, energy and angular momentum transport in pre-main sequence stars. Detection of giant cells already in our Sun is very difficult (LaBonte *et al.* 1981, Lisle *et al.* 2004) since they are merged within stronger signals like granulation or supergranulation which have to be removed correctly. They further demand to substract the global differential rotation of the Sun, the limb effect and disentangle possible line shifts due to the solar magnetic field in active regions. Full sphere simulations resolving only the largest scales of turbulent convection down to supergranulation with mean flows, combined with a specific wavelet analysis (Starck *et al.* 2006) are suitable tools to probe these giant cells and study their properties.

2. Convective structures and giant cells properties

Four different models with an e-folding increasing radial density contrast from 13 to 272 are computed with a common outer edge at $R = 0.98R_*$ and a fixed surface density scale height $H_\rho = 10$ Mm. The initial stellar state is obtained from the CESAM code (Morel 1997). Typical run for these simulations lasts 3000 days, i.e. 10 convective overturning time and the rms Reynolds number in the mid layer varies between 100 and 700.

The different aspect ratio $R_*/\Delta R$ between models (from 14 to 2.2 with the increasing density contrast) has in appearance little influence on the convective patterns close to the stellar surface as shown in Figure 1 for the radial velocity. Thus, provided that the level of turbulence is sufficient, the small-scale convection patterns are really linked to the local density scale height which is common to all models and equal to 10 Mm. Hence the difficulty to extract information about the underlying giant cells from the surface. The 3D structure of the convective patterns are similar whatever the latitude is for the $\Delta\rho = 13, 37$ models whereas the $\Delta\rho = 100, 272$ models show convective cells extended through the whole convective zone and parallel to the rotation axis at high latitude i.e. convective structures modify their orientation to accomodate for the Coriolis force in these models where the Rossby number is below unity.

After having detected a large scale signal with our wavelet analysis (see Bessolaz & Brun 2011 for details), a time correlation analysis similar to the work of Miesch *et al.*

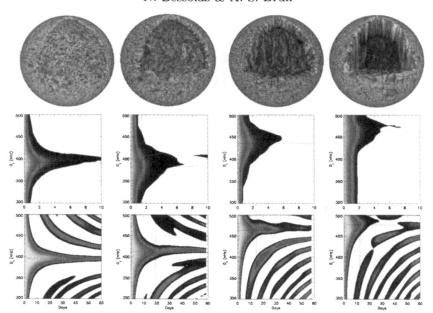

Figure 1. *Top* : 3D rendering (done with SDVision, Pomarede *et al.* 2008) of the radial velocity pattern cropped in one quadrant in the North hemisphere to highlight the convective structures in depth for the different models (by increasing the density contrast from left to right). *Bottom* : Autocorrelation function computed close to the surface ($R = 0.97R_*$) and for the latitudinal band [0-20°] according to the tracking rate for all models both for the full image (upper figures) and only the largest scale (lower figures). Notice the different much larger temporal range for the large scale analysis.

(2008) is performed both on the full convective flow and on the reconstructed largest spatial scales of the wavelet analysis. This helps to quantify the lifetime of these large scale structures detected over two stellar periods for each model (see Fig. 1). The optimal tracking velocity Ω_{opt} which maximize the autocorrelation function (acf) defines the lifetime of giant cells for a threshold of 0.5 of the acf function. For the full images including a large range of scales, the lifetime is really similar for all models around 1.5-2 days whereas large scale structures have lifetimes greater than 10 days. The lifetime of these large scale structures seems also to decrease with the density contrast from $\tau = 60$ days till $\tau = 12$ days because they undergo much distortions due to the stronger differential rotation in these high density contrast models. Giant cells are elongated in the North/South direction with a varying latitudinal extension linked to the depth of the convection zone. Finally, they have a proper tracking rate really greater than the local differential rotation and the discrepancy increases with the density contrast.

Acknowledgements : We acknowledge funding by ERC through grant STARS2 #207430.

References

Bessolaz, N. & Brun, A. S. 2011, *ApJ*, 728, 115
Brun, A. S., Miesch, M. S., & Toomre, J. 2004, *ApJ*, 614, 1073
Labonte, B. J., Howard, R., & Gilman, P. A. 1981, *ApJ*, 250, 796
Lisle, J. P., Rast, M. P., & Toomre, J. 2004, *ApJ*, 608, 1167
Miesch, M. S., Brun, A. S., DeRosa, M. L., & Toomre, J. 2008, *ApJ*, 673, 557
Morel, P. 1997, *A&AS*, 124, 597
Pomarede *et al.* 2008, Astronomical Society of the Pacific Conference Series, vol. 386, 327
Starck, J.-L., Moudden, Y., Abrial, P., & Nguyen, M. 2006, *A&A*, 446, 1191

Astrophysical Dynamics: From Stars to Galaxies
Proceedings IAU Symposium No. 271, 2010
N.H. Brummell, A.S. Brun, M. S. Miesch & Y.Ponty eds.

© International Astronomical Union 2011
doi:10.1017/S1743921311017820

The Effect of Small Scale Motion on an Essentially-Nonlinear Dynamo

Benjamin M. Byington[1], Nicholas H. Brummell[1] & Steven M. Tobias[2]

[1]Department of Applied Mathematics, University of California Santa Cruz, 1156 High St,
Santa Cruz, CA 95064
[2]Department of Applied Mathematics, University of Leeds, Leeds LS2-9JT, UK

Abstract. A dynamo is a process by which fluid motions sustain magnetic fields against dissipative effects. Dynamos occur naturally in many astrophysical systems. Theoretically, we have a much more robust understanding of the generation and maintenance of magnetic fields at the scale of the fluid motions or smaller, than that of magnetic fields at scales much larger than the local velocity. Here, via numerical simulations, we examine one example of an "essentially nonlinear" dynamo mechanism that successfully maintains magnetic field at the largest available scale (the system scale) without cascade to the resistive scale. In particular, we examine whether this new type of dynamo at the system scale is still effective in the presence of other smaller-scale dynamics (turbulence).

Keywords. magnetic fields - MHD - turbulence - dynamos.

1. Results and Discussion

The dynamo under consideration operates in an "essentially nonlinear" manner, in the sense described in a paper by Cline, Brummell & Cattaneo, ApJ, 599: 1449 - 1468, 2003. Such dynamos do not have a typical kinematic phase characterized by negligible Lorentz forces and exponentially growing eigenfunctions. Rather, they require a critical initial magnetic field strength to operate, and then the dynamo generating velocity is a function of the magnetic field itself. Figures 1a and 2a show the dynamic evolution of one realization of this type of dynamo. Here, B_x is shown as a diagnostic of dynamo activity. After some initial transience, the system relaxes into a nearly periodic steady state, with strong magnetic structures repeatedly created by the velocity shear and rising via a magnetic buoyancy mechanism. Notice that the magnetic field exists on the scale of the system here without cascading to smaller scales, even though the magnetic Reynolds number (based on the shear velocity) is of order $R_m \sim 3000$.

Here, we attempt to assess the effects of turbulence at scales much smaller than the magnetic structures on this dynamo, by applying small scale perturbations to the temperature field. Various spectral signatures and time scales have been considered, but here we will discuss only white noise (δ-correlated in space and time). The system proves to be highly sensitive to the presence of noise, in the sense that very small perturbations are sufficient to kick the solution into a different nonlinear state (Figure 1b & 2b). Temperature fluctuations of magnitude 5×10^{-7} (cf. a mean temperature of order unity) are sufficient to kick the system into a similar but more energetic and chaotic state with a higher frequency of structure creation. The new dynamics remain if the noise is subsequently turned off, indicating that the new state appears to be the more stable of the two.

This new dynamo state however is very robust to the presence of noise. Varying levels of noise were applied to the system, spanning many orders of magnitude (Figure 3), up to

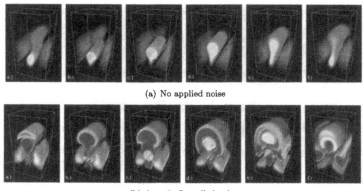

(a) No applied noise

(b) Amp 5e-7 applied noise

Figure 1. Volume Renderings of B_x over one period

levels comparable in strength to the background field. Stronger perturbations increases the energy at smaller wavelengths, and increased the level of dissipation at small scales, but even the strongest perturbations failed to disrupt dynamo action.

In conclusion, the system (at these parameters) appears to have two nonlinear states. The first is unstable, giving way to the second with even small applications of noise. However, the second state appears stable and remains even when perturbed by unphysically large amounts of noise. Of particular importance is that we have found a system-scale, essentially nonlinear dynamo that appears to be robust, operating even in the presence of strong small scale perturbations. Future work will examine other parameters and extend the model to include more realistic turbulence instead of ad hoc applications of noise.

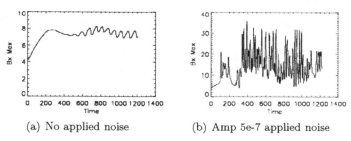

(a) No applied noise   (b) Amp 5e-7 applied noise

Figure 2. Maximal B_x for several periods

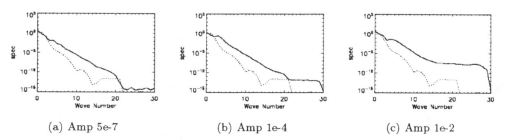

(a) Amp 5e-7   (b) Amp 1e-4   (c) Amp 1e-2

Figure 3. Spectrum of horizontal slice of B_x through a typical magnetic structure for several levels of noise. Dashed line: no applied noise.

Astrophysical Dynamics: From Stars to Galaxies
Proceedings IAU Symposium No. 271, 2010
N.H.Brummell, A.S. Brun, M.S. Miesch & Y.Ponty, eds.
© International Astronomical Union 2011
doi:10.1017/S1743921311017832

Influence of Magnetic Helicity in MHD

Simon Candelaresi, Fabio Del Sordo and Axel Brandenburg

NORDITA, AlbaNova University Center, Roslagstullsbacken 23, SE-10691 Stockholm, Sweden
and
Department of Astronomy, Stockholm University, SE 10691 Stockholm, Sweden

Abstract. Observations have shown that the Sun's magnetic field has helical structures. The helicity content in magnetic field configurations is a crucial constraint on the dynamical evolution of the system. Since helicity is connected with the number of links we investigate configurations with interlocked magnetic flux rings and one with unlinked rings. It turns out that it is not the linking of the tubes which affects the magnetic field decay, but the content of magnetic helicity.

Keywords. Sun: magnetic fields

Magnetograms of the Sun's surface have shown (Pevstov et al. 1995) that the field lines in the active regions are twisted. From observations of the chromosphere and the corona it was conjectured (Leka et al. 1996) that the magnetic field in sunspots gets twisted before it emerges. Extrapolation of the three-dimensional structure of magnetic field lines have shown twist in the field (Gibson et al. 2002). Since twisting is connected to helicity we can say that the Sun's magnetic field shows helical patches.

For two flux tubes which are not twisted and do not intersect each other the magnetic helicity is related to the number of mutual linking via the formula (Moffatt 1969)

$$H = \int_V \mathbf{A} \cdot \mathbf{B} \, dV = 2n\phi_1\phi_2,$$

where H is the magnetic helicity, $\mathbf{B} = \nabla \times \mathbf{A}$ is the magnetic field in terms of the vector potential $\mathbf{A}$, ϕ_1 and ϕ_2 are the magnetic fluxes through the tubes and n is the linking number. Since H is a conserved quantity in ideal MHD the linking number is also conserved.

In ideal MHD and in presence of magnetic helicity the spectral magnetic energy is bounded from below through the realizability condition (Moffatt 1969)

$$M(k) \geqslant k|H(k)|/2\mu_0 \quad \text{with} \quad \int M(k) \, dk = \langle \mathbf{B}^2 \rangle/2\mu_0, \quad \int H(k) \, dk = \langle \mathbf{A} \cdot \mathbf{B} \rangle$$

and the magnetic permeability μ_0 and where $\langle . \rangle$ denotes volume integrals. It is the aim of our work (Del Sordo et al. 2010) to study the dynamical evolution of a system with interlocked flux rings with and without helicity.

The setups under consideration consist of three flux tubes with constant magnetic flux. For two configurations the rings are interlocked while for one they are separated. In one of the interlocked configurations we change the direction of the flux such that the magnetic helicity becomes zero (Fig. 1). The inner ring has a radius which is 1.2 times larges then for the outer rings. We solve the full resistive MHD equations for an isothermal and incompressible medium.

For the interlocked configuration with finite magnetic helicity the magnetic energy decays with a power law of $t^{-1/2}$, while both non-helical configurations decay much faster with a power law of $t^{-3/2}$ (Fig. 2).

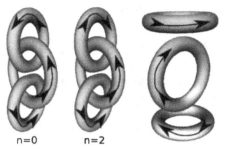

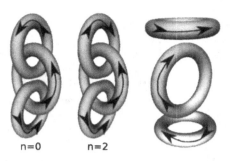

Figure 1. The initial triple ring configuration. Interlocked rings with $n = 0$ (left) and $n = 2$ (center) and non interlocked configuration with $n = 0$ (right). The arrows indicate the direction of the magnetic field.

Figure 2. Magnetic energy decay normalize for the initial time for linking number 2 (soli line), 0 (dashed line) and the non-interlocke case (dotted line).

One can see (Fig. 3) that the linked structure is conserved for the finite helicity case. For even later times the linking gets transformed into twisting.

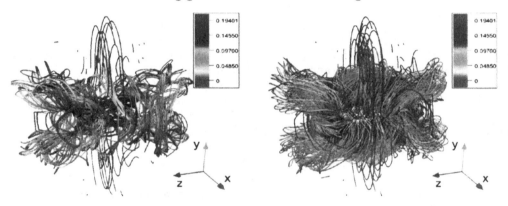

Figure 3. Magnetic field configuration after 4 Alfvén times for the case of linking number zero (left) and finite linking number (right). The colors represent the magnitude of the magnetic field.

Due to the realizability condition the decay of magnetic energy is slowed down by the magnetic helicity which decays slowly. Since helicity is decaying slowly and is almost conserved also the number of linkings is almost conserved. This is why the linking is transformed into twisting of the fields which then contributes to the helicity. The test run with non-interlocked rings shows that it is the helicity content of the system and not the actual number of linkings which affects the decay.

References

Moffatt, H. K., J. Fluid Mech. **35**, 117 (1969).
Pevtsov, A. A., Canfield, R. C. & Metcalf, T. R. 1995, ApJ, 440, L109
Leka, K. D., Canfield, R. C., McClymont, A. N., van Driel-Gesztelyi, L., ApJ **462**, 547 1996
Gibson, S. E., Fletcher, L., Del Zanna, G., et al., ApJ **574**, 1021 2002
Del Sordo, F., Candelaresi, S., Brandenburg, A., Phys. Rev. E **81**, 036401 2010

Astrophysical Dynamics: From Stars to Galaxies
Proceedings IAU Symposium No. 271, 2010
N.H.Brummell, A.S.Brun, M.S. Miesch & Y.Ponty, eds.
© International Astronomical Union 2011
doi:10.1017/S1743921311017844

Large resistivity in numerical simulations of radially self-similar outflows

Miljenko Čemeljić[1], Nektarios Vlahakis[2] and Kanaris Tsinganos[2]

[1]Academia Sinica Institute of Astronomy and Astrophysics and Theoretical Institute for Advanced Research in Astrophysics, P.O. Box 23-141, Taipei, Taiwan
email: miki@tiara.sinica.edu.tw

[2]IASA and Section of Astrophysics, Astronomy and Mechanics, Department of Physics, University of Atens, Panepistemiopolis 15784, Zografos, Athens, Greece
email: vlahakis,tsingan@phys.uoa.gr

Abstract. We investigate conditions in a radially self-similar outflow in the regime of large resistivity. Using the PLUTO code, we performed simulations with proper choice of boundary conditions, relaxed at the footpoints of critical surfaces in the flow. We investigate outflow propagation in a high-resistive disk corona, and compare it to the results with small or vanishing resistivity.

Keywords. magnetic fields (magnetohydrodynamics): MHD, methods: numerical, stars: winds, outflows.

1. Initial and boundary conditions

General classes of self-consistent ideal-MHD solutions have been constructed in Vlahakis & Tsinganos (1998). In Gracia *et al.* 2006 ideal-MHD solutions with modified

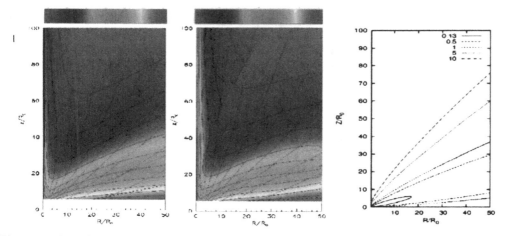

Figure 1. Initial state in our NIRVANA simulations for ideal-MHD and for simulations with not too large resistivity is shown in the leftmost panel. Logarithmic color grading and solid lines represent density, magnetic field lines are shown in dashed lines, and critical surfaces are shown in dotted lines. The density, magnetic field and velocity are slightly modified from the analytical solutions for numerical reasons. In the middle panel is shown the final, stationary state in low resistivity simulations, which does not differ significantly from the initial state. In the rightmost panel is shown the value of $0.5\beta(V R/V_0 R_0)$ for the analytical solution, which gives critical value of resistivity that corresponds to Rb. We introduced a new number, describing influence of the magnetic diffusivity η on the energy transport, which can be written in terms of Rm and plasma beta as Rb=Rmβ/2.

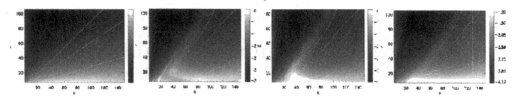

Figure 2. To avoid influence of the outer boundary, we triple the computational domain in R direction, and results are taken before the reflected wave can return from the outer R-boundary to our domain of interest, RxZ=50x100. We show results for large resistivity, one order of magnitude above the critical resistivity found in previous study. Boundary conditions are not over-specified and solutions are different for different b.c. relaxed, what could infer on the outcome of simulations, and we study it closer. Here we show the initial density and critical surfaces, and results after the relaxation for Vx, Vy, Bz boundary conditions relaxed, left to right respectively, for $\eta=1.5$.

analytical initial conditions were obtained, and followed with extension in the resistive-MHD in Cemeljic *et al.* 2008-both using the NIRVANA code (version 2.0, Ziegler (1998)).

Solutions with small and moderate resistivity show smooth departure from the ideal-MHD solutions as shown in Fig. 1, up to critical resistivity, for which solutions changed abruptly — see Fig. 2. This critical resistivity was 10^3 times larger than numerical resistivity in simulations. Here we investigate the case with large resistivity, solving the resistive-MHD equations with the diffusive terms added in the induction and energy equations, by the PLUTO code by Mignone *et al.* 2007. Anomalous resistivity often introduced in simulations is usually few orders of magnitude larger than microscopic resistivity, and it is instructive to investigate how a code treats such problem.

2. Results

We show, for the first time, results of resistive simulations above the critical resistivity found in our previous study, in comparison with the analytical solution of the closely related ideal-MHD problem. Understanding those results, and disentangling the physical and numerical causes for differences, is essential for understanding the simulations of launching and propagation of resistive astrophysical jets.

References

Čemeljić, M., Gracia, J., Vlahakis, N., & Tsinganos, K. 2008, *MNRAS*, 389, 1022
Gracia, J., Vlahakis, N., & Tsinganos, K. 2006, *MNRAS*, 367, 201
Mignone, A., Bodo, G., Massaglia, S., Matsakos, T., Tesileanu, O., Zanni, C., & Ferrari, A. 2007, *AP&SS*, 287, 129
Vlahakis, N. & Tsinganos, K. 1998, *MNRAS*, 298, 777
Ziegler, U. 1998, *Co.Ph.Com*, 109, 111

Astrophysical Dynamics: From Stars to Galaxies
Proceedings IAU Symposium No. 271, 2010
N.H.Brummell, A.S.Brun, M.S. Miesch & Y.Ponty, eds.

The Effect of Scattering on the Temperature Stratification of 3D Model Atmospheres of Metal-Poor Red Giants

Remo Collet[1], Wolfgang Hayek[2,1,3], and Martin Asplund[1]

[1] Max Planck Institut für Astrophysik, Karl-Schwarzschild-Str. 1, D–85741 Garching, Germany
email: `remo,hayek,asplund@mpa-garching.mpg.de`

[2] Research School of Astronomy & Astrophysics, Cotter Road, Weston ACT 2611, Australia

[3] Institute for Theoretical Astrophysics, Sem Sælands Vei, N–0315 Oslo, Norway

Abstract. We study the effects of different approximations of scattering in 3D radiation-hydrodynamics simulations on the photospheric temperature stratification of metal-poor red giant stars. We find that assuming a Planckian source function and neglecting the contribution of scattering to extinction in optically thin layers provides a good approximation of the effects of coherent scattering on the photospheric temperature balance.

Keywords. stars: atmospheres, convection, hydrodynamics, radiative transfer, scattering, methods: numerical

1. Introduction

Three-dimensional model atmospheres of metal-poor late-type stars predict lower surface temperatures than their 1D counterparts. This leads to important differences between 3D and 1D spectroscopic analyses whenever temperature-sensitive features are used to determine elemental abundances (Asplund *et al.* 1999; Asplund & García Pérez 2001; Collet *et al.* 2006, 2007). 3D models by various groups agree qualitatively, but not quantitatively on how cool the upper atmospheric layers in metal-poor stars actually are (Bonifacio *et al.* 2009; Bonifacio 2010). It has been suggested that such discrepancies and, in particular, the very low photospheric temperatures may be due to the approximated treatment of scattering in the radiative transfer. In this study, we explore the role of scattering on the predicted surface temperatures in 3D model atmospheres of metal-poor red giants.

2. Method

We use the 3D radiation-hydrodynamics code BIFROST (Gudiksen *et al.*, in prep.) to construct surface convection simulations of a red giant star ($T_{\rm eff} = 5100$ K, $\log g\,[{\rm g\,cm^{-2}}]=$ 2.2, [Fe/H]$= -3$). We account for the effects of radiative cooling by solving the radiative transfer equation at each time-step. For the treatment of scattering, we consider three cases: (i) no contribution of continuum scattering to extinction in optically thin layers (Asplund *et al.* 1999; Collet *et al.* 2007); (ii) scattering treated as pure absorption everywhere; (iii) proper iterative solution of the radiative transfer equation in the case of a source function with coherent scattering term (Hayek *et al.* 2010).

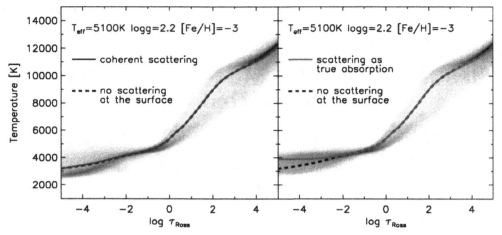

Figure 1. *Left panel, grey shaded area*: temperature distribution as a function of Rosseland optical depth in a snapshot of the 3D metal-poor red giant surface convection simulation computed for a source function with coherent scattering term. *Continuous blue line*: corresponding mean temperature stratification. *Right panel, grey shaded area*: temperature distribution in a snapshot of the 3D simulation assuming a Planckian source function and scattering as pure absorption. *Dashed lines, both panels*: mean temperature stratification for the simulation assuming a Planckian source function for the radiative transfer and no contribution of scattering to extinction in optically thin layers.

3. Results

The left panel of Fig. 1 shows the temperature distribution with optical depth in a 3D model metal-poor red giant atmosphere constructed by solving the radiative transfer equation in the case of a source function with coherent scattering term. The predicted temperature stratification is in very good agreement with the one obtained by simply assuming a Planckian source function and neglecting the contribution of scattering to the total extinction at the surface. Conversely, treating scattering as true absorption everywhere in the simulation leads to a significantly hotter temperature stratification.

To summarize, we have verified that, at least for 3D metal-poor red giant surface convection simulations, the assumption of a Planckian source function and of no scattering extinction in optically thin layers well reproduces the effects of coherent scattering on the photospheric temperature stratification. Consequently, we exclude that the cool temperature stratifications predicted by the 3D metal-poor red giant simulations of Collet *et al.* (2007) may be due to their approximated treatment of scattering.

References

Asplund, M., Nordlund, Å., Trampedach, R., & Stein, R. F. 1999, *A&A*, 346, L17

Asplund, M. & García Pérez, A. E. 2001, *A&A*, 372, 601

Bonifacio, P. 2010, in: K. Cunha, M. Spite, & B. Barbuy (eds.), *Chemical Abundances in the Universe: Connecting First Stars to Planets*, Proc. IAU Symposium No. 265 (Rio de Janeiro), p. 81

Bonifacio, P., Spite, M., Cayrel, R., *et al.* 2009, *A&A*, 501, 519

Collet, R., Asplund, M., & Trampedach, R. 2006, *ApJ (Letters)*, 644, L121

Collet, R., Asplund, M., & Trampedach, R. 2007, *A&A*, 469, 687

Hayek, W., Asplund, M., Carlsson, R., *et al.* 2010, *A&A*, in press

Astrophysical Dynamics: From Stars to Galaxies
Proceedings IAU Symposium No. 271, 2010
N.H.Brummell, A.S.Brun, M.S. Miesch & Y.Ponty, eds.

Vorticity from irrotationally forced flow

Fabio Del Sordo and Axel Brandenburg

NORDITA, Roslagstullsbacken 23, SE-10691 Stockholm, Sweden; and
Department of Astronomy, Stockholm University, SE 10691 Stockholm, Sweden

Abstract. In the interstellar medium the turbulence is believed to be forced mostly through supernova explosions. In a first approximation these flows can be written as a gradient of a potential being thus devoid of vorticity. There are several mechanisms that could lead to vorticity generation, like viscosity and baroclinic terms, rotation, shear and magnetic fields, but it is not clear how effective they are, neither is it clear whether the vorticity is essential in determining the turbulent diffusion acting in the ISM. Here we present a study of the role of rotation, shear and baroclinicity in the generation of vorticity in the ISM.

Keywords. Galaxies: magnetic fields – ISM: bubbles

The study of the interstellar medium (ISM) is strictly connected with that of supernovae explosions, being this one of the most important phenomena in determining its dynamic. Such events act on scales up to ~ 100 pc injecting enough energy to sustain turbulence with velocities of ~ 10 km/s. An understanding of the generation of turbulence and vorticity in the ISM is a basic requirement in order to formulate a theory for the production of interstellar magnetic fields. The first step to perform numerical simulations of these explosion is to assume the presence of pure potential forcing. Indeed a supernova explosion can be seen at first glance as a pure spherical expansion that is driving turbulence in the ISM. In our simulations we use the PENCIL CODE, http://pencil-code.googlecode.com/ We simulate spherical expansions following the work of Mee & Brandenburg (2006). We analyze flows that are only weakly supersonic and use a constant and uniform viscosity in an unstratified medium. We solve the Navier-Stokes equations in the presence of viscosity and with a potential forcing $\nabla \phi$ where ϕ is given by

$$\phi(\boldsymbol{x}, t) = \phi_0 \, N \exp \left\{ [\boldsymbol{x} - \boldsymbol{x}_{\mathrm{f}}(t)]^2 / R^2 \right\}. \qquad (0.1)$$

Here $\boldsymbol{x} = (x, y, z)$ is the position vector, $\boldsymbol{x}_{\mathrm{f}}(t)$ is the random forcing position, R is the radius of the Gaussian, and N is a normalization factor. We consider two forms for the time dependence of $\boldsymbol{x}_{\mathrm{f}}$. First, we take $\boldsymbol{x}_{\mathrm{f}}$ such that the forcing is δ-correlated in time. Second, we include a forcing time $\delta t_{\mathrm{force}}$ that defines the interval during which $\boldsymbol{x}_{\mathrm{f}}$ remains constant. We study then the effect of rotation on such environment. Rotation causes the action of the Coriolis force, $2\boldsymbol{\Omega} \times \boldsymbol{u}$, in the evolution equation for the velocity. We investigate flows with Reynolds numbers up to 150. We find that there is a clear production of small scale vorticity when we use both a finite and δ-correlated forcing. In this last case we observe some spurious vorticity also at very low Coriolis number, due probably to some numerical artifact. We then consider the effects of shear on the potential flow. In the presence of linear shear with $\boldsymbol{u}^S = (0, Sx, 0)$, the evolution equation for the departure from the mean shear attains additional terms, $-\boldsymbol{u}^S \boldsymbol{\nabla} \cdot \boldsymbol{u} - \boldsymbol{u} \cdot \boldsymbol{\nabla} \boldsymbol{u}^S$. Also in this case we observe the production of vorticity but we do not find clear results for small values of the shear. We have thus observed a production of vorticity due to rotation and shear, but these effects are not able to produce enough vorticity under the physical condition of the ISM, like those described by Beck et al. (1996). When we

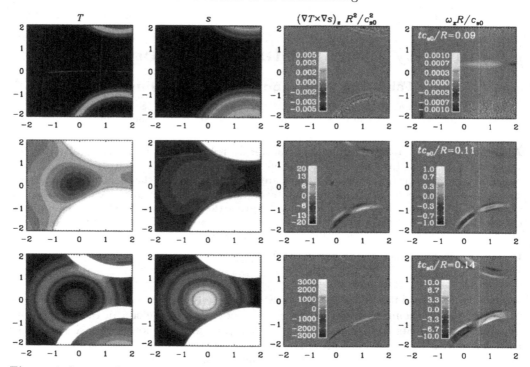

Figure 1. Images of temperature, entropy, baroclinic term ($\nabla T \times \nabla s)_z$ and normalized vertical vorticity for a two-dimensional run with $\delta t_{\text{force}} c_{s0}/R = 0.1$ at an instant shortly before the second expansion wave is launched (top row), and shortly after the second expansion wave is launched (second and third row). In the second and third row the vorticity production from the baroclinic term is clear, while in the top row, $(\nabla T \times \nabla s)_z$ and ω_z are just at the noise level of the calculation. Shock surfaces are well localized and the zones of maximum production of vorticity are those in which the fronts encounter each other. Adapted from Del Sordo & Brandenburg (2010).

relax the isothermal condition the system is then also under the action of a baroclinic term. This is proportional to the cross product of the gradients of pressure and density and emerges when taking the curl of the pressure gradient term, $\rho^{-1}\nabla p$. The results we obtain are shown in Figure 1.

It turns then out that the baroclinic term is more efficient in the production of the vorticity. Moreover the biggest amount of vorticity is observed when shock fronts encounter each other.

Speaking about possible dynamo action, as pointed out by Brandenburg & Del Sordo (2009), the presence of vorticity does not seem to affect the diffusion of magnetic fields differently than a complete irrotational turbulence. Nevertheless the vorticity plays an important role in dynamo processes so it is still important to address the problem of the generation of the vorticity investigating the possible role of other effects.

References

Beck, R., Brandenburg, A., Moss, D., Shukurov, A., & Sokoloff, D. 1996, *ARA&A*, 34, 155
Brandenburg, A., & Del Sordo, F. 2009, in Turbulent diffusion and galactic magnetism, ed. Highlights of Astronomy, Vol. **15** (E. de Gouveia Dal Pino), (in press) CUP, arXiv:0910.0072
Del Sordo, F. & Brandenburg, A. 2010, *A&A*, submitted, arXiv:1008:5281
Mee, A. J. & Brandenburg, A. 2006, *MNRAS*, 370, 415

Astrophysical Dynamics: From Stars to Galaxies
Proceedings IAU Symposium No. 271, 2010
N.H.Brummell, A.S.Brun, M.S. Miesch & Y.Ponty, eds.
© International Astronomical Union 2011
doi:10.1017/S174392131101787X

Multi-season photometry of the newly-discovered roAp star HD75445

C. A. Engelbrecht[1], F. A. M. Frescura[2], B. S. Frank[2] and P. R. Nicol[1]

[1] Department of Physics, University of Johannesburg, P.O. Box 524, Auckland Park, Johannesburg 2006, South Africa
email: chrise@uj.ac.za

[2] School of Physics, University of the Witwatersrand, Private Bag 3, WITS, Johannesburg 2050, South Africa
email: fabio.frescura@wits.ac.za

Abstract. HD75445 was recently announced by Kochukhov *et al.* (2009) to be a low-amplitude roAp star, based on spectroscopic measurements. We present putative pulsation frequencies of HD75445 determined from 22 hours of Johnson B photometry obtained in 2008, 2009 and 2010. We present the first photometric periodicities detected in this star. We make a marginal detection of one of Kochukhov *et al.* (2009)'s spectroscopic periods, along with a range of confidently detected periodicities covering the low-frequency end of the roAp instability spectrum and the high-frequency end of the Delta Scuti instability spectrum.

Keywords. stars: variables: roAp, delta Scuti, stars: HD75445

1. Introduction

Kochukhov *et al.* (2009) reported HD75445 as a low-amplitude roAp star, based on high-resolution spectroscopy including Nd II and Nd III lines. They reported periods of 9.20, 9.01 and 8.37 minutes respectively, all present at extremely low amplitudes. They also determined an effective temperature of 7700 K for HD75445, while Kochukhov & Bagnulo (2006) determined $\log (L/L_{sun}) = 1.17$ and $(M/M_{sun}) = 1.81$ for HD75445. These numbers imply that HD75445 just falls within the hotter half of detected roAp stars (see, for example, Théado *et al.* (2009)) and is expected to display pulsation frequencies at the higher end of the range of pulsation frequencies seen in roAp stars (see, for example, Dupret *et al.* (2008)). We detect only one high-frequency roAp mode (and this is only a marginal detection), accompanied by a set of much stronger lower-frequency modes extending all the way into the Delta Scuti domain.

2. Observations and Analysis

We used PMT photometry to observe HD75445 in the Johnson B filter for between 90 minutes and 320 minutes on each of 4 nights in February 2010, 3 nights in January 2009 and one night in January 2008, using the 0.5 m telescope at the Sutherland station of the South African Astronomical Observatory. We marginally confirm one of the spectroscopically determined periods reported in Kochukhov *et al.* (2009), at approximately 8.6 minutes. However, we also find strong periodicities ranging from 18 to 23.5 minutes (none between 8.6 and 18 minutes), i.e. populating the lower-frequency end of the currently recognised roAp pulsation spectrum. The subset of these detections that are common to all three observing seasons appear in Table 1. Moreover, we detect strong

Table 1. Five periodicities common to all three seasons (2008, 2009 and 2010) of our photometry of HD75445, plus the 8.6 minute period only detected in the 2009 dataset.

Number	Freq.(μHz)	Period (min.)	Amp. (mmag)	SNR
f_1	411	40.5	0.5	15
f_2	438	38.0	0.8	11
f_3	708	23.5	0.7	11
f_4	756	22.1	0.7	14
f_5	922	18.1	0.4	5
f_6	1950	8.6	0.4	4

Notes:
[1] The signal-to-noise ratio (SNR) was determined using the Period04 package described by Lenz & Breger (2005). The detection of the 18.1 minute period is a bit questionable in the 2008 data, where a strong peak is found at 15 ± 1 minutes, rather than 18. The detected Delta Scuti-type perodicities longer than 45 minutes only appear in the 2009 data and are not shown in this table.

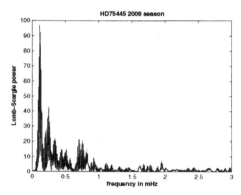

Figure 1. Lomb-Scargle periodogram of 2009 data.

signals at the short end of the Delta Scuti instability spectrum (six firmly established periodicities, ranging from 32 to 146 minutes) in the 2009 dataset. Two of these appear in all three observing seasons and are also listed in Table 1. The appearance of roAp and Delta Scuti pulsations in the same star was first reported only very recently (Balona *et al.* (2010)), using satellite data. Our results present the first such detection using ground-based photometry. Figure 1 displays the Lomb-Scargle periodogram for the observations obtained in 2009 (representing 12.4 hours of photometry in total, i.e. more than half the total duration of the multi-season photometry). The strength of the detected Delta Scuti-type signals is clearly seen.

Acknowledgements: [1] We thank the South African Astronomical Observatory for generous amounts of observing time and the South African National Research Foundation and the University of Johannesburg for financial support. CAE thanks Chris Koen for valuable comments and Peter Martinez for suggesting HD75445 as an observing target.

References

Balona, L. A., and 11 co-authors 2010, arXiv:1006.4013v1
Dupret, M.-A., Théado, S., & Noels, A. 2008, *J. Phys. Conf. Ser.*, 118, 012052
Kochukhov, O., Bagnulo, S., Lo Curto, G., & Ryabchikova, T. 2009, *A & A*, 493, L45-48
Kochukhov, O. & Bagnulo, S. 2006, *A & A*, 450, 763
Lenz, P. & Breger, M. 2005, *CoAst*, 146, 53
Théado, S., Dupret, M.-A., Noels, A. & Ferguson, J.W. 2009 *A& A*, 493, 159

Astrophysical Dynamics: From Stars to Galaxies
Proceedings IAU Symposium No. 271, 2010
N.H.Brummell, A.S.Brun, M.S. Miesch & Y.Ponty, eds.

© International Astronomical Union 2011
doi:10.1017/S1743921311017881

The Density and the Velocity Power Spectrum in Thermally Bi-stable Flows

Adriana Gazol[1], and Jongsoo Kim[2]

[1]Centro de Radioastronomía y Astrofísica UNAM,
A. P. 3-72, c.p. 58089, Morelia, Michoacán, México
email: a.gazol@crya.unam.mx

[2]Korea Astronomy and Space Science Institute,
61-1, Hwaam-Dong, Yuseong-Ku, Daejeon 305-348, Korea
email: jskim@kasi.re.kr

Abstract. We present numerical results concerning the behavior of the density and the velocity power spectrum in turbulent thermally bistable flows for different Mach numbers.

Keywords. turbulence, instabilities, ISM: kinematics and dynamics, ISM: structure

---

The turbulent nature of the interstellar medium (ISM) is believed to play a crucial role on the determination of its density and velocity structure (e.g. Elmegreen & Scalo 2004; Scalo Elmegreen (2004)). In the atomic interstellar gas, turbulence coexists with the isobaric mode of thermal instability (TI) (Field 1965), which do also play a crucial role on the determination of the structure and the dynamics of this kind of gas. The density and the velocity spectra of the atomic gas are thus the result of the interplay between these two physical ingredients.

In this paper we present results from a numerical study on the behavior of the density and the velocity power spectra in thermally bistable flows. We have analyzed a set of five 3D simulations where turbulence is randomly driven in Fourier space at a fixed wavenumber and with different Mach numbers M (with respect to the gas at 10K) ranging from 0.2 to 4.5. For details concerning the numerical model see (Gazol & Kim 2010).

The density power spectrum becomes shallower as M increases (see fig. 1). This behavior is the same as the one reported for isothermal simulations (Kim & Ryu 2005), however the slope values we obtain are much shallower than those obtained for the isothermal case. This fact and the trend we observe are interpreted as a consequence of the simultaneous turbulent compressions, thermal instability generated density fluctuations, and the weakening of thermal pressure force in diffuse gas. This behavior is consistent with the fact that observationally determined spectra exhibit different slopes in different regions. The values of the spectral indexes resulting from our simulations are consistent with observational values. For a detailed discussion on the interpretation as well as on comparison with observations, see (Gazol & Kim 2010).

We do also explore the behavior of the velocity power spectrum, which becomes steeper as M increases (see figure 2), and which for low values of M is also shallower than the one obtained for isothermal simulations. The spectral index goes from a value much shallower than the Kolmogorov one for $M = 0.2$ to a value steeper than the Kolmogorov one for $M = 4.5$. The flattening of the spectrum for low M could be due to the increased relevance of the velocities generated by the development of TI, but a detailed analysis of the velocity power spectrum for thermally bistable flows is needed. For a discussion on this issue see (Gazol & Kim 2010).

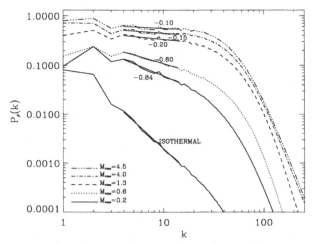

Figure 1. Density power spectra P_ρ for simulations with different M. For $M = 0.2$ the spectra resulting from an isothermal run is also displayed. Solid lines represent least-squares fits over the range $4 \leqslant k \leqslant 14$ and numbers indicate the slopes of these fits.

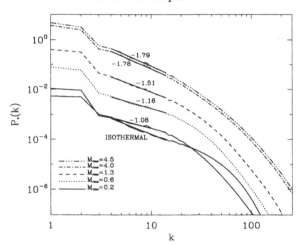

Figure 2. Velocity power spectra P_v for simulations with different M. For $M = 0.2$ the spectra resulting from an isothermal run is also displayed. Solid lines represent least-squares fits over the range $4 \leqslant k \leqslant 14$ and numbers indicate the slopes of these fits.

In conclusion, the presence of TI does significantly affect the statistics of turbulence in thermally bi-stable flows. Different turbulence conditions, in particular different Mach numbers, can lead to considerably different slopes of the density and the velocity power spectrum of the atomic interstellar gas.

References

Elmegreen, B. G., Scalo J. 2004, *ARAA*, 42, 211
Field, G. B. 1965, *ApJ*, 142, 531
Gazol, A. & Kim, J. 2010 *ApJ*, 630, 911
Kim, J. & Ryu, D. 2005, *ApJ*, 630, L45
Scalo, J. & Elmegreen, B. G. 2004, *ARAA*, 42, 275

Astrophysical Dynamics: From Stars to Galaxies
Proceedings IAU Symposium No. 271, 2010
N.H.Brummell, A.S.Brun, M.S. Miesch & Y.Ponty, eds.

© International Astronomical Union 2011
doi:10.1017/S1743921311017893

The star capture model for fueling quasar accretion disks

Gareth F. Kennedy[1], Jordi Miralda-Escudé[2,1] & Juna A. Kollmeier[3]

[1]Institut de Ciències del Cosmos, University of Barcelona
[2]Institució Catalana de Recerca i Estudis Avançats
[3]Carnegie Observatories, Pasadena

Abstract. Although the powering mechanism for quasars is now widely recognized to be the accretion of matter in a geometrically thin disk, the transport of matter to the inner region of the disk where luminosity is emitted remains an unsolved question. Miralda-Escudé & Kollmeier (2005) proposed a model whereby quasars are fuelled when stars are captured by the accretion disk as they plunge through the gas. Such plunging stars can then be destroyed and deliver their mass to the accretion disk.

Here we present the first detailed calculations for the capture of stars originating far from the accretion disk near the zone of influence of the central black hole. In particular we examine the effect of adding a perturbing mass to a fixed stellar cusp potential on bringing stars into the accretion disk where they can be captured. The work presented here will be discussed in detail in an upcoming publication Kennedy *et al.* (2010).

Keywords. galaxies: formation, galaxies: kinematics and dynamics, galaxies: nuclei, quasars: general

1. Model and results

The model consists of a stellar cluster and an accretion disk both centered on a $10^8 M_\odot$ massive black hole (MBH). To examine the effect of the disk on the orbits of stars at the zone of influence (ZOI) of the MBH all stars are initially set up to have the same binding energy as a star on a circular orbit at the zone of influence (E_{ZOI}). A spherical cusp potential of the form of a harmonic oscillator is added with the frequency chosen to give the same zone of influence for the MBH as a stellar velocity dispersion of 200 km/s.

The disk model is an accretion α-disk as described in Goodman (2003) modified at $r = 500$ Schwarzschild radii (R_S) to fall off as r^{-2} to limit the disk mass to approximately $2 \times 10^6 M_\odot$ inside R_{ZOI}. For comparison this region contains $10^8 M_\odot$ of stars, therefore the disk mass can be neglected when integrating particle orbits.

The interaction between the stars and the disk is modelled by a single velocity impulse when the particles cross the plane of the disk, at radial distance r_d, of the form given in Miralda-Escudé & Kollmeier (2005).

To model the particle motions we developed a special purpose parallel code that uses the Bulirsch-Stoer method for the numerical integration of the orbits. This code has more than sufficient accuracy to ensure that stars cannot migrate into the loss cone of the disk by numerical errors.

The possible final states for a star are that it is absorbed into the disk, it is directly captured by the MBH, or it remains in an uncaptured orbit. Stars absorbed into the disk are assumed to eventually be destroyed and their material accreted onto the MBH. The absorption of a star is said to occur when the orbital angular momentum of the star being aligned to the disk while the apocentre is within the disk. A capture occurs whenever the star passes within $4R_S$ of the MBH (the condition for capture in a Schwarzschild black

(a) Without perturbing mass   (b) With perturbing mass

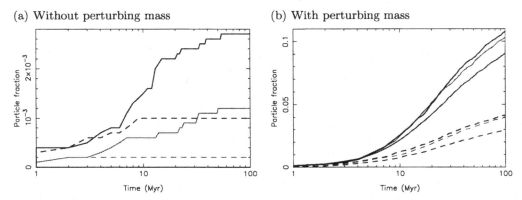

Figure 1. Left panel shows the fraction of particles captured by the MBH directly (*dashed*) and via absorption by the disk (*solid*) for the three stellar cusp potentials. Right panel shows the fractions for the same initial conditions but with a perturbing mass of $10^6 M_\odot$ at the zone of influence.

hole when the Newtonian orbit is extrapolated, in the limit of high eccentricity). This is often due to the relative inclination at the first plunge being high which leads to the star being brought close to the MBH before the angular momentum has time to align with the disk.

A sample of results from Kennedy *et al.* (2010) is shown in Figure 1. For the cases presented here 10^4 test particles were integrated, all of them with a binding energy equivalent to a circular orbit at the zone of influence for the MBH in the assumed stellar cusp. We explore three models for the gravitational potential of the stellar cusp: spherically symmetric (*red*), oblate (*green*) and triaxial (*blue*). Panel (a) shows the fraction of stars absorbed into the disk (*solid lines*) or directly captured by the MBH (*dashed lines*). The case with no perturbing masses added to the fixed potential is shown in panel (a), while panel (b) shows the effect of adding a $10^6 M_\odot$ point mass on a circular orbit at the zone of influence in the plane of the gas disk. Similar results are also achieved for perturbing masses on inclined and/or elliptical orbits.

2. Conclusions and Implications

We find that the addition of a $10^6 M_\odot$ perturbing mass greatly enhances the rate at which the loss cone of the disk is replenished. In fact the case with the perturber shown in Figure 1 (b) is very close to the most efficient full loss cone case (not shown). In addition, approximately twice as many stars that are brought into the loss cone of the disk will end up being absorbed into the disk rather than directly captured by the MBH independent of the presence of a perturbing mass. The mass transferred into the disk from the stars is expected to be higher still since we have neglected mass loss of the stars due to stellar winds or stripping of the envelope as they pass through the disk. In summary we find this method very promising to bring material into the accretion disk by maintaining a full loss cone for absorption of stars in the accretion disk. The full details of this work will appear in an upcoming publication Kennedy *et al.* (2010).

References

Goodman, J 2003 *MNRAS*, 339, 937
Kennedy, G. F., Miralda-Escudé, J & Kollmeier, J 2010, *in preparation*
Miralda-Escudé, J. & Kollmeier, J. A. (2005) *ApJ*, 619, 30

Astrophysical Dynamics: From Stars to Galaxies
Proceedings IAU Symposium No. 271, 2010
N.H.Brummell, A.S.Brun, M.S. Miesch & Y.Ponty, eds.

Particle dynamics in the stellar magnetosphere by gravitational collapse

Volodymyr Kryvdyk[1]

National Taras Shevchenko University of Kyiv, 64, Volodymyrs'ka St.,01601 Kyiv, Ukraine
email: kryvdyk@univ.kiev.ua

Abstract. The particle dynamics and in the stellar magnetosphere during gravitational collapse is investigated. The formations of relativistic jets and the generation of the radiation bursts in the stellar magnetosphere by gravitational collapse are considered. As follows from results, the stars on the stage of gravitational collapse must be powerful sources of the relativistic jets and the non-thermal radiation. These jets will formed in the polar caps of collapsing stars magnetospheres, when the stellar magnetic field increases during collapse and the charged particles will be accelerate. These jets will generate the non-thermal radiation. The radiation flux grows with decreasing stellar radius and can be observed in the form of radiation burst in wide band wave- from radio to gamma-ray. These bursts radiation can be observed as gamma- and X- rays bursts.

Keywords. gravitational collapse, stellar magnetosphere, gamma- and X- rays bursts

1. Particles acceleration and their nonthermal emission in magnetosphere

Particles acceleration and generation of electromagnetic radiation in magnetosphere to result from processes Kryvdyk (1999), Kryvdyk (2004), Kryvdyk & Agapitov (2005a),and Kryvdyk & Agapitov (2005b) 1) Magnetic fields of collapsing stars increase during the collapse. 2) These variable magnetic fields accelerate the charged particles in the magnetospheres. 3) These particles emit the electromagnetic waves in the wide frequency band, from gamma rays to radio waves. 4)The radiation flux grows during collapse and to reaches a maximum on the final stage of collapse. This radiation can be observed as bursts in the all-frequency band, from radio to gamma ray. The burst duration completes with the collapse time. The radiation flux from collapsing on the final stage of collapse exceeds the initial flux a million times. 5) Thus the collapsing stars can be the powerful sources of the non-thermal radiation when before a supernova flares, the star compress and emit bursts.

Model particles generation in magnetosphere of collapsing stars give on figure 1.

2. Conclusion

1.Charged particles will accelerate in the magnetosphere of collapsing star to relativistic energy. 2. Relativistic polar jets will formed during collapse (figure 2). 3. These jets emit the electromagnetic radiation in all frequency bands. 4. This radiation can be observed as bursts from radio to gamma ray.

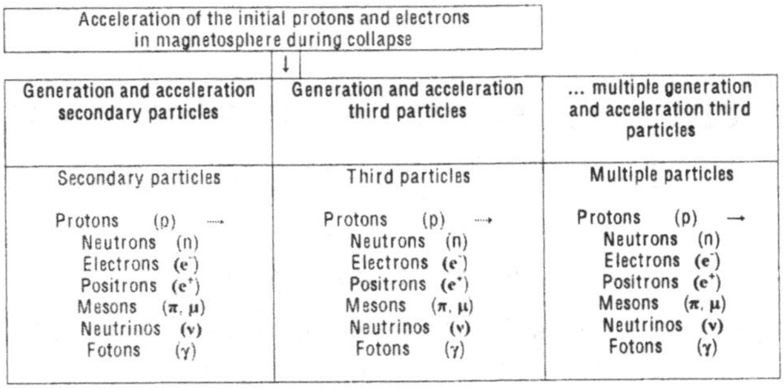

Acceleration of the initial protons and electrons in magnetosphere during collapse		
Generation and acceleration secondary particles	Generation and acceleration third particles	... multiple generation and acceleration third particles
Secondary particles	Third particles	Multiple particles
Protons (p) → Neutrons (n) Electrons (e⁻) Positrons (e⁺) Mesons (π, μ) Neutrinos (ν) Fotons (γ)	Protons (p) → Neutrons (n) Electrons (e⁻) Positrons (e⁺) Mesons (π, μ) Neutrinos (ν) Fotons (γ)	Protons (p) → Neutrons (n) Electrons (e⁻) Positrons (e⁺) Mesons (π, μ) Neutrinos (ν) Fotons (γ)

Figure 1. Model particle generation in magnetosphere.

Figure 2. Jets in magnetosphere of collapsing stars

References

Kryvdyk, V. 1999, *MRAS*, 309, 593

Kryvdyk, V. 2004, *Adv. Sp. Res.*, 33, 484

Kryvdyk, V. & Agapitov, A. 2005a, *ASP Conf. Ser.*, 334, 277

Kryvdyk, V. & Agapitov, A. 2005b, *Adv. Sp. Res.*, 330, 415

Kryvdyk, V. 2005, *Springer Proc. Phys.*, 99, 215

Astrophysical Dynamics: From Stars to Galaxies
Proceedings IAU Symposium No. 271, 2010
N.H.Brummell, A.S.Brun, M.S. Miesch & Y.Ponty, eds.

© International Astronomical Union 2011
doi:10.1017/S1743921311017911

Can turbulent reconnection be fast in 2D?

K. Kulpa-Dybeł[1], G. Kowal[1,2], K. Otmianowska-Mazur[1], A. Lazarian [3] and E. Vishniac[4]

[1]Astronomical Observatory, Jagiellonian University, ul Orla 171, 30-244 Kraków, Poland
[2]Núcleo de Astrofsica Teorica, Universidade Cruzeiro do Sul-Rua Galvão Bueno 868,
CEP 01506-000 São Paulo, Brazil
[3]Department of Astronomy, University of Wisconsin, 475 North Charter Street,
Madison, WI 53706, USA
[4]Department of Physics and Astronomy, McMaster University, 1280 Main Street West,
Hamilton, ON L8S 4M1, CANADA

Abstract. Turbulent reconnection is studied by means of two-dimensional (2D) compressible magnetohydrodynamical numerical calculations. The process of homogeneous turbulence is set up by adding two-dimensional random forcing implemented in the spectral space at small wave numbers with no correlation between velocity and forcing. We apply the initial Harris current sheet configuration together with a density profile calculated from the numerical equilibrium of magnetic and gas pressures. We assume that there is no external driving of the reconnection. The reconnection develops as a result of the initial vector potential perturbation. We use open boundary conditions. Our main goal is to find the dependencies of reconnection rate on the uniform resistivity. We present that the reconnection speed depends on the Lindquist number in 2D in the case of low as well as high resolution. When we apply more powerful turbulence the reconnection is faster, however the speed of reconnection is smaller than in the case of our three-dimensional numerical simulations.

Keywords. magnetic fields, turbulence, methods: numerical

1. Introduction

Magnetic reconnection plays a major role in various astrophysical phenomena such as star formation, solar explosions or galactic dynamo. For instance, in order for astrophysical dynamos to function smoothly the reconnection should have speeds close to Alfvén speed V_A. This is called fast reconnection, meaning that it does not depend on resistivity or depends on the resistivity logarithmically. The generally accepted Sweet-Parker (Parker 1957, Sweet 1958) model of magnetic reconnection applied to astrophysical bodies gives very slow reconnection rate which depends on the Lundquist number S as $V_A \sim S^{-1/2}$. Lazarian & Vishniac (1999) (LV99) proposed that in 3D a stochastic magnetic field component can dramatically enhance reconnection rates. This model has been recently confirmed numerically (including the insensitivity to the Lundquist number) by Kowal *et al.* (2009). Because the claim in LV99 model is that 3D effects are essential for fast reconnection, we are interested to test the effects of dimensionality on the reconnection rate. What is more, in a number of earlier studies of 2D reconnection it was conjectured that magnetic reconnection could become fast in the presence of turbulence.

2. Dependence on the uniform resistivity

In our work we studied the dependence of the reconnection rate on the uniform resistivity η_u in low (1024×2048) and high (4096×8192) resolution numerical simulations. We run several simulations with different values of the uniform resistivity η_u and with

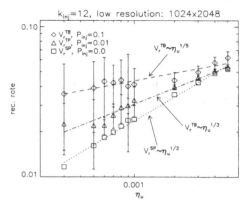
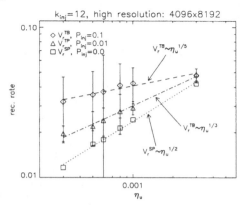

Figure 1. Dependence of the reconnection rate on the uniform resistivity η_u for models with and without turbulence, V_r^{TB} (diamonds and triangles) and V_r^{SP} (squares), respectively. Turbulence is injected at the scale $k = 12$ with $P_{inj} = 0.1$ and $P_{inj} = 0.01$. Simulations are performed with low (left panel) and high (right panel) resolution. For the Sweet-Parker reconnection (squares) the variance is negligible and not shown.

two values of the power of turbulence $P_{inj} = 0.1$ and $P_{inj} = 0.1$. For the Sweet-Parker model (Fig. 1, left panel) the obtained dependence of reconnection pace on the uniform resistivity coincide in 2D and 3D and agrees with the theoretical prediction, i.e., $V_r^{SP} \sim \eta_u^{1/2}$. Kowal et al. (2009) showed that in 3D the turbulent reconnection process is not constrained by Ohmic resistivity. In contrast to their work, we find that adding turbulence to the system leads to a weaker dependence between the reconnection rate and the uniform resistivity. Moreover, the dependence between reconnection rate and uniform resistivity is stronger for lower values of P_{inj}. Namely, for $P_{inj} = 0.1$ and $P_{inj} = 0.01$ we see that the reconnection rate scales as $\sim \eta_u^{1/5}$ and $\sim \eta_u^{1/3}$, respectively. For a more detailed description of these results we refer the reader to Kulpa-Dybeł et al. (2010).

To verify the influence of the numerical resolution on the recconection rate we made several simulations with different values of the resolution. The rest of the input parameters have the same value in all these models. In Fig. 1 we show the comparison of reconnection rates obtained for low and high resolution simulations. We can see that in both cases the reconnection rate depends on the uniform resistivity. This indicates that increasing the numerical resolution does not lead to a stable state of reconnection and does not change our results.

Acknowledgements

This work was supported by Polish Ministry of Science and Higher Education through grants: 92/N-ASTROSIM/2008/0 and 3033/B/H03/2008/35. The computations presented here have been performed on the GALERA supercomputer in TASK in Gdańsk.

References

Kowal, G., Lazarian, A., Vishniac, E. T., & Otmianowska-Mazur, K. 2009, *ApJ*, 700, 63
Kulpa-Dybeł, K., Kowal, G., Otmianowska-Mazur, K., Lazarian, A., & Vishniac, E. 2010 *A&A*, 514, 26
Lazarian, A. & Vishniac, E. T. 1999, *ApJ*, 517, 700
Parker, E. N. 1957, *J. Geophys. Res*, 62, 509
Sweet, P. A. 1958, *Electromagnetic Phenomena in Cosmical Physics* Conf. Proc. IAU Symposium 6, (Cambridge, UK:Cambridge University Press), 123

Astrophysical Dynamics: From Stars to Galaxies
Proceedings IAU Symposium No. 271, 2010
N.H. Brummell, A.S. Brun, M.S. Miesch & Y. Ponty, eds.

© International Astronomical Union 2011
doi:10.1017/S1743921311017923

Cosmological magnetic field seeds produced by the Weibel instabilities

M. Lazar*, R. Schlickeiser and T. Skoda

Institut für Theoretische Physik IV, Ruhr-Universität Bochum, D-44780 Bochum, Germany
*email: mlazar@tp4.rub.de

Abstract. The source of the cosmological magnetic field is still unknown because the widely invoked dynamo processes are only able to regenerate and amplify some initial magnetic field seeds. In the hot and highly ionized intergalactic matter such magnetic field seeds can easily be produced by the (electro-)magnetic instabilities of Weibel type. Here we discuss suplementary mechanisms that can make these Weibel created fields to evolve at large scales presently observed in galaxies and clusters and can also enhance these magnetic field seeds after the dissipation.

Keywords. cosmology: large-scale structure of universe, plasmas, instabilities, magnetic fields

1. Introduction

Dynamo mechanism presently widely invoked in cosmological magnetic field genesis need some seeds to work with. Here we show that such seed fields can originate from the Weibel instabilities (Weibel 1959). These instabilities are driven by various kinetic anisotropies, e.g., heating flows, shocks and interpenetrating plasma shells, and generate long-lived quasistatic magnetic fields providing plausible explanations for the origin of cosmological magnetic field (Schlickeiser & Shukla 2003) or the magnetic boosts in astrophysical sources (GRBs, SNRs) of nonthermal radiation (Medvedev & Loeb 1999).

2. Weibel instabilities

The instabilities of Weibel-type operate in initially unmagnetized plasma systems: the filamentation instability driven by a relative motion of different plasma shells (beam-plasma, counterstreams), Weibel instability driven by a thermal anisotropy, or the cumulative Weibel/filamentation instability driven by counterstreaming plasmas with intrinsic kinetic anisotropies. Any magnetic pertubation generates currents of charged particles, which reinforce the perturbation (Ampere's law) to grow to large amplitudes. It saturates due to the magnetic trapping of plasma particles, when the magnetic bounce frequency $(\Omega^2 \equiv ek_{max}vB_{max}/(mc))$ increases to a value comparable to the linear growth rate prior to the saturation $\Omega_{max} \simeq \Gamma_{max}$ (Davidson 1972). The mode with the largest growth rate, $\Gamma_{max}(k_{max})$, dominates and sets the characteristic length scale of the magnetic field fluctuations, $\lambda \sim k_{max}^{-1}$ at the saturation. Particles free streaming across the excited magnetic field lines is suppressed once the particle's gyroradius $\rho = v/\Omega_{max} = eB_{max}/(mc)$ becomes comparable to the length scale of the magnetic field fluctuations, k_{max}^{-1}, yielding (Medvedev & Loeb 1999) for the maximum $B_{max} \simeq mvck_{max}/e = mc\Gamma_{max}/e$.

Saturated magnetic field have been derived for the filamentation instability $B_{max} = v_0\sqrt{4\pi nm/\gamma_0}$, the Weibel instability ($v_{th} = v_{th,2} >> v_{th,1}$) $B_{max} = \sqrt{2\pi nm}\, v_{th}$ and for the cumulative Weibel/filamentation instability $B_{max} = \sqrt{4\pi nm(v_{th}^2 + 2v_0^2)}$, and these magnetic fields are consistent with the observations: $B = \eta B_{max} \sim 10^{-8} - 10^{-6}$ G in clusters, $\sim 10^{-7} - 10^{-5}$ G in intergalactic medium, $\sim 10^{-7}$ G in galaxies, for a conversion factor $\eta = 0.01$-0.1 (Schlickeiser & Shukla 2003, Lazar *et al.* 2009).

By comparison to the cosmological magnetic field presently measured on scales of galaxies or clusters ($\gtrsim 1$ kpc), the coherence length scales of the Weibel created fields (of the order of plasma skin depth, $c/\omega_{pp} < 10^5$ km), are very small.

3. Evolution of the Weibel magnetic field to large cosmological scales

The Weibel fields do not decay much after the saturation (Silva *et al.* 2003), and these fields may become coherent up to very large scales expected to reach kpc scales through inverse cascade processes (Cho & Vishniac 2000). Inverse cascades produce in plasma turbulence, and trasfer energy towards large scales. In a weak-turbulence model, inverse cascades are described by the theory of nonlinear wave interactions. For the Weibel field, the main channels of resonant wave-wave conversion after the saturation were identified according to long-time evolution in numerical experiments, indicating an inverse cascade with small-scale structures merging into large scale coherent fields (Lazar *et al.* 2010).

For a long term evolution to large cosmological time and space scales these magnetic fields dissipates by many order of magnitudes. However, recent numerical experiments have shown that the weak seeds are again enhanced by the turbulent flow motions (Ryu *et al.* 2008). Intense fields are obtained at large coherence scales of the order of dominant eddies (curvature radius of typical cosmological shocks). Thus, the turbulence excited during the process of large-scale structure formation can amplify very weak strengths of the *intergalactic magnetic field* (after the dissipation), e.g., $B \geqslant 10^{-10}$ G, and reach coherence length ~ 100 kpc, fitting well the presently observed magnetic fields.

Amplification of the galactic magnetic field seeds can be attributed to turbulent flows and dynamos observed and predicted in convective zones of the stars and planets, stellar wind and supernova explosions, star forming regions, interstellar medium (Beck 2007), or by the ($\alpha - \Omega$, mean field) dynamos in stars (the Sun), accretion disks (protostellar disks, close binaries, AGNs). Intergalactic magnetic field can be enhanced by turbulent dynamo in accretion disks and motion of galaxies through IGM, or by ($\alpha - \Omega$, mean field) dynamos in spiral galaxies, giant molecular clouds (Ryu *et al.* 2008).

4. Conclusions

The instabilities of Weibel-type can potentially create magnetic seed fields and their evolution to large cosmological scales is sustained by the amplification mechanisms and the inverse cascades in plasma turbulence predicted in galaxies and intergalactic medium.

References

Beck, R. 2007 *A&A*, 470, 539

Schlickeiser, R. & Shukla, P. K. 2003, *ApJ (Letters)*, 599, L57

Lazar, M., Schlickeiser, R., Wielebinski, R., & Poedts, S. 2009, *ApJ*, 693, 1133

Weibel, E. S. 1959, *Phys.Rev.Lett.*, 2, 83

Davidson, R. C., Hammer, D. A., Haber, I., & Wagner, C. E. 1972, *Phys.Fluids*, 15, 317

Medvedev, M. & Loeb, A. 1999, *ApJ*, 526, 697

Ryu, D., Kang, H., Cho, J. & Das, S. 2008, *Science*, 320, 909

Silva, L. O., Fonseca, R. A., Tonge, J. W., Dawson, J. M., Mori, W. B., & Medvedev, M. V. 2003, *ApJ (Letters)*, 596, L121

Cho, J. & Vishniac, E. T. 2000, *ApJ*, 539, 237

Lazar, M., Lapenta, G., Schlickeiser, R., & Poedts, S. 2010, *Phys. Plasmas*, in preparation

Astrophysical Dynamics: From Stars to Galaxies
Proceedings IAU Symposium No. 271, 2010
N.H. Brummell, A.S. Brun, M.S. Miesch & Y. Ponty, eds.

Rapid mass segregation in young star clusters *without* substructure?

C. Olczak[1,2,3], R. Spurzem[4,2,3], Th. Henning[1]

[1] Max-Planck-Institut für Astronomie, Heidelberg, Germany
email: olczak@mpia.de

[2] National Astronomical Observatories of China, Chinese Academy of Sciences, Beijing, China

[3] The Kavli Institute for Astronomy and Astrophysics at Peking University, Beijing, China

[4] Astronomisches Rechen-Institut, Universität Heidelberg, Heidelberg, Germany

Abstract. The young star clusters we observe today are the building blocks of a new generation of stars and planets in our Galaxy and beyond. Despite their fundamental role we still lack knowledge about the initial conditions under which star clusters form and the impact of these often harsh environments on the formation and evolution of their stellar and substellar members.

We present recent results showing that mass segregation in realistic models of young star clusters occurs very quickly for subvirial spherical systems without substructure. This finding is a critical step to resolve the controversial debate on mass segregation in young star clusters and provides strong constraints on their initial conditions. The rapid concentration of massive stars is usually associated with strong gravitational interactions early on during cluster evolution and the subsequent formation of multiple systems and ejection of stars.

Keywords. stellar dynamics, methods: n-body simulations, galaxies: star clusters

1. Introduction

It is commonly accepted that star formation does usually not occur in isolation but that a large majority of young stars – up to 90 % – are part of a cluster (e.g. Lada & Lada 2003). One of the most widely discussed aspect of the dynamical evolution of young star clusters is that of mass segregation (e.g. Raboud & Mermilliod 1998; Bonnell & Davies 1998; Gouliermis *et al.* 2004; Chen *et al.* 2007; Vesperini *et al.* 2009; Xin-Yue *et al.* 2009; Ascenso *et al.* 2009). A major question is whether the observed degree of mass segregation has developed dynamically (via two-body encounters) or whether it can be only explained via primordial mass segregation (i.e. inherent to the star formation process).

Though it has been shown that mass segregation in a stellar system with a mass spectrum (e.g. Spurzem & Takahashi 1995; Khalisi *et al.* 2007) occurs rapid enough to account for the observed mass segregation, recent investigations have focused on how substructure in stellar systems can accelerate this process (McMillan *et al.* 2007; Allison *et al.* 2009; Moeckel & Bonnell 2009, see also (Aarseth & Hills 1972)). However, all these works have simulated the evolution under subvirial initial conditions.

We thus investigate numerically the *pure* effect of subvirial conditions in young star clusters and infer to which degree they can speed-up mass segregation.

2. Numerical Setup

We have performed numerical simulations of isolated, non-rotating, moderately populated star clusters with 1000 stars initially, a Kroupa IMF in the range $0.08 \, M_\odot$ to $50 \, M_\odot$ Kroupa (2001), a density profile $\rho \propto r^{-2}$ and varying initial virial ratios Q. We

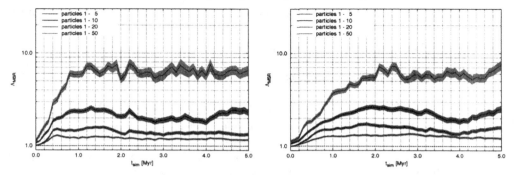

Figure 1. Evolution of mass segregation in a star cluster with $Q = 0.1$ (left) and $Q = 0.3$ (right) represented by Λ_{MST} of four different stellar groups (from top to bottom: 5, 10, 20, and 50 most massive stars). The shaded regions indicate the associated uncertainties. One dynamical crossing time corresponds to $0.19\,\mathrm{Myr}$ for $Q = 0.1$ and $0.28\,\mathrm{Myr}$ for $Q = 0.3$.

use the NBODY6-GPU code Aarseth (2003) to model the dynamics of these numerical models and analyze the degree of mass segregation, Λ_{MST}, via a minimum spanning tree algorithm as described by Allison *et al.* (2009).

3. Results

We find that mass segregation occurs rapidly in star clusters with subvirial initial conditions (see Fig. 1). If $Q = 0.1$, the maximum degree of segregation is reached in less than 1 Myr (or roughly five crossing times). The five most massive stars show strong mass segregation with $\Lambda_{MST} \approx 7$. The first evidence of significant mass segregation, $\Lambda_{MST} \gtrsim 2$, occurs even within 1-2 crossing times. For less cold initial conditions ($Q = 0.3$) the evolution is about two times slower yet still very rapid and equally pronounced.

The concentration of massive stars is accompanied by strong gravitational interactions that involve early on the formation of multiple systems and ejection of (massive) stars. This counteracts further mass segregation and quickly limits its maximum degree.

Compared with the time scale on which mass segregation occurs in the simulations of McMillan *et al.* (2007) and Allison *et al.* (2009) the present setup leads to an equally rapid evolution. This indicates that in fact subvirial initial conditions might be the dominant dynamical driver of rapid mass segregation in young star clusters.

References

Aarseth, S., 2003, *Gravitational N-body Simulations* (Cambridge, Cambridge University Press, 2003, 430 p.).
Aarseth, S. J., & J. G. Hills, 1972, A&A **21**, 255.
Allison, R. J., S. P. Goodwin, R. J. Parker, R. de Grijs, S. F. Portegies Zwart, & M. B. N. Kouwenhoven, 2009, ApJ **700**, L99.
Ascenso, J., J. Alves, & M. T. V. T. Lago, 2009, A&A **495**, 147.
Bonnell, I. A., & M. B. Davies, 1998, MNRAS **295**, 691.
Chen, L., R. de Grijs, & J. L. Zhao, 2007, AJ **134**, 1368.
Gouliermis, D., S. C. Keller, M. Kontizas, E. Kontizas, & I. Bellas-Velidis, 2004, A&A **416**, 137.
Khalisi, E., P. Amaro-Seoane, & R. Spurzem, 2007, MNRAS **374**, 703.
Kroupa, P., 2001, MNRAS **322**, 231.
Lada, C. J. & E. A. Lada, 2003, ARA&A **41**, 57.
McMillan, S. L. W., E. Vesperini, & S. F. Portegies Zwart, 2007, ApJ **655**, L45.
Moeckel, N. & I. A. Bonnell, 2009, MNRAS, 13700908.0253.
Raboud, D. & J.-C. Mermilliod, 1998, A&A **333**, 897.
Spurzem, R. & K. Takahashi, 1995, MNRAS **272**, 772.
Vesperini, E., S. L. W. McMillan, & S. Portegies Zwart, 2009, ApJ **698**, 615.
Xin-Yue, E., J. Zhi-Bo & F. Yan-Ning, 2009, Chinese Astronomy and Astrophysics **33**, 139.

Astrophysical Dynamics: From Stars to Galaxies
Proceedings IAU Symposium No. 271, 2010
N.H. Brummell, A.S. Brun, M.S. Miesch & Y. Ponty, eds.

© International Astronomical Union 2011
doi:10.1017/S1743921311017947

Supergranulation and its Activity Dependence

U. Paniveni[1,2], V. Krishan[1,3], J. Singh[1] and R. Srikanth[4,3]

[1] Indian Institute of Astrophysics, Bangalore-560034,India
[2] NIE Institute of technology, Hootagalli Industrial
Area, Mysore-570018, India
[3] Raman Research Institute, Sadashiva nagar
Bangalore-560080, Karnataka, India
[4] Poornaprajna Institute of Scientific Research,
Devanahalli, Bangalore-562 110, India

Abstract. We study the complexity of supergranular cells using the intensity patterns obtained at the Kodaikanal solar observatory during the solar maximum. Our data consists of visually identified supergranular cells, from which a fractal dimension D is obtained according to the relation $P \propto A^{D/2}$ where A is the area and P is the perimeter of the cells. We find a difference in the fractal dimension between the active and the quiet region cells which is conjectured to be due to the magnetic activity level.

Keywords. Sun: granulation —-turbulence

---

1. Introduction

Heat flux transport is chiefly by convection rather than by photon diffusion in the convection zone of all cool stars such as the Sun, the thickness of convection zone being 30% of the solar radius below the photosphere (Noyes 1982). Convection is revealed predominantly on two scales– on the typical scale of 1″-2″ as granulation, and on the typical scale of 30″-40″ as supergranulation. The typical lifetime of supergranular cell is around 24 hours.The horizontal motions of the supergranular cells transport magnetic flux from the central upflow region in the cell to the edge, where the resultant production of excess heat at the chromospheric level traces out the supergranular network (Kosovichev 2007).Supergranules are characterized by typical horizontal speeds of $0.3 - 0.4$ km s^{-1}. Fractal analysis is a valuable mathematical tool to quantify the complexity of geometric structures and thus gain insight into the underlying dynamics. Fractal analysis was first adopted by Roudier and Muller (1986) to study the complexity of solar granulation. In the context of supergranulation, fractal analysis can shed light on the turbulence of magneto-convective processes that generate the magnetic structures (Stenflo and Holzreuter 2003;Lawrence,Ruzmaikin and Cadavid 1993). For our purpose, fractal dimension D is characterized by the area-perimeter relation of the structures (Mandelbrot 1977). Self-similarity, meaning the same degree of complexity regardless of the scale at which the structures are observed, is expressed by a linear relationship between $\log P$ and $\log A$ over some range of scales. A fractal analysis helps quantify the supergranular irregularity, which can shed light on the nature of solar turbulence (Paniveni *et al.* 2005).

2. Data Analysis

We analysed intensity data, consisting of Ca II K filtergrams ($\lambda = 3934$ Å) of the Sun, obtained between 16th May 2001 and 26th August 2001, during the solar maximum phase of the 23rd Solar cycle at the solar observatory, Kodaikanal. The images have

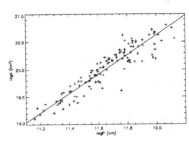 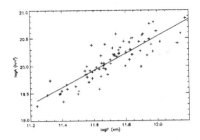

Figure 1. (a) Left : Plot of the natural logarithm of the supergranular area (in sq.km) against the natural logarithm of perimeter (km) in the Active Region (b) Right : Plot of the natural logarithm o the supergranular area (in sq.km) against the natural logarithm of perimeter (km) in the Quiet Region

a resolution of about $2''$, which is twice the granular scale. Only cells lying within $60°$ angular distance from the disc centre were selected in order to minimize projection effects. Cells close to the sunspot or plage region were identified as active region cells and cells away from these were noted as quiet region cells. We analysed 152 active region cells and 87 quiet region cells. We chose a fiducial y-direction on the cell and performed intensity profile scans along the x-direction for all the pixel positions on the y-axis. In each scan, the cell extent is taken to be marked by two juxtaposed 'crest'(separated by a 'trough') as expected in the intensitygrams. This set of data points was used to determine the area and perimeter of a given cell and of the spectrum of all selected supergranules. The area-perimeter relation is used to evaluate the fractal dimension.

3. Results and Discussion

The $\log(A)$ vs $\log(P)$ relation is linear as shown in the Figure1(a) for the active region. A correlation co-efficient of 0.94 indicates strong correlation. Fractal dimension D calculated as 2/slope is found to be 1.12 ± 0.07.The $\log(A)$ vs $\log(P)$ relation is linear as shown in the Figure1(b) in the quiet region. A correlation co-efficient of 0.88 indicates strong correlation. Fractal dimension D calculated as 2/slope is found to be 1.25 ± 0.14. The linear relations of both Figure1(a) and Figure1(b) suggest that supergranules are self-similar and can be regarded as fractal objects. According to Mandelbrot (1975), an isosurface has a fractal dimension given by D_I = (Euclid dimension) - 1/2 (exponent of the variance). The pressure variance $\langle p^2 \rangle$ is proportional to the square of the velocity variance i.e. $\langle p^2 \rangle \propto r^{4/3}$ (Batchelor 1953).The fractal dimension of an isobar is found to be $D_p = 2 - (1/2 \times 4/3) = 1.33$. Our data furnishes an average fractal dimension of nearly $D = 1.21$ which indicates that the supergranular network is close to being an isobar.

References

Batchelor, G. K., *The theory of Homogeneous Turbulence* (Cambridge University Press 1953).

Kosovichev, A. G. in *Dynamic Sun*, chapter 5; ed. B. N. Dwivedi (CUP, 2007)

Lawrence, J. K., Ruzmaikin, A. A.,Cadavid, A. C.Astrophysical Journal 417,805 (1993).

Mandelbrot, B., 1975, J. Fluid Mech., 72, part2, 401-416.

Mandelbrot, B. 1977, Fractals (San Francisco: Freeman).

Noyes, R. W., *The Sun, Our star* (Harvard University press 1982).

Paniveni, U., Krishan, V., Singh, J., & Srikanth, R., 2005, Solar Physics, 231, 1-10.

Roudier, Th. & Muller, R., 1986, Solar Physics,107, 11.

Stenflo, J.O., and Holzreuter, R. in Current Theoretical Models and Future High Resolution Solar Observations: Preparing for ATST, ed. A.A. Pevtsov & H. Uitenbroek, ASP Conf.Ser., 286, 169 (2003).

Astrophysical Dynamics: From Stars to Galaxies
Proceedings IAU Symposium No. 271, 2010
N.H. Brummell, A.S. Brun, M.S. Miesch & Y. Ponty, eds.

© International Astronomical Union 2011
doi:10.1017/S1743921311017959

3D MHD modelling of a chromosphere above the sun's convective zone

Rui Pinto[1] and Sacha Brun[1]

[1]Service d'Astrophysique - CEA Saclay/DSM/Irfu,
91191 Gif-sur-Yvette Cedex, France,
email: rui.pinto@cea.fr

Abstract. We report on the extension of the ASH code to include an atmospheric stable layer (*i.e* not convective). This layer is meant to model the sun's chromosphere within the anelastic approximation limits while coping with the wide range of densities, time and spatial scales between $r = 0.7\ R_\odot$ and $r = 1.03\ R_\odot$. Convective overshoot into the stable atmospheric layer is observed in a region $\sim 0.01\ R_\odot$ thick, exciting waves which propagate upwards into the atmosphere.

1. Introduction

The physical connections between the sun's convection zone (CZ) and the atmospheric layers are reputedly a hard problem to study. The chromosphere separates plasma of very different length and time-scales, as well as of different MHD regimes (*e.g* $\beta < 1$ and $\beta > 1$). In consequence, most atmospheric and sub-photospheric phenomena are studied separately. The most remarkable exceptions are small-scale numerical investigations (typically, with $10-40$ Mm tall and wide domains) of the sub-photospheric and chromospheric layers (*e.g* Martínez-Sykora, *et al.* 2008). We perform global-scale 3D simulations including the CZ and the sun's chromosphere. Our goal is to study flux emergence issues consistently and without treating the photosphere as an impenetrable boundary (or any type of boundary condition whatsoever). This is a necessary condition to correctly capture the dynamics of the transition between the CZ and the chromosphere, together with the MHD transport processes taking place there. Our study differs from the *small-scale* simulations cited above in that it aims at capturing the physics of *global* phenomena such as sunspot origin and evolution, magnetic transport between loop footpoints (Grappin, *et al.* 2008), large-scale flows.

2. Numerical model and results

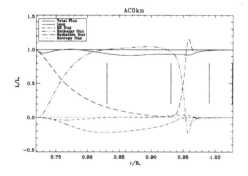

Figure 1. Energy flux balance in the radial direction in "units of solar luminosity". Note the layer of negative enthalpy flux ($0.95 - 0.96\ R_\odot$), which indicates convective overshoot into the photosphere. Note also that all the star's luminosity is transported radiatively at the top. The vertical grey bars indicate the positions of the spherical shells in Fig. 2.

We use the ASH 3D MHD anelastic stellar code (Clune *et al.* 1999; Brun *et al.* 2004). The initial state was built from a CESAM solar profile of the CZ (Morel, 1997; Brun,

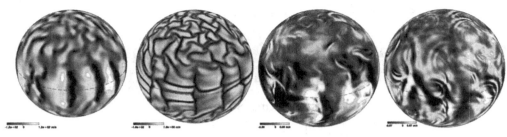

Figure 2. Radial velocity V_r at spherical surfaces with different radii. From left to right, top to bottom: $r = 0.83,\ 0.93,\ 0.99,\ 1.03\ R_\odot$ (*cf.* Fig. 1).

et al. 2002) with a transition to a nearly isothermal chromosphere in hydrostatic equilibrium. The chromospheric temperature is explicitly set to $\sim 4\times$ the standard 6000 K for numerical stability. We do not include a coronal *Transition Region* (TR), but rather extend our chromosphere up to $1.03\ R_\odot$. The minimum of H_ρ is 10 Mm. Before considering any magnetic effects, we let the convection settle in a purely hydrodynamical CZ and waited until the whole system relaxed. At this point, the system reaches a steady state (in a statistical sense; the main properties remain unchanged despite any local fluctuations). The conditions for the goodness of the anelastic approximation are verified at all times. The transition from the CZ to the chromosphere is clearly visible in fig. 1, which shows the radial energy flux balance (averaged over θ and ϕ) in the relaxed state. The enthalpy flux vanishes there, and becomes negative in the overshoot layer. The radiative flux assumes all the energy transport afterwards (in the convectively stable atmosphere). Fig. 2 shows the radial velocity V_r over spherical cuts taken at different radii (signalled by the vertical grey lines in Fig. 1). The first spherical shell sits well within the CZ, the second one corresponds to a layer just below the photosphere and the two remaining ones are chromospheric layers. Convective overshoot excite waves which propagate into the chromosphere, while any traces of the underlying granular motions fade away. The atmospheric layer shows a prograde differential rotation, but rotating slower and more uniformly than the CZ/photosphere. We added at this point a global magnetic field and let it evolve in the CZ. The issues concerning its evolution (*e.g* ohmic diffusion, advection into the atmosphere, back-reaction on the flows) will be addressed in a forthcoming paper.

3. Conclusions

We developed a 3D model of the CZ and solar chromosphere using the ASH code. Some simplifications were made in order to achieve both physical correctness (from the anelastic approximation point of view) and numerical stability. In particular, the chromosphere is warmer than the solar one and it is rooted at a lower radius. The coronal TR was not included in the model, and the chromosphere was extend up to $1.03\ R_\odot$. We judge this as an essential step for the comprehension of flux emergence scenarios (*cf.* Jouve &Brun, 2009), before any finer modelling becomes feasible. These issues will be discussed in a future paper.

Acknowledgements: We acknowledge funding by ERC through grant STARS2 207430.

References

Brun, A.-S., *et al.* 2002, A&A, 391, 725

Brun, A.-S., *et al.* 2004, ApJ, 612, 1073

Clune, T., *et al.* 1999, Par. Computing, 25, 361

Grappin, R., *et al.* 2008, A&A, 490, 353

Jouve, L., *et al.* 2009, ApJ, 701, 1300

Martínez-Sykora, *et al.* 2008, ApJ, 679, 871

Miesch, et al. 2000, ApJ, 532, 593

Morel, P. 1997, A&AS, 124, 597

Astrophysical Dynamics: From Stars to Galaxies
Proceedings IAU Symposium No. 271, 2010
N.H. Brummell, A.S. Brun, M.S. Miesch & Y. Ponty, eds.

Solar wind, mass and momentum losses during the solar cycle

R. Pinto[1], S. Brun[1], L. Jouve[2] and R. Grappin[3]

[1] Service d'Astrophysique - CEA Saclay/DSM/Irfu, 91191 Gif-sur-Yvette Cedex, France
[2] DAMTP, Wilberforce Road, CB30WA Cambridge, UK
[3] LUTH Obs. Paris-Meudon, France & LPP Ecole Polytechnique, Palaiseau, France
email: rui.pinto@cea.fr

Abstract. We study the connections between the sun's convection zone evolution and the dynamics of the solar wind and corona. We input the magnetic fields generated by a 2.5D axisymmetric kinematic dynamo code (STELEM) into a 2.5D axisymmetric coronal MHD code (DIP). The computations were carried out for an 11 year cycle. We show that the solar wind's velocity and mass flux vary in latitude and in time in good agreement with the well known time-latitude assymptotic wind speed diagram. Overall sun's mass loss rate, momentum flux and magnetic breaking torque are maximal near the solar minimum.

1. Introduction

The subject of the evolution of the corona and solar wind's properties during the solar cycle has been studied mostly based on observations of surface magnetic fields and/or coronal imaging. We follow a different approach here and focus on the connections between atmospheric and sub-surface phenomena and their variation during the solar cycle. Two 2.5D MHD models were used: STELEM for the convection zone dynamo (Jouve & Brun, 2007) and DIP for the wind and corona (Grappin, *et al.* 2000). A time series of dynamo generated field's were used as source fields in the DIP code. The solar wind flow develops in the domain until the system relaxes to stable steady state. Successive realisations of the procedure allow us to build a general picture of the evolution of the sun's coronal environment in response to the cyclic evolution of the dynamo source field.

2. Results

Fig. 1 shows the first 3 $R_\odot$ of the northern hemisphere at four different instants ($t = 0, 3, 3.5, 10$ years). Orange and blue shades trace different **B**-field polarities in the open-field regions. The coronal (poloidal) magnetic field is globally dipolar faraway from the surface, except during the polarity reversal. Close to the surface, the magnetic topology is much more complex. The polarity reversal happens *quickly* in the corona, even if the underlying **B**-field evolves *slowly*. For $r \gg R_\odot$, $|\mathbf{B}| \approx B_r$, as the solar wind "opens up" the field lines. Furthermore, B_r is nearly independent of the latitude faraway from the sun. The magnetic field decomposes mainly in low-order multipolar components at the minimum, while higher order components appear at the maximum. The *multiple* current sheets shown in Fig. 1 may be interpreted as a highly warped current sheet in the non-axisymmetric corona. Fast solar wind originates essentially from high latitude regions, while the slow wind flows mostly aside the *streamers* at lower latitudes. Exceptions may occur, though. Wind speed and flux tube expansion factor are well correlated at all latitudes. From solar minimum to solar maximum (Fig. 2), the global $|\mathbf{B}|$ increases, the mean Alfvén radius $\langle r_A \rangle$ decreases, the total surface open flux decreases, the coronal hole/streamer boundaries approach the poles and $\dot{M}$ falls. Angular momentum flux and breaking torque τ weaken, with $\tau < 0$, all the time (Matt & Pudritz, 2008; Weber & Davis, 1967).

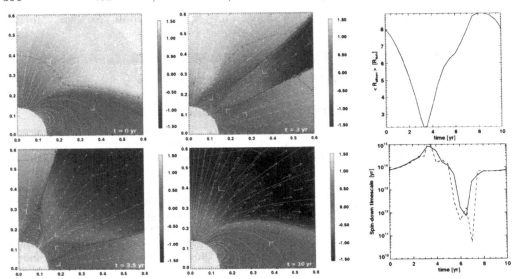

Figure 1. Snapshots of the evolution of the corona during the solar cycle (only the first $\sim 3\ R_\odot$ and northern hemisphere are shown). White lines are magnetic field lines. The colorscale represents the wind flow velocity (Mach number) projected onto the signed magnetic field. This quantity traces the B-field's polarity in the open field regions. Red/orange means positive polarity, while green/blue means negative polarity. The large arrowheads show the local **B**-field orientation. The black contours trace Mach numbers $0, 1$.

Figure 2. Mean Alfvén radius $\langle r_A \rangle$ (top) and magnetic breaking time-scale $\tau = J_\odot/\dot{J}$ (bottom) during the solar cycle. $J_\odot$ is the sun's angular momentum (Gilman, *et al.* 1989); $\dot{J}$ is the wind's angular momentum loss rate.

3. Conclusions

The surface magnetic field, coronal topology and assymptotic wind speed computed are all in good agreement with observations (*e.g* Wang & Sheeley, 2006). Overall sun's mass loss rate is dominated by outward flows originating at low latitudes. Their predominance depends on the time-varying magnetic field geometry and how it restrains the solar wind from "opening-up" outflow channels. The global angular momentum flux is maximum near the solar minimum, and so is the resulting magnetic breaking torque. Note that this is the moment when the total photospheric surface that is magnetically connected open flux regions (where the solar wind flows) is maximal. At the same time the coronal magnetic field is at it's *least multipolar* configuration, meaning that it decays the slowest with radial distance and the Alfvén radius (*lever arm* length) is naturally the highest. Variations in the radial and transverse large scale magnetic gradients may be pertinent to the study of the triggering of eruptive phenomena. Future work will include a more detailed treatment of the chromospheric dense layers.

Acknowledgements: We acknowledge funding by ERC through grant STARS2 207430.

References

Grappin, R., Lorat, J., & Buttighoffer, A. 2000, Astronomy and Astrophysics, 362, 342

Jouve, L. & Brun, A. S. 2007, Astronomy and Astrophysics, 474, 239

Gilman, P., Morrow, C. & DeLuca, E. 1989, The Astrophysical Journal, 338, 528

Matt, S. & Pudritz, R. E. 2008, The Astrophysical Journal, 678, 1109

Wang, Y. & Sheeley, N. R. 2006, Astrophysical Journal, 653, 708

Weber, E. J. & Davis, L. 1967, Astrophysical Journal, 148, 217

Astrophysical Dynamics: From Stars to Galaxies
Proceedings IAU Symposium No. 271, 2010
N.H. Brummell, A.S. Brun, M.S. Miesch & Y. Ponty, eds.

© International Astronomical Union 2011
doi:10.1017/S1743921311017972

Modelling turbulent fluxes due to thermal convection in rectilinear shearing flow

Radoslaw Smolec[1], Günter Houdek[1] and Douglas Gough[2]

[1]Institute of Astronomy, University of Vienna, A-1180 Vienna, Austria
email: radek.smolec@univie.ac.at, guenter.houdek@univie.ac.at

[2]Institute of Astronomy and Department of Applied Mathematics and Theoretical Physics,
University of Cambridge, Cambridge CB3 0HA, UK
email: douglas@ast.cam.ac.uk

Abstract. We revisit a phenomenological description of turbulent thermal convection along the lines proposed originally by Gough (1965) in which eddies grow solely by extracting energy from the unstably stratified mean state and are subsequently destroyed by internal shear instability. This work is part of an ongoing investigation for finding a procedure to calculate the turbulent fluxes of heat and momentum in the presence of a shearing background flow in stars.

Keywords. convection, turbulence, hydrodynamics

Introduction

Convection models based on the mixing-length approach still represent the main method for computing the turbulent fluxes in stars with convectively unstable regions. In such regions the pulsational stability of the star is affected not only by the radiative heat flux but also by the modulation of the convective heat flux and by direct mechanical coupling of the pulsation with the convective motion via the Reynolds stresses. Time-dependent formulations of the mixing-length approach for radial pulsation have been proposed by, for example, Gough (1965, 1977a) and Unno (1967). In a first step towards a generalization to nonradially pulsating stars, Gough & Houdek (2001) adopted Gough's formulation, incorporating into it a treatment of the influence of a shearing background flow. In this generalized framework of the mixing-length formalism, in which turbulent convective eddies grow according to linearized theory and are subsequently broken up by internal shear instability, there is a consequent reduction in the mean amplitude of the eddy motion, and a corresponding reduction in the heat flux.

In order to test and calibrate the formalism, it is preferable first to compare its predictions with existing results of hopefully more reliable investigations, such as experiments or numerical simulations. Here we extend our earlier work (Gough & Houdek 2001), comparing the functional forms of our mean temperature and shear profiles with those of direct numerical simulations (DNS; Domaradzki & Metcalfe 1988, DM88) of Rayleigh-Bénard convection in air (Prandtl number, $\sigma = 0.71$) with a strongly shearing background flow. We consider a plane-parallel layer of fluid confined between rigid horizontal perfectly conducting boundaries of infinite extent at fixed temperatures, the lower being hotter than the upper by ΔT. The upper boundary moves horizontally with constant velocity ΔU, and we assume, in accordance with the Boussinesq approximation, that the shear, E, in the mean (plane Couette) flow does not vary over the scale of an eddy.

Turbulent fluxes and mean equations in the presence of a shear

We follow the basic procedure by Gough & Houdek (2001) to solve the linearized equations describing the dynamics in a statistically stationary flow of a viscous Boussinesq fluid confined between two horizontal planes. In Cartesian co-ordinates (x, y, z) the

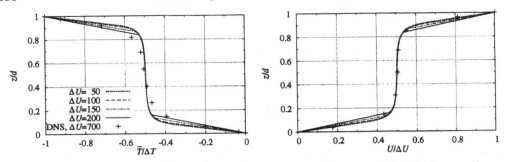

Figure 1. Mean temperature $\overline{T}/\Delta T$ (left panel) and mean velocity $U/\Delta U$ (right panel) as a function of height z are compared with DNS data (crosses) for different ΔU values.

equations are nondimensionalized using the vertical distance between the planes, d, and the thermal diffusion time across d as units of space and time. Linearized modes of convection are obtained by expansion about $E := d_z U = 0$ (where $d_z \equiv d/dz$) to second order in E. The resulting expressions for the eigenfunctions of the fluctuating temperature, T', and turbulent velocity field, $\boldsymbol{u} = (u, v, w)$, are then used to compute the turbulent fluxes of heat, $\overline{wT'}$, and momentum (Reynolds stresses), $\overline{\rho u_i u_j}$, in the manner of Gough (1977a). A horizontal bar denotes statistical average (average over the horizontal plane). In our dimensionless formulation, the sum of the radiative and convective fluxes is independent of z and equal to the Nusselt number of the layer, $N = F_r(z) + F_c(z)$. In the diffusion approximation the radiative flux is equal to the mean temperature gradient, $F_r(z) = -d_z \overline{T}$. The thermal mean equation is solved simultaneously with (the only nontrivial (x-) component of) the mean momentum equation: $d_z \overline{\rho u w} - \sigma d_{zz} U = 0$ (where $d_{zz} \equiv d^2/dz^2$). Note that the Reynolds stress $\overline{\rho u w}$ distorts the shear, and consequently the x-component of the mean flow, U, is not a linear function of height z.

Results and conclusions

Fig. 1 shows the normalized mean vertical velocity profiles, $U/\Delta U$ and mean temperature profiles, $\overline{T}/\Delta T$, for four values of ΔU: 50, 100, 150 and 200. Our results are compared with DNS data (DM88), which assume $\Delta U = 700$. The mean profiles are in reasonable agreement with the DNS data, but for smaller values of ΔU in our model computations. Best agreement with the DNS data is obtained with $\Delta U = 150$. Two factors may be responsible for the discrepancy in the values of ΔU: (i) to maintain greatest simplicity, we adopted horizontal-stress-free boundary conditions for the eddies, whereas in the simulations the bounding planes were taken to be rigid – this may perhaps account for up to a factor of about three in the ΔU differences between the model ($\Delta U=150$) and the simulation ($\Delta U=700$); (ii) nonlocal effects may also contribute to the remaining differences between our results and the DNS data. We plan to investigate these issues, extending our model in the manner of Gough (1977b) to accommodate nonlocal behaviour.

This research is supported by the Austrian FWF grant No. AP 2120521. DOG is grateful to the Leverhulme Foundation for an Emeritus Fellowship.

References

Domaradzki, J. A. & Metcalfe, R. W. 1988, *J. Fluid Mech.*, 193, 499

Gough, D. O. 1965, *Geophys. Fluid Dyn. II, Woods Hole Oceanographic Institution*, p. 49

Gough, D. O. 1977a, *ApJ*, 214, 196

Gough, D. O. 1977b, *Problems of stellar convection, Springer-Verlag, Berlin*, p. 15

Gough, D. O. & Houdek, G. 2001, *ESASP*, 464, 637

Unno W. 1967, *PASJ*, 19, 140

Astrophysical Dynamics : From Stars to Galaxies
Proceedings IAU Symposium No. 271, 2010
N.H. Brummell, A.S. Brun, M.S. Miesch & Y. Ponty, eds.

Magnetic confinement of the solar tachocline: influence of turbulent convective motions

Antoine Strugarek[1], Allan Sacha Brun[1] and Jean-Paul Zahn[2,1]

[1]Laboratoire AIM Paris-Saclay, CEA/Irfu Université Paris-Diderot CNRS/INSU,
F-91191 Gif-sur-Yvette (email: antoine.strugarek@cea.fr)

[2]LUTh, Observatoire de Paris-Meudon, France

Abstract. We present the results of 3D simulations, performed with the ASH code, of the nonlinear, magnetic coupling between the convective and radiative zones in the Sun, through the tachocline. Contrary to the predictions of Gough & McIntyre (1998), a fossil magnetic field, deeply buried initially in the solar interior, will penetrate into the convection zone. According to Ferraro's law of iso-rotation, the differential rotation of the convective zone will thus expand into the radiation zone, along the field lines of the poloidal field.

Keywords. Sun: magnetic fields - stars: evolution - stars : rotation

1. Two scenarios for the tachocline confinement

Spiegel & Zahn (1992) were the first to develop the concept of the tachocline, the thin region that links in the Sun the differential rotation of the convection zone with the quasi uniform rotation of the radiation zone. They showed that, due to thermal diffusion, this differential rotation would spread far into the deep interior. Since this is not observed, Spiegel & Zahn (1992) invoked an anisotropic turbulence that would erode the latitudinal gradient of angular velocity. Gough & McIntyre (1998) questioned this anisotropic turbulent momentum transport, quoting examples from geophysical studies to rebut this scenario. They proposed that the confinement of the tachocline is achieved by a fossil magnetic field buried in the radiation zone. However, numerical simulations carried out by Garaud (2002) and Brun & Zahn (2006) did not confirm this magnetic scenario. Revisiting GM98, it appeared that the meridional circulation of the convection zone was a key factor in confining the magnetic field inside the radiation zone, and that the simulations should allow for radial flow between these two zones. As a matter of fact, when implementing an *ad-hoc* meridional flow in the convection zone, Sule *et al.* (2005) and Garaud & Garaud (2008) managed to numerically recover the GM98 results. To circumvent these boundary conditions issues, we present here the first 3D MHD numerical simulations that couple the solar radiation zone with the convection zone, both regions being treated in the most realistic way. These simulations were performed with the ASH code (for Anelastic Spherical Harmonics, Brun *et al.* (2004)).

Evolution of a magnetic field of fossil origin

We start the simulation with a relaxed convection zone, and with a dipolar magnetic field confined in the radiation zone, in equipartition of energy with the rotation in that zone (*see* figure 1(a). As the model evolves, we observe the development in the tachocline of a strong toroidal magnetic field, which is due to shearing of the poloidal field by the differential rotation; it is shown in figure 1(c). In that figure the instantaneous meridional circulation (i.e. the azimuthally averaged meridional velocity) is drawn in black. There is no sign of polar confinement of the magnetic field as predicted by Wood & McIntyre (2010). The amplitude of the meridional circulation, near the poles, varies from $18\,\mathrm{m.s}^{-1}$ at the base of the convective zone to $0.06\,\mathrm{m.s}^{-1}$ at the base of the tachocline, thus losing 3 orders of magnitude in 0.03 solar radius. Although our tachocline is somewhat too thick, diffusivity values have been selected to approach as much as possible the solar parameters in such a global 3D simulation.

As the magnetic field evolves, it is at the same time advected, sheared and diffused through the tachocline into the convective zone (Fig. 1(d)). Magnetic pumping certainly exists in the equatorial region, but is not efficient enough to confine the magnetic field there. As seen in figure 1(b), magnetic field lines open into the convective zone. According to Ferraro's law, the differential rotation of the convective zone spreads into the radiative zone through the Maxwell tensor, and it achieves that much faster than if it were transported solely by thermal diffusion.

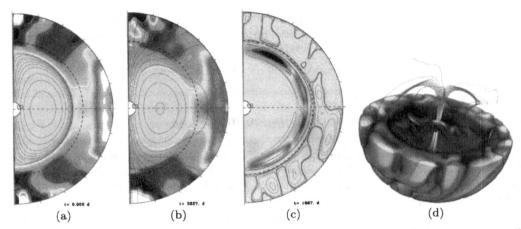

(a)   (b)   (c)   (d)

Figure 1. Meridional snapshots of azimuthal averages of angular velocity (colored background) and magnetic field lines (black); (a) initial field configuration, (b) after around 180 convective turnover times. Figure (c) shows the magnetopause (GM98) at the same epoch, with the instantaneous meridional circulation in black lines. Figure (d) displays a snapshot of magnetic field lines, and of the azimuthal velocity field v_φ in colored volume; it illustrates the three-dimensional character of the penetration into the convection zone of the magnetic field.

Our simulations demonstrate that the velocity field in the convection zone cannot be reduced to a steady meridional flow, when the scope is to describe the penetration into that zone of an interior field of fossil origin. Instead of being deflected by this laminar flow, the field pervades the whole convection zone, and it imprints the differential rotation on the radiation zone. Since this is not observed, we conclude that the magnetic scenario must be refined. More detailed results are to appear soon (Strugarek *et al.* (2010)).

We acknowledge funding by ERC through grant STARS2 207430.

References

Brun, A. S., Miesch, M. S., & Toomre, J. 2004, ApJ Series, 614, 1073

Brun, A. S. & Zahn, J.-P. 2006, A&A, 457, 665

Garaud, P. 2002, MNRAS, 329, 1

Garaud, P. & Garaud, J.-D. 2008, MNRAS, 391, 1239

Gough, D. O. & McIntyre, M. E. 1998, Nature, 394, 755

Miesch, M. S. 2003, ApJ, 586, 663

Spiegel, E. A. & Zahn, J.-P. 1992, A&A, 265, 106

Strugarek, A., Brun, A. S., & Zahn, J. P. 2010, submitted to A&A

Sule, A., Rudiger, G. & Arlt, R. 2005, A&A, 437, 1061

Wood, T. & McIntyre, M. 2010, submitted to NJP

Astrophysical Dynamics: From Stars to Galaxies
Proceedings IAU Symposium No. 271, 2010
N.H. Brummell, A.S. Brun, M.S. Miesch & Y. Ponty, eds.

© International Astronomical Union 2011
doi:10.1017/S1743921311017996

Variation in convective properties across the HR diagram

Joel Tanner, Sarbani Basu, Pierre Demarque and Frank Robinson

Yale University, New Haven, Connecticut, USA

Abstract. We perform 3D radiative hydrodynamic simulations to study convection in low-mass main-sequence stars with the aim of improving stellar models. Comparing models from a $0.90M_\odot$ evolutionary track with 3D simulations reveals distinct differences between simulations and mixing length theory. The simulations show obvious structural differences throughout the superadiabatic layer where convection is inefficient at transporting energy. The discrepancy between MLT and simulation changes as the star evolves and the dynamical effects of turbulence increase. Further, the simulations reveal a T-tau relation that is sensitive to the strength of the turbulence, which is in contrast to 1D stellar models that use the same T-tau relation across the HR diagram.

Keywords. stars: atmospheres, stars: evolution

The convective transport of energy in stars plays an important role in accurately determining stellar structure, particularly in the near-surface layers. Commonly used treatments of convection in stellar modeling such as mixing length theory (MLT) (Böhm-Vitense, 1958), are a large source of uncertainty, especially when applied to transition regions between radiative and convective energy transport. While MLT works very well in deep layers, it breaks down near the surface where radiation carries a significant fraction of the total energy flux.

A 3D simulation is characterized by its surface gravity, effective temperature and chemical composition. We get the surface gravity, stellar flux and initial stratification for each simulation from 1D stellar evolution models. The simulation domain is located at the top of the convection zone, with the top and bottom of the domain located at approximately 2 and 9 pressure scale heights above and below the photosphere, respectively. The domain is small enough that curvature and radial variation in gravity can be safely ignored. The vertical walls are periodic and the horizontal walls are free slip and impenetrable (closed box).

The code is a finite-difference code on a staggered grid and simultaneously solves the mass-, momentum-(Navier-Stokes) and energy conservation equations consistently coupled with the radiative transfer equations. The subgrid-scale model employed is the one from Smagorinsky (1963). The radiation flux is computed using the 3D Eddington approximation (Unno & Spiegel, 1966).

Including turbulence in stellar models alters the structure of the outer layers. We find that 3D simulations produce turbulent pressure that can be in excess of 10% of the gas pressure, which is an effect that is not present in MLT. In particular, the superadiabaticity is larger and occurs closer to the surface in turbulent simulations than in static 1D models.

Fig. 1 compares the SAL from a 3D turbulent simulation to the SAL computed with a 1D stellar model. The 1D model underestimates both the the height and breadth of the superadiabatic peak at all stages of evolution. Further, the difference between the MLT and simulated SALs are greater in the more evolved models where the turbulent pressure is larger.

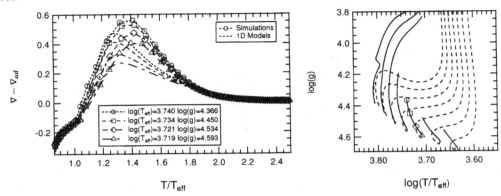

Figure 1. Comparison of the superadiabaticity in 1D models and 3D simulations. The 1D models correspond to different evolutionary states on a $0.90 M_\odot$ track. Locations of the 1D models in the log(g)-log(T) are shown in the left panel with pre- and post- main sequence evolutionary tracks for reference.

Fig. 2 shows the turbulent pressure at different stages of evolution. During the evolution, the effective temperature increases slightly, but there is a dramatic decrease in the surface gravity. Both effects contribute to higher turbulent pressure, which reaches nearly 14% of the gas pressure.

Although convection is driven in the unstable regions, convective overshoot affects the structure above the photosphere. This leads to altered T-τ relations in the presence of convective overshoot. The right panel of Fig. 2 shows the differences between the 1D Eddington T-τ and the simulated T-τ relations. Contrary to what is assumed in 1D stellar modeling, the simulated T-τ relations change as the star evolves, and can differ from the Eddington relation by more than 15%.

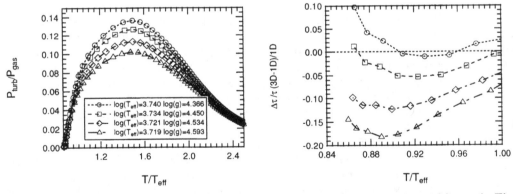

Figure 2. Turbulent pressure in simulations of the SAL for models along a $0.90 M_\odot$ track. The turbulence alters the 1D Eddington $T - \tau$ relation as the star evolves. Differences between the simulated and model $T - \tau$ relations are shown in the right panel.

References

Böhm-Vitense, E., 1958, S. Astrophys., 46, 108

Smagorinsky, J., 1963, Mon. Weather Rev., 91, 99

Unno, W. & Spiegel, E. A., 1966, PASJ, 18, 85

Astrophysical Dynamics: From Stars to Galaxies
Proceedings IAU Symposium No. 271, 2010
N.H. Brummell, A.S. Brun, M.S. Miesch & Y. Ponty, eds.

© International Astronomical Union 2011
doi:10.1017/S174392131101800X

Using simulations of solar surface convection as boundary conditions on global simulations

Regner Trampedach and Kyle Augustson

JILA, University of Colorado, 440 UCB, Boulder, CO 80309-0440, U.S.A.
email: trampeda@lcd.colorado.edu, augustso@lcd.colorado.edu

Abstract. Direct numerical simulations of convective stellar envelopes, are divided between two different physical regimes, that are rather difficult to reconcile — at least with the computational power of present-day computers. This paper outlines an attempt at bridging the gap between surface and interior simulations of convection.

Keywords. convection, hydrodynamics, radiative transfer, Sun: interior

---

1. The Dichotomy

Global scale simulations Have spherical geometry, include rotation, and have no lateral boundaries. They do have top and bottom boundaries, which for lack of better alternatives, are hard impenetrable and heating at the bottom and cooling at the top. This realm is optically deep and radiative transfer is well into the diffusion limit. The bulk of the solar convection zone can be well approximated by not only an ideal gas but even a perfect gas. The simplified physics is key to making such simulations feasible.

If the relevant features are resolved in the simulations, they can realistically describe the large scale convection, meridional flows, overshoot at the bottom of convective envelopes, dynamo action and probably also magnetic cycles.

Surface simulations, on the other hand, are small boxes straddling the photosphere. Here, the radiative transfer needs to be solved explicitly, for several directions, and the effect of millions of spectral lines has to be accounted for. This region also covers the major ionization zones so that at least the ionization and dissociation balances of H and He has to be included in a realistic simulation of stellar convection.

The resolution has to be large enough to resolve the granules and the steep photospheric temperature gradient. A realistic simulation of stellar convection also has an open bottom boundary, so as not to recycle the entropy deficiency, turbulence and vorticity of the downdrafts (which have been cooled at the photosphere) into the upflows, weakening the contrast between the two flows. Such simulations can be directly compared to all imaginable observations of the Sun and stars (?),and can greatly improve our interpretations of these observations.

The scale of convection increases rapidly, when going inward from the photosphere, giving rise to a fundamental scale-separation between interior and surface.

2. Giant Cells

Apart from the scale, the surface of the global scale simulations looks exactly like the photosphere of the surface simulations, as shown in the middle and right-hand-side panels, respectively, of Fig. 1. This granulation pattern is caused by the very rapid cooling in the photosphere or from the artificial solid cooling slab put on top of the global simulations.

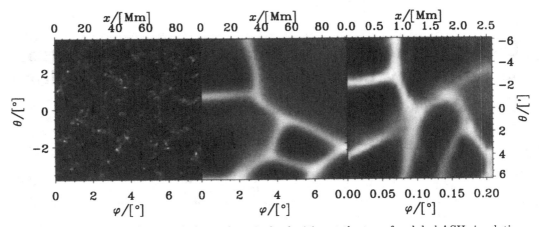

Figure 1. Comparing the morphology of vertical velocities at the top of a global ASH simulation of the solar convective envelope by ? (middle), with a surface simulation (left) at the same depth of $0.979\,R_\odot = 14.6\,\text{Mm}$ below the photosphere. The right panel shows the photospheric velocities of the surface simulation, enlarged by a factor of 37 to match the top of the ASH simulation.

This, however, looks nothing like the location in the large surface simulation by ?, that corresponds to the top of the global simulations, as shown in the left panel of Fig. 1.

A lot of the difference between the middle and left panels of Fig. 1, is due to the factor of 22 in physical resolution between the two simulations, but even resampling the surface simulation to 45×45-points, would not result in the middle panel. The loosely connected network of downflows with many isolated downflows in between, cannot be reproduced with the closed boundary, cooling layer of the global simulations.

Below about a Mm from the photosphere, the flux is carried out by a number of discrete, entropy deficient, over-dense downdrafts — not a homogeneous cooling layer. We therefore propose to open the top boundary of the global simulations for such downdrafts, stochastically reproducing, e.g., the correlation of entropy and velocity of the downdrafts of the surface simulations. This procedure would be subject to conservation of mass at the top boundary and the approximate constancy of convective flux.

The curved spherical segment (CSS) code for simulations of wedges of convection zones ? is going to be used as a testbed for these new boundary conditions. This code can be seen as an intermediate step between the surface and the deep interior, as it is fully compressible, only covers a fraction of the radial and horizontal extent, but does employ spherical geometry, only includes radiation in the diffusion approximation, and has a resolution and covers time-scales in-between the two other cases. This fast and highly parallelizable code is ideal for these kind of experiments and in itself can be used for analyzing the effect of the boundary conditions on, e.g., the near surface shear-layer.

Astrophysical dynamics: From Stars to Galaxies
Proceedings IAU Symposium No. IAUS271, 2010
N.H. Brummell, A.S. Brun, M.S. Miesch & Y. Ponty, eds.

© International Astronomical Union 2011
doi:10.1017/S1743921311018011

Numerical simulations of the circumstellar medium of massive binaries

Allard Jan van Marle[1] and Rony Keppens[1,2,3]

[1] Centre for Plasma Astrophysics, K.U. Leuven,
Celestijnenlaan 200B, B-3001, Heverlee, Belgium
email: [AllardJan.vanMarle] [Rony.Keppens]@wis.kuleuven.be
[2] Sterrenkundig Instituut, University of Utrecht,
Postbus 80000, NL-3508TA, Utrecht, the Netherlands
[3] FOM Institute for Plasma Physics Rijnhuizen,
P.O. Box 1207 NL-3430 BE Nieuwegein, the Netherlands

Abstract. We have made 3-D models of the collision of binary star winds and followed their interaction over multiple orbits. This allows us to explore how the wind-wind interaction shapes the circumstellar environment. Specifically, we can model the highly radiative shock that occurs where the winds collide. We find that the shell that is created at the collision front between the two winds can be highly unstable, depending on the characteristics of the stellar winds.

Keywords. methods: numerical, binaries: general, circumstellar matter, mass loss, stars: winds, outflows

1. Introduction

Massive stars emit powerful stellar winds that shape the environment around their progenitor stars. Recent advances in both hardware and software have made it possible to simulate the interactions of stellar winds emitted by binary stars. Such simulations were shown by Pittard (2009) and Pittard & Parkin (2010) for close binaries. Here we add to these simulations by showing two wide orbit (P=1 yr) binaries. One for a Wolf-Rayet + O-star binary, the second an LBV + O-star binary.

2. Astrophysical hydrodynamics for binary environments

We use the MPI-AMRVAC code (Meliani *et al.* 2007) to solve the usual equations of hydrodynamics (conservation of mass, momentum) with the energy equation given by:

$$\frac{\partial e}{\partial t} + \nabla \cdot (e\mathbf{v}) + \nabla \cdot (p\mathbf{v}) = -n^2 \Lambda(T), \qquad (2.1)$$

with $\mathbf{v}$ the velocity, p the pressure and e the total energy density. The last equation includes the effect of radiative cooling $n^2 \Lambda(T)$, which depends on local ion density, temperature and metallicity. The cooling curve $\Lambda(T)$ for gas at solar metallicity is obtained from Mellema & Lundqvist (2002). For our binary simulations we use a Cartesian grid, centered around the centre of mass of the binary orbit. At each timestep we calculate

Type	Mass $[M_\odot]$	$\dot{M}_A$ $[M_\odot/yr]$	$\dot{M}_B$ $[M_\odot/yr]$	V_A [km/s]	V_B [km/s]
LBV + O	50 + 20	1×10^{-4}	5×10^{-7}	200	2000
WR + O	50 +20	5×10^{-6}	5×10^{-7}	1500	2000

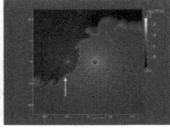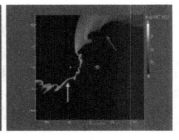

Figure 1. Circumstellar morphology for a Wolf-Rayet + O star binary (left) and the density (centre) and temperature (right) for a LBV+O star binary.

the location of each star and fill spheres, centered on these points, with free-streaming stellar wind material. The parameters for both simulations are given in the table below.

The density of the wind-shaped environment of a hydrogen rich Wolf-Rayet + O star binary, moving counter clockwise in orbit, after one complete orbit is shown in the left-hand image of fig. 1. The strong, high density Wolf-Rayet star (blue) wind clearly dominates. The interaction region is smooth and shows no sign of instabilities. This is due to the relatively ineffective radiative cooling. The shell formed by the colliding gas remains thick, preventing the formation of thin-shell instabilities.

The central and right hand plots of fig. 1 shows the circumstellar density and temperature respectively for a similar binary (same period and stellar masses) during a different evolutionary phase. The high mass star (blue symbol) now has a high density, relatively slow, Luminous Blue Variable (LBV) type wind. Due to the high density the radiative cooling allows the shell to be compressed leading to the formation of two types of thin-shell instabilities. The type of instabilities depends on the nature of the shock. Between the stars and at the front of the bowshock (white arrow) the interaction is mostly radiative, so the thermalized zone is very thin. As a result the shell feels ram-pressure from both sides and forms non-linear thin-shell instabilities (Vishniac 1994). This effect is even more pronounced in the region exactly between the two stars, where the shock is strongest due to the head-on collision of the two winds, as was found in 2D simulations by Stevens *et al.* (1992). Further downstream (red arrow) the thermalized zone becomes quite thick. Therefore, the shell feels thermal pressure on one side and ram-pressure on the other, leading to the formation of linear thin-shell instabilities (Vishniac 1983).

Acknowledgements

A.J.v.M. acknowledges support from NSF grant AST-0507581, from the FWO, grant G.0277.08 and K.U.Leuven GOA/09/009.

References

Meliani, Z., Keppens, R., Casse, F., & Giannios, D. 2007, *mnras*, 376, 1189
Mellema, G. & Lundqvist, P. 2002, *A&A*, 394, 901
Pittard, J. M. 2009, *mnras*, 396, 1743
Pittard, J. M. & Parkin, E. R. 2010, *mnras*, 403, 1657
Stevens, I. R., Blondin, J. M., & Pollock, A. M. T.. 1992, *ApJ*, 386, 265
Townsend, R. H. D.. 2009, *ApJS*, 181, 391
Vishniac, E. T. 1983, *ApJ*, 274, 152
Vishniac, E. T. 1994, *ApJ*, 428, 186

Astrophysical Dynamics: From Stars to Galaxies
Proceedings IAU Symposium No. 271, 2010
N.H.Brummell, A.S.Brun, M.S. Miesch & Y.Ponty, eds.

© International Astronomical Union 2011
doi:10.1017/S1743921311018023

Recurrent flux emergence from dynamo-generated fields

Jörn Warnecke[1,2] and Axel Brandenburg[1,2]

[1] Nordita, AlbaNova University Center,
Roslagstullsbacken 23, SE-10691 Stockholm, Sweden
email: joern@nordita.org

[2] Department of Astronomy, AlbaNova University Center,
Stockholm University, SE 10691 Stockholm, Sweden

Abstract. we investigate the emergence of a large-scale magnetic field. This field is dynamo-generated by turbulence driven with a helical forcing function. Twisted arcade-like field structures are found to emerge in the exterior above the turbulence zone. Time series of the magnetic field structure show recurrent plasmoid ejections.

Keywords. Sun: magnetic fields, Sun: coronal mass ejections (CMEs)

1. Introduction

The magnetic field at the visible surface of the Sun is known to take the form of bipolar regions and the field continues in an arch-like fashion. These formations appear usually as twisted loop-like structures. These loops can be thought of as a continuation of more concentrated flux ropes in the bulk of the solar convection zone. Twisted magnetic fields are produced by a large-scale dynamo mechanism that is generally believed to be the motor of solar activity (Parker 1979). One such dynamo mechanism is the α effect that produces a large-scale poloidal magnetic field from a toroidal one. In order to study the emergence of helical magnetic fields from a dynamo, we consider a model that combines a direct simulation of a turbulent large-scale dynamo with a simple treatment of the evolution of nearly force-free magnetic fields above the surface of the dynamo. In the context of force-free magnetic field extrapolations this method is also known as the stress-and-relax method (Valori *et al.* 2005). Above the solar surface, we expect the magnetic fields to drive flares and coronal mass ejections through the Lorentz force. In the present paper we highlight some of the main results of our earlier work (Warnecke & Brandenburg 2010).

2. The Model

The equation for the velocity correction in the Force-Free Model is similar to the usual momentum equation, except that there is no pressure, gravity, or other driving forces on the right-hand side, so we just have

$$\frac{D\boldsymbol{U}}{Dt} = \boldsymbol{J} \times \boldsymbol{B}/\rho + \boldsymbol{F}_{\mathrm{visc}}, \tag{2.1}$$

where $\boldsymbol{J} \times \boldsymbol{B}$ is the Lorentz force, $\boldsymbol{J} = \boldsymbol{\nabla} \times \boldsymbol{B}/\mu_0$ is the current density, μ_0 is the vacuum permeability, $\boldsymbol{F}_{\mathrm{visc}}$ is the viscous force, and ρ is here treated as a constant the determines the strength of the velocity correction. Equation (2.1) is solved together with the induction equation. In the lower layer the velocity is excited by a forcing function

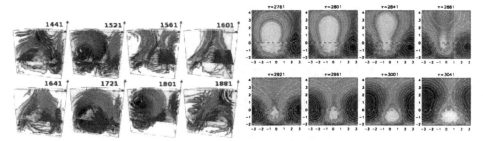

Figure 1. *Left panel*: Time series of arcade formation and decay. Field lines are colored by their local field strength which increases from pink to green. The plane shows B_z increasing from red (positive) to pink (negative). The normalized time τ is giving in each panel. *Right panel*: Time series of the formation of a plasmoid ejection. Contours of $\langle A_x \rangle_x$ are shown together with a color-scale representation of $\langle B_x \rangle_x$; dark blue stands for negative and red for positive values. The contours of $\langle A_x \rangle_x$ correspond to field lines of $\langle \boldsymbol{B} \rangle_x$ in the yz plane. The dotted horizontal lines show the location of the surface at $z = 0$. Adapted from Warnecke & Brandenburg (2010).

and the density is evolving using the continuity equation. The forcing function consists of random plane helical transversal waves with an average forcing wavenumber k_f.

3. Results

The magnetic field grows first exponentially and then shows subsequent saturation that is typical for forced turbulent dynamo action. In the turbulent layer the magnetic field reaches around 78% of the equipartition field strength, B_{eq}. The dynamo generates a large-scale field whose vertical component has a sinusoidal variation in y. After some time the magnetic field extends well into the exterior where it tends to produce an arcade-like structure, as seen in the left panel of Figure 1. The arcade opens up in the middle above the line where the vertical field component vanishes at the surface. This leads to the formation of anti-aligned field lines with a current sheet in the middle. The dynamical evolution is seen clearly in a sequence of field line images in the left hand panel of Figure 1, where anti-aligned vertical field lines reconnect above the neutral line and form a closed arch with plasmoid ejection above. This arch then changes its connectivity at the foot points in the sideways direction (here the y direction), making the field lines bulge upward to produce a new reconnection site with anti-aligned field lines some distance above the surface. Field line reconnection is best seen for two-dimensional magnetic fields, because it is then possible to compute a flux function whose contours correspond to field lines in the corresponding plane. In the present case the large-scale component of the magnetic field varies only little in the x direction, so it makes sense to visualize the field averaged in the x direction. The right panel of Figure 1 shows clearly the recurrent reconnection events with subsequent plasmoid ejection. The dynamics of the magnetic field in the exterior is indeed found to mimic open boundary conditions at the interface between the turbulence zone and the exterior at $z = 0$. In particular, it turns out that a twisted magnetic field generated by a helical dynamo beneath the surface is able to produce flux emergence in ways that are reminiscent of that found in the Sun.

References

Parker, E. N. 1979, Cosmical magnetic fields (Clarendon Press, Oxford)
Valori, G., Kliem, B., & Keppens, R. 2005, A&A, 433, 335
Warnecke, J. & Brandenburg, A. 2010, A&A, DOI: 10.1051/0004-6361/201014287

Astrophysical dynamics – from stars to galaxies
Proceedings IAU Symposium No. 271, 2010
N.H. Brummell, A.S. Brun, M.S. Miesch & Y. Ponty, eds.

© International Astronomical Union 2011
doi:10.1017/S1743921311018035

Magnetic confinement in the solar interior

Toby S. Wood

Baskin School of Engineering, University of California Santa Cruz
1156 High Street, Santa Cruz, CA 95064
email: tsw25@soe.ucsc.edu

Abstract. The observed uniform rotation of the Sun's radiative interior can be explained by the presence of a global-scale interior magnetic field, provided that the field remains confined below the convection zone. In high latitudes, such magnetic confinement is possible by means of persistent downwelling, driven by the convection zone's turbulent stresses.

Keywords. MHD, Sun: abundances, Sun: interior, Sun: magnetic fields, Sun: rotation

1. Introduction

Within the solar convection zone turbulent stresses drive differential rotation (Fig. 1) and "gyroscopically pump" mean meridional circulations (MMCs) that burrow into the radiative interior (Spiegel & Zahn 1992). These MMCs transport angular momentum below the base of the convection zone, yet the differential rotation goes over into uniform rotation within the radiative interior, via a thin shear layer called the "tachocline".

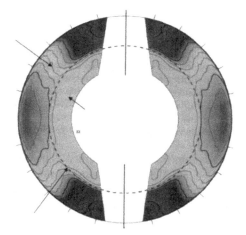

Figure 1. The Sun's differential rotation, adapted from Schou *et al.* (1998). The radiative interior rotates approximately uniformly with angular velocity $\Omega_i = 2.7 \times 10^{-6}\,\mathrm{s}^{-1}$, or 435 nHz. Within the convection zone, the angular velocity increases with colatitude through 350, 400, 450 nHz (heavy contours) to a maximum just under 470 nHz at the equator.

The uniform interior rotation can be explained by the presence of a global-scale interior magnetic field $\mathbf{B}_i$, in accordance with Ferraro's Law. To be effective, the field lines must directly couple low and high latitudes within the radiative interior; we refer to such a field as "confined". In order for $\mathbf{B}_i$ to remain confined over the Sun's lifetime, there needs to be some mechanism that can counteract the diffusion of $\mathbf{B}_i$ into the convection zone (Gough & McIntyre 1998, hereafter GM98).

For simplicity we suppose that $\mathbf{B_i}$ is roughly an axial dipole. In low latitudes, where the field lines are nearly horizontal, we assume that $\mathbf{B_i}$ is confined by downward "magnetic flux pumping" in the convective overshoot layer (e.g. Tobias *et al.* 2001; Kitchatinov & Rüdiger 2008). Downward pumping of this kind is believed to act only on the horizontal components of the magnetic field, and is therefore less plausible as a confinement mechanism in high latitudes. However, the tachocline's MMCs are expected to be downwelling in high latitudes, suggesting that $\mathbf{B_i}$ can be held in advective-diffusive balance across a thin "magnetic confinement layer" at the bottom of the tachocline (GM98).

2. The magnetic confinement layer

We seek to model the confinement layer in a neighborhood of the north pole (Fig. 2). We compute axisymmetric, steady-state solutions to the nonlinear, Boussinesq MHD equations using a finite-difference code.

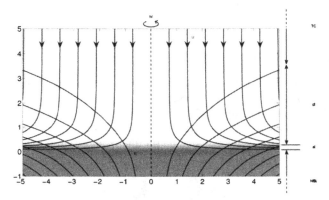

Figure 2. A vertical cross-section through the magnetic confinement layer near the north pole. The streamlines with arrows show the downwelling MMC responsible for confining the magnetic field $\mathbf{B}$. Over the Sun's lifetime, helium settles out of the convection zone and establishes a stable compositional gradient within the outer part of the radiative interior. The helium settling layer (shaded) is almost impermeable, and so the downwelling MMC turns equatorward within a thin "helium sublayer". For solar magnetic and helium diffusivities the sublayer is thinner than the confinement layer by about an order of magnitude (Wood & McIntyre 2010).

We find that downwelling of magnitude $\sim 10^{-5}\,\mathrm{cm\,s^{-1}}$ is able to confine a broad range of interior field strengths ($|\mathbf{B_i}| \lesssim 500\mathrm{G}$) across a confinement layer of thickness $\sim 1\,\mathrm{Mm}$. The retrograde Coriolis torque from the equatorward flow is balanced almost exactly by a prograde Lorentz torque from the confined magnetic field, with viscosity playing almost no role. The solutions are magnetostrophic to excellent approximation.

Our results also show the importance of compositional stratification for the structure of the confinement layer. Further details, and a discussion of the implications for solar lithium depletion, will appear in a future publication (Wood & McIntyre 2010).

References

Gough, D. O. & McIntyre, M. E. 1998, *Nature*, 394, 755
Kitchatinov, L. L. & Rüdiger, G. 2008, *AN*, 329, 372
Schou, J., *et al.* 1998, *ApJ*, 505, 390
Spiegel, E. A. & Zahn, J.-P. 1992, *A&A*, 265, 106
Tobias, S. M., Brummell, N. H., Clune, T. L., & Toomre, J. 2001, *ApJ*, 549, 1183–1203
Wood, T. S. & McIntyre, M. E. 2010, *J. Fluid Mech.*, *submitted*

Author Index

Subject Index